Photosynthesis and Respiratory Cycles During Environmental Stress Response in Plants

Photosynthesis and Respiratory Cycles During Environmental Stress Response in Plants

Edited by

Aryadeep Roychoudhury, PhD

APPLE ACADEMIC PRESS

First edition published 2023

Apple Academic Press Inc.
1265 Goldenrod Circle, NE,
Palm Bay, FL 32905 USA

760 Laurentian Drive, Unit 19,
Burlington, ON L7N 0A4, CANADA

CRC Press
6000 Broken Sound Parkway NW,
Suite 300, Boca Raton, FL 33487-2742 USA

4 Park Square, Milton Park,
Abingdon, Oxon, OX14 4RN UK

Library and Archives Canada Cataloguing in Publication

Title: Photosynthesis and respiratory cycles during environmental stress response in plants / edited by Aryadeep Roychoudhury, PhD.
Names: Roychoudhury, Aryadeep, editor.
Description: First edition. | Includes bibliographical references and index.
Identifiers: Canadiana (print) 20220275262 | Canadiana (ebook) 20220275297 | ISBN 9781774911839 (hardcover) | ISBN 9781774911846 (softcover) | ISBN 9781003315162 (ebook)
Subjects: LCSH: Plants—Photorespiration. | LCSH: Plants—Respiration. | LCSH: Plants—Effect of stress on.
Classification: LCC QK891 .P56 2022 | DDC 572/.462—dc23

Library of Congress Cataloging-in-Publication Data

CIP data on file with US Library of Congress

ISBN: 978-1-77491-183-9 (hbk)
ISBN: 978-1-77491-184-6 (pbk)
ISBN: 978-1-00331-516-2 (ebk)

About the Editor

Aryadeep Roychoudhury, PhD

Aryadeep Roychoudhury, PhD, is Assistant Professor at the Department of Biotechnology, St. Xavier's College (Autonomous), Kolkata, West Bengal, India. He is involved in active research in the field of abiotic stress responses in plants with perspectives of physiology, molecular biology, and cell signaling under diverse stress conditions. He has over 20 years of research experience in the concerned discipline. Dr. Roychoudhury is currently handling several government-funded projects on abiotic stress responses in rice and supervising five PhD students as Principal Investigator. To date, he has published over 200 articles in peer-reviewed journals and chapters in books of international and national repute. He has edited many books from reputed publishers and has also guest edited several special issues. He is a regular reviewer of articles in high-impact international journals, a life member of several scientific associations and societies, and the recipient of the Young Scientist Award 2019, conferred by the International Foundation for Environment and Ecology, at the University of Allahabad, Prayagraj, Uttar Pradesh, India. His name is included in the Stanford University's List of World's Top 2% Scientists.

Dr. Roychoudhury received his BSc (Hons.) in Botany from Presidency College, Kolkata, and his MSc in Biophysics and Molecular Biology from the University of Calcutta, West Bengal, India. He did his PhD at the Bose Institute, Kolkata, under Jadavpur University, Kolkata, India.

Contents

Contributors

Awatif M. Abdulmajeed
Biology Department, Faculty of Science, University of Tabuk, Umluj 46429, Saudi Arabia

Taghreed S. Alnusaire
Biology Department, College of Science, Jouf University, Sakaka 2014, Saudi Arabia

M. S. Amritha
Plant Physiology and Biochemistry Division, Department of Botany, University of Calicut, C.U. Campus P.O., Thenhipalam 673635, Kerala, India

Muhammad Imran Ashraf
Institute of Soil and Environmental Sciences, University of Agriculture, Faisalabad, Punjab, Pakistan

Muhammad Ashar Ayub
Institute of Soil and Environmental Sciences, University of Agriculture, Faisalabad, Punjab, Pakistan; E-mail: muhammadasharayub@gmail.com

Sharbani Bahali
Department of Biotechnology, Centurion University of Technology and Management, Odisha, India

Sabyasachi Banerjee
Gupta College of Technological Sciences, Ashram More, G.T. Road, Asansol 713101, West Bengal, India

Monika Bansal
SIILAS Campus, Jaipur National University, Jaipur 302025, Rajasthan, India

Preetha Bhadra
Department of Biotechnology, M.S. Swaminathan School of Agriculture, Centurion University of Technology and Management, Odisha761211, India

Biswajit Bose
Department of Botany, School of Basic Sciences, Shri Ramasami Memorial (SRM) University, Sikkim, East Sikkim, India

Sankhadip Bose
Department of Pharmacognosy & Phytochemistry, Bengal School of Technology, A College of Pharmacy, Chuchura, Hooghly, West Bengal, India; E-mail: sankha.bose@gmail.com

Kolluru Viswanatha Chaitanya
Department of Microbiology and Food Science Technology, GITAM School of Science, GITAM Deemed to be University, Visakhapatnam 530045, India

Sudip Choudhury
Department of Chemistry, Assam University, Silchar 788011, India; E-mail: sudip.choudhury@aus.ac.in, sudipch1@gmail.com

Somali Dhal
Amity Institute of Biotechnology, Amity University Kolkata, Major Arterial Road (South-East); Action Area II, Newtown, Kolkata 700135, West Bengal, India

Amara Farooq
Institute of Soil and Environmental Sciences, University of Agriculture, Faisalabad, Punjab, Pakistan

Sunayana Goswami
Department of Chemistry, Assam University, Silchar 788011, India

Eiji Hirasawa
Department of Biology and Geosciences, Osaka City University, Osaka, Japan

Syed Sarfraz Hussain
Kauser Abdulla Malik School of Life Sciences (KAM-SLS), Forman Christian College (A Chartered University), Ferozepur Road, Lahore 54600, Pakistan

Muhammad Irfan
Kauser Abdulla Malik School of Life Sciences (KAM-SLS), Forman Christian College (A Chartered University), Ferozepur Road, Lahore 54600, Pakistan

Awadhesh Kumar
ICAR-National Rice Research Institute, Cuttack, Odisha, India

Roopa Kumari
Department of Chemistry, Ramjas College, University of Delhi, New Delhi, India

Milan Kumar Lal
ICAR-Central Potato Research Institute, Shimla, Himachal Pradesh, India

Sagar Maitra
Department of Agronomy and Agroforestry, M.S. Swaminathan School of Agriculture Centurion University of Technology and Management, Gajapati 761211, Odisha, India

Sudip Kumar Mandal
Dr. B. C. Roy College of Pharmacy and Allied Health Sciences, Bidhannagar, Durgapur, West Bengal, India

Debajyoti Mukherjee
R&D Head, Herbal Extraction, Keva Fragrances Pvt. Ltd., Mumbai, India

Naveed Ul Mushtaq
Department of Bioresources, School of Biological Sciences, University of Kashmir, Srinagar 190006, India

Allah Nawaz
Institute of Soil and Environmental Sciences, University of Agriculture, Faisalabad, Punjab, Pakistan; E-mail: an_uaf99@yahoo.com

Lopamudra Nayak
ICAR-National Rice Research Institute, Cuttack, Odisha, India

Sarangadhar Nayak
ICAR-National Rice Research Institute, Cuttack, Odisha, India

Harshata Pal
Amity Institute of Biotechnology, Amity University Kolkata, Major Arterial Road (South-East); Action Area II, Newtown, Kolkata 700135, West Bengal, India; E-mail: hpal@kol.amity.edu

Saurav Paul
Department of Chemistry, Assam University, Silchar 788011, India

Suchismita Prusty
Department of Biotechnology, Centurion University of Technology and Management, Odisha, India

Jos T. Puthur
Plant Physiology and Biochemistry Division, Department of Botany, University of Calicut, C.U. Campus P.O., Thenhipalam 673635, Kerala, India; E-mail: jtputhur@yahoo.com

Sameer H. Qari
Biology Department, Aljumum University College, Umm Al-Qura University, Makkah, P.O. 2203, Saudi Arabia

K. P. Aswathi Raj
Plant Physiology and Biochemistry Division, Department of Botany, University of Calicut, C.U. Campus P.O., Thenhipalam 673635, Kerala, India

Sanchayita Rajkhowa
Department of Chemistry, Jorhat Institute of Science & Technology, Jorhat, Assam, India; E-mail: sanchayita.rajkhowa@gmail.com

Abdul Qayyum Rao
National Centre of Excellence in Molecular Biology (NCEMB), 87-West Canal Bank Road, University of the Punjab, Lahore 54600, Pakistan

Aadil Rasool
Department of Bioresources, School of Biological Sciences, University of Kashmir, Srinagar 190006, India

L. Jeyanthi Rebecca
Department of Agricultural Biotechnology, Bharath Institute of Higher Education and Research, Selaiyur, Chennai 600073, India; E-mail: jeyanthi.bt@bharathuniv.ac.in

Reiaz Ul Rehman
Department of Bioresources, School of Biological Sciences, University of Kashmir, Srinagar 190006, India; E-mail: rreaizbiores@gmail.com

Atasi Routray
Department of Biotechnology, Centurion University of Technology and Management, Odisha, India

Sindhujit Roy
Department of Biotechnology, SRM University, Chennai, India

Anuradha Roy Choudhury
Department of Chemistry, Assam University, Silchar 788011, India

Aryadeep Roychoudhury
Department of Biotechnology, St. Xavier's College (Autonomous), 30, Mother Teresa Sarani, Kolkata 700016, West Bengal, India; E-mail: aryadeep.rc@gmail.com

Ayman E. L. Sabagh
Department of Agronomy, Faculty of Agriculture, University of Kafrelsheikh, Egypt

Ranjan Kumar Sahoo
Department of Biotechnology, Centurion University of Technology and Management, Odisha, India; E-mail: ranjan.sahoo@cutm.ac.in

Masina Sairam
Department of Agronomy and Agroforestry, M.S. Swaminathan School of Agriculture, Centurion University of Technology and Management, Gajapati 761211, Odisha, India; E-mail: sagar.maitra@cutm.ac.in

Seerat Saleem
Department of Bioresources, School of Biological Sciences, University of Kashmir, Srinagar 190006, India

Jyotirmoy Sarma
Department of Chemistry, The Assam Kaziranga University, Jorhat, Assam, India

Wasifa Hafiz Shah
Department of Bioresources, School of Biological Sciences, University of Kashmir, Srinagar 190006, India

Ahmed Ali Shahid
National Centre of Excellence in Molecular Biology (NCEMB), 87-West Canal Bank Road, University of the Punjab, Lahore 54600, Pakistan

Bujun Shi
School of Agriculture, Food & Wine, Waite Campus University of Adelaide SA 5064, Australia

Abbas Shoukat
Institute of Soil and Environmental Sciences, University of Agriculture, Faisalabad, Punjab, Pakistan

Kadambot HM Siddique
The UWA Institute of Agriculture, The University of Western Australia, Perth WA 6001, Australia

Ankur Singh
Department of Biotechnology, St. Xavier's College (Autonomous), 30, Mother Teresa Sarani, Kolkata 700016, West Bengal, India

Mona H. Soliman
Botany and Microbiology Department, Faculty of Science, Cairo University, Giza 12613, Egypt
Biology Department, Faculty of Science, Taibah University, Al-Sharm, Yanbu El-Bahr, Yanbu 46429, Kingdom of Saudi Arabia; E-mail: hmona@sci.cu.edu.eg

Challa Surekha
Department of Biochemistry and Bioinformatics, GITAM School of Science, GITAM Deemed to be University, Visakhapatnam 530045, India; E-mail: schalla@gitam.edu

Inayatullah Tahir
Department of Botany, School of Biological Sciences, University of Kashmir, Srinagar 190006, India

T. T. Dhanya Thomas
Plant Physiology and Biochemistry Division, Department of Botany, University of Calicut, C.U. Campus P.O., Thenhipalam 673635, Kerala, India

Wajid Umar
Doctoral School of Environmental Science, the Hungarian University of Agricultural and Life Sciences, Gödöllő, Hungary

Shabir H Wani
Mountain Research Center for Field Crops, Khudwani, Sher-e-Kashmir University of Agricultural Sciences and Technology of Kashmir, Jammu and Kashmir, India;
E-mail: shabirhwani@skuastkashmir.ac.in

Muhammad Zohaib
Institute of Soil and Environmental Sciences, University of Agriculture, Faisalabad, Punjab, Pakistan

Naqshe Zuhra
Institute of Soil and Environmental Sciences, University of Agriculture, Faisalabad, Punjab, Pakistan

Abbreviations

ABA	abscisic acid
ABS/CSm	absorption flux per cross section
ABS/RC	absorption flux per reaction center
ADP	adenosine diphosphate
Ag+	silver
Al	aluminium
AlaAT	alanine aminotransferase
AM	arbuscular mycorrhizal
AMF	arbuscular mycorrhizal fungi
AMP	adenosine mono phosphate
ANTs	adenine nucleotide translocators
AOX	alternative oxidase
APX	ascorbate peroxidase
AS	antenna system
ATPs	adenosine triphosphates
ATSDR	Agency for Toxic Substances and Disease Registry
BS	bundle sheath
BSCs	bundle sheath cells
CA	carbonic anhydrase
cAMP	3',5'-cyclic adenosine monophosphate
CaMV	cauliflower mosaic virus
CAB2	Chl a/b binding protein 2
CACs	citric acid cycle
CAT	catalase
CBB	Calvin–Benson–Bassham
CBC	Calvin–Benson Cycle
CCM	carbon concentrating mechanisms
Cd	cadmium
Chl*	chlorophyll
Chla	chlorophyll a
Chlb	chlorophyll b
CAM	Crassulacean acid metabolism
CAPP	cold atmospheric pressure plasma
CKs	cytokinins

CMV	cucumber mosaic virus
CP	chloroplast protrusion
CS	citrate synthase
Cyt b6f	cytochrome b6f
DAMPs	damage-associated molecular patterns
DEG	differentially expressed unigenes
DHAP	dihydroxyacetone phosphate
DIo/CSm	dissipated energy flux per cross section
DIo/RC	dissipated energy flux per reaction center
E	transpiration rate
EED	excess energy dissipation
EFM	elementary flux mode
EL	electrolyte leakage
ERD 10	early response to dehydration 10
ET	ethylene
ETC	electron transport chain
ETo/CSm	electron transport flux per cross section
ETo/RC	electron transport flux per reaction center
F-6-P	fructose-6-phosphate
F-1,6-P2	fructose-1,6-bisphosphate
FAD	flavin adenine dinucleotide
FADH2	flavin adenine dinucleotide reduced
FBA-3	fructose-bisphosphate aldolase 3
Fm	maximum Chl a fluorescence
Fo	initial Chl a fluorescence
FV	fluorescence yield
Fv	variable Chl a fluorescence
Fv/Fm	potential or maximum quantum yield of PSII
G-6-P	glucose-6-phosphate
GA	gibberellins
GABA	gamma-aminobutyric acid
GAP	glyceraldehyde-3-phosphate
GAPDH	glyceraldehyde-3-phosphate dehydrogenase
GDP	guanosine diphosphate
GOGAT	glutamine oxoglutarate aminotransferase
GPI	glucose-6-phosphate isomerase
GPX	glutathione peroxidase
GR	glutathione reductases
gs	stomatal conductance

GST	glutathione S-transferase
GTP	guanosine triphosphate
HIF-α	hypoxia-inducible factor 1-alpha
HK	hexokinase
HM	heavy metal
HNT	high night temperature
HOCl	hypochlorous acid
HSF	heat shock factors
HSP	heat shock proteins
HR	hypersensitive response
IAA	auxins
IDH	isocitrate dehydrogenase
IR	ionizing radiations
JA	jasmonic acid
KDH	α-ketoglutarate dehydrogenase
LEA	late embryogenesis protein
LHC or LH	light-harvesting antenna complex
LHC	light-harvesting complex
LSU	large subunit
MAVS	mitochondrial antiviral-signaling protein
MC	mesophyll cells
MCA	metabolic control analysis
MDA	malondialdehyde
MDAR	monodehydroascorbate reductases
MDH	malate dehydrogenase
ME	malic enzyme
MG	methylglyoxal
mTCA	mitochondrial tricarboxylic acid
NAD^+	nicotinamide adenine dinucleotide
NADH	nicotinamide adenine dinucleotide hydrogen
NADP	nicotinamide adenine dinucleotide phosphate
NADPH	nicotinamide adenine dinucleotide phosphate hydrogen
NADP-MDH	NADP-malate dehydrogenase
NADP-ME	NADP-malic enzyme
NAD-ME	NAD-malic enzyme
NO	nitric oxide
NPQ	nonphotochemical quenching
NUE	nitrogen use efficiency
OAA	oxaloacetate

O_3	ozone
OPP	oxidative pentose phosphate
OXPHOS	oxidative phosphorylation system
PAs	polyamines
PCD	programmed cell death
PEP	phosphoenolpyruvate
PEPC	phosphoenolpyruvate carboxylase
PFK	phosphofructokinase
3-PG	3-phosphoglycerate
PGK	phosphoglycerate kinase
PGM	phosphoglucomutase
PHD	prolyl hydroxylases
PHYB	phytochrome B
PIABS	performance index on absorption basis
PK	pyruvate kinase
PLD	phospholipase D
Pn	photosynthesis
POX	peroxidase
PPDK	pyruvate orthophosphate dikinase
PPN	plant parasitic nematodes
PPRs	pattern recognition receptors
PPT	phosphate translocator
PR	pathogen related
PRK	phosphoribulokinase
PS I	photosystem I
PS II	photosystem II
PTOX	plastid terminal oxidase
RbcS	RuBisCO small subunit
RC	reaction center
RCA	RuBisCO activase
RC/CSm	reaction center per cross section
RC/CSo	number of reaction centers per excited cross section
RLR	retinoic acid-inducible protein like receptor
RNS	reactive nitrogen species
RONSS	reactive oxygen, nitrogen, and sulphur species
ROS	reactive oxygen species
RSS	reactive sulphur species
RuBisCO	ribulose-1,5-biphosphate carboxylase
RuBP	ribulose-1,5-bisphosphate

RWC	relative water content
SA	salicylic acid
SCMV	sugarcane mosaic virus
SDH	succinate dehydrogenase
SOD	superoxide dismutase
SPS	sucrose phosphate synthase
TCA	tricarboxylic acid cycle
TFs	transcription factors
TMV	tobacco mosaic virus
TP	triose phosphates
TPI	triosephosphate isomerase
TRo/CSm	trapping flux per cross section
TRo/RC	trapping flux per reaction center
UV-B	ultraviolet B radiation
WUE	water use efficiency

Preface

Climate change imposes various forms of environmental stresses, such as salinity, drought, heat, cold, high light intensity, UV radiation, and heavy metals, which are hostile to plants due to their sessile nature. Climate change is also responsible for decreasing plant growth and productivity, leading to massive loss of crops worldwide.

Agricultural production is estimated to be enhanced by 100–110% by 2050 in order to meet the per capita caloric demands and provide food security to the ever-increasing human population. The reactive oxygen species (ROS) and reactive nitrogen species formed under such situations cause structural alterations of various biomolecules and enzyme proteins and hence cause loss in their functional activity. Photosynthesis and respiration are the two most vital and tightly regulated physiological processes that govern plant productivity by regulating carbon sequestration from the atmosphere, carbon assimilation, and energy generation through ATP production by photophosphorylation and oxidative phosphorylation. Both of these processes involve multistep enzymatic pathways with conversion of one intermediate to the other. Any perturbation in the enzyme structure due to environmental stress interrupts the photosynthetic and respiratory cycles, together with cellular energy homeostasis, thereby retarding plant growth and development. It is imperative that re-establishment of the energy balance requires efficient maintenance of photosynthesis and respiration machinery during stress through elaborate stress-adaptive strategies.

Abiotic stresses, either in single or in combinations, exert severe consequences by affecting the photosynthetic efficiency of plants. Photosynthesis enables plants to produce carbohydrates, acting as primary sources of energy to the animal world. The photosynthetic apparatus comprises chiefly of the primary pigment, chlorophyll for light absorption, along with accessory pigments like carotenoids, which transfer light energy to chlorophyll molecules, photosystems (photosytems I and II), and light harvesting complexes for undergoing photochemical reactions to release oxygen that supports respiration for the animal world and also generates highly reducing compounds, NADPH and ATP, the energy currency to support life.

Pheophytins with high-reduction potential serves as a primary electron acceptor for photosystem II. Carotenoids also function as potent antioxidants because they can quench singlet oxygen formed due to high light intensity, thereby helping in photoprotection of photosynthetic machinery. The atmospheric CO_2 is fixed as carbohydrates through the C3 cycle, also known as the Calvin cycle, by using NADPH as a reducing agent, in light-independent enzymatic reaction steps. Most of the photosynthetic components, including the pigments and enzymes, are severely affected by the different stress factors due to the formation of ROS like superoxide radicals, hydroxyl radicals, and H_2O_2 as well as cytotoxic compounds like malondialdehyde and methylglyoxal. Heavy metals like cadmium react with sulfhydryl sites of the enzymes δ-aminolevulinic acid (ALA) dehydratase and porphobilinogen deaminase. As a result, ALA and porphyrins, which are the intermediates in the biosynthetic pathway, start accumulating. The accumulation of ALA is known to generate ROS in plants, further leading to oxidative damages. Therefore, it is highly important to decipher and investigate the impact of diverse environmental stresses on the different constituent components of photosynthetic machinery in order to have an in-depth understanding of the factors, leading to decreased photosynthetic rate and reduced yield.

In addition to C3 cycle, plants also exhibit additional type of photosynthesis, namely C4 cycle (Hatch and Slack pathway). Tropical plants such as maize, sorghum, and sugarcane survive in warmer climate and have specially developed C4 pathways in order to exhibit better photosynthetic performance, increased water, nitrogen, and radiation use efficiency; and better allocation of resources under drought situations, owing to a more potential carbon-concentrating or fixation mechanism by enhancing carboxylase activity of RuBisCo (ribulose 1,5-bisphosphate carboxylase). Hence, the ROS-mediated damages are far lesser in C4 plants than in C3 plants, where drastic reduction of electron-transport chain (ETC) occurs under stress conditions. Genetic engineering of C3 plants with overexpressed C4 enzymes appear to be an optimistic strategy to increase photosynthetic capacity and fitness under stressful environment. Moreover, many genes and enzymes implicated in photosynthetic pathways have been shown to be upregulated or downregulated in response to diverse stress cues and therefore serve as important candidates to develop stress-resilient photosynthetic machinery to sustain crop productivity.

Aerobic metabolism has unavoidable consequence of the production of ROS, like superoxide radicals and H_2O_2. The ETC in mitochondria is one of the major sites of ROS production in plant cells. The cytosolic caspase-like proteases are activated due to the release of functionally active cytochrome during ROS production. Thus, mitochondria execute a major function in the programmed cell death during stress conditions. Apart from the well-organized mitochondrial-stress-regulated enzymes like heat shock proteins, alternative oxidases, and NADPH dehydrogenases, a set of other proteins have been identified. Since the mitochondrial carrier protein family constitutes one of the important stress-responsive regulatory proteins, it is imperative that stress conditions cause alterations in metabolite transport across the mitochondrial inner membrane. Rotenone and piericidin A, the known inhibitors of mitochondrial complex I, have the ability to enhance stress tolerance. Proteome changes are coordinated by the large respiratory subunits of ETC, thereby altering respiration in response to oxidative stress conditions. Data from different studies indicate that oxidative stress largely affects the Krebs cycle because of inhibition of an array of Krebs cycle enzymes, leading to perturbation in the carbon flux. Environmental stress has been reported to modulate the activity of the enzymes participating in the Krebs cycle. Citrate accumulation plays a crucial role during stress, since it lowers ROS production by inducing alternative oxidase. Disaccharides (sucrose, trehalose), raffinose family oligosaccharides (RFOs), and fructans are the three major types of water-soluble carbohydrates that are essentially involved in plant stress responses. Based on such observations, mitochondria and respiratory cycles are assumed to play a pivotal role in the regulation of tolerance against multiple abiotic stresses in plants.

Both higher photosynthetic efficiency and proper maintenance of respiratory cycles under stress situations are possible by applying potent growth regulators including phytohormones, osmoprotectants, antioxidants, salicylic acid, brasssinosteroids, melatonin, strigolactones, polyamines, and mineral nutrients. In addition, proper understanding of the genetic regulation of the biochemical pathways in response to stress also provide opportunities to genetically engineer the pathways for enhanced stress tolerance through improved photosynthesis and respiration, with higher yield and productivity.

While tracing the footprints of research on photosynthesis and respiration in the context of global climate change, environmental pollution, and abiotic stresses, I found that there is dearth of reliable books as well as a loop hole of proper knowledge between book content and ongoing research trends. There is no single book that brings together the photosynthetic and respiratory mechanisms in plants under hostile situations and the proper mitigating strategies to continue uninterrupted photosynthesis and respiration under such situations in a comprehensive and elaborate way. Since photosynthesis and respiration are the two main physiological processes for sugar biosynthesis and mobilization for driving all other vital functions, I feel the extreme necessity to have a compiled comprehensive book on this aspect that will deliver sound knowledge in the related disciplines to scientists, researchers, and academicians, motivating them for future research.

The chapters included in this book are contributed by renowned scientists, researchers, and global leaders in this field and focus on the effect of environmental stressors on photosynthetic pigments, photosystems, activities of photosynthetic enzymes and protein complexes, PSII photochemistry, carbon fixation pathways, photosynthetic efficiency, glycolytic and Krebs cycle pathways, and ATP production and ETC of plants. The recovery of photosynthesis and respiration through application of phytohormones, signaling molecules, and other protective agents has also been emphasized. Genetic engineering to enhance photosynthetic efficiency has also been highlighted.

Altogether, the book will not only enable us to understand the nature of extensive damages brought to the photosynthetic and respiratory processes but will also provide a blueprint in designing proper mitigating strategies to sustain these two vital processes under challenging environment. This will play an important role in modulating energy signaling and balance. Overall this book considerably bridges the gap between research that is already conducted and future trends in photosynthesis and respiration research. I sincerely hope that this book will serve as a proper documentation of research undertaken so far and future trends to be followed to address this problem.

Aryadeep Roychoudhury
Kolkata, West Bengal, India

CHAPTER 1

Introduction to Plant Responses to Environmental Stress

PREETHA BHADRA[1], SAGAR MAITRA[2*], and MASINA SAIRAM[2]

[1]*Department of Biotechnology, M.S. Swaminathan School of Agriculture, Centurion University of Technology and Management, Odisha 761211, India*

[2]*Department of Agronomy and Agroforestry, M.S. Swaminathan School of Agriculture, Centurion University of Technology and Management, Odisha 761211, India*

[*]*Corresponding author. E-mail: sagar.maitra@cutm.ac.in*

ABSTRACT

Anthropogenic activities contribute to different environmental stresses, such as water scarcity or excess water, salinity, high and low temperatures, heavy metal pollution, and solar radiation which cause considerable damages in plants. Inhibited growth of plants and decreased crop productivity have led to global food insecurity. Plant response to these environmental pressures can be visualized in the form of physiological and molecular mechanisms. Photosynthesis and carbon assimilation are affected in crops that are influenced generally by environmental disturbances. The downregulation of Calvin–Benson–Bassham cycle is influenced by stomatal conductance, CO_2 diffusion through stomata, diminished activity of ribulose bisphosphate carboxylase/oxygenase action, which thus diminishes the CO_2 concentration, subsequently bringing about decreased biomass and crop productivity. Other responses involve the combination of stress hormones, for example,

Photosynthesis and Respiratory Cycles during Environmental Stress Response in Plants. Aryadeep Roychoudhury (Ed.)

abscisic acid regulates stomatal movement, and the formation of reactive oxygen species causing oxidative damage. There is an intricate organization of collaboration among metabolites and enzymes that regulate plant adaptations to diverse stresses. This chapter focuses on the major abiotic stresses adversely affecting plant growth, productivity, and survival.

1.1 INTRODUCTION

Plant environment, physiology, and natural chemistry have been evaluated widely since time immemorial (Zargar et al., 2015; Zhang et al., 2015). Changes in ecological conditions may be characterized, based on the reactions that a living cell goes through, by restricting certain activities in presence of any challenge (Yadav et al., 2004; Wang et al., 2012; Zhang et al., 2013). For straightforwardness, we characterize a restricting variable as one that decides the pace of development or biochemical response, and that an adjustment in its level can bring about an adjustment of the plant metabolic activities. Stresses are consequently characterized as ecological disorders that outcome awkwardness in metabolism and require biochemical as well as metabolic changes (Eckardt and Pell, 1995; Tournaire et al., 2003; Velikova et al., 2005; Iglesias-Ferna´ndez and Matilla, 2016; Shahid et al., 2017). Schematically, the grouping of occasions related to the reaction of living cells to a natural variation might be portrayed as follows: (1) environmental change, (2) sensing mechanism, (3) adaptation, and (4) new steady state. Plants in an open field face combined vagaries in the natural environment. The progressions occur in two diverse time scales, namely, circadian cycle, which senses light and temperature in a 24 h cycle; and occasional cycle that fluctuates as per the environmental conditions and geological area of the specific environment where the green development occurs (Chatterjee et al., 2017; Dizengremel et al., 2012). In thick algal societies, another cycle is forced by the concentrated blending framework that basically brings about a light-dim cycle, which varies as far as part of seconds when contrasted with the hours or months in the other two cycles. Microalgae have surely evolved different systems for detecting and adapting to variations in their conditions. Such adjustments are intended to balance productively the retention of excitation energy along with the creation of diminishing force (NADPH) and compound energy (ATP) for supporting cellular development. Failure to perform this is due to overexcitation of the photosynthetic apparatus that may bring about the production of reactive oxygen species prompting oxidative damages (Baker, 1996; Carmo-Silva et al. 2010; Brauer et al., 2016).

Proper functioning of chloroplast includes communications with other cell organelles, trade of metabolites, support of phytohormones, reactive oxygen species (ROS), and serious cross-talk with the core (anterograde and retrograde flagging). Chloroplasts, both during biogenesis and senescence, react to the natural variations by adjusting with proper alterations. Plastids have a plastid-encoded RNA polymerase (PEP) and phage-type RNA polymerase (NEP). Numerous TOC receptors empower the activity of various import pathways with various substrate inclinations. These may take part in the separation of plastids. As of late, advancement in chloroplast development is managed by the rebuilding of the protein import apparatus. A RING-type ubiquitin E3 ligase has been identified, targeting the parts of the plastid protein import hardware. Plastid improvement is under the immediate control of the ubiquitin-proteasome framework.

The Calvin–Benson–Bassham (CBB) cycle is dedicated to CO_2 assimilation and sugar production in oxyphototrophs. Phosphoribulokinase (PRK) is a fundamental catalyst of the CBB cycle, catalyzing ATP-mediated transformation of ribulose-5-phosphate (Ru5P) to ribulose-1,5-bisphosphate. The oxyphototrophic PRK is redox-controlled and can be additionally managed by the reversible relationship with both glyceraldehyde-3-phosphate dehydrogenase (GAPDH) and oxidized chloroplast protein CP12. The subsequent GAPDH/CP12/PRK complex is focal in the regulation of the CBB cycle (Jordan et al., 1996; James et al., 2006; Hancock et al., 2014; Gupta et al., 2018). Environmental stresses faced by plants may be characterized as the state when abiotic factors disturb the plants to accomplish their growth, development, and regenerative processes. Abiotic stress influences the development, photosynthesis, and other vital activities of plants. Accordingly, an outline related to the impact of environmental burdens on the Calvin cycle of photosynthesis as well as connected proteins is introduced in this chapter. Under challenging situations, plants try to adjust by hindering metabolic exercises, particularly biosynthesis, that influences general development and yield (Lawlor and Tezara, 2009; Osakabe and Osakabe, 2012; Mustafa and Komatsu, 2016).

1.2 WATER STRESS

Environmental pattern is rapidly changing because of increasing earth temperature, and therefore, crop development is confronted with genuine ecological burdens like dry spell, salinity, temperature, UV, ozone (O_3), and flooding (Fig. 1.1). Flooding pressure is one of the significant concerns brought about by sudden precipitation within a short span of time. Excess water pressure in the farming area may create waterlogging or submergence stress. Waterlogging is brought about by the abundance of water in cultivating regions, so that the entire root region as well as certain portions of shoot is inundated, while submergence implies to the situation when an entire plant is covered with water (Parry et al., 2008; Qureshi et al., 2013). The excess soil moisture stress is because of poor seepage and concentrated water system. Considering the losses due to excess water stress, many harvest raisers have zeroed in on the advancement of flood lenient assortments or recognizable proof of flood-tolerant crops and cultivars (Fig. 1.1). Along these lines, a few flooding resilience assortments or promotions have been created, while then again, explicit hereditary and biochemical systems have additionally been recognized in rice and other crops (Schnarrenberger et al., 1995; Reddy et al., 2004; Yan et al., 2006).

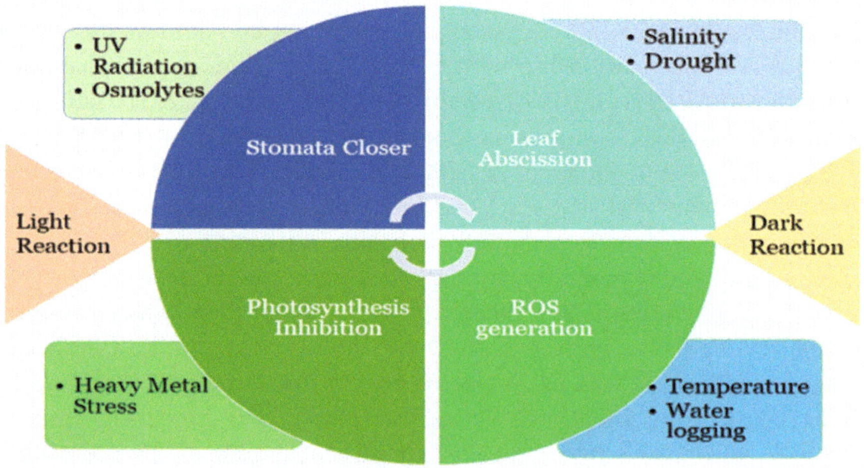

FIGURE 1.1 Different forms of abiotic stress.

1.2.1 MORPHOLOGICAL RESPONSE

Under submerged conditions, plants face oxygen insufficiency and react morphologically to maximize oxygen uptake. Aerenchyma tissue may develop in the root cortical cells that provide oxygen ceaselessly from shoot to roots. Besides, plants undergo morphological changes in roots to endure such pressure (Vu et al., 1984; Zeng et al., 2008). The unusual roots are invigorated by prior root primordia situated at the shoot region. Roots, situated at the soil surface, are stretched out in woody and herbaceous species. Shoot parts likewise show some morphological changes as in *Rumex palustris* during flooding pressure. The increments in the petiole length and leaf regions over the water level are other such responses. Shoots also undergo hypertrophy that resembles white elastic tissue, which are normally noticeable in stems, close to water surfaces. Hypertrophic tissues are auxiliary aerenchyma and are outside the phellogens. The principal job of hypertrophic tissues in the shoot is the improvement of oxygen development in plant tissues to create energy through mitochondrial activity.

1.2.2 WATERLOGGING TOLERANCE STRATEGY

Waterlogging in a crop field is characterized by the submergence of plant parts. Under such circumstance, excess moisture condition in the rhizosphere is very common. Under submergence, the majority of the shoots is presented to air in different crops such as maize, wheat, and soybean, and hence is presented to water stagnation because of excess rainfall (Ruelland et al., 2009; Ueda et al., 2015; Sharma et al., 2017). Waterlogging stress instigates genuine yield loss, relying upon the crop development stage (reproductive stage showed more yield loss than vegetative stage). To endure waterlogging, plants bring about morphological or metabolic changes. In the case of soybean, unusual rooting occurs as compared to shoot formation that is near the water surface during waterlogging.

1.3 SALT STRESS

Productivity of all crops is affected by high salt levels; yet, it contrasts extraordinarily among species and among cultivars. The formative developmental phase of plant and the external factors likewise adjust plant reactions against salinity. The abundance of salt not only impacts growth and developmental processes such as poor seed germination, plant height, root length, and yield output but also affects the food web as a whole. This occurs because of salt-incited oxidative damage and reduced photosynthesis rate. In spite of the adverse effects, plants can endure and complete the crop cycle in high salinity conditions (Kim et al., 2005; Lee et al., 2007; Mutava et al., 2015) due to the evolution of organic, synthetic, and physiological systems. Among the salt tolerance systems, salt overly sensitive pathways, hyperosmotic sensors, and Na and K carriers are the mainstream regulators. In addition, osmoregulation and hormonal variations have been recognized as natural salinity resilience mechanism of plants. Rhizospheric colonization of the growth-promoting rhizobacteria (PGPR) can also improve plant resilience against salinity. Generating transgenic salt-tolerant plants is another effective approach, which however has limitation because salt tolerance is a multigenic trait (Denman et al., 2007; Fernandez et al., 2012; Hu et al., 2015).

1.4 HEAVY METAL STRESS

The presence of heavy metals in farmlands causes various abnormalities in the developmental processes of plants and affects distinct biochemical and physiological cycles. Among the various metals, Mo, Cu, Co, Mn, Ni, Zn, and Fe are needed in minor quantities by plants and are viewed as fundamental micronutrients. Some of them act as cofactors for a few essential metabolic enzymes; however, when present in excess, they exert a destructive effect on plants. Arsenic (As), cadmium (Cd), chromium (Cr), and lead (Pb) are nonessential to plants and are regarded as insignificant elements (Fig. 1.2).

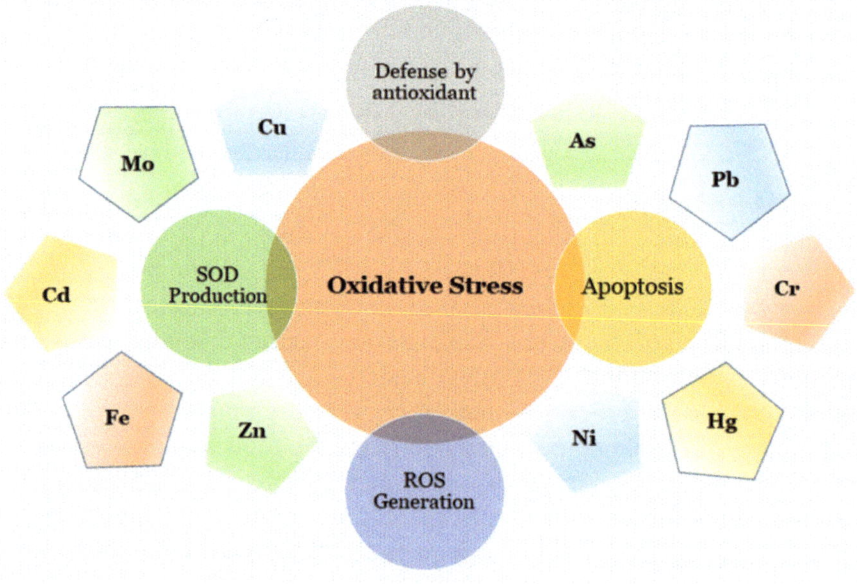

FIGURE 1.2 Heavy metals causing plant stress.

1.5 RADIATION STRESS

During the previous few decades, pulverization of timberlands and burning of petroleum derivatives prompted the aggravation of carbon dioxide, chlorofluorocarbons, methane, nitrous oxide, and such greenhouse gases, which enhance solar radiation to arrive on the earth (Xiong et al., 2006; Chakrabarty et al., 2009; Tripathi et al., 2016). This active radiation is retained and reradiated by the above ozone-depleting harmful substances, causing global warming and creating an enhancement in temperature on earth surface. Further, the amount of ozone-depleting substances is as yet expanding and causing climatic aberrations. Environmental change contemplates are vital in farming and natural science. Reactions of the plants to different ecological disturbances are represented by complex subatomic and biochemical sign transduction measures, coordinating to regulating resilience in plants. Stress conditions created by radiation include (UV-B) irradiance, which induces a varied cluster of plant reactions including variations in developmental rates and productivity. Different reports have shown the harming impact of UV-B to DNA, proteins, and lipids, causing ROS accumulation by perturbing the decrease of NADPH formation during photosynthetic electron transport chain. The induction of various phenolic compounds (stilbenes and flavonoids) and vitamin D is stimulated by UV-B (Pankovic et al., 1999; Feller et al., 2008; Cechin et al., 2018) as defense mechanism.

1.6 CONCLUSION AND FUTURE PERSPECTIVES

Plants are consistently affected by different ecological burdens that incorporate abiotic and biotic factors. Stresses due to abiotic factors are common in the present scenario of climate change because of a few anthropogenic factors like contamination, global warming, and depletion of ozone. It was seen that up to 70% of the yield efficiency is influenced by natural elements (Asada, 1994). The physiological activities of plants inclusive of unequivocal reactions are likewise being upset under stress situations. So, to fulfill the developing need for food around the world, viable, novel, and strong methodologies ought to be adopted. The application of phytohormones could be an intelligent game plan, which can improve the yield. The phytohormones are incredible endogenous substances for fitting functional reactions in plants. Further investigations in the field of proteomics, transcriptomics, metabolomics, and novel cell imaging procedures would upgrade our comprehension of photosynthetic mechanism under abiotic stresses. The different strides in RuBisCO debasement, and the enzyme regulations associated with photosynthetic cycles still need to be clarified under environmental stresses. Nanotechnology is also regarded as one of the encouraging ways to deal with the improvement of environmental stress resilience. Nanoparticle application has been found to improve abiotic stress resistance by lowering ROS as well as improving photosynthesis. However, positive or adverse consequences of nanoparticles largely rely on various factors, namely, their size, surface charge, etc. Future research should further explore beneficial outcomes of nanoparticles on different physiological regulations like osmolyte synthesis, and maintaining ionic homeostasis under abiotic stresses.

KEYWORDS

- **environmental stress**
- **antioxidants**
- **plant responses**
- **reactive oxygen species**
- **oxidative damages**

REFERENCES

Asada, K. Mechanisms for Scavenging Reactive Molecules Generated in Chloroplasts under Light Stress. In *Photoinhibition of Photosynthesis: From Molecular Mechanisms to the Field*; Baker, N. R.; Bowyer, J. R.; Bios Scientific Publishers Ltd: Oxford, **1994,** pp 129–142.

Armada, E.; Rolda´n, A.; Azcon, R. Differential Activity of Autochthonous Bacteria in Controlling Drought Stress in Native Lavandula and Salvia Plants Species under Drought Conditions in Natural Arid Soil. *Microb. Ecol.* **2014,** *67*, 410420.

Ashraf, M.; Harris, P. J. C. Photosynthesis under Stressful Environments: An Overview. *Photosynthetica* **2013**, *51*, 163190.

Asner, G. P.; Alencar, A. Drought Impacts on the Amazon Forest: The Remote Sensing Perspective. *New Phytol.* **2010,** *187* (3), 569578.

Atkin, O. K.; Macherel, D. The Crucial Role of Plant Mitochondria in Orchestrating Drought Tolerance. *Ann. Bot.* **2009,** *103*, 581597.

Baker, N. R. Photoinhibition of Photosynthesis. In *Light as an Energy Source and Information Carrier in Plant Physiology*; Jennings, P. R., Zucchelli, F., Ghetti, F., Colombetti, G.; NATO ASI Series, Series A: Life Science, Vol. 287; Plenum Press: New York, 1996.

Bahrami-Rad, S.; Hajiboland, R. Effect of Potassium Application in Drought-Stressed Tobacco (*Nicotiana rustica* L.) Plants: Comparison of Root with Foliar Application. *Ann. Agric. Sci.* **2017,** *62*, 121130.

Balestrini, R.; Lumini, E.; Borriello, R.; Bianciotto, V. Plant-Soil Biota Interactions. In *Soil Microbiology, Ecology and Biochemistry*; 2015; p 311338.

Bartels, D.; Sunkar, R. Drought and Salt Tolerance in Plants. *CRC Crit. Rev. Plant Sci.* **2005,** *24*, 2358.

Brauer, M.; Freedman, G.; Frostad, J.; van Donkelaar, A.; Martin, R. V.; Dentener, F. Ambient Air Pollution Exposure Estimation for the Global Burden of Disease 2013. *Environ. Sci. Technol.* **2016,** *50*, 7988.

Carmo-Silva, A. E.; Keys, A. J.; Andralojc, P. J.; Powers, S. J.; Arrabac¸a, M. C.; Parry, M. A. J. Rubisco Activities, Properties, and Regulation in Three Different C4 Grasses under Drought. *J. Exp. Bot.* **2010,** *61* (9), 23552366.

Cechin, I.; Gonzalez, G. C.; Corniani, N.; Terezinha de, F. F. The Sensitivity of Sunflower (*Helianthus annuus* L.) Plants to UV-B Radiation Is Altered by Nitrogen Status. *Cie^ncia Rural, Santa Maria* **2018,** *48* (02), e20170369.

Chakrabarty, D.; Trivedi, P. K.; Misra, P.; Tiwari, M.; Shri, M.; Shukla, D. Comparative Transcriptome Analysis of Arsenate and Arsenite Stresses in Rice Seedlings. *Chemosphere* **2009,** *74*, 688702.

Chatterjee, S.; Sharma, S.; Gupta, D. K. Arsenic and Its Effect on Major Crop Plants: Stationary Awareness to Paradigm with Special Reference to Rice Crop. In *Arsenic Contamination in the Environment: The Issues and Solutions*; Gupta, D. K., Chatterjee, S., Eds.; Springer International Publishing AG, 2017; pp 123143. https://doi.org/10.1007/978-3-319-54356-7-6.

Denman, K. L.; Brasseur, G.; Chidthaisong, A.; Ciais, P.; Cox, P. M.; Dickinson, R. E. Couplings between Changes in the Climate System and Biogeochemistry. In *Climate Change 2007: The Physical Science Basis. Contribution of Working Group I to the Fourth Assessment Report of the Intergovernmental Panel on Climate Change*; Solomon, S., Qin, D., Manning, M., Chen, Z., Marquis, M., Averyt, K. B., Tignor, M., Miller, H. L., Eds.; Cambridge University Press: Cambridge, 2007; pp 499–587.

Dizengremel, P.; Vaultier, M. N.; Thiec, D. L.; Cabane, M.; Bagard, M.; Ge´rant, D. Phosphoenolpyruvate Is at the Crossroads of Leaf Metabolic Responses to Ozone Stress. *New Phytol.* **2012,** *195*, 512–517.

Duman, F.; Ozturk, F.; Aydin, Z. Biological Responses of Duckweed (*Lemna minor* L.) Exposed to the Inorganic Arsenic Species As(III) and As(V): Effects of Concentration and Duration of Exposure. *Ecotoxicology.* **2010,** *19*, 983–993.

Eckardt, N.; Pell, E. Oxidative Modification of Rubisco from Potato Foliage in Response to Ozone. *Plant Physiol. Biochem.* **1995,** *33*, 273–282.

Faller, P.; Kienzler, K.; Krieger-Liszkay, A. Mechanism of Cd21 Toxicity: Cd21 Inhibits Photoactivation of Photosystem II by Competitive Binding to the Essential Ca21 Site. *Biochim. Biophys. Acta. Bioenerg.* **2005,** *1706*, 158–164.

Fang, Y.; Xiong, L. General Mechanisms of Drought Response and Their Application in Drought Resistance Improvement in Plants. *Cell. Mol. Life Sci.* **2015,** *72*, 673–689.

Feller, U.; Anders, I.; Demirevska, K. Degradation of Rubisco and Other Chloroplast Proteins under Abiotic Stress. *Gen. Appl. Plant Physiol.* **2008,** *34*, 5–18.

Fernandez-Garcia, N.; Romojaro, F.; Olmos, E.; Estrella, E.; Bolarin, M. C.; Flores, F. B. Understanding the Mechanisms of Chilling Injury in Bell Pepper Fruits Using the Proteomic Approach. *J. Proteomics* **2012,** *75*, 5463–5478.

Finnegan, P. M.; Chen, W. Arsenic Toxicity: The Effects on Plant Metabolism. *Front. Physiol.* 2012, 3 (182). https://doi.org/10.3389/fphys.2012.00182.

Gupta, D. K.; Chatterjee, S., Eds. *Arsenic Contamination in the Environment: The Issues and Solutions*; Springer International Publishing: AG Switzerland, 2007. https://doi.org/10.1007/978-3-319-54356-7.

Gupta, B. K.; Sahoo, K. K.; Ghosh, A.; Tripathi, A. K.; Anwar, K.; Das, P. Manipulation of Glyoxalase Pathway Confers Tolerance to Multiple Stresses in Rice. *Plant Cell Environ.* **2018,** *41* (5), 1186–1200. https://doi.org/10.1111/pce.12968.

Hancock, R. D.; Morris, W. L.; Ducreux, L. J.; Morris, J. A.; Usman, M.; Verrall, S. R. Physiological, Biochemical and Molecular Responses of the Potato (*Solanum tuberosum* L.) Plant to Moderately Elevated Temperature. *Plant Cell Environ.* **2014,** *37*, 439–450.

Hu, X.; Wu, L.; Zhao, F.; Zhang, D.; Li, N.; Zhu, G. Phosphoproteomic Analysis of the Response of Maize Leaves to Drought, Heat and Their Combination Stress. *Front. Plant Sci.* **2015,** *6*, 1–21.

Iglesias-Ferna′ndez, R.; Matilla, A. J. Flooding Stress and O2-shortage in Plants. In *Water Stress and Crop Plants: A Sustainable Approach*; Ahmad, P., Ed., Vol. 1; About Wiley Online Library Chapter 41; 2016. https://doi.org/10.1002/9781119054450.ch41.

James, R. A.; Munns, R.; von Caemmerer, S.; Trejo, C.; Miller, C.; Condon, T. A. G. Photosynthetic Capacity Is Related to the Cellular and Subcellular Partitioning of Naþ, Kþ and Cl- in Salt-Affected Barley and Durum Wheat. *Plant Cell Environ.* **2006,** *29*, 2185–2197.

Jordan, B. R.; He, J.; Chow, W. S.; Anderson, J. M. Changes in mRNA Levels and Polypeptide Subunits of Ribulose- 1, 5-Bisphosphate Carboxylase in Response to Supplemental UV-B Radiation. *Plant Cell Environ.* **1996,** *15*, 91–98.

Kim, S. Y.; Lim, J. H.; Park, M. R.; Kim, Y. J.; Park, T.; Seo, Y. W. Enhanced Antioxidant Enzymes Are Associated with Reduced Hydrogen Peroxide in Barley Roots under Saline Stress. *J. Biochem. Mol. Biol.* **2005,** *38*, 218–224.

Kosova, K.; Vitamvas, P.; Prasil, I. T.; Renaut, J. Plant Proteome Changes under Abiotic Stress—Contribution of Proteomics Studies to Understanding Plant Stress Response. *J. Proteome* **2011,** *74*, 1301–1322.

Lawlor, D. W. Limitations to Photosynthesis in Water-Stressed Leaves: Stomatal vs. Metabolism and the Role of ATP. *Ann. Bot.* **2002,** *89*, 871–885.

Lawlor, D. W.; Tezara, W. Causes of Decreased Photosynthetic Rate and Metabolic Capacity in Water Deficient Leaf Cells: A Critical Evaluation of Mechanisms and Integration of Processes. *Ann. Bot.* **2009,** *103*, 561–579.

Lee, D. G.; Ahsan, N.; Lee, S. H.; Kang, K. Y.; Bahk, J. D.; Lee, I. J. A Proteomic Approach in Analyzing Heat-Responsive Proteins in Rice Leaves. *Proteomics* **2007,** *7*, 3369–3383.

Mustafa, G.; Komatsu, S. Toxicity of Heavy Metals and Metal-Containing Nanoparticles on Plants. *Biochim. Biophys. Acta.* **2016,** *1864*, 932–944.

Mutava, R. N.; Prince, S. J. K.; Syed, N. H.; Song, L.; Valliyodan, B.; Chen, W. Understanding Abiotic Stress Tolerance Mechanisms in Soybean: A Comparative Evaluation of Soybean Response to Drought and Flooding Stress. *Plant Physiol. Biochem.* **2015,** *86*, 109–120.

Nouri, M. Z.; Moumeni, A.; Komatsu, S. Abiotic Stresses: Insight Into Gene Regulation and Protein Expression in Photosynthetic Pathways of Plants. *Int. J. Mol. Sci.* **2015,** *16*, 20392–20416.

Osakabe, K.; Osakabe, Y. Plant Light Stress. In *Encyclopaedia of Life Sciences*; Robinson, S. A., Ed.; Nature Publishing Group: London, 2012.

Pankovic, D.; Sakac, Z.; Kevresan, S.; Plesnicar, M. Acclimation to Long-Term Water Deficit in the Leaves of Two Sunflower Hybrids: Photosynthesis, Electron Transport and Carbon Metabolism. *J. Exp. Bot.* **1999,** *50*, 127–138.

Parry, M. A. J.; Keys, A. J.; Madgwick, P. J.; Carmo-Silva, A. E.; Andralojc, P. J. Rubisco Regulation: A Role for Inhibitors Centre. *J. Exp. Bot.* **2008,** *59*, 1569–1580.

Pell, E. J.; Landry, L. G.; Eckardt, N. A.; Click, R. E. Effects of Gaseous Air Pollutants on Ribulose Bisphosphate Carboxylase/Oxygenase: Effects and Implications. In *Plant Response to the Gaseous Environment: Molecular, Metabolic and Physiological Aspects*; Alscher, R. G., Wellburn, A. R., Eds.; Chapman and Hall: London, 1994; pp 239–254.

Qureshi, M. I.; Abdin, M. Z.; Ahmad, J.; Iqbal, M. Effect of Long-Term Salinity on Cellular Antioxidants, Compatible Solute and Fatty Acid Profile of Sweet Annie (*Artemisia annua* L.). *Phytochemistry*. **2013,** *95*, 215–223.

Reddy, A. R.; Chaitanya, K. V.; Vivekanandan, M. Drought-Induced Responses of Photosynthesis and Antioxidant Metabolism in Higher Plants. *J. Plant Physiol.* **2004,** *161*, 1189–1202.

Rı´o Segade, S.; Vilanova, M.; Giacosa, S.; Perrone, I.; Chitarra, W.; Pollon, M. Ozone Improves the Aromatic Fingerprint of White Grapes. *Sci. Rep.* **2017,** *24*, 16301. https://doi.org/10.1038/s41598-017-16529-5.

Ruelland, E.; Vaultier, M. N.; Zachowski, A.; Hurry, V. Cold Signalling and Cold Acclimation in Plants. In *Advances in Botanical Research*; Jean-Claude, K., Michel, D., Eds., vol. 49. Academic Press, 2009; pp 35–150.

Schnarrenberger, C.; Flechner, A.; Martin, W. Enzymatic Evidence Indicating a Complete Oxidative Pentose Phosphate in the Chloroplasts and an Incomplete Pathway in the Cytosol of Spinach Leaves. *Plant Physiol.* **1995,** *108*, 609–614.

Shahid, M.; Shamshad, S.; Rafiq, M.; Khalid, S.; Bibi, I.; Niazi, N. K. Chromium Speciation, Bioavailability, Uptake, Toxicity and Detoxification in Soil-Plant System: A Review. *Chemosphere* **2017,** *178*, 513–533.

Sharma, S.; Chatterjee, S.; Kataria, S.; Joshi, J.; Datta, S.; Vairale, M. G. A Review on Responses of Plants to UV-B Radiation Related Stress. In *UV-B Radiation: From Environmental Stressor to Regulator of Plant Growth*; Singh, V. P., Singh, S., Prasad, S. M., Parihar, P., Eds.; John Wiley Sons: New York, 2017; pp 75–97. https://doi.org/10.1002/9781119143611, Ch. 5.

Tournaire-Roux, C.; Sutka, M.; Javot, H.; Gout, E.; Gerbeau, P.; Luu, D. T. Cytosolic pH Regulates Root Water Transport during Anoxic Stress through Gating of Aquaporins. *Nature*. **2003,** *425*, 393–397.

Tripathi, A. K.; Pareek, A.; Singla-Pareek, S. L. A NAP-Family Histone Chaperone Functions in Abiotic Stress Response and Adaptation. *Plant Physiol.* **2016,** *171*, 2854–2868.

Ueda, Y.; Siddique, S.; Frei, M. A Novel Gene, Ozone-Responsive Apoplasticprotein1, Enhances Cell Death in Ozone Stress in Rice. *Plant Physiol.* **2015,** *169*, 873–889.

Velikova, V.; Tsonev, T.; Pinelli, P.; Alessio, G. A.; Loreto, F. Localized Ozone Fumigation System for Studying Ozone Effects on Photosynthesis, Respiration, Electron Transport

Rate and Isoprene Emission in Field-Grown Mediterranean Oak Species. *Tree Physiol.* **2005,** *25*, 1523–1532.

Vu, C. V.; Allen, L. H.; Garrard, L. A. Effects of UV-B Radiation (280–320 nm) on Ribulose-1, 5 Bisphosphate Carboxylase in Pea and Soybean. *Environ. Exp. Bot.* **1984,** *24*, 131–143.

Wang, H. B.; Xie, F.; Yao, Y. Z.; Zhao, B.; Xiao, Q. Q.; Pan, Y. H. The Effects of Arsenic and Induced-Phytoextraction Methods on Photosynthesis in Pteris Species with Different Arsenic-Accumulating Abilities. *Environ. Exp. Bot.* **2012,** *75*, 298–306.

Xin, Z.; Browse, J. Cold Comfort Farm: The Acclimation of Plants to Freezing Temperatures. *Plant Cell Environ.* **2000,** *23*, 893–902.

Xiong, Z.; Zhao, F.; Li, M. Lead toxicity in Brassica pekinensis Rupr.: effect on nitrate assimilation and growth. *Environ. Toxicol.***2006,** *21* (2), 147–153.

Yadav, R. S.; Hash, C. T.; Bidinger, F. R.; Devos, K. M.; Howarth, C. J. Genomic Regions Associated with Grain Yield and Aspects of Post Flowering Drought Tolerance in Pearl Millet across Environments and Tester Background. *Euphytica.* **2004,** *136*, 265–277.

Yan, S. P.; Zhang, Q. Y.; Tang, Z. C.; Su, W. A.; Sun, W. N. Comparative Proteomic Analysis Provides New Insights Into Chilling Stress Responses in Rice. *Mol. Cell. Proteomics.* **2006,** *5*, 484–496.

Zargar, S. M.; Kurata, R.; Inaba, S.; Oikawa, A.; Fukui, R.; Ogata, Y. Quantitative Proteomics of Arabidopsis Shoot Microsomal Proteins Reveals a Cross-Talk Between Excess Zn and Iron Deficiency. *Proteomics.* **2015,** *15*, 1196–1201.

Zeng, Z. Z.; Wei, J.; Wei, R. H.; Zhu, X. Q. Investigation of Frost Bitten Sugarcane in Liujiang County. In *Proceedings of Annual Meeting of Guangxi Society of Sugarcane Technologists*; 2008; pp 189–192.

Zhang, L.; Zhang, L.; Sun, J.; Zhang, Z.; Ren, H.; Sui, X. Rubisco Gene Expression and Photosynthetic Characteristics of Cucumber Seedlings in Response to Water Deficit. *Sci. Hortic.* **2013,** *161*, 81–87.

Zhang, B. Q.; Yang, L. T.; Li, Y. R Physiological and Biochemical Characteristics Related to Cold Resistance in Sugarcane. *Sugar Tech.* **2015,** *17*, 49–58.

CHAPTER 2

Abiotic Stress-Mediated Regulation of Photosynthesis and Modulations in Photosynthetic Apparatus: Impact on Photosynthetic Genes and Enzyme Functioning

AWATIF M. ABDULMAJEED[1], SAMEER H. QARI[2], TAGHREED S. ALNUSAIRE[3], and MONA H. SOLIMAN[4,5*]

[1]*Biology Department, Faculty of Science, University of Tabuk, Umluj 46429, Saudi Arabia*

[2]*Biology Department, Aljumum University College, Umm Al-Qura University, Makkah, P.O. 2203, Saudi Arabia*

[3]*Biology Department, College of Science, Jouf University, Sakaka, 2014, Saudi Arabia*

[4]*Botany and Microbiology Department, Faculty of Science, Cairo University, Giza 12613, Egypt*

[5]*Biology Department, Faculty of Science, Taibah University, Al-Sharm, Yanbu El-Bahr, Yanbu 46429, Kingdom of Saudi Arabia*

[*]*Corresponding author. E-mail: hmona@sci.cu.edu.eg*

ABSTRACT

Environmental stresses alter the growth and productivity of crop plants by affecting the key metabolic pathways. Among these pathways, photosynthesis

Photosynthesis and Respiratory Cycles during Environmental Stress Response in Plants. Aryadeep Roychoudhury (Ed.)

holds a key position contributing immensely to overall development and productivity. The tolerance levels to different abiotic stresses vary considerably from species to species. Abiotic stresses significantly alter photosynthetic functioning by upsurging the production of toxic reactive oxygen production leading to disruption of the photosynthetic components like electron transport, photosystems I and II, adenosine triphosphate (ATP) generation, and carbon fixation by overall influence on the enzymes regulating the carbon fixation process. Both C3 and C4 systems display differential behavior in response to abiotic stresses. The alteration in the stomatal and nonstomatal attributes ultimately affects the photosynthesis considerably. Designing plants with improved photosynthetic functioning is imperative with focus on the antioxidant machinery of the chloroplast, PSII repair mechanisms, redox homeostasis, and the associated signaling components. The present review addresses the influence of abiotic stresses on photosynthetic functioning and the mechanisms contributing to amelioration of the damaging effects.

2.1 INTRODUCTION

Plants confront various environmental stress factors during their life cycle, which include both biotic as well as abiotic stresses. Stresses impart a significant decline in the growth, enzyme functioning, nutrient uptake and assimilation, photosynthesis, protein synthesis, and synthesis of metabolites (Baslam et al., 2020; Begum et al., 2019). To alleviate the damaging effects of stresses, plants upregulate several tolerance mechanisms that may be common or stress specific. The decline in photosynthetic efficiency in plants under stress has been reported widely (Ashraf and Harris, 2013; Khan et al., 2014; Iqbal et al., 2015; Kaur et al., 2017; Yadav et al., 2018; Elkelish et al., 2019; Ahanger et al., 2019, 2020; Soliman et al., 2019, 2020; Qin et al., 2021). Stress-induced decline in photosynthesis has cumulative effect on the key photosynthetic attributes including stomatal and nonstomatal components. Reduced stomatal conductance, intercellular CO_2 concentrations, increased reactive oxygen species (ROS) production, and downregulation of Ribulose-1,5-bisphosphate carboxylase activity, PS I and PS II functioning, and chlorophyll biosynthesis are the key damaging effects of stress that ultimately contribute to photosynthetic arrest in plants (Xia et al., 2006; Kalaji et al., 2016; Fatma et al., 2016; Ahmad et al., 2019, 2020). The general effects of stress factors on plant growth include declined root and shoot growth, limited access to soil mineral elements, membrane dysfunctioning, reduced enzyme activity, photoinhibition, etc. Besides having other reasons for these mentioned damaging effects, it is believed that the key contributors include the excessive generation of reactive oxygen and nitrogen species. ROS include toxic species like superoxide, hydrogen peroxide, singlet oxygen, and hydroxyl radicals (Begum et al., 2020; Qin et al., 2021; Ahanger et al., 2021), while reactive nitrogen species include nitric oxide, etc. These reactive radicals are produced in different cellular organelles including chloroplast, mitochondria, peroxisomes, etc. The excess concentrations of these reactive radicals are toxic to metabolism; however, they can serve as signaling molecules at optimal concentrations (Baslam et al., 2020). The present chapter reviews the effects of different environmental stresses on growth, photosynthesis, and the enzymes of photosynthesis. In addition, the role of tolerance mechanisms, protective compounds, and the influence of different molecules in the regulation of photosynthesis through modulation of tolerance mechanisms have been reviewed.

2.2 IMPACT OF DROUGHT STRESS ON PHOTOSYNTHESIS

Drought stress is another key environmental stress factor damaging plant growth and productivity (Ahanger and Agarwal, 2017a; Begum et al., 2019, 2020). Drought stress arises due to shortage of water and the global climate change has made the situation even graver. Prime effects of drought stress include restricted root and shoot growth, reduced mineral uptake and assimilation, degradation of enzymes, lipids and nucleic acids, photosynthetic inhibition, water loss, and altered osmoregulation (Ohashi et al., 2006; Ahanger et al., 2014; Li et al., 2017; Zhao et al., 2020). Limited availability of water directly influences the uptake and water use efficiency, hence affecting the major cellular metabolic pathways (Zhao et al., 2020). Reduced growth and photosynthesis under drought stress have been attributed to excess generation of ROS in the key cellular organelles including chloroplasts, leading to damage to D1 protein, and photosynthetic electron transport gets hampered (Wang et al., 2018). Xerophytic plants have developed certain key adaptations to endure drought periods, which include absorption of enough water to complete the life cycle quickly to escape drought periods, reduction in water loss due to transpiration by causing a significant reduction in number of stomata, stomatal closure, and morphological modification like production of thick cuticle to control water loss (Lopes et al., 2011; Swamy and Kumar, 2013; Basu et al., 2016). Drought has been reported to cause a significant decline in plant height, biomass accumulation, turgor loss, and leaf number (Semerci et al., 2017). Drought resistance and tolerance vary with the species and the cultivar levels have been reported by Sun et al. (2013) and Ahanger and Agarwal (2017a). Deficiency of soil water causes tissue water imbalance, and in response to this reduced water potential, transpiration rates are cut down by the early closure of stomata. Stomatal closure has been considered as the prime response of plants to drought (Bastos et al., 2011; Ying et al., 2013; Wang et al., 2018; Kapoor et al., 2020). Stomatal mechanisms are very important for maintaining the tissue water balance for protecting major cellular processes; however, severe drought conditions impact carbon assimilation and hence the production of carbon assimilates via significant decline in photosynthesis (Tenhunen et al., 1990; Cornic, 2000; Feller et al., 2016). Drought mediated decline in carbon assimilation ultimately reduces the availability of carbon-containing metabolites for further assimilation and regulation of metabolic processes (McDowell and Sevanto, 2010). Wang et al. (2018) demonstrated that drought-mediated alteration in photosynthetic stomatal attributes like photosynthetic rate, stomatal conductance, transpiration rate, and intercellular CO_2 concentration as well as nonstomatal attributes like maximum photochemical efficiency (F_v/F_m), actual photochemical efficiency of PSII (Φ_{PSII}), photochemical quenching coefficient (q_P), and coefficient of photochemical fluorescence quenching assuming interconnected PSII antennae (q_L), exhibit strong positive correlation with water potential (ψ_w). Drought-mediated decline in photosynthetic CO_2 assimilation reflects in stunted growth and respiration (Pinheiro and Chaves, 2011). Photosynthesis under drought is strongly affected by leaf area index and hence productivity (Xu et al., 2020).

Drought stress results in differential expression of genes involved in abiotic stress response and pathways related to catalytic activity, methyltransferase activity, transferase activity, and superoxide metabolic process. In addition, the expression of genes encoding proteins belonging to helix-loop-helix, WRKY, NAM (no apical meristem), ATAF1/2, and CUC2 (cup-shaped cotyledon) (NAC), fatty acyl-CoA reductase (FAR1), B3, myeloblastosis (MYB)-related and basic leucine zipper families have a key role in drought-rehydration mechanism (Zhou et al., 2021).

To avoid the drought-mediated photo inhibition, CO_2 supply must be maintained, which otherwise declines due to drought-induced stomatal closure. However, the effect of other attributes besides the stomatal components on photosynthesis cannot be neglected (Hajiboland et al., 2017). The reduced production of RuBisCO restricted allocation of key mineral ions like N and Mg to photosynthetic apparatus; membrane disruption and excess ROS production are the other key reasons in addition to the stomatal and nonstomatal components responsible for restricted photosynthesis (Parry et al., 2002; Guliyev et al., 2008; Ahanger and Agarwal, 2017a; Ahanger et al., 2017, 2021; Kapoor et al., 2020). Besides this, the key factor reducing the rate of photosynthesis and carbon fixation is the damaged PSII structure and functioning (Tattini et al., 2014).

Responses to drought are regulated by the modulations in the endogenous phytohormones. It has been reported that concentration of abscisic acid (ABA) is regulated by modulating the activity of key enzymes including 9-cis-epoxy carotenoid dioxygenase and δ-arabino-1,4-lactone oxidase (Bao et al., 2016). Recently, Wang et al. (2021) demonstrated active involvement of ABA and jasmonic acid (JA) in drought stress tolerance through upregulating the genes involved in their synthesis; JA functions upstream of ABA. Jasmonates have active roles in key signal transduction pathways, including those of gene networks, signaling intermediates and proteins, regulatory proteins, enzymes, and molecules acting for protection of plants against the toxic effects of abiotic stresses (Ali and Baek, 2020). Drought-induced decline in photosynthesis is balanced actively by the influence of ABA on stomatal conductance and transpiration (Guo et al., 2008; reviewed in Vishwakarma et al., 2017). In drought-stressed *Solanum pennellii,* Egea et al. (2018) demonstrated significant upregulation of genes linked with synthesis and signaling of jasmonate and ethylene that might have significantly regulated the stomatal and nonstomatal photosynthetic attributes. Jasmonate-induced synthesis of secondary metabolites has been reported to mediate regulation of photosynthetic functioning in *Ipomoea batatas* under drought stress (Yoshida et al., 2020). Salicylic acid application modulates leaf protein synthesis, resulting in significant protection to photosynthesis (Khalvandi et al., 2021). These reports imply that phytohormones have key roles in the regulation of photosynthesis in plants through modulation of key metabolic pathways.

Restricted water availability affects the continuous generation of $NADP^+$ ultimately causing leakage of electron and hence electron transport chain gets over reduced (Nishiyama and Murata, 2014). In addition, drought affects the oxygen-evolving center of PSII sternly, leading to significant drop in the PSII reaction center and declined chlorophyll fluorescence (Skotnica et al., 2000). Drought stress affects photosynthesis not only by hampering the stomatal movements but also due to the degradation of key proteins like that of the RuBisCO enzyme, hence declining its regeneration, ultimately affecting the photosynthetic CO_2 assimilation and impairment in PSII reaction center (Reddy et al., 2004; Demirevska et al., 2009). Drought stress results in the degradation of thylakoid membranes of chloroplasts, negatively affecting the photosynthetic pigment and photosynthetic rate (Shao et al., 2016). The kinetics of chlorophyll fluorescence is also altered reflected in reduced PSII functioning (Zhang et al., 2011). The effect of desiccation-mediated decline in tissue water content is obvious on PSII electron transport through increased plastoquinone reduction. Different transcription factor family mediates the regulation of gene expression (Saibo et al., 2009; Joshi et al., 2016), for example, protochlorophyllide oxidoreductase has a key role in chlorophyll biosynthesis. Transcription factor DREB1A binds with dehydration responsive element and its upregulation positively affects drought tolerance (Kudo et al., 2017). Overexpression of DREB improves chlorophyll and photosynthesis through reduction in the oxidative stress parameters (Guo et al., 2019).

2.3 SALINITY STRESS AND PHOTOSYNTHESIS

Salinity stress is another important environmental stress factor affecting photosynthesis and plant growth significantly, hence restricting the yield productivity of major crop plants. Salinity stress results due to excess salt accumulation (Ahanger and Agarwal, 2017b; Elkelish et al., 2019; Soliman et al., 2020). Increasing salinization of soil worldwide is declining the yield and productivity (Landi et al., 2017). More than 6% of global land area is salt affected (Yuan et al., 2016). Salinity stress substantially affects key metabolic and cellular processes like photosynthesis, energy metabolism, protein synthesis, and lipid metabolism (Song et al., 2016; Elkelish et al., 2019; Sui et al., 2018). Excess accumulation of ions within the soil solution imparts ionic and osmotic stress, resulting in a significant decline in water balance (Munns, 2002; Qados, 2011). Salinity stress initiates the accumulation of ROS, thereby triggering oxidative stress and damaging the key macromolecules (Saffan, 2008; Ahanger et al., 2020). Like other stresses, salinity also damages the photosynthetic functioning by affecting the chloroplast structural integrity. One of the key reasons for the salinity-mediated decline in photosynthesis is oxidative damage resulting from excess ROS accumulation (Fatma et al., 2016). Under severe salinity stress, plant death can occur. Physiological and biochemical responses to salinity stress include altered photosynthesis, sucrose synthesis, ion homeostasis, and hormonal metabolism. Among the key strategies for improving salinity stress tolerance are (1) salt exclusion to maintain the ion homeostasis, and ensure that salt concentrations are maintained relatively low in shoot than root, and (2) increasing sugar content by protecting photosystem structures, increasing photosynthetic performance and increasing synthesis of sucrose with concomitant decline in its degradation (Yang et al., 2020). There is considerable difference between glycophytes and halophytes in their response to salinity stress tolerance. Membrane transporters and the signaling pathways like salt overly sensitive act as prime sensors for the salinity stress (Isayenkov and Maathuis, 2019).

Salinity declines photosynthesis and the export of photosynthate to vegetative sinks significantly (Suwa et al., 2006). Decline in gas exchange and photosynthesis under salinity stress shows considerable variations in sensitive and tolerant plant cultivars (Najar et al., 2018; Yang et al., 2020). Salinity declines stomatal density, chlorophyll synthesis, and hence photosynthesis gets affected (Kwon et al., 2019). In NAD-malic enzyme and phosphoenolpyruvate carboxykinase type C4 plants, Omoto et al. (2010) reported that salinity stress declines chlorophyll content and chloroplast structure. Swelling of thylakoids, disruption of thylakoid envelopes in mesophyll cell chloroplasts, and a significant reduction in the stromal and starch areas in bundle sheath chloroplast occur. Moreover, salinity resulted in increased granal development reflected as relatively higher number of thylakoid per granum, granal indices, and appressed thylakoid density, compared with untreated plants (Omoto et al., 2010). The protective roles of key pathways like xanthophyll cycle, water–water cycle, and photorespiration are believed to be promising in the protection of photosynthetic machinery under salinity stress (Chaves et al., 2009; Acosta-Motos et al., 2017). Photorespiration prevents overreduction of chloroplast electron transport chain by acting as sink for the excess reducing power (Halliwell and Gutteridge, 2000). Water–water cycle also protects the effects of excess radicals by mediating the photoreduction of molecular oxygen to water in PSI by the electrons generated in PSII from water (Asada, 1999). Excess ROS can inhibit the activity of enzymes, APX, SOD, Calvin cycle, and photorespiration (Asada, 1999). Water–water cycle maintains linear flow of electrons for the production of ATP necessary for the photorespiratory pathway, thereby assisting in dissipation of excess photon energy. Xanthophyll cycle is another mechanism involved in the dissipation of excess excitation in which carotenoid violaxanthin is converted to zeaxanthin by the enzyme violaxanthin de-epoxidase (Zhang et al., 2012). Xanthophyll cycle dissipates excess excitation energy in PSII antenna in the form of heat through nonphotochemical quenching (NPQ) (Zhang et al., 2012; Acosta-Motos et al., 2017). Zeaxanthin acts as antioxidant, and plants accumulating higher content of zeaxanthin show better alleviation of salinity stress-induced photoinhibition and photooxidative damage through quick neutralization of toxic singlet oxygen radicals in the thylakoid membranes (Zhang et al., 2012). Xanthophyll cycle contributes to photoprotection by direct quenching of overexcitation by products of the light phase of xanthophyll cycles. Other indirect participation of xanthophyll cycle carotenoids in photoprotection include (1) indirect quenching of overexcitation by aggregation-dependent light-harvesting complexes quenching (2) light-driven mechanisms in light-harvesting complex II, and (3) a model based on charge transfer quenching between chlorophyll a and zeaxanthin (Latowski et al., 2011). Qiu et al. (2003) observed a linear relationship between the de-epoxidation state of the xanthophyll cycle and maximal efficiency of PSII, Fv/Fm, and NPQ. Changes in actual efficiency of PS II photochemistry, NPQ, content of xanthophylls, and kinetics of de-epoxidation were studied in ABA-fed and non-ABA-fed leaves of rice and cabbage under NaCl stress (Zhu et al., 2011). Salinity stress causes progressive decrease in actual PSII photochemistry, higher reduction state of PSII, and significant increase in NPQ, which was alleviated by ABA treatments by inducing lowered reduction state of PSII and increased NPQ capacity. This resulted in increased photosynthetic electron transport and energy dissipation with reduced accumulation of excess light. This exogenous ABA treatment enhanced the size of xanthophyll cycle pool, de-epoxidation of the xanthophyll cycle components as well as an increase in level of NPQ by altering the kinetics of de-epoxidation of the xanthophyll cycle (Zhu et al., 2011).

Salinity drastically declines the photosynthesis by inhibiting the activity of Calvin cycle enzymes. In salt-stressed cucumber, Shu et al. (2014) demonstrated a significant decline in maximum carboxylase activity of ribulose-1,5-bisphosphate carboxylase/oxygenase, the maximal velocity of RuBP regeneration (Jmax), utilization capacity of triose-phosphate, and carboxylation efficiency, resulting in decline in net photosynthesis, actual photochemical efficiency of PSII, and hence decline in growth. Besides this, salinity has been found to upregulate the transcriptional levels of ribulose-1,5-bisphosphate carboxylase, glyceraldehyde-3-phosphate dehydrogenase, and phosphoribokinase with significant downregulation of ribulose-1,5-bisphosphate carboxylase/oxygenase large subunit and small subunit, fructose-1,6-biphosphate aldolase, triose-3-phosphate isomerase, fructose-1,6-bisphosphate phosphatase, and 3-phosphoglyceric acid kinase (Shu et al., 2014). Salinity stress imparts reproducible and significant effect on photosynthesis by enhancing ROS generation and influencing the functioning of antioxidant system and the expression of proteins coding for photosynthetic Calvin–Benson cycle, photosynthetic electron transport, redox homeostasis, defense, energy, and metabolism (Meng et al., 2016).

2.4 TEMPERATURE STRESS AND PHOTOSYNTHESIS

Both high and low temperatures are adverse to plant growth. Temperature stresses (low and high) induce disruption of plant growth by altering membrane functioning, cellular homeostasis, and enzyme functioning (Mathur et al., 2014). Stunted growth, accumulation of ROS, reduced mineral assimilation, and loss of structural and functional integrity of organelles due to temperature exposures have been reported in several crop plants (Lu et al., 2017; Hu et al., 2020). It is believed that the functioning of both photosystems (PSI and PSII), the reductive pathways of CO_2 assimilation, chlorophylls, and the electron transport systems have direct influence on photosynthesis and overall growth performance of the plant (Ashraf and Hafeez, 2004). High temperature imparts considerable restrictions on the photosynthetic functioning of plants. Photosynthetic machinery is very sensitive to high temperatures and temperature-mediated photoinhibition leads to direct decline of other cellular functions (Mathur et al., 2014; Zhang et al., 2017). It has been observed that decline in the production of RuBisCO is among the key factors for the subsequent damaging effects of heat stress on photosynthetic functioning (Jajoo and Allakhverdiev, 2017).

Photosynthesis is one of the sensitive pathways of plants that is intensively affected by high-temperature stress. Targets of temperature stress in plant photosynthesis include PSII, ribulose-1, 5-bisphosphate carboxylase/oxygenase (RuBisCO), Cytochrome b559, and plastoquinone (Mathur et al., 2014). Exposure of tall fescue (*Festuca arundinacea*) to high temperature affected photosynthesis, PSII functioning, and the antioxidant activity (Cui et al., 2006). Declined photosynthesis, under high-temperature stress, results from altered stomatal and nonstomatal attributes, increased photoinhibition, and reduction in the RuBisCO activity. Nonstomatal alterations in high-temperature stress are manifested as lower variable chlorophyll fluorescence yield (Fv), maximum photochemical efficiency of PSII (Fv/Fm), and efficiency of open reaction center in light (ΦPSII) (Cui et al., 2006). Song et al. (2014) reported significant differential expression of photosynthesis-related genes controlling carbon fixation and stomatal conductance in poplar. Moreover, damaged photosystems, reduced electron transport, H_2O_2

accumulation, and activation of glycolate pathway were among the other effects of high-temperature stress (Song et al., 2014).

Heat stress has been reported to result in over reduction of PSII acceptor site, damage donor site of PSII, reduce energetic connectivity of PSII units, and hence the performance of PSII is reduced. Higher temperature (37 °C) declines PSII activity with increase in PSI activity and the increase in photosynthetic O_2 evolution is synchronized with the increase in PSI activity (Zhang and Liu, 2016). Low-temperature stress also declines photosynthesis through considerable effects on the stomatal and nonstomatal attributes, in addition to the structural disintegration of thylakoids and chloroplasts (Zhang et al., 2020a, 202b). In heat stressed wheat seedlings, mesophyll ultrastructure is altered, stomatal limitation is induced, and RuBisCO activity is reduced affecting plant growth and grain filling significantly (Feng et al., 2014). High and low temperatures have been reported to decline the import of precursor of small subunit of RuBisCO into plastids, which has been attributed to downregulation of Toc159. This impaired binding of proteins and import into the chloroplast is due to the downregulation of TOC complex, TIC complex, HSP93, and stomatal processing peptidase (Dutta et al., 2009). Plants tend to adjust the photosynthetic metabolism in according to the changes in temperatures and the optimal temperature requirements. The main factor includes the alterations in the optimum photosynthetic temperature as influenced by the external temperature (Berry and Bjorkman, 1980).

Low-temperature stress is also equally damaging to that of high temperature and proves to be the limiting factor declining the optimal productivity potential by considerably declining the photosynthetic activities and the associated metabolic pathways (Sane et al., 2003; Savitch et al., 2011). It has been reported that low-temperature stress disrupts growth and photosynthesis of plants by significantly declining the key stomatal and nonstomatal attributes of photosynthesis like transpiration, intercellular CO_2, stomatal conductance, carbon assimilation cycle, and electron transport (Hou et al., 2016; Savitch et al., 2011). Therefore, the identification of the key regulatory effects of the major macromolecules synthesized endogenously or applied exogenously is crucial. There have been significant evidences that exogenous application of protectant molecules can be beneficial in strengthening the key tolerance mechanisms aimed to provide sound protection to photosynthetic system under temperature stress (Hasanuzzaman et al., 2013; Kurepin et al., 2015; Li et al., 2018). Gai et al. (2020) demonstrated mitigation of photosynthetic and growth decline in *Glycine max* due to exogenous application of α-oxoglutarate. Exogenous proline protects growth and photosynthesis by upregulating the antioxidant system considerably (Mohammadrexakhani et al., 2019). Low-temperature stress reduces the intercellular CO_2 levels by inducing stomatal closure, hence declining the activity of RuBisCO (Partelli et al., 2009). Temperature stress drastically declines chlorophyll biosynthesis by affecting the expression of enzymes mediating the biosynthetic steps of chlorophyll (Tewari and Tripathy, 1998) and hence the light harvesting in plastids is influenced (Efeoglu and Terzioglu, 2009; Ashraf and Harris, 2013). Huang et al. (2017) demonstrated reduced gene expression of chlorophyll biosynthetic pathway under high-temperature stress. High-temperature stress triggers increased cyclic electron transport and loss of thylakoid membrane integrity, so that leakage of key constituents is evident (Schrader et al., 2004; Sharkey, 2005; Velikova et al., 2011; Niu and Xiang, 2018).

The increased electron flow under temperature extremes leads to maintenance of energy gradient across the thylakoid membranes along with ATP homeostasis for maximally preventing any probable damage to photosynthetic apparatus (Sharkey and Zhang, 2010). Damage to photosystems due to extremes temperature regimes is mainly due to the increased membrane fluidity of thylakoids making way for easy displacement of light-harvesting complex of PSII. High-temperature exposure disrupts the water-oxidizing complex, light-harvesting complex, and PSII reaction center (Martin and Ort, 1982; Havaux, 1993; Lípová et al., 2010; Popova et al., 2019). Such damaging effects of temperature regimes on the photosynthetic apparatus have been largely attributed to decline in uptake of ions or their leakage from the photosynthetic apparatus, making PSII pigment–protein complex weak, and also increases the release of extrinsic polypeptides from the thylakoid membranes (Wise et al., 2004; Niu and Xiang, 2018). Similarly, there are numerous studies of photo-damages observed after chilling (low-temperature exposure) (Hou et al., 2016). Low temperature triggers inhibition of PSII functioning and restricts the sinks for excitation energy, that is, photorespiration and CO_2 fixation (Huner et al., 1998), besides affecting the repair mechanisms.

Both high (Cen and Sage, 2005; Benomar et al., 2019) and lower (Partelli et al., 2009) temperatures influence photosynthesis by declining the RuBisCO activity and its generation and this enzyme inactivation under temperature stress is also due to the loss of redox homeostasis (Zhang and Portis, 1999). Content of RuBisCO, RuBisCO activase, maximal CO_2 carboxylation rate, and stomatal conductance have key roles in thermal acclimation of net photosynthesis (Benomar et al., 2019). Carboxylation activity of RuBisCO is influenced by the temperature regime exposure (Crafts-Brandner and Law, 2000). In addition, high temperature inhibits the RuBisCO activase-mediated activation of RuBisCO in cotton and wheat (Feller et al., 1998). A strong correlation exists between RuBisCO activase and RuBisCO content in plants (Yanori and von Caemmerer, 2009). High-temperature stress causes decline in the activation of RuBisCO, which is often related to its thermal liability (Salvucci and Crafts-Brandner, 2004). During stressful conditions, photorespiration can assist in dissipation of excess reducing equivalents and energy directly using ATP, NAD(P)H or reduced ferredoxin (Fd), or indirectly via alternative oxidase (Voss et al., 2013; Huang et al., 2015). Accumulation of ROS and photorespiration has been reported in rice under cold stress (Cheng et al., 2007). Heat stress drastically affects photosynthesis by declining the D1 protein turnover and it has been reported that higher expression of D1 protein turnover-related genes including *PsbA*, *STN8*, *PBCP*, *Deg1*, *Deg2*, *Deg5*, *Deg8*, *FtsH1/5,* and *FtsH2/8* assists in alleviation of temperature-induced damages (Li et al., 2016). Degradation of D1 under stress is attributed to the migration of FtsH protease from stroma of thylakoid to grana (Yamamoto et al., 1999). Exogenous spermidine alleviates the damaging effects of heat stress on the stomatal and nonstomatal attributes of photosynthesis in tall fescue, improves the expression of *psbA* and *psbB* genes encoding for PSII core proteins (Zhang et al., 2017). Further, spermidine application significantly improved the partial potentials at the steps of energy bifurcations (PI_{ABS} and PI_{total}) and the quantum yields and efficiencies (φP_0, δR_0, φR_0, and γRC). Besides, the specific energy fluxes per Q_A^- reducing PSII reaction center is also reduced. In *Solanum lycopersicum* L., high-temperature stress declines photosynthetic carbon assimilation at multiple sites in PSII and PSI by interfering with intercellular CO_2, stomatal conductance, and Fv/Fm; the donor site limitation of PSI exhibited an increase with reduction in the acceptor site limitation of PSI (Lu et al., 2017). PSII structural and functional integrity is hampered by heat stress (Jajoo and Allakhverdiev, 2017). It has been revealed by genetic studies that photo-inhibition due to high- and low-temperature stress may be due to altered fatty acid saturation of the thylakoid lipids (Barber, 1995). Temperature stress-mediated inhibition of CO_2 fixation affects thylakoid energization like NPQ, electrochromic absorption shift, and light scattering, reflecting the reduced potential to utilize the fixed energy into Calvin cycle (Law and Crafts-Brandner, 1999).

2.5 HEAVY METAL(LOID)S AND PHOTOSYNTHESIS

Heavy metals and metalloids are among the toxic environmental stress factors affecting plant photosynthesis, thereby influencing the plant growth and productivity (Khan et al., 2015; Ahmad et al., 2020, 2021). Both natural as well as anthropogenic factors contribute toward the accumulation of toxic metal ions in the soil, thereby posing damaging effects on plant growth, and the animal and human health (Gratão et al., 2005; Khan et al., 2015; Kumar et al., 2019). Heavy metal emissions from volcanoes and dust blown by winds are the natural sources, whereas anthropogenic contributors include excessive use of pesticides and the inorganic fertilizers particularly phosphate, industrial processes like smelting, metal-plating, mining, and dyeing significantly increase the concentrations of heavy metals in soil (Singh et al., 2013; Karahan et al., 2020). Agricultural soil contaminated with heavy metals affect considerably the growth of plants. Uptake and accumulation of heavy metals exhibit considerable variations with plant species, besides other soil characteristics like pH, aerations, etc. Within the plant cells and tissues, the biochemical and metabolic processes are adversely altered, which are evident at both cellular and molecular levels. Toxic metal ions replace essential mineral elements from the key cellular biomolecules, compete for the functional groups and the enzyme active sites, alter the transcription and damage membrane structural and functional integrity (Huang et al., 2017). Like any other stress factor, heavy metals also induce ROS accumulation, resulting in oxidative damages to major biomolecules and also interfere with the electron transport mechanisms within the cellular organelles like chloroplast and mitochondria (Rascio and Navari-Izzo, 2011; Huang et al., 2017).

In tobacco, cadmium inhibits the chlorophyll synthesis by downregulating the expression of key genes, whereas Zn does not show much effect due to increased chlorophyllide a oxygenase activity. Heavy metals affect carotenoid metabolism by significantly downregulating the expression of violaxanthin de-epoxidase and zeaxanthin epoxidase, which contribute to NPQ. Cadmium significantly downregulates the expression of PSII donor site proteins like PPD3, PPD6, OEE1, OEE2-1, OEE2-2, OEE2-3, and OEE3-2, receptor-site proteins (D1, D2, CP43, CP47, Cyt b559α, Cyt b559β, PsbL, PsbQ, PsbR, Psb27-H1, and Psb28) and core proteins of the PSI reaction center (psaA, psaB, psaC, psaD, psaE-A, PsaE-B, psaF, psaG, psaH-1, psaK, psaL, psaN, and psaOL). However, zinc only affects the PPD6, OEE3-2, PsbL, PsbQ, Psb27-H1, psaL, and psaOL expression (Zhang et al., 2020a, 2020b). Cadmium potentially inhibits chlorophyll synthesis and downregulates the expression of photosynthesis-related proteins and subunits as well as suppressing the xanthophyll cycle and NADPH dependent cyclic electron flow, in addition to affecting the Fd-dependent nitrogen metabolism and ROS scavenging, thereby blocking the electron transport and causing photoinhibition of PSII and PSI (Zhang et al., 2020a, 2020b). Photoinhibition due to heavy metals results mainly due to excess accumulation of ROS, declined activity of PSII, PSI, and carboxylase efficiency (Huihui et al., 2020). Heavy metals affect both stomatal and nonstomatal photosynthetic attributes and damage the oxygen-evolving complex located on PSII, besides causing significant decline in electronic pressure on PSII acceptor site and PSI (Huihui et al., 2020). The intensity of photoinhibition varies with the metal ion and the concentration of the ion (Huihui et al., 2020; Zhang et al., 2020a, 2020b).

Leaf necrosis resulting after exposure to heavy metals is an evident symptom of reduction in the chloroplast size and density. It has been reported that plants exposed to heavy metals exhibit a significant alteration in the ultrastructural features of chloroplasts, and the visible effects include inflated thylakoids and changed chloroplast appearance, declined number of chloroplast, and increased chlorophyll degradation (Per et al., 2016; Kapoor et al., 2019). The accumulation of plastoglobuli enhances (Ying et al., 2010; Kapoor et al., 2019), while accumulation of starch is lowered (Ying et al., 2010). Wang et al. (2011) observed reduction in stacking of grana due to loss of parallel arrangement in the thylakoid lamellae, which was reflected as grana swelling. Similarly, treatment of chromium (Appenroth et al., 2003) and cadmium (Parmar et al., 2013) results in altered lamellar arrangement and reduces the number of grana, dysfunctioning of thylakoid and chloroplast membranes, and reduction in starch with increased plastoglobuli. Among the prime reasons for reduced photosynthesis are surge in ROS, and loss of membrane functioning of chloroplasts and thylakoids. Exposure to lead (Pb) affects the chloroplast functioning by altering the lipid composition in thylakoid membranes (Sharma and Dubey, 2005). Increased peroxidation of lipids in heavy-metal exposed plants results in loss of structural order of chloroplasts (Remans et al., 2010). The lipid damaging effects of heavy metals have been reported to be obvious as upregulation of lipoxygenase activity, resulting in speedy oxidation of polyunsaturated fatty acids for the formation of free radicals (Nahar et al., 2016; Hameed et al., 2012). Heavy metal induced upregulation of lipoxygenase significantly reduces the key lipid components of chloroplast and thylakoid membranes (Parmar et al., 2013). Exposure to metal ions like cadmium and aluminum induces lipid peroxidation exhibited as increased lipoxygenase activity (Nahar et al., 2016, 2017).

In addition of the effects on the membrane structure and functioning, the damaging effects of heavy metals include the declined pigment biosynthesis, concomitant with increased degradation (Parmar et al., 2013). Heavy metals like Cd compete with the sulfhydryl sites of the key enzymes of chlorophyll biosynthesis pathway like δ-aminolevulinic acid dehydratase, thereby bringing down the synthesis of chlorophyll (Skrebsky et al., 2008). Heavy metal stress-induced accumulation of ALA and porphyrins triggers excess generation of ROS in plants, further leading to oxidative damages (Noriega et al., 2007). Heavy metals inhibit the activity of enzymes mediating the biosynthesis of chlorophyll (Ganesh et al., 2008).

The influence of heavy metals on the synthesis of chlorophyll, photosynthesis, and assimilate production can be majorly attributed to significant decline in the uptake and assimilation of the key mineral elements including magnesium, sulfur, nitrogen, phosphorous, calcium, potassium (Singh et al., 2013; Khan et al., 2015; Asgher et al., 2014), and other key microelements (Medynska-Juraszek et al., 2020). Heavy metals bring down the rate of photosynthesis by replacing the key mineral components, Mg, in the porphyrin head of the chlorophyll molecule (Bertrand and Poirier, 2005). The pathway for the conversion of light into chemical energy occurs in the membrane thylakoid of chloroplasts. PSII is among the principal components and is a membrane protein complex having proteins associated intrinsically as well extrinsically that mediate water oxidation and the reduction of plastoquinone (Hasni et al., 2015). Treatment of Cd alters the donor sites of PSII by replacing Ca in oxygen-evolving complex (Pagliano et al., 2006). It has been reported that Pb potentially dissociates the extrinsic polypeptides at the donor site of PSII and replaces Ca, thereby affecting oxygen evolution (Bernier and Carpentier, 1995; Msilini et al., 2011). Yruela et al. (2000) showed the release of intrinsic proteins and the proteins from inner antenna of PSII due to excess copper.

The rate of electron transport from QA to QB is reduced due to Cd stress (Geiken et al., 1998). Heavy metal toxicity reduces the light-harvesting capacity of plants as has been reported with chromium that is believed to interact with the heme group of cytochrome, thereby altering the electron transport (Dixit et al., 2002). Exposure to Cd induces disorder of key proteins of LHCII, that is, Lhcb1 and Lhcb2 (Parmar et al., 2013). The activity and synthesis of RuBisCO are declined due to heavy metal stress (Per et al., 2016). Access to CO_2 and its assimilation are adversely affected by the heavy metals. Enzymes mediating the carbon assimilation in C3 and C4 are down-regulated due to metal stress (Sharma and Dubey, 2005).

2.6 LIGHT AND PHOTOSYNTHESIS

Photosynthesis depends on light and any modulations in the nature and intensity of light would directly affect photosynthesis and plant development (Bukhov, 2004; Wang et al., 2009; Hogewoning et al., 2010; Wang et al., 2017; Liu and van Iersel, 2021). The key regulatory reactions, reflecting the photosynthetic machinery to respond to changes in the irradiance by changing the activities of the light and dark photosynthetic phases within seconds or minutes, determine the photosynthetic efficiency. Over-reduction of electron transport chain is regulated by negative feedback mode reflected in declined generation of ROS. On the contrary, during dark reactions, these regulatory responses include positive feedback related mostly to redox regulation of activities of several enzymes that are much sensitive to reduction status of PSI acceptor site (Bukhov, 2004). Light-mediated regulation of photosynthesis includes regulation of stomatal functioning, leaf growth, pigment synthesis, structural stability of chloroplast, gene expression of D1 protein, and carbon assimilation under visible and UV light (Zheng et al., 2008). Excess light results in the generation of harmful oxygen radicals, thereby inducing photoinhibition with significant reduction in primary productivity of plants (Gururani et al., 2015). In addition, extreme low light also inhibits photosynthesis by restricted energy availability, while high light intensity imparts photoinhibition by overloading the light-harvesting complexes (Pascual et al., 2017). It has been confirmed that low light availability induced leaf etiolation due to the conditioned arrest of photosynthetic process (Solymosi and Schoefs, 2010). In this connection, the insufficient light negatively influences photosynthetic efficiency, declines carbon and nitrogen metabolism by inducing oxidative stress, besides declining the stomatal conductance and intercellular CO_2 concentrations (Apel and Hirt, 2004; Wang et al., 2013; Liu et al., 2014; Fu et al., 2017). On the other hand, high intense visible light reduces photosynthesis by triggering photoinhibition, photo-inactivation, photolability, and solarization.

Grape plants grown under red and blue light exhibited increased chlorophyll and photosynthesis during early growth period; however, at later growth stage, significant damage to photosynthetic apparatus was observed (Shuai et al., 2016). *Capsicum annum* plants grown under red and blue light showed differential responses in terms of photosynthesis and carbon assimilation with red light resulting in reduced CO_2 assimilation by affecting PSII activity and electron transport, thereby reducing the biomass accumulation significantly as compared with blue light, which caused inactivation of photosystems (Li et al., 2020a). In addition, blue light treatments upregulate the expression and activities of RuBisCO, fructose-1, 6-bisphosphatase and glyceraldehyde-phosphate dehydrogenase, and the enzymatic components contributing to synthesis of RuBP synthesis (Li et al., 2020a). Recently, Yavari et al. (2021) also demonstrated significant alteration in the photosynthetic functioning and the expression of key genes like *GSH2* and *PSBA*, while downregulated *NPQ1* and *FNR2* in Arabidopsis under various light treatments. Blue light responses in plants are regulated by photoreceptor, phototropin, reflecting in phototropism, chloroplast and stomatal movements, and expression of light-regulated genes. Exposure of *Chlamydomonas reinhardtii* to low-fluence white light increased the transcript levels of glutamate-1-semialdehyde aminotransferase, phytoene desaturase, and light-harvesting polypeptides like LHCBM6 (Im et al., 2006). Red light also upregulates glutamate-1-semialdehyde aminotransferase and LHCBM6 confirming the involvement of red light photoreceptor in the regulation of photosynthetic genes in *Chlamydomonas reinhardtii.* Besides, phototropin also modulates the transcript levels of PSI and PSII polypeptides confirming that phototropin: (1) is involved in blue-light induced alterations in transcript accumulation, and (2) is involved in signaling synchronizing synthesis of chlorophylls, carotenoids, chlorophyll-binding proteins and other components of photosynthetic machinery (Im et al., 2006). Inactivation of blue light signalling components like CRY1, CRY2, and HY5D declined the synthesis of chlorophyll and the expression of chlorophyll biosynthesis genes (Doroshenko et al., 2019).

Kreslavski et al. (2016) demonstrated that UV-A exposure declines photosynthesis by affecting pigment synthesis, and maximal photochemical and effective quantum yield of PSII in phytochrome-B deficient *Arabidopsis* plants; however, preillumination with red and far-red light significantly mitigates the decline suggesting the important role of red and far-red light reversible phytochromes in improving resistance of PSII against UV-A. This study also suggested the key role of phytochrome B in transcriptional regulation of antioxidant enzymes and transcription factor genes (Kreslavski et al., 2016).

Wang et al. (2009), while examining the effects of different light qualities, namely, purple, blue, green, yellow, red, and white light on *Cucumis sativus* observed significant decline in growth and photosynthesis under all light sources, compared with white light-grown plants. Photosynthetic parameters including CO_2 assimilation, quantum yield of PSII, and electron transport rate declined more significantly in green, yellow, and red light-grown seedlings, whereas purple and blue light enhanced stomatal conductance and initial and total RuBisCO activity, besides increasing the transcription of ten genes encoding the key enzymes of Calvin cycle. In addition, yellow, green, and red light resulted in inactivation of photosystems, stomatal closure, and downregulation of Calvin cycle enzymes like *rbc* L and *rca* (Wang et al., 2009). Duration of high intense light exposures creates high oxidizing potential inside PSII reaction center, leading to significant damage to D1 protein (Chen et al., 2012). In barley, UV-B promotes dephosphorylation of all the thylakoid phosphoproteins, and dephosphorylation of reaction center proteins D1 and D2 is accompanied by degradation (Barbato et al., 1999). It has been shown that primary damage occurs at the catalytic Mn cluster of water oxidation often sensitized by absorption of UV by Mn (III) and Mn(IV) ions ligated by organic residues. Such damage to PSII due to UV exposure can be repaired by the *de novo* synthesis of the D1 and D2 protein subunits, which is facilitated by visible light. Repair of PSII is retarded due to lack of DNA repair as reported in the photolyase-deficient cyanobacterial mutant (Sicora et al., 2006). The mitochondrial activity gets adversely affected by the light intensity, besides declining the photochemical efficiency by damaging PSI and PSII reflected as dissipation of energy in the form of fluorescence (Faseela and Puthur, 2018). Plant photosynthetic attributes including chlorophyll and carotenoid content, quantum efficiency of PSII, electron transport, and net photosynthesis are affected by high light exposures (Faseela and Puthur, 2018; Yi et al., 2020; Yavari et al., 2021). Decline in photosynthesis and the palisade tissue layer due to yellow light exposure is regulated by genes as revealed by RNA sequencing, and differentially expressed genes are closely associated with light stimulus and high light response, leaf and root growth and development, photosynthesis, carbohydrate and secondary metabolite metabolism, and hormone and antioxidant metabolism (Liu et al., 2020). Extreme light results in photooxidation of chlorophyll and high-light stress induces the degradation of chlorophyll by chlorophyll reductase (Sato et al., 2015). Alterations in the key physiological processes like photosynthesis, stomatal conductance, and transpiration are the immediate responses of plants to high light (Cowie et al., 2016). Exposure to low light has been reported to deactivate fructose-1,6-bisphosphatase and sedoheptulose-1,7-bisphosphatase, reflecting that the activation of photosynthesis after low light exposure depends on the reactivation of fructose-1,6-bisphosphatase and sedoheptulose-1,7-bisphosphatase (Li et al., 2020b).

2.7 FUNGICIDES/PESTICIDES AND PHOTOSYNTHETIC REGULATION

Fungicides and pesticides are used to control the damaging effects of fungi and pests on crop growth and hence productivity. However, it should be mentioned that these chemicals can be toxic to normal physiology and biochemistry of plants when applied at higher concentrations, and can result in death of plants (Kumar et al., 2010; Dias, 2012; Sharma et al., 2019). Herbicide application reduces photosynthesis by influencing the C_4 cycle enzymes, phosphoenolpyruvate carboxylase, malate dehydrogenase, RuBisCO, and pyruvate orthophosphate dikinase (Nemat Alla et al., 2007). Application of fungicide reduces photosynthesis by declining chlorophyll synthesis and PSII activity, thereby declining photoassimilate production (Petit et al., 2012). The key fungicides include the contact fungicides destroying the chloroplast, affecting PSII activity, and restricting the biosynthesis of chlorophyll, and systemic fungicides like anilides, benzimidazoles, and pyrimidine that are phytotoxic; however, azoles are believed to stimulate photosynthesis. Fungicides affect photosynthesis by inducing perturbation and interaction with specific target sites (Petit et al., 2012). Combination of fungicide and pesticide influences the photosynthetic functioning by affecting CO_2 exchange, besides imparting an obvious effect on the dark respiration and carbon balance (Untiedt and Blanke, 2004). Fungicide application modifies both photosynthetic activity and chlorophyll fluorescence (Untiedt and Blanke, 2004; Xia et al., 2006).

Copper, sulfur, or fludioxonil-based compounds are used against plant disease through nonsystemic control programs. In agricultural practices, copper is used as a pesticide (e.g., Bordeaux mixture) as well as algicide, being an essential nutrient and component of various proteins involved in both photosynthetic (plastocyanin) and the respiratory (cytochrome oxidase) electron transport chain (Baron et al., 1995). However, excess concentration proves strongly phytotoxic for crops observed as pronounced photoinhibition. Declined photosynthesis is associated with a parallel decrease in stomatal conductance (Vinit-Dunant et al., 2002). Both fungicides and pesticides alter growth and development by interfering with the chlorophyll synthesis and photosynthesis (Parween et al., 2016). Significant decline in the CO_2 fixation,

PSII activity, and photosynthesis due to dimethoate has been observed to be dose dependent (Mishra et al., 2008). Free radical generation is triggered and electron transport between QA and QB is blocked due to pesticide application (Devine and Shukla, 2000). The stability of D1 protein is affected and degradation of chlorophyll is triggered due to excess ROS accumulation (Kleczkowski, 1994). Light-dependent photosynthetic processes are affected by pesticide application, resulting in significant reduction in the generation of ATP and $NADP^+$ (Nadasy et al., 2000).

2.8 CONCLUSION AND FUTURE PERSPECTIVES

Environmental stresses (biotic and abiotic) damage plant growth and developmental processes by affecting the key metabolic functioning like photosynthesis. The structural and functional stability of photosynthetic apparatus are altered drastically due to excess generation of toxic ROS. Enzymes regulating the carbon metabolism are downregulated due to stresses. Growth retardation, alteration in redox homeostasis, and oxidative damages are the prime effects of stresses, inducing a significant decline in photosynthetic performance. In addition, the loss of structural and functional stability of the chloroplast reported by many workers justifies the damaging effects of the environmental stress factors on plant photosynthesis. Therefore, making efficient use of the available techniques for improving the photosynthetic stability of crop plants for better productivity is essential. Future research will definitely involve the identification of quantitative trait loci (QTLs) and exploitation of gene and gene products for enhancing the tolerance mechanisms for improved photosynthesis.

KEYWORDS

- **abiotic stress**
- **photosynthesis**
- **photosystem II**
- **ribulose 1,5-bisphosphate carboxylase**
- **carbon assimilation**

REFERENCES

Acosta-Motos, J. R.; Ortuno, M. F.; Bernal-Vicente, A.; Diaz-Vivancos, P.; Sanchez-Blanco, M. J.; Hernandez, J. A. 2017. Plant Responses to Salt Stress: Adaptive Mechanisms. *Agronomy* **2017,** *7*, 18. DOI: 10.3390/agronomy7010018.

Ahanger, M. A.; Agarwal, R. M. Potassium Up-Regulates Antioxidant Metabolism and Alleviates Growth Inhibition under Water and Osmotic Stress in Wheat (*Triticum aestivum* L). *Protoplasma* **2017a,** *254* (4), 1471–1486.

Ahanger, M. A.; Agarwal, R. M. Salinity Stress Induced Alterations in Antioxidant Metabolism and Nitrogen Assimilation in Wheat (*Triticum aestivum* L) as Influenced by Potassium Supplementation. *Plant Physiol. Biochem.* **2017b,** *115*, 449–460.

Ahanger, M. A.; Mir, R. A.; Alyemeni, M. N.; Ahmad, P. Combined Effects of Brassinosteroid and Kinetin Mitigates Salinity Stress in Tomato through the Modulation of Antioxidant and Osmolyte Metabolism. *Plant Physiol. Biochem.* **2020,** *147*, 31–42.

Ahanger, M. A.; Qi, M.; Huang, Z.; Xu, X.; Begum, N.; Qin, C.; Zhang, C.; Ahmad, N.; Mustafa, N. S.; Ashraf, M.; Zhang, L. Improving Growth and Photosynthetic Performance of Drought Stressed Tomato by Application of Nano-Organic Fertilizer Involves Up-Regulation of Nitrogen, Antioxidant and Osmolyte Metabolism. *Ecotoxicol. Environ. Saf.* **2021,** 216. https://doi.org/10.1016/j.ecoenv.2021.112195.

Ahanger, M. A.; Qin, C.; Begum, N.; Maodong, Q.; Dong, X. X.; El-Esawi, M.; El-Sheikh, M. A.; Alatar, A. A.; Zhang, L. Nitrogen Availability Prevents Oxidative Effects of Salinity on Wheat Growth and Photosynthesis by Up-Regulating the Antioxidants and Osmolytes Metabolism, and Secondary Metabolite Accumulation. *BMC Plant Biol.* **2019,** *19*, 479. https://doi.org/10.1186/s12870-019-2085-3.

Ahanger, M. A.; Tomar, N. S.; Tittal, M.; Argal, S.; Agarwal, R. M. Plant growth under Water/salt Stress: ROS Production; Antioxidants and Significance of Added Potassium under Such Conditions. *Physiol. Mol. Biol. Plants* **2017,** *23* (4), 731–744.

Ahmad, P.; Ahanger, M. A.; Alam, P.; Alyemeni, M. N.; Wijaya, L.; Ali, S.; Ashraf, M. Silicon (Si) Supplementation Alleviates NaCl toxicity in Mung Bean [*Vigna radiata* (L.) wilczek] through the Modifications of Physiobiochemical Attributes and Key Antioxidant Enzymes. *J. Plant Growth Regul.* **2019,** *38* (1), 70–82.

Ahmad, P.; Alam, P.; Balawi, T. H.; Altalayan, F. H.; Ahanger, M. A.; Ashraf, M. Sodium Nitroprusside (SNP) Improves Tolerance to Arsenic (As) Toxicity in *Viciafaba* through the Modifications of Biochemical Attributes, Antioxidants, Ascorbate-Glutathione Cycle and Glyoxalase Cycle. *Chemosphere* **2020**. DOI: 10.1016/j.chemosphere.2019.125480.

Ahmad, P.; Alyemeni, M. N.; Wijaya, L.; Ahanger, M. A.; Ashraf, M.; Alam, P.; Paray, B. A.; Rinklebe, J. Nitric Oxide Donor, Sodium Nitroprusside, Mitigates Mercury Toxicity in Different Cultivars of Soybean. *J. Hazard. Mater.* **2021,** *408*, 124852. https://doi.org/10.1016/j.jhazmat.2020.124852.

Ali, M. S.; Baek, K. H. Jasmonic Acid Signaling Pathway in Response to Abiotic Stresses in Plants. *Int. J. Mol. Sci.* Jan **2020,** *21* (2), 621. DOI: 10.3390/ijms21020621.

Apel, K.; Hirt, H. Reactive Oxygen Species: Metabolism, Oxidative Stress, and Signal Transduction. *Annu. Rev. Plant Biol.* **2004,** *55*, 373–399.

Appenroth, K. J.; Keresztes, Á.; Sárvári, É.; Jaglarz, A.; Fischer, W. Multiple Effects of Chromate on *Spirodela polyrhiza*: Electron Microscopy and Biochemical Investigations. *Plant Biol.* **2003,** *5* (3), 315–323.

Asada, K. The Water–Water Cycle in Chloroplasts: Scavenging of Active Oxygen and Dissipation of Excess Photons. *Ann. Rev. Plant Physiol. Plant Mol. Biol.* **1999,** *50*, 601–639.

Ashraf, M.; Hafeez, M. Thermotolerance of Pearl Millet and Maize at Early Growth Stages: Growth and Nutrient Relations. *Biol. Planta* **2004,** *48* (1), 81–86.

Ashraf, M.; Harris, P. Photosynthesis under Stressful Environments: An Overview. *Photosynthetica* **2013,** *51* (2), 163–190.

Bao, G.; Zhuo, C.; Qian, C.; Xiao, T.; Guo, Z.; Lu, S. Co-expression of NCED and ALO Improves Vitamin C Level and Tolerance to Drought and Chilling in Transgenic Tobacco and Stylo Plants. *Plant Biotechnol. J.* **2016,** *14* (1), 206–214.

Barbato, R.; Bergo, E.; Giacometti, G. M. Effects of Ultraviolet-B Light on Photosystem II Phosphoproteins in Barley Wild Type and Its Chlorophyll *b*-Less Mutant Chlorina f2. *J. Photochem. Photobiol. B: Biol.* **1999,** *48* (2–3), 189–193.

Barber. J. Molecular Basis of the Vulnerability of Photosystem II to Damage by Light. *Funct. Plant Biol.* **1995,** *22* (2), 201–208.

Baron, M.; Arellano, J. B.; Gorge, L. Copper and Photosystem II: A Controversial Relationship. *Physiol. Plant* **1995,** *94*, 174–180.

Baslam, M.; Mitsui, T.; Hodges, M.; Priesack, E.; Herrit, M. T.; Aranjuelo, I.; Sanz-Saez, A. Photosynthesis in a Changing Global Climate: Scaling Up and Scaling Down in Crops. *Front. Plant Sci.* **2020**. https://doi.org/10.3389/fpls.2020.00882.

Bastos, E. A.; Sebastião, P. d. N.; Everaldo, M. d. S.; Francisco, R. F. F.; Reinaldo, L. G. Identification of Cowpea Genotypes for Drought Tolerance. *Rev. Ciên. Agron.* **2011,** *42* (1), 100–107.

Basu, S.; Ramegowda, V.; Kumar, A.; Pereira, A. Plant Adaptation to Drought Stress. *F1000Res*. **2016**. DOI: 10.12688/f1000research.7678.1.

Begum, N.; Ahanger, M. A.; Zhang, L. AMF Inoculation and Phosphorus Supplementation Alleviates Drought Induced Growth and Photosynthetic Decline in *Nicotiana tabacum* by Up-Regulating Antioxidant Metabolism and Osmolyte Accumulation. *Environ. Exp. Bot.* **2020,** *176*. https://doi.org/10.1016/j.envexpbot.2020.104088.

Begum, N.; Qin, C.; Ahanger, M. A.; Raza, S.; Khan, M. I.; Ahmed, N.; Ashraf, M.; Zhang, L. Role of Arbuscular Mycorrhizal Fungi in Plant Growth Regulation: Implications in Abiotic Stress Tolerance. *Front. Plant Sci.* **2019**. DOI: 10.3389/fpls.2019.01068.

Benomar, L.; Moutaoufik, M. T.; Elferjani, R.; Isabel, N.; DesRochers, A.; ElGuellab, A.; Khlifa, R.; Hassania, L. A. I. Thermal Acclimation of Photosynthetic Activity and RuBisCO Content in Two Hybrid Poplar Clones. *PLoS One* **2019,** *14* (2), e0206021. DOI: 10.1371/journal.pone.0206021.

Bernier, M.; Carpentier, R. The Action of Mercury on the Binding of the Extrinsic Polypeptides Associated with the Water Oxidizing Complex of Photosystem II. *FEBS Lett.* **1995,** *360* (3), 251–254.

Berry, J.; Bjorkman, O. Photosynthetic Response and Adaptation to Temperature in Higher Plants. *Annu. Rev. Plant. Physiol. Plant Mol. Biol.* **1980,** *31* (1), 491–543.

Bertrand, M.; Poirier, I. Photosynthetic Organisms and Excess of Metals. *Photosynthetica* **2005,** *43* (3), 345–353.

Bukhov, N. G. Dynamic Light Regulation of Photosynthesis (A Review). *Russian J. Plant Physiol.* **2004,** *51*, 742–753.

Cen, Y. P.; Sage, R. F. The Regulation of Rubisco Activity in Response to Variation in Temperature and Atmospheric CO_2 Partial Pressure in Sweet Potato. *Plant Physiol* Oct **2005,** *139* (2), 979–990.

Chaves, M. M.; Flexas, J.; Pinheiro, C. Photosynthesis under Drought and Salt Stress: Regulation Mechanisms from Whole Plant to Cell. *Ann. Bot.* **2009,** *103*, 551–560.

Chen, L.; Jia, H.; Tian, Q.; Du, L.; Gao, Y.; Miao, X.; Liu, Y. Protecting Effect of Phosphorylation on Oxidative Damage of D1 Protein by Down-Regulating the Production of Superoxide Anion in Photosystem II Membranes under High Light. *Photosynth. Res.* **2012,** *112* (2):141–148.

Cornic, G. Drought Stress Inhibits Photosynthesis by Decreasing Stomatal Aperture – Not by Affecting ATP Synthesis. *Trends Plant Sci.* **2000,** *5* (5), 187–188.

Cowie, B. W.; Byrne, M. J.; Witkowski, E. T. F.; Venter, N. Exacerbation of Photosynthetic Damage through Increased Heat-Light Stress Resulting from *Gargaphia decoris* sap-Feeding. *Biol. Control* **2016,** *94*, 82–89.

Crafts-Brandner, S. J.; Law, R. D. Effect of Heat Stress on the Inhibition and Recovery of the Ribulose-1,5–Bisphosphate Carboxylase/Oxygenase Activation State. *Planta* **2000,** *212* (1), 67–74.

Cui, L.; Li, J.; Fan, Y.; Xu, S.; Zhang, Z. High Temperature Effects on Photosynthesis, PSII Functionality and Antioxidant Activity of Two *Festuca arundinacea* Cultivars with Different Heat Susceptibility. *Bot. Studi.* **2006,** *47*, 61–69.

Demirevska, K.; Zasheva, D.; Dimitrov, R.; Simova-Stoilova, L.; Stamenova, M.; Feller, U. Drought Stress Effects on Rubisco in Wheat: Changes in the Rubisco Large Subunit. *Acta Physiol. Plant* **2009,** *31*, 1129. https://doi.org/10.1007/s11738-009-0331-2.

Devine, M. D.; Shukla, A. Altered Target Sites as a Mechanism of Herbicide Resistance. *Crop Prot.* **2000,** *19*, 881–889.

Dias, M. C. Phytotoxicity: An Overview of the Physiological Responses of Plants Exposed to Fungicides. *J. Bot.* **2012,** ID 135479. https://doi.org/10.1155/2012/135479.

Dixit, V.; Pandey, V.; Shyam, R. Chromium Ions Inactivate Electron Transport and Enhance Superoxide Generation in Vivo in Pea (*Pisum sativum* L. cv. Azad) Root Mitochondria. *Plant Cell Environ.* **2002,** *25* (5), 687–693.

Doroshenko, A. S.; Danilova, M. N.; Medvedeva, A. S.; Kusnetsov, V. V. Influence of Blue-Light Signaling Components on the Regulation of Cytokinin-Dependent *Arabidopsis thaliana* Seedlings' Greening. *Russian J. Plant Physiol.* **2019,** *66*, 864–871.

Dutta S, Mohanty S, Tripathy BC. 2009. Role of Temperature Stress on Chloroplast Biogenesis and Protein Import in Pea. Plant Physiology. **2009,** *150* (2): 1050–1061.

Efeoglu, B.; Terzioglu, S. Photosynthetic Responses of Two Wheat Varieties to High Temperature. *Eur. Asia J. BioSci.* **2009,** *3*, 97–106.

Egea, I.; Albaladejo, I.; Meco, V.; Morales, B.; Sevilla, A.; Bolarin, M. C.; Flores, F. B. The Drought-Tolerant Solanum Pennellii Regulates Leaf Water Loss and Induces Genes Involved in Amino Acid and Ethylene/Jasmonate Metabolism under Dehydration. *Sci. Rep.* **2018,** *8*, 2791. DOI: 10.1038/s41598-018-21187-2.

Elkelish, E. E.; Soliman, M. H.; Alhaithloul, H. A.; El-Esawi, M. A. Selenium Protects Wheat Seedlings against Salt Stress-Mediated Oxidative Damage by Up-Regulating Antioxidants and Osmolytes Metabolism. *Plant Physiol. Biochem.* **2019,** *137*, 144–153.

Faseela, P.; Puthur, J. T. The Imprints of the High Light and UV-B Stresses in *Oryza sativa* L. 'Kanchana' Seedlings Are Differentially Modulated. *J. Photochem. Photobiol. B* **2018,** *178*, 551–559.

Fatma, M.; Masood, A.; Per, T. S.; Khan, N. A. Nitric Oxide Alleviates Salt Stress Inhibited Photosynthetic Performance by Interacting with Sulfur Assimilation in Mustard. *Front. Plant Sci.* **2016,** *7*, 521. DOI: 10.3389/fpls.2016.00521.

Feller, U.; Crafts-Brandner, S. J. C.; Salvucci, M. E. Moderately High Temperatures Inhibit Ribulose-1, 5-Bisphosphate Carboxylase/Oxygenase (Rubisco) Activase-Mediated Activation of Rubisco. *Plant Physiol.* **1998,** *116*, 539–546.

Feng, B.; Liu, P.; Li, G.; Dong, S. T.; Wang, F. H.; Kong, L. A.; Zhang, J. W. Effect of Heat Stress on the Photosynthetic Characteristics in Flag Leaves at the Grain-Filling Stage of Different Heat-Resistant Winter Wheat Varieties. *J. Agron. Crop Scie.* **2014,** *200* (2), 143–155.

Fu, Y.; Li, H.; Yu, J.; Liu, H.; Cao, Z.; Manukovsky, N. S.; Liu, H. Interaction Effects of Light Intensity and Nitrogen Concentration on Growth, Photosynthetic Characteristics and Quality of Lettuce (*Lactuca sativa* L. Var. youmaicai). *Sci. Horti.* **2017,** *214*, 51–57.

Gai, Z.; Liu, L.; Zhang, J.; Liu J.; Cai L. Effects of Exogenous α-Oxoglutarate on Proline Accumulation, Ammonium Assimilation and Photosynthesis of Soybean Seedling (*Glycine max(L.) Merr.*) Exposed to Cold Stress. *Sci. Rep.* **2020,** *10*, 17017. https://doi.org/10.1038/s41598-020-74094-w.

Ganesh, K. S.; Baskaran, L.; Rajasekaran, S.; Sumathi, K.; Chidambaram, A. L.; Sundaramoorthy, P. Chromium Stress Induced Alterations in Biochemical and Enzyme Metabolism in Aquatic and Terrestrial Plants. *Colliods Surf. B* **2008,** *63* (2), 159–163.

Geiken, B.; Masojidek, J.; Rizzuto, M.; Pompili, M. L.; Giardi, M. T. Incorporation of [35S] Methionine in Higher Plants Reveals That Stimulation of the D1 Reaction Centre II Protein Turnover Accompanies Tolerance to Heavy Metal Stress. *Plant Cell Environ.* **1998,** *21* (12), 1265–1273.

Gratão, P. L.; Polle, A.; Lea, P. J.; Azevedo, R. A. Making the Life of Heavy Metal-Stressed Plants a Little Easier. *Funct. Plant. Biol.* **2005,** *32* (6), 481–494.

Guliyev, N.; Bayramov S.; Babayev, H. Effect of Water Deficit on Rubisco and Carbonic Anhydrase Activities in Different Wheat Genotypes. In *Photosynthesis. Energy from the Sun*; Allen, J. F., Gantt E., Golbeck, J. H., Osmond, B., Eds.; Springer: Dordrecht, 2008. https://doi.org/10.1007/978-1-4020-6709-9-315.

Guo, Q.; Lu, N.; Sun, Y.; Lv, W.; Luo, Z.; Zhang, H.; Ji, Q.; Yang, Q.; Chen, S.; Zhang, W.; Li, Y. Heterologous Expression of the DREB Transcription Factor *AhDREB* in *Populus tomentosa* Carrière Confers Tolerance to Salt Without Growth Reduction under Greenhouse Conditions. *Forests* **2019,** *10* (3), 214. https://doi.org/10.3390/f10030214.

Guo, X. L.; Ma, Y. Y.; Liu, Z. H.; Liu, B. H. Effects of Exterior Abscisic Acid on Calcium Distribution of Mesophyll Cells and Calcium Concentration of Guard Cells in Maize Seedlings. *Agric. Sci. China* **2008,** *7*, 438–446.

Gururani, M. A.; Venkatesh, J.; Tran, L. S. Regulation of Photosynthesis during Abiotic Stress-Induced Photoinhibition. *Mol. Plant* **2015,** *8* (9), 1304–1320.

Hajiboland, R.; Cheraghvareh, L.; Poschenrieder, C. Improvement of Drought Tolerance in Tobacco (*Nicotiana rustica* L.) Plants by Silicon. *J. Plant Nutr.* **2017,** *40* (12), 1661–1676.

Halliwell, B.; Gutteridge, J. M. C. *Free Radicals in Biology and Medicine*; Oxford University Press: London, 2000.

Hameed, A.; Qadri, T. N.; Mahmooduzzafar, S. T.O. Plant Tolerance and Fatty Acid Profile in Responses to Heavy Metals. In *Abiotic Stress Responses in Plants*; Ahmad, P., Prasad, M., Eds.; Springer: New York, 2012. https://doi.org/10.1007/978-1-4614-0634-1-20.

Hasanuzzaman, M.; Nahar, K.; Alam, M. M.; Roychowdhury, R.; Fujita, M. Physiological, Biochemical, and Molecular Mechanisms of Heat Stress Tolerance in Plants. *Int. J. Mol. Sci.* **2013,** *14* (5), 9643–9684.

Hasni, I.; Yaakoubi, H.; Hamdani, S.; Tajmir-Riahi, H. A.; Carpentier, R. Mechanism of Interaction of Al3+ with the Proteins Composition of Photosystem II. *PLoS ONE* **2015,** *10* (3), e0120876.

Havaux, I. Rapid Photosynthetic Adaptation to Heat Stress Triggered in Potato Leaves by Moderately Elevated Temperatures. *Plant Cell Environ.* **1993,** *16* (4), 461–467.

Hogewoning, S. W.; Trouwbiorst, G.; Maljaars, H.; Poorter, H.; van Ieperen, W.; Harbinson, J. Blue Light Dose–Responses of Leaf Photosynthesis, Morphology, and Chemical Composition of *Cucumis sativus* Grown under Different Combinations of Red and Blue Light. *J. Exp. Bot.* **2010,** *61* (11), 3107–3117.

Hou, W.; Sun, A.; Chen, H.; Yang, F.; Pan, J.; Guan, M. Effects of Chilling and High Temperatures on Photosynthesis and Chlorophyll Fluorescence in Leaves of Watermelon Seedlings. *Biol. Planta* **2016,** *60* (1), 148–154.

Hu, S.; Ding, Y.; Zhu, C. Sensitivity and Responses of Chloroplasts to Heat Stress in Plants. *Front. Plant Sci.* **2020.** https://doi.org/10.3389/fpls.2020.00375.

Huang, W.; Ma, H. Y.; Huang, Y.; Li, Y.; Wang, G. L.; Jiang, Q.; Wang, F.; Xiong, A. S. Comparative Proteomic Analysis Provides Novel Insights into Chlorophyll Biosynthesis in Celery under Temperature Stress. *Physiol Planta* **2017,** *161* (4), 468–485.

Huang, M.; Zhu, H.; Zhang, J.; Tang, D.; Han, X.; Chen, L.; Du, D.; Yao, J.; Chen, K.; Sun, J. Toxic Effects of Cadmium on Tall Fescue and Different Responses of the Photosynthetic Activities in the Photosystem Electron Donor and Acceptor Sides. *Sci. Rep.* **2017,** *7*, 14387. https://doi.org/10.1038/s41598-017-14718-w.

Hui, D.; Yu, C. L.; Deng, Q.; Dzantor, E. K.; Zhou, S.; Dennis, S.; Sauve, R.; Johnson, T. L.; Fay, P. A.; Shen, W.; Luo, Y. Effects of Precipitation Changes on Switchgrass Photosynthesis, Growth, and Biomass: A Mesocosm Experiment. *PLoS ONE* **2018,** *13* (2), e0192555.

Huihui, Z.; XIn, L.; Zisong, X, Yue W, Zhiyuan T, Meijun A, Yuechui Z, Wenxu Z, Nan X, Guangyu S. 2020. Toxic effects of heavy metals Pb and Cd on mulberry (Morus alba L.) seedling leaves: Photosynthetic function and reactive oxygen species (ROS) metabolism responses. Ecotoxicol Environ Safety. **2020,** 195:110469. DOI: 10.1016/j.ecoenv.2020.110469.

Huner, N. P.; Öquist, G.; Sarhan, F. Energy Balance and Acclimation to Light and Cold. *Trends Plant Sci.* **1998,** *3* (6), 224–230.

Im, C. S.; Eberhard, S.; Huang, K.; Beck, C. F.; Grossman, A. R. Phototropin Involvement in the Expression of Genes Encoding Chlorophyll and Carotenoid Biosynthesis Enzymes and LHC Apoproteins in *Chlamydomonas reinhardtii*. *Plant J.* **2006,** *48* (1), 1–16.

Iqbal, N.; Umar, S.; Khan, N. A. Nitrogen Availability Regulates Proline and Ethylene Production and Alleviates Salinity Stress in Mustard (*Brassica juncea*). *J Plant Physiol.* **2015,** *178*, 84–91.

Isayenkov, S. V.; Maathuis, F. J. M. Plant Salinity Stress: Many Unanswered Questions Remain. *Front. Plant Sci.* **2019**. https://doi.org/10.3389/fpls.2019.00080.

Jajoo, A.; Allakhverdiev, S. I. High-Temperature Stress in Plants: Consequences and Strategies for Protecting Photosynthetic Machinery. In *Plant Stress Physiology*, 2nd ed.; CAB International: Oxfordshire, 2017; pp 138–154.

Joshi, R.; Wani, S. H.; Singh, B.; Bohra, A.; Dar, Z. A.; Lone, A. A.; Pareek, A.; Singla-Pareek, S. L. Transcription Factors and Plants Response to Drought Stress: Current Understanding and Future Directions. *Front. Plant Sci.* **2016,** *7*, 1029. DOI: 10.3389/fpls.2016.01029.

Kalaji, H. M.; Jajoo, A.; Oukarroum, A.; Brestic, M.; Zivcak, M.; Samborska, I. A.; Cetner, M. D.; Łukasik, I.; Goltsev, V.; Ladle, R. J. Chlorophyll a Fluorescence as a Tool to Monitor

Physiological Status of Plants under Abiotic Stress Conditions. *Acta Physiol. Planta* **2016,** *38* (4), 102.

Kapoor, D.; Bhardwaj, S.; Landi, M.; Sharma, A.; Ramakrishnan, M.; Sharma, A. The Impact of Drought in Plant Metabolism: How to Exploit Tolerance Mechanisms to Increase Crop Production. *Appl. Sci.* **2020,** *10*, 5692. DOI: 10.3390/app10165692.

Kapoor, D.; Singh, M.; Kaur, S.; Bhardwaj, R.; Zheng, B.; Sharma, A. Modulation of the Functional Components of Growth, Photosynthesis, and Anti-Oxidant Stress Markers in Cadmium Exposed *Brassica juncea* L. *Plants* **2019,** *8* (8), 260.

Karahan, F.; Ozyigit, I. I.; Saracoglu, I. A.; Yalcin, I. E.; Ozyigit, A. H.; Ilcim, A. Heavy Metal Levels and Mineral Nutrient Status in Different Parts of Various Medicinal Plants Collected from Eastern Mediterranean Region of Turkey. *Biol Trace Elem Res* **2020,** *197*, 316–329.

Kaur, P.; Bali, S.; Sharma, A.; Vig, A. P.; Bhardwaj, R. Effect of Earthworms on Growth, Photosynthetic Efficiency and Metal Uptake in *Brassica juncea* L. Plants Grown in Cadmium-Polluted Soils. *Environ. Sci. Pollut. Res.* **2017,** *24* (15),13452–13465.

Khalvandi, M.; Siosemardeh, A.; Roohi, E.; Keramati, S. Salicylic Acid Alleviated the Effect of Drought Stress on Photosynthetic Characteristics and Leaf Protein Pattern in Winter Wheat. *Heliyon.* Jan **2021,** *7* (1), e05908. DOI: 10.1016/j.heliyon.2021.e05908.

Khan, M. I. R.; Asgher, M.; Khan, N. A. Alleviation of Salt-Induced Photosynthesis and Growth Inhibition by Salicylic Acid Involves Glycine Betaine and Ethylene in Mungbean (*Vignaradiata* L.). *Plant Physiol. Biochem.* **2014,** *80*, 67–74.

Kleczkowski, L. A. Inhibitors of Photosynthetic Enzymes/Carriers and Metabolism. *Annu. Rev. Plant Physiol. Plant Mol. Biol.* **1994,** *45* (1), 339–367.

Kreslavski, V. \D.; Schmitt, F. J.; Keuer, C.; Friedrich, T.; Shirshikova, G. N.; Zharmukhamedov, S. K.; Kosobryukhov, A. A.; Allakhverdiev. Response of the Photosynthetic Apparatus to UV-A and Red Light in the Phytochrome B-Deficient *Arabidopsis thaliana* L. *hy3* Mutant. *Photosynthetica* **2016,** *54*, 321–333.

Kudo, M.; Kidokoro, S.; Yoshida, T.; Mizoi, J.; Todaka, D.; Fernie, A. R.; Shinozaki, K.; Yamaguchi-Shinozaki, K. Double Overexpression of DREB and PIF Transcription Factors Improves Drought Stress Tolerance and Cell Elongation in Transgenic Plants. *Plant Biotech J* **2017,** *15* (4), 458–471.

Kumar, J. I. N.; Bora, A.; Amn, M. K. Chronic Toxicity of a Triazole Fungicide Tebuconazole on the Growth and Metabolic Activities of Heterocystous, Nitrogen-Fixing Paddy Field Cyanobacterium, Westiellopsis prolifica Janet. *J. Microbiol. Biotechnol.* **2010,** *20* (7), 1134–1139.

Kurepin, L. V.; Ivanov, A. G.; Zaman, M.; Pharis, R. P.; Allakhverdiev, S. I.; Hurry, V.; Huner, N. P. A. Stress-Related Hormones and Glycinebetaine Interplay in Protection of Photosynthesis under Abiotic Stress Conditions. *Photosynth. Res.* **2015,** *126*, 221–235.

Kwon, O. K.; Mekapogu, M.; Kim, K. S. Effect of Salinity Stress on Photosynthesis and Related Physiological Responses in Carnation (*Dianthus caryophyllus*). *Hortic. Environ. Biotechnol.* **2019,** *60*, 831–839.

Latowski, D.; Kuczynska, P.; Strzałka, K. Xanthophyll Cycle—a Mechanism Protecting Plants against Oxidative Stress. *Redox Rep.* **2011,** *16* (2), 78–90.

Law, R. D.; Crafts-Brandner, S. J. Inhibition and Acclimation of Photosynthesis to Heat Stress Is Closely Correlated with Activation of Ribulose-1,5-bisphosphate Carboxylase/Oxygenase. *Plant Physiol.* **1999,** *120* (1), 173–182.

Li, J.; Cang, Z.; Jiao, F.; Bai, X.; Zhang, D.; Zhai, R. Influence of Drought Stress on Photosynthetic Characteristics and Protective Enzymes of Potato at Seedling Stage. *J.Saudi Soc.Agric.Sci.* **2017,** *16* (1), 82–88.

Li, S-L.; Li, Z-G.; Yang, L-T.; Li, Y-R.; He, Z-L. Differential Effects of Cold Stress on Chloroplasts Structures and Photosynthetic Characteristics in Cold-Sensitive and Cold-Tolerant Cultivars of Sugarcane. *Sugar Tech.* **2018,** *20* (1), 11–20.

Li, Y.; Xin, G.; Liu, C.; Shi, Q.; Yang, F.; Wei, M. Effects of Red and Blue Light on Leaf Anatomy, CO_2 Assimilation and the Photosynthetic Electron Transport Capacity of Sweet Pepper (*Capsicum annuum* L.) Seedlings. *BMC Plant Biol.* **2020a,** *20,* 318. https://doi.org/10.1186/s12870-020-02523-z.

Li, Y. T.; Li, Y.; Li, Y. N.; Liang, Y.; Sun, Q.; Li, G.; Liu, P.; Zhang, Z. S.; Gao, H. Y. Dynamic Light Caused Less Photosynthetic Suppression, Rather Than More, under Nitrogen Deficit Conditions Than under Sufficient Nitrogen Supply Conditions in Soybean. *BMC Plant Biol.* **2020b,** *20,* 339. https://doi.org/10.1186/s12870-020-02516-y.

Lípová, L.; Krchňák, P.; Komenda, J.; Ilík, P. Heat-Induced Disassembly and Degradation of Chlorophyll-Containing Protein Complexes in Vivo. *Biochim. Biophys. Acta* **2010,** *1797*\ (1), 63–70.

Liu, J.; van Iersel, M. W. Photosynthetic Physiology of Blue, Green, and Red Light: Light Intensity Effects and Underlying Mechanisms. *Front. Plant Sci.* **2021.** https://doi.org/10.3389/fpls.2021.619987.

Liu, Q-H.; Wu, X.; Chen, B-C.; Ma J-Q.; Gao J. Effects of Low Light on Agronomic and Physiological Characteristics of Rice Including Grain Yield and Quality. *Rice Sci.* **2014,** *21* (5), 243–251.

Liu, X.; Chen, Z.; Jahan, M. S.; Wen, Y.; Yao, X.; Ding, H.; Guo, S.; Xu, Z. RNA-Seq Analysis Reveals the Growth and Photosynthetic Responses of Rapeseed (*Brassica napus* L.) under Red and Blue LEDs with Supplemental Yellow, Green, or White Light. *Hort. Res.* **2020,** *7,* 206. https://doi.org/10.1038/s41438-020-00429-3.

Lopes, M. S.; Araus, J. L.; van Heerden, P. D. R.; Foyer, C. H. Enhancing Drought Tolerance in C(4) Crops. *J. Exp. Bot.* **2011,** *62* (9), 3135–3153.

Lu, T.; Meng, Z.; Zhang, G.; Qi, M.; Sun, Z.; Liu, Y.; Li, T. Sub-high Temperature and High Light Intensity Induced Irreversible Inhibition on Photosynthesis System of Tomato Plant (*Solanum lycopersicum* L.). *Front. Plant Sci.* **2017,** *8,* 365. DOI: 10.3389/fpls.2017.00365.

Martin, B.; Ort, D. R. Insensitivity of Water-Oxidation and Photosystem II Activity in Tomato to Chilling Temperatures. *Plant Physiol.* **1982,** *70* (3), 689–694.

Mathur, S.; Agarwal, D.; Jajoo, A. Photosynthesis: Response to High Temperature Stress. *J.Photochem. Photobiol.B Biol.* **2014,** *137,* 116–126.

McDowell, N. G.; Sevanto, S. The Mechanisms of Carbon Starvation: How, When, or Does It Even Occur At All? *New Phytol.* **2010,** *186* (2), 264–266.

Medynska-Juraszek, A.; Rivier, P. A.; Rasse, D.; Joner, E. J. Biochar Affects Heavy Metal Uptake in Plants through Interactions in the Rhizosphere. *Appl. Sci.* **2020,** *10,* 5105. DOI: 10.3390/app10155105.

Meng, F.; Luo, Q.; Wang, Q.; Zhang, X.; Qi, Z.; Xu, F.; Lei, X.; Cao, Y.; Chow, W. S.; Sun, G. Physiological and Proteomic Responses to Salt Stress in Chloroplasts of Diploid and Tetraploid Black Locust (*Robinia pseudoacacia* L.). *Sci. Rep.* **2016,** *6,* 23098. https://doi.org/10.1038/srep23098.

Mishra, V.; Srivastava, G.; Prasad, S. M.; Abraham, G. Growth, Photosynthetic Pigments and Photosynthetic Activity during Seedling Stage of Cowpea (*Vigna unguiculata*) in Response to UV-B and Dimethoate. *Pestic Biochem. Physiol.* **2008,** *92* (1), 30–37.

Msilini, N.; Zaghdoudi, M.; Govindachary, S.; Lachaal, M.; Ouerghii, Z.; Carpentier, R. Inhibition of Photosynthetic Oxygen Evolution and Electron Transfer from the Quinone Acceptor Q_A to Q_B by Iron Deficiency. *Photosynth. Res.* **2011,** *107*, 247–256.

Munns, R. Comparative Physiology of Salt and Water Stress. *Plant Cell Environ.* **2002,** *25*, 239–250.

Nadasy, E.; Lehoczky, E.; Lukacs, P.; Adam, P. Influence of Different Pre-Emergent Herbicides on the Growth of Soybean Varieties. *Zeitschr Pflanzenkr Pflanzensch* **2000,** *17*, 635–639.

Nahar, K.; Hasanuzzaman, M.; Suzuki, T.; Fujita, M. Polyamines-Induced Aluminum Tolerance in Mung Bean: A Study on Antioxidant Defense and Methylglyoxal Detoxification Systems. *Ecotoxicology* **2017,** *26* (1), 58–73.

Nahar, K.; Hasanuzzaman, M.; Alam, M. M.; Rahman, A.; Suzuki, T.; Fujita, M. Polyamine and Nitric Oxide Crosstalk: Antagonistic Effects on Cadmium Toxicity in Mungbean Plants through Up-regulating the Metal Detoxification, Antioxidant Defense and Methylglyoxal Detoxification Systems. *Ecotoxicol. Environ. Saf.* **2016,** *126*, 245–255.

Nemat Alla, M. M.; Hassan, N. M.; El-Bastawisy, Z. M. Differential Influence of Herbicide Treatments on Activity and Kinetic Parameters of C4 Photosynthetic Enzymes from *Zea mays*. *Pestic. Biochem. Physiol.* **2007,** *89* (3), 198–205.

Nishiyama, Y.; Murata, N. Revised Scheme for the Mechanism of Photoinhibition and Its Application to Enhance the Abiotic Stress Tolerance of the Photosynthetic Machinery. *Appl. Microbiol. Biotechnol.* **2014,** *98* (21), 8777–8796.

Niu, Y.; Xiang, Y. An Overview of Biomembrane Functions in Plant Responses to High-Temperature Stress. *Front. Plant Sci.* **2018,** *9*, 915. DOI: 10.3389/fpls.2018.00915.

Noriega, G. O.; Balestrasse, K. B.; Batlle, A.; Tomaro, M. L. Cadmium Induced Oxidative Stress in Soybean Plants Also by the Accumulation of δ-Aminolevulinic Acid. *Biometals* **2007,** *20* (6), 841–851.

Ohashi, Y.; Nakayama, N.; Saneoka, H.; Fujita, K. Effects of Drought Stress on Photosynthetic Gas Exchange, Chlorophyll Fluorescence and Stem Diameter of Soybean Plants. *Biol. Plant* **2006,** *50*, 138–141.

Omoto, E.; Taniguchi, M.; Miyake, H. Effects of Salinity Stress on the Structure of Bundle Sheath and Mesophyll Chloroplasts in NAD-Malic Enzyme and PCK Type C_4 Plants. *Plant Prod.Sci.* **2010,** *13* (2), 169–176.

Pagliano, C.; Raviolo, M.; Dalla Vecchia, F.; Gabbrielli, R.; Gonnelli, C.; Rascio, N.; Barbato, R.; La Rocca, N. Evidence for PSII Donorside Damage and Photoinhibition Induced by Cadmium Treatment on Rice (*Oryza sativa* L.). *J. Photochem. Photobiol. B* **2006,** *84* (1), 70–78.

Parmar, P.; Kumari, N.; Sharma, V. Structural and Functional Alterations in Photosynthetic Apparatus of Plants under Cadmium Stress. *Bot Stud.* **2013,** *54* (1), 45. DOI: 10.1186/1999-3110-54-45.

Parry, M. A. J.; Andralojc, P. J.; Khan, S.; Lea, P. J.; Keys, A. J. Rubisco Activity: Effects of Drought Stress. *Ann Bot.* Jun 15, **2002,** *89* (7), 833–839.

Partelli, F. L.; Vieira, H. D.; Viana, A. P.; Batista-Santos, P.; Rodrigues, A. P.; Leitao, A. E.; Ramalho, J. C. Low Temperature Impact on Photosynthetic Parameters of Coffee Genotypes. *Pesquisa Agropecuária Brasileira* **2009,** *44* (11). http://dx.doi.org/10.1590/S0100-204X2009001100006.

Parween, T.; Jan, S.; Mahmooduzzafar, S.; Fatma, T.; Siddiqui, Z. H. Selective Effect of Pesticides on Plant—A Review. *Crit. Rev. Food Sci. Nutr.* **2016,** *56* (1), 160–179.

Pascual, J.; Rahikainen, M.; Kangasjärvi, S. *Plant Light Stress*; Wiley: Chichester, 2017. https ://doi.org/10.1002/97804-70015-902. a0001 319.pub3.

Per, T. S.; Khan, N. A.; Masood, A.; Fatma, M. Methyl Jasmonate Alleviates Cadmium-Induced Photosynthetic Damages through Increased S-Assimilation and Glutathione Production in Mustard. *Front. Plant Sci.* **2016,** *7*, 1933. DOI: 10.3389/fpls.2016.01933.

Petit, A. N.; Fontaine, F.; Vatsa, P.; Clement, C.; Vaillaant-Gaveau, N. Fungicide Impacts on Photosynthesis in Crop Plants. *Photosynth. Res.* **2012,** *111*, 315–326.

Popova, A. V.; Dobrev, K.; Velitchkova, M.; Ivanov, A. G. Differential Temperature Effects on Dissipation of Excess Light Energy and Energy Partitioning in *lut2* Mutant of *Arabidopsis thaliana* under Photoinhibitory Conditions. *Photosynth. Res.* **2019,** *139*, 367–385.

Qados, A. M. S. A. Effect of Salt Stress on Plant Growth and Metabolism of Bean Plant *Vicia faba* (L.). *J. Saudi Soc. Agric.Sci.* **2011,** *10* (1), 7–15.

Qin, C.; Ahanger, M. A.; Lin, B.; Huang, Z.; Zhou, J.; Ahmed, N.; Ai, S.; Mustafa, NSA.; Ashraf, M.; Zhang, L. Comparative Transcriptomic Analysis Reveals the Regulatory Effects of Acetylcholine on Salt Tolerance of *Nicotiana benthamiana*. *Phytochemistry* **2021,** *181*, 112582. https://doi.org/10.1016/j.phytochem.2020.112582.

Qiu, N.; Lu, Q.; Lu, C. Photosynthesis, Photosystem II Efficiency and the Xanthophyll Cycle in the Salt-Adapted Halophyte *Atriplex centralasiatica*. *New Phytol.* **2003,** *159* (2), 479–486.

Rascio, N.; Navari-Izzo, F. Heavy Metal Hyperaccumulating Plants: How and Why Do They Do It? And What Makes Them So Interesting? *Plant Sci.* **2011,** *180* (2), 169–181.

Reddy, A. R.; Chaitanya, K. V.; Vivekanandan, M. Drought-Induced Responses of Photosynthesis and Antioxidant Metabolism in Higher Plants. *J. Plant Physiol.* **2004,** *161* (11), 1189–1202.

Remans, T.; Opdenakker, K.; Smeets, K.; Mathijsen, D.; Vangronsveld, J.; Cuypers, A. Metal-Specific and NADPH Oxidase Dependent Changes in Lipoxygenase and NADPH Oxidase Gene Expression in *Arabidopsis thaliana* Exposed to Cadmium or Excess Copper. *Funct. Plant Biol.* **2010,** *37* (6), 532–544.

Saffan, S. E. Effect of Salinity and Osmotic Stresses on Some Economic Plants. *Res. J. Agric. Biol. Sci.* **2008,** *4* (2), 159–166.

Saibo, N. J. M.; Lourenco, T.; Oliveira, M. M. Transcription Factors and Regulation of Photosynthetic and Related Metabolism under Environmental Stresses. *Ann.Bot.***2009,** *103*, 609–623.

Salvucci, M. E.; Crafts-Brandner, S. J. Inhibition of Photosynthesis by Heat Stress: The Activation State of Rubisco as a Limiting Factor in Photosynthesis. *Physiol. Planta.* **2004,** *120* (2), 179–186.

Sane, P. V.; Ivanov, A. G.; Hurry, V.; Huner, N. P. A.; Öquist, G. Changes in the Redox Potential of Primary and Secondary Electron Accepting Quinones in Photosystem II Confer Increased Resistance to Photoinhibition in Low-Temperature-Acclimated *Arabidopsis*. *Plant Physiol.* **2003,** *132*, 2144–2151.

Sato, R.; Ito, H.; Tanaka, A. Chlorophyll *b* Degradation by Chlorophyll *b* Reductase under High-Light Conditions. *Photosynth. Res.* **2015,** *126* (2–3), 249–259.

Savitch, L. V.; Ivanov, A. G.; Gudynaite-Savitch, L.; Huner, N. P. A.; Simmonds, J. Cold Stress Effects on PSI Photochemistry in *Zea mays*: Differential Increase of FQR-Dependent Cyclic Electron Flow and Functional Implications. *Plant Cell Physiol.* **2011,** *52*, 1042–1054.

Schrader, S. M.; Wise, R. R.; Wacholtz, W. F.; Ort, D. R.; Sharkey, T. D. Thylakoid Membrane Responses to Moderately High Leaf Temperature in Pima Cotton. *Plant Cell Environ.* **2004,** *27*, 725–735.

Semerci, A.; Cicek, N.; Karahan, F. A.; Özyürek, E.; Arici, Y. K.; Ekmekci, Y. Some Growth and Chlorophyll Fluorescence Parameters of Black and Hybrid Poplar Clones under Water Stress. *Turk. J. Agric. For.* **2017,** *41* (5), 348–356.

Shao, R. X.; Xin, L. F.; Zheng, H. F, Li, L. L.; Ran, W. L.; Mao, J.; Yang, Q. H. Changes in Chloroplast Ultrastructure in Leaves of Drought-Stressed Maize Inbred Lines. *Photosynthetica* **2016,** *54*, 74–80.

Sharkey, T. D. Effects of Moderate Heat Stress on Photosynthesis: Importance of Thylakoid Reactions, Rubisco Deactivation, Reactive Oxygen Species, and Thermotolerance Provided by Isoprene. *Plant Cell Environ.* 2**0005,** *28* (3), 269–277.

Sharkey, T. D.; Zhang, R. High Temperature Effects on Electron and Proton Circuits of Photosynthesis. *J. Integr. Plant Biol.* **2010,** *52* (8), 712–722.

Sharma, A.; Yuan, H.; Kumar, V.; Ramakrishnan, M.; Kohli, S. K.; Kaur, R.; Thukral, A. K.; Bhardwaj, R.; Zheng, B. Castasterone Attenuates Insecticide Induced Phytotoxicity in Mustard. *Ecotoxicol. Environ. Saf.* **2019,** *179*, 50–61.

Sharma, P.; Dubey, R. S. Lead Toxicity in Plants. *Braz. J. Plant Physiol.* **2005,** *17* (1), 35–52.

Shu, S.; Chen, L.; Lu, W.; Sun, J.; Guo, S.; Yuan, Y.; Li, J. Effects of Exogenous Spermidine on Photosynthetic Capacity and Expression of Calvin Cycle Genes in Salt-Stressed Cucumber Seedlings. *J. Plant Res.* **2014,** *127* (6), 763–73.

Shuai, W.; Xiaodi, W.; Xiangbin, S.; Baoling, W.; Xiaocui, Z.; Haibo, W.; Fengzhi, L. Red and Blue Lights Significantly Affect Photosynthetic Properties and Ultrastructure of Mesophyll Cells in Senescing Grape Leaves. *Hort.Plant J.* **2016,** *2* (2), 82–90.

Sicora, C.; Szilárd, A.; Sass, L.; Turcsányi, E.; Máté, Z.; Vass, I. UV-B and UV-A Radiation Effects on Photosynthesis at the Molecular Level. In: *Environmental UV Radiation: Impact on Ecosystems and Human Health and Predictive Models. Nato Science Series: IV: Earth and Environmental Sciences*; Ghetti, F.; Checcucci, G.; Bornman, J. F.; Eds.; Vol. 57; Springer: Dordrecht, 2006. https://doi.org/10.1007/1-4020-3697-3-10.

Singh, H. P.; Mahajan, P.; Kaur, S.; Batish, D. R.; Kohli, R. K. Chromium Toxicity and Tolerance in Plants. *Environ. Chem. Lett.* **2013,** *11* (3), 229–254.

Skotnica, J.; Matoušková, M.; Nauš, J.; Lazár, D.; Dvořák, L. Thermoluminescence and Fluorescence Study of Changes in Photosystem II Photochemistry in Desiccating Barley Leaves. *Photosynth. Res.* **2000,** *65* (1), 29–40.

Skrebsky, E. C.; Tabaldi, L. A.; Pereira, L. B.; Rauber, R.; Maldaner, J.; Cargnelutti, D.; Gonçalves, J. F.; Castro, G. Y.; Shetinger, M. R.; Nicoloso, F. T. Effect of Cadmium on Growth, Micronutrient Concentration, and δ-aminolevulinic Acid Dehydratase and Acid Phosphatase Activities in Plants of *Pfaffia glomerata. Braz. J. Plant Physiol.* **2008,** *20* (4), 285–294.

Soliman, M.; Alhaithloul, H. A.; Hakeem, K. R.; Alharbi, B. M.; El-Esawi, M.; Elkelish, A. Exogenous Nitric Oxide Mitigates Nickel-Induced Oxidative Damage in Eggplant by Up-regulating Antioxidants, Osmolyte Metabolism, and Glyoxalase Systems. *Plants* **2019,** *8*, 562. DOI: 10.3390/plants8120562.

Soliman, M.; Elkelish, A.; Souad, T.; Alhaithloul, H.; Farooq, M. Brassinosteroid Seed Priming with Nitrogen Supplementation Improves Salt Tolerance in Soybean. *Physiol. Mol. Biol. Plants.* **2020,** *26* (3), 501–511.

Solymosi, K.; Schoefs, B. Etioplast and Etio-Chloroplast Formation under Natural Conditions: The Dark Side of Chlorophyll Biosynthesis in Angiosperms. *Photosynth. Res.* **2010,** *105* (2), 143–166.

Song, J.; Zhou, J.; Zhao, W.; Xu, H.; Wang, F.; Xu, Y.; Wang, L.; Tian, C. Effects of Salinity and Nitrate on Production and Germination of Dimorphic Seeds Applied Both through the Mother Plant and Exogenously during Germination in *Suaeda salsa. Plant Species Biol.* **2016,** *31*, 19–28.

Song, Y.; Chen, Q.; Ci, D.; Shao, X.; Zhang, D. Effects of High Temperature on Photosynthesis and Related Gene Expression in Poplar. *BMC Plant Biol.* **2014,** *14*, 111. https://doi.org/10.1186/1471-2229-14-111.

Sui, N.; Wang, Y.; Liu, S.; Yang, Z.; Wang, F.; Wan, S. Transcriptomic and Physiological Evidence for the Relationship between Unsaturated Fatty Acid and Salt Stress in Peanut. *Front. Plant Sci.* **2018,** *9*, 7.

Sun, J.; Gu, J.; Zeng, J.; Han, S.; Song, A.; Chen, F.; Fang, W.; Jiang, J.; Chen, S. Changes in Leaf Morphology, Antioxidant Activity and Photosynthesis Capacity in Two Different Drought-Tolerant Cultivars of Chrysanthemum during and after Water Stress. *Sci. Hort.* **2013,** *161*, 249–258.

Swamy, B. P. M.; Kumar, A. Genomics-Based Precision Breeding Approaches to Improve Drought Tolerance in Rice. *Biotechnol. Adv.* **2013,** *31* (8), 1308–1318.

Tattini, M.; Velikova, V.; Vickers, C.; Brunetti, C.; Di Ferdinando, M.; Trivellini, A.; Fineschi, S.; Agati, G.; Ferrini, F.; Loreto, F. Isoprene Production in Transgenic Tobacco Alters Isoprenoid, Non-Structural Carbohydrate and Phenylpropanoid Metabolism, and Protects Photosynthesis from Drought Stress. *Plant Cell Environ.* **2014,** *37* (8), 1950–1964.

Tenhunen, J. D.; Serra, A. S.; Harley, P. C.; Dougherty, R. L.; Reynolds, J. F. Factors Influencing Carbon Fixation and Water Use by Mediterranean Sclerophyll Shrubs during Summer Drought. *Oecologia* **1990,** *82*, 381–393.

Tewari, A. K.; Tripathy, B. C. Temperature-Stress-Induced Impairment of Chlorophyll Biosynthetic Reactions in Cucumber and Wheat. *Plant Physiol.* **1998,** *117* (3), 851–858.

Untiedt, R.; Blanke, M. M. Effects of Fungicide and Insecticide Mixtures on Apple Tree Canopy Photosynthesis, Dark Respiration and Carbon Economy. *Crop Protection* **2004,** *23* (10), 1001–1006.

Velikova, V.; Varkonyi, Z.; Szabo, M.; Maslenova, L.; Nogues, I.; Peeva, V.; Busheva, M.; Garab, G.; Sharkey, T. D.; Loreto, F. Increased Thermostability of Thylakoid Membranes in Isoprene-Emitting Leaves Probed with Three Biophysical Techniques. *Plant Physiol.* **2011,** 157 (2), 905–916.

Vinit-Dunant, F.; Epron, D.; Alaoui-Sosse, B.; Badot, P-M. Effects on Copper on Growth and Photosynthesis of Mature and Expanding Leaves in Cucumber Plants. *Plant Sci.* **2002,** *163*, 53–58.

Vishwakarma, K.; Upadhyay, N.; Kumar, N.; Yadav, G.; Singh, J.; Mishra, R. K.; Kumar, V.; Verma, R.; Upadhyay, R. G.; Pandey, M.; Sharma, S. Abscisic Acid Signaling and Abiotic Stress Tolerance in Plants: A Review on Current Knowledge and Future Prospects. *Front. Plant Sci.* **2017,** *8*, 161. DOI: 10.3389/fpls.2017.00161.

Voss, I.; Sunil, B.; Scheibe, R.; Raghavendra, A. S. Emerging Concept for the Role of Photorespiration as an Important Part of Abiotic Stress Response. *Plant Biol.* **2013,** *15* (4), 713–722.

Wang, F.; Chen, F.; Cai, Y.; Zhang, G.; Wu, F. Modulation of Exogenous Glutathione in Ultrastructure and Photosynthetic Performance against Cd Stress in the Two Barley Genotypes Differing in Cd Tolerance. *Biol. Trace Element Res.* **2011,** *144* (1–3), 1275–1288.

Wang, H.; Gu, M.; Cui, J.; Shi, K.; Zhou, Y.; Yu, J. Effects of Light Quality on CO_2 Assimilation, Chlorophyll-Fluorescence Quenching, Expression of Calvin Cycle Genes and Carbohydrate Accumulation in *Cucumis sativus. J. Photochem. Photobiol.B: Biol.* **2009,** *96* (1), 30–37.

Wang, L.; Deng, F.; Ren, W. J.; Yang, W. Y. Effects of Shading on Starch Pasting Characteristics of Indica Hybrid Rice (*Oryza sativa* L.). *PLoS ONE* **2013,** *8* (7), e68220.

Wang, X.; Li, Q.; Xie, J.; Huang, M.; Cai, J.; Ting, Z.; Dai, T.; Jiang, D. Abscisic Acid and Jasmonic Acid Are Involved in Drought Priming-Induced Tolerance to Drought in Wheat. *Crop J.* **2021,** *9* (1), 120–132.

Wang, Y.; Tong, Y.; Chu, H.; Chen, X.; Guo, H.; Yuan, H.; Yan, D.; Zheng, B. Effects of Different Light Qualities on Seedling Growth and Chlorophyll Fluorescence Parameters of *Dendrobium officinale. Biologia* **2017,** *72* (7), 735–744.

Wang, Z.; Li, G.; Sun, H.; Ma, L.; Guo, Y.; Zhao, Z.; Gao, H.; Mei, L. Effects of Drought Stress on Photosynthesis and Photosynthetic Electron Transport Chain in Young Apple Tree Leaves. *Biol. Open.* **2018,** *7* (11), bio035279. DOI: 10.1242/bio.035279.

Wise, R.; Olson, A.; Schrader, S.; Sharkey, T. Electron Transport Is the Functional Limitation of Photosynthesis in Field-Grown Pima Cotton Plants at High Temperature. *Plant Cell Environ.* **2004,** *27* (6), 717–724.

Xia, X. J.; Huang, Y. Y.; Wang, L.; Huang, L. F.; Yu, Y. L.; Zhou, Y. H.; Yu, J. Q. Pesticides-Induced Depression of Photosynthesis Was Alleviated by 24-Epibrassinolide Pretreatment in *Cucumis sativus* L. *Pestic. Biochem. Physiol.* **2006,** *86* (1), 42–48.

Xu, Q.; Ma, X.; Lv, T.; Bai, M.; Wang, Z.; Niu, J. Effects of Water Stress on Fluorescence Parameters and Photosynthetic Characteristics of Drip Irrigation in Rice. Water **2020,** *12*, 289. DOI: 10.3390/w12010289.

Yadav, P.; Kaur, R.; Kanwar, M. K.; Sharma, A.; Verma, V.; Sirhindi, G.; Bhardwaj, R. Castasterone Confers Copper Stress Tolerance by Regulating Antioxidant Enzyme Responses, Antioxidants, and Amino Acid Balance in *B. juncea* Seedlings. *Ecotoxicol. Environ. Saf.* **2018,** *147*, 725–734.

Yamamoto, H.; Miyake, C.; Dietz, K. J.; Tomizawa, K.; Murata, N.; Yokota, A. Thioredoxin Peroxidase in the Cyanobacterium *Synechocystis* sp. PCC 6803. *FEBS Lett.* **1999,** *447* (2–3), 269–273.

Yang, Z.; Li, J. L.; Liu, L. N.; Xie, Q.; Sui, N. Photosynthetic Regulation under Salt Stress and Salt-Tolerance Mechanism of Sweet Sorghum. *Front. Plant Sci.* **2020,** *15*. https://doi.org/10.3389/fpls.2019.01722.

Yanori, W.; von Caemmerer, S. Effect of Rubisco Activase Deficiency on the Temperature Response of CO_2 Assimilation Rate and Rubisco Activation State: Insights from Transgenic Tobacco with Reduced Amounts of Rubisco Activase. *Plant Physiol.* **2009,** *151*, 2073–2082.

Yavari, N.; Tripathi, R.; Wu, B. S.; MacPherson, S.; Singh, J.; Lefsrud, M. The Effect of Light Quality on Plant Physiology, Photosynthetic, and Stress Response in *Arabidopsis thaliana* Leaves. *Plos ONE* **2021**. https://doi.org/10.1371/journal.pone.0247380.

Ying, R. R.; Qiu, R. L.; Tang, Y. T.; Hu, P. J.; Qiu, H.; Chen, H. R.; Shi, T. H.; Morel, J. L. Cadmium Tolerance of Carbon Assimilation Enzymes and Chloroplast in Zn/Cd Hyperaccumulator *Picris divaricata. J. Plant Physiol.* **2010,** *167* (2), 81–87.

Ying, Y.; Yue, Y.; Huang, X.; Wang, H.; Mei, L.; Yu, W.; Zheng, B.; Wu, J. Salicylic Acid Induces Physiological and Biochemical Changes in Three Red Bayberry (*Myric rubra*) Genotypes under Water Stress. *Plant Growth Regul.* **2013,** *71*, 181–189.

Yoshida, C. H. P.; Pacheco, A. C.; Lapaz, A. M.; Gorni, P. H.; Vitolo, H. F.; Bertoli, S. C. Methyl Jasmonate Modulation Reduces Photosynthesis and Induces Synthesis of Phenolic Compounds in Sweet Potatoes Subjected to Drought. *Bragantia* **2020,** *79* (3). http://dx.doi.org/10.1590/1678-4499.20200203.

Yuan, F.; Leng, B.; Wang, B. Progress in Studying Salt Secretion from the Salt Glands in Recretohalophytes: How Do Plants Secrete Salt? *Front. Plant Sci.* **2016,** *7*, 997. DOI: 10.3389/fpls.2016.00977.

Zhang, F.; Lu, K.; Gu, Y.; Zhang, L.; Li, W.; Li, Z. Effects of Low-Temperature Stress and Brassinolide Application on the Photosynthesis and Leaf Structure of Tung Tree Seedlings. *Front. Plant Sci.* **2020**. https://doi.org/10.3389/fpls.2019.01767.

Zhang, H.; Xu, Z.; Guo, K.; Huo, Y.; He, G.; Sun, H.; Guan, Y.; Xu, N.; Yang, W, Sun, G. Toxic Effects of Heavy Metal Cd and Zn on Chlorophyll, Carotenoid Metabolism and Photosynthetic Function in Tobacco Leaves Revealed by Physiological and Proteomics Analysis. *Ecotoxicol. Environ.Saf.* **2020,** *202* (1), 110856. https://doi.org/10.1016/j.ecoenv.2020.110856.

Zhang, L.; Hu, T.; Amombo, E.; Wang, G.; Xie, Y.; Fu, J. The Alleviation of Heat Damage to Photosystem II and Enzymatic Antioxidants by Exogenous Spermidine in Tall Fescue. *Front. Plant Sci.* **2017,** *8*, 1747. DOI: 10.3389/fpls.2017.01747.

Zhang, L. T.; Zhang, Z. S.; Gao, H. Y.; Xue, Z. C.; Yang, C.; Meng, X. L.; Meng, Q. W. Mitochondrial Alternative Oxidase Pathway Protects Plants against Photoinhibition by Alleviating Inhibition of the Repair of Photodamaged PSII through Preventing Formation of Reactive Oxygen Species in Rumex K-1 Leaves. *Physiol. Planta.* **2011,** *143* (4), 396–407.

Zhang, N.; Portis, A. R. Jr. Mechanism of Light Regulation of Rubisco: A Specific Role for the Larger Rubisco Activase Isoforms Involving Reductive Activation by Thioredoxin-f. *Proc Natl Acad Sci USA* **1999,** *96* (16), 9438–9443.

Zhang, Q. Y.; Wang, L. Y.; Kong, F. Y.; Deng, Y. S.; Li, B.; Meng, Q. W. Constitutive Accumulation of Zeaxanthin in Tomato Alleviates Salt Stress-Induced Photoinhibition and Photooxidation. *Physiol. Plant.* **2012,** *146*, 363–373.

Zhang, L.; Liu, J. Effects of Heat Stress on Photosynthetic Electron Transport in a Marine Cyanobacterium *Arthrospira* sp. *J. Appl. Phycol.* **2016,** *28*, 757–763.

Zhao, W.; Liu, L.; Shen, Q.; Yang, J.; Han, X.; Tian, F.; Wu, J. Effects of Water Stress on Photosynthesis, Yield, and Water Use Efficiency in Winter Wheat. *Water* **2020,** *12*, 2127. DOI: 10.3390/w12082127.

Zheng, J.; Hu, M. J.; Guo, Y. P. Regulation of Photosynthesis by Light Quality and Its Mechanism in Plants. *Ying Yong Sheng Tai Xue Bao.* **2008,** *19* (7), 1619–1624.

Zhou, J.; Chen, S.; Shi, W.; David-Schwartz, R.; Li, S.; Yang, F.; Lin, Z. Transcriptome Profiling Reveals the Effects of Drought Tolerance in Giant Juncao. *BMC Plant Biol.* **2021,** *21*, 2. https://doi.org/10.1186/s12870-020-02785-7.

Zhu, S. Q.; Chen, M. W.; Ji, B. H.; Jiao, D. M.; Liang, J. S. Roles of Xanthophylls and Exogenous ABA in Protection against NaCl-Induced Photodamage in Rice (*Oryza sativa* L) and Cabbage (*Brassica campestris*). *J. Exp. Bot.* **2011,** *62* (13), 4617–4625.

CHAPTER 3

Redesigning Plant Photosynthesis Using High-Throughput Techniques for Enhanced and Sustainable Food Production under Environmental Stresses

SYED SARFRAZ HUSSAIN[1*], MUHAMMAD IRFAN[1], ABDUL QAYYUM RAO[2], BUJUN SHI[3], AHMED ALI SHAHID[2], and KADAMBOT H M SIDDIQUE[4]

[1]*Kauser Abdulla Malik School of Life Sciences (KAM-SLS), Forman Christian College (A Chartered University), Ferozepur Road, Lahore 54600, Pakistan*

[2]*National Centre of Excellence in Molecular Biology (NCEMB), 87-West Canal Bank Road, University of the Punjab, Lahore 54600, Pakistan*

[3]*School of Agriculture, Food & Wine, Waite Campus University of Adelaide SA 5064, Australia*

[4]*The UWA Institute of Agriculture, The University of Western Australia, Perth WA 6001, Australia*

[*]*Corresponding author. E-mail: syedsarfraz_2000@yahoo.com*

ABSTRACT

It seems logical to talk about plants with an increased productivity for the fast increasing world population. Crop plant development showing tolerance

Photosynthesis and Respiratory Cycles during Environmental Stress Response in Plants. Aryadeep Roychoudhury (Ed.)

to various abiotic stresses has got immense potential and seems a promising strategy that may help to satisfy population demands for nutritious food. Currently, plant biotechnology is responsible to develop environmental stress tolerance and higher-yielding crop plants with increased nutritional value. Even though plant growth and development are governed by a variety of genetic, biochemical, environmental, and molecular processes, photosynthesis is a complex mechanism responsible for providing the energy required for crop productivity. The successful improvement in photosynthetic efficiency of crops, especially the leading cereal crops, has substantially increased global food production even without increasing cultivation areas. However, the availability of high throughput tools and technologies such as dynamic and sophisticated genome editing techniques for improving cultivars will play a vital role, especially in increasing photosynthetic efficiency. This technique represents a promising strategy to tackle yield-related problems, such as improvement in photosynthetic efficiency and abiotic stress tolerance of plants and may serve as a foundation for the second green revolution in the future. This chapter aims to discuss recent advances, current challenges, and future perspectives for targeting improved photosynthesis for enhanced and sustainable food, fiber, and fuel production under different environmental stresses.

3.1 INTRODUCTION: PHOTOSYNTHESIS, PLANT PRODUCTIVITY, AND ABIOTIC STRESSES

The human population is growing by 1.2% per year (Kaneda, 2016), and the current global population which stands at 7.6 billion is expected to touch 9.8 billion by mid of the 21st century, namely, 2050 (United Nations, 2017). Modern technologies and strategies to increase crop productivity under optimal and suboptimal conditions are important to achieve food security for the ever-increasing population. Crop productivity needs to increase by at least 50% to keep pace with the needs of the global population (Alexandratos and Bruinsma, 2012). However, this will be difficult using the available plant biodiversity and crop production technologies. Several environmental factors, such as natural disasters and anthropogenic activities like soil depletion, afforestation, and loss of arable land, are damaging plant health, resulting in irreversible crop yield losses (Hussain et al., 2012; Gururani et al., 2015; Dann and Leister, 2017; Hrmova and Hussain, 2021). Abiotic stress is defined as any environmental constraint reducing plant growth, development, and yield below optimum levels.

Studies have shown that abiotic stresses often reduce photosynthesis, leading to low plant productivity (Dhar et al., 2014; Mu et al., 2016, 2017; Sharma et al., 2016a, 2019; Fahad et al., 2017; Kaur et al., 2017; Demmig-Adams et al., 2018; Kohli et al., 2018; Paunov et al., 2018; Soares et al., 2018; Yadav et al., 2018; Zandalinas et al., 2018). Abiotic stresses such as drought, extreme temperature fluctuations, salinity, and other associated stresses can severely damage macromolecules, especially photosynthetic apparatus (photosystems I and II), electron transport chain, chlorophyll biosynthesis, and others (Efeoglu and Terzioglu, 2009; Ashraf and Harris, 2013; Nishiyama and Murata, 2014; Nouri et al., 2015; Kalaji et al., 2016; Sharma et al., 2016b; Yamori, 2016). However, plants have several strategies to counteract the adverse effects of these abiotic factors. Plants have evolved some common mechanisms during most stress conditions, while others are stress- and plant-specific (Eberhard et al., 2008; Foyer et al., 2012). However, it is noteworthy that plant and abiotic stress interactions are complex and dynamic in nature (Eberhard et al., 2008; Foyer et al., 2012). It is also known that most stress-related effects in plants are partially or completely reversible. However, it depends on the factors like the extent of damage and plant organ/system affected (Maayan et al., 2008; Ambavaram et al., 2014).

A number of studies have contributed significantly for developing a current understanding of photosynthetic process in plants over many decades (Foyer and Noctor, 2005; Baker, 2008; Foyer et al., 2012; Nishiyama and Murata, 2014; Cortleven et al., 2014; Tikkanen and Aro, 2014; Tikkanen et al., 2014; Gururani et al., 2015; Ort et al., 2015). Photosynthesis is a specialized plant process responsible for generating energy in the form of organic molecules (Pinheiro and Chaves, 2011; Ashraf and Harris, 2013; Morales et al., 2020). Photosynthesis fuels several metabolic processes directly involved in converting light energy into chemical energy (Pinheiro and Chaves, 2011; Ashraf and Harris, 2013; Chen et al., 2018; Demmig-Adams et al., 2018). Photosynthesis is a complex mechanism that comprises multistep redox reactions and involves various metabolic pathways, complex signaling, and regulatory system (Foyer and Noctor, 2005; Baker, 2008; Foyer et al., 2012; Nishiyama and Murata, 2014; Cortleven et al., 2014; Gururani et al., 2015). Several regulatory components particularly phytohormones, transcription factors, phosphatases, and mitogen-activated protein kinases, control and critically regulate many genes associated with stress (Foyer and Shigeoka, 2011; Osakabe et al., 2014; Gururani et al., 2015). Being naturally complex, plant productivity is a collaborative contribution of multiple physiological, biochemical, genetic, and molecular processes (Hrmova and Hussain, 2021). Stable photosynthetic performance is vital for normal plant growth and productivity (Dann and Leister, 2017; Nguyen et al., 2018; Sharma et al., 2020; Muhammad et al., 2021).

It is imperative to develop strategies to improve photosynthetic capacity for high crops under different abiotic stresses (Hussain et al., 2011, 2012; Hrmova and Hussain, 2021). Transgenic approaches are promising strategies for developing crop plants with high-stress tolerance to various stresses (Capell et al., 2004; Ashraf and Foolad, 2007; Pinheiro and Chaves, 2011; Hussain et al., 2012, 2014b; Song et al., 2014; Dusenge et al., 2019). Numerous studies have demonstrated the advantages of this approach, as transgenic plants harbor important stress-inducible genes directly involved in metabolic functioning, modeling, redesigning photosynthesis, and stress tolerance (Kathuria et al., 2007; Bartels and Hussain, 2008; Hussain et al., 2012; Long et al., 2015; Ort et al., 2015; Simkin et al., 2019; Weber and Bar-Even, 2019).

It is expected that various environmental stresses will continue to negatively impact photosynthesis and, ultimately, crop plant productivity. Photosynthetic plasticity is another way that plants can adapt to changing environment to maintain high photosynthetic efficiency. Natural biodiversity provides ways to exploit this plasticity, which can be explored using current high-throughput biotechnological and genomic techniques (Nouri et al., 2015). There are several ways to enhance photosynthetic capacity other than those dictated by natural variation. For example, current high-throughput phenotyping platforms offer extensive opportunities in crop breeding programs to monitor potential traits under different abiotic stresses. Plants exploit over 3000 genes to carry out photosynthesis successfully, and modern gene expression approaches are helpful to quantify transcripts and locate when and where a gene is turned on/off. These studies can effectively address several vital aspects of the plant photosynthesis process in the future. Genome editing technologies, such as transcription activator-like effector nucleases (TALENs) (Bedell et al., 2012; Li et al., 2012) and the CRISPR/Cas system, already play a significant role in precise genome engineering in plants (Cong et al., 2013; Feng et al., 2014) that could be used to generate targeted variations to enhance photosynthetic capacity of crop plants (Scheben and Edwards, 2017; Scheben et al., 2017). Plant breeding programs, aimed at achieving the above-mentioned goals, can take advantages of these technologies.

Another recent development is the integration of innovative ideas for redesigning photosynthesis triggered by new developments in in silico modeling and computational and synthetic biology to understand photosynthetic processes. Such computer-based programs can help realize the outcomes of potential redesigns of photosynthesis (Ort et al., 2011; Ort et al., 2015).

3.2 ABIOTIC STRESSES AND PHOTOSYNTHESIS-ASSOCIATED MICRORNA (MIRNA) PROFILING AND FUNCTIONAL CHARACTERIZATION

The discovery of sRNAs, especially miRNAs, and their effect on gene regulation and expression is one of the most exciting revolutions of the last decade (Lelandais-Briere et al., 2010; Hussain et al., 2014b). Numerous studies have demonstrated the importance of miRNAs (19–24 nt) in biological and metabolic regulatory processes, transforming our concept of several biological mechanisms, including gene regulation (Trindale et al., 2009; Cuperus et al., 2011; Khraiwesh et al., 2012; Hussain et al., 2014a; Ferdous et al., 2015), and developmental processes in plants, such as tolerance to environmental stresses (Zhang and Li 2013; Hussain et al., 2014b; Ferdous et al., 2015; Shriram et al., 2016; Alptekin et al., 2017; Liu et al., 2017a; Pan et al., 2018; Song et al., 2018).

Furthermore, several advanced technologies, such as digital sequencing, computer-based mining, and molecular experimental approaches, are available for genome-wide miRNA profiling and molecular characterization of miRNAs in plants (Achard et al., 2004; Juarez et al., 2004; Sunkar and Zhu, 2004; Lauter et al., 2005; Jones-Rhoades et al., 2006; Nikovics et al., 2006; Wang et al., 2007). These remarkable developments lead the molecular biologists toward exploring and exploiting the stress miRNAs in plants (Xie et al., 2007; Buhtz et al., 2008; Pant et al., 2008; Subramanian et al., 2008; Jagadeeswaran et al., 2009; Lelandais-Briere et al., 2009; Yao et al., 2007; Joshi et al., 2010; Zhao et al., 2010; Schreiber et al., 2011). A number of reports have revealed that hundreds of plant miRNAs target many genes involved in critical biological processes such as growth and development, and metabolic pathways (Sunkar et al., 2006; Laporte et al., 2007; Wang and Li 2007; Berger et al., 2009; Chitwood et al., 2009; Chuck et al., 2009; Hsieh et al., 2009; Husbands et al., 2009; Meng et al., 2009; Liu and Chen, 2009; Todesco et al., 2010).

High photosynthetic efficiency leads to improved crop yields and constitutes an integral part of a promising strategy. Some evidences are available on the potential involvement of small RNA in the photosynthesis process in several plants, algae, or cyanobacteria. Recently, thorough transcriptome-based analyses in many model cyanobacteria identified and isolated hundreds of regulatory sRNA candidates (Billis et al., 2014; Kopf et al., 2014a, 2014b, 2015a, 2015b; Pfreundt et al., 2014; Voigt et al., 2014; Wu et al., 2015, 2016). RNA sequencing and other high-throughput techniques, such as microarray, computer-aided prediction, and experimental verification have been used to discover potential sRNA candidates involved in photosynthesis or allied processes in cyanobacteria (Steglich et al., 2008; Georg et al., 2009; Voss et al., 2009; Ionescu et al., 2010; Gierga et al., 2012; Billis et al., 2014; Kopf et al., 2014a, 2014b; Pfreundt et al., 2014; Voigt et al., 2014).

miRNA affects global gene expression in several ways. A single miRNA can target multiple genes, or multiple miRNAs can synergistically regulate a single gene. It is well documented that the Arabidopsis genome has more than 184 miRNAs, which are computationally or experimentally validated to directly or indirectly regulate over 600 genes, which also include at least 225 known targets (Griffiths-Jones et al., 2008; Alves et al., 2009; Hussain et al., 2014b). The same is true for other plants, including cereals. The identification of hundreds of plant-associated miRNA genes demonstrates that miRNAs have vital roles to play in biologically challenging and complex responses under stress (Comai and Zhang, 2012; Khraiwesh et al., 2012; Sun, 2012). Several studies have used strategies, such as generation of mutants (loss- and gain-of-function), miRNA overexpression, or generated transgenic plants expressing miRNA-resistant versions to identify target plant genes (Li et al., 2010; Khan et al., 2011; Bustos-Sanmamed et al., 2013; Turner et al., 2013; Yang et al., 2013; Zhou et al., 2013; Hussain et al., 2014b; Wang et al., 2015; Zhao et al., 2016).

All abiotic stresses primarily target plant photosynthesis. Plant photosynthetic efficiency is negatively affected by environmental stress, and various abiotic stresses negatively impact photosynthetic capacity and also CO_2 assimilation, leading to the progressive reduction of plant productivity. Numerous studies highlighted the involvement of miRNAs in photosynthesis and associated processes, including miR397, miR857, miR397a, miR397b, miR398, miR408 (Trindale et al., 2009; Zhou et al., 2010; Kantar et al., 2011; Kulcheski et al., 2011; Zhang et al., 2011, 2014a, 2014b; Zhang and Li 2013; Hussain et al., 2014b; Wang et al., 2014; Zhao et al., 2015, 2016; Pilon, 2017; Pan et al., 2018; Song et al., 2018). Transgenic model and crop plants carrying miR408 were generated to explore the performance of miR408 in three different transgenic plants (Zhang et al., 2017). The study used molecular and biochemical analyses to conclusively show that miR408 has high expression levels in transgenic plants, which significantly contributed to their high photosynthetic efficiency. Furthermore, the transgenic rice plants carrying miR408 had increased grain weight and grain yield in small-scale field trials, compared to control plants (Zhang et al., 2017). A study on transgenic Arabidopsis overexpressing miR408 revealed that miR408 has a positive impact toward Arabidopsis biomass and yield (Song et al., 2018). These two studies conclusively showed that miR408 is a potential candidate gene for developing plants with enhanced photosynthetic capacity, biomass, and yield using genetic manipulation.

3.3 OVEREXPRESSION OF PHOTOSYNTHESIS-ASSOCIATED REGULATORY GENES

Photosynthesis is one of the complex processes that is involved in plant physical (growth and development) and reproductive (biomass) processes (Ort et al., 2015, 2017; Simkin et al., 2019). Several research opportunities for enhancing photosynthetic capacity in terrestrial plants have been targeted (Orr et al., 2017; Nowicka et al., 2018; Kubis and Bar-Even, 2019; Nowicka, 2019), and reports showed encouraging results (Long et al., 2015; Foyer et al., 2017; Nowicka et al., 2018; Wallace et al., 2018; Eva et al., 2019; Fernie and Yan, 2019). One such opportunity is the overexpression of different photosynthesis-associated genes (both structural and regulatory). The key enzymes of the Calvin–Benson Cycle (CBC) have significant negative effects on the rate of carbon assimilation and plant development. Transgenic plants expressing genes such as SBPase (sedoheptulose-1,7-bisphosphatase), FBPA (fructose-1,6-bisphosphate aldolase), FBPase (fructose-1,6-bisphosphatases), or phosphoribulokinase revealed reductions in final biomass yield (Kossmann et al., 1992; Haake et al., 1998, 1999; Harrison et al., 1998, 2001; Paul et al., 2000; Sahrawy et al., 2004; Lawson et al., 2006; Feng et al., 2007; Rojas-Gonzalez et al., 2015). Overexpression of SBPase in tobacco, Arabidopsis, tomato, and wheat increased biomass yield and, more importantly, seed yield by 30–40% (Lefebvre et al., 2005; Rosenthal et al., 2011; Simkin et al., 2015, 2017, 2019; Driever et al., 2017). Overexpression of FBPA (another CBC enzyme) increased biomass yield by 70–120% in transgenic plants under elevated CO_2 level (Uematsu et al.,

2012). In contrast, Khozaei et al. (2015) reported that overexpression of some CBC enzymes had a negative effect on transgenic plant development. Similarly, transgenic plants carrying genes such as betaine aldehyde dehydrogenase, chloroplast small heat-shock protein, plastidial protein synthesis elongation factor, glycine betaine, sucrose nonfermenting1-related protein kinase 2, Na^+/H^+ antiporter, and tobacco aquaporin improved photosynthetic performance under different environmental stresses (Aharon et al., 2003; He et al., 2005; Wang et al., 2005; Su et al., 2006; Brini et al., 2007; Park et al., 2007; Fu et al., 2008; Zhang et al., 2010; Fan et al., 2012).

TFs are the master switches responsible for gene expression changes. TFs have gained immense importance in plant function, including plant growth and development, and ensure normal plant growth and development under environmental stresses (Nakano et al., 2006; Hrmova and Lopato, 2014; Wang et al., 2016). TFs are extremely important for normal plant growth, with a large portion of plant genomes comprising of genes encoding TFs. Genome sequences of several plants are available and over 10% of all identified genes represent TFs including Arabidopsis (1922), rice (1611), sorghum (2450), and maize (3337) (Hrmova and Hussain, 2021). These TFs have been classified into several families, depending on the structure and properties of their respective binding domains. Wang et al. (2011) described 60 families of known TFs. Classification criteria are based on several characteristics including primary sequences, DNA binding motifs, binding domain 3D structures, oligomerization patterns, and post-translational modifications (Hrmova and Hussain, 2021). Similarly, several studies used members of the MYB, ERF, MYC, NAC, bZIP, WRKY TF families via transgenic approach to explore their roles in abiotic stress tolerance (Schwechheimer et al., 1998; Singh et al., 2002; Saibo et al., 2009; Hussain et al., 2011; Hrmova and Hussain, 2021). Currently, molecular biologists pay more attention to determining the molecular mechanism of gene expression and elucidating the role of different transcription factors (TFs) for a particular trait (Hrmova and Hussain, 2021). The manipulation of gene regulatory elements using transgenic approaches has had promising results under various environmental stresses. Hussain et al. (2011) and Hrmova and Hussain (2021) reviewed several TFs that directly or indirectly confer tolerance in plants under different abiotic stresses. ABP9 (ABRE binding protein 9), a bZIP TF, specifically binds to the ABRE2 motif. A model transgenic plant carrying ABP9 TF was developed to investigate its potential role under drought and heat stress, alone or combined (Zhang et al., 2008). Transgenic Arabidopsis plants carrying ABP9 with constitutive expression significantly increased their photosynthetic capacity under heat and drought stresses (Zhang et al., 2008), which was mainly attributed to increased ABA content, enhanced efficiency for dissipating excess light, photosynthetic pigment composition, high carbon-use efficiency, and high water use efficiency (Zhang et al., 2008). ABP9 also manages the positive regulation of some genes responsible for protecting PSII. This multifunctional contribution of ABP9 highlights its importance for improving plant photosynthesis under drought and heat stress (Zhang et al., 2008).

Similarly, several studies have highlighted the involvement of HY5 (LONG HYPOCOTYL 5), a bZIP-type TF, in *CAB* gene regulation, and expression by light (Chattopadhyay et al., 1998; Maxwell et al., 2003; Saibo et al., 2009). Furthermore, HY5 TF controls the expression of Chl *a/b* binding protein 2 (CAB2) (Maxwell et al., 2003) and is involved in the regulation and expression of several genes, including *RbcS1A* (a gene for RuBisCO small subunit) (Chattopadhyay et al., 1998, Lee et al., 2007). Overexpression of OsMYB4 in *Arabidopsis thaliana* accumulated glycine betaine. Transgenic plants showed high-stress tolerance and increased photosynthesis, compared with control plants under salinity stress (Mattana et al., 2005). Glycine betaine positively stabilizes the RuBisCO structure under high salt stress (Sakamoto and Murata, 2002, Yang et al., 2005, Khafagy et al., 2009). Therefore, OsMYB4 TF contributes indirectly to maintaining and regulating photosynthetic genes under salt stress (Ashraf and Harris, 2013). Another study demonstrated that overexpression of BpMYB106 TF in transgenic birch (*Betula platyphylla* Suk.) upregulated several genes involved in photosynthesis and the oxidative phosphorylation pathway (Zhou and Li 2016). RNA-seq profiling also demonstrated that expression of BpMYB106 results in the activation of several photosynthesis-related genes in transgenic birch through its interaction with the MYB2 element of promoters. Consequently, transgenic birch plants had enhanced photosynthetic efficiency, water use efficiency, growth rate, and plant height (Zhou and Li 2016). The above results demonstrate that TFs are invaluable tools for the success of next-generation crop plants (Hymus et al., 2013; Kobayashi et al., 2013).

3.4 GENOME EDITING AND ARTIFICIAL TRANSCRIPTION FACTORS FOR TRANSCRIPTIONAL REGULATION

Genome editing is a widespread technique used by plant scientists, which focuses on four systems: meganucleases, ZFNs (zinc-finger nucleases),

TALENs (transcription activator-like effector nucleases), and CRISPR/Cas (clustered regularly interspaced short palindromic repeat (CRISPR)/CRISPR-associated nuclease protein). Plant gene modifications are possible using these tools (Shukla et al., 2009; Cermak et al., 2015; Nishihara et al., 2018; Li et al., 2020). Studies have shown that the CRISPR/Cas9 approach is comparatively simple and reliable for increasing photosynthetic efficiency in plants by manipulating directly or indirectly some photosynthesis-related traits (Baltes and Voytas, 2015; Schiml and Puchta, 2016; Weeks et al., 2016). CRISPR-Cas is one of the easiest, rapid, and reliable technologies for genome modification in major crop plants, including corn, potato, wheat, tomato, rice, sorghum, soybean, and barley (Feng et al., 2013; Jiang et al., 2013; Miao et al., 2013; Upadhyay et al., 2013; Brooks et al., 2014; Wang et al., 2014; Xu et al., 2014; Zhang et al., 2014b; Zhou et al., 2014, 2015; Cai et al., 2015; Ito et al., 2015; Jacobs et al., 2015; Li et al., 2015; Shan et al., 2015; Sun et al., 2016; Svitashev et al., 2015; Du et al., 2016; Feng et al., 2016; Zhu et al., 2016; Tang et al., 2017; Zhang et al., 2019; Sheng et al., 2020; Usman et al., 2020).

The CRISPR/dCas9-based approach allowed scientists to utilize the perfect targeting efficiency of dCas9 (dead Cas9) using regulatory and reporter proteins for gene repression or activation and several other downstream applications in plants (Guo et al., 2015; Hilton et al., 2015; Woo et al., 2015; Zhang et al., 2016; Dreissig et al., 2017; Li et al., 2017a, 2017b; Liang et al., 2017; Lowder et al., 2017; Xue and Acar, 2018; Zezulin and Musunuru, 2018; Shelake et al., 2019a, 2019b; Veillet et al., 2019). TALEN and ZFN (sequence-specific nucleases: SSNs) have emerged as multipurpose tools that can easily target sequence-specific genetic loci, leading to DSB (Bogdanove and Voytas, 2011; Jinek et al., 2012; Qi 2015; Schmidt et al., 2019; Mackelprang and Lemaux, 2020). Target specificity of these SSNs depends on customizable DNA binding domains, which play a significant role in engineering artificial transcription factors (ATFs) to study gene knockout or gene knockin expression profiles (Jinek et al., 2012; Schiml and Puchta, 2016; Knott and Doudna, 2018; Chen et al., 2019; Kim and Kim, 2019). Several publications have been reported using ATFs for transcriptional regulation (Gupta et al., 2012; Reyon et al., 2012; Sanjana et al., 2012). Zinc-finger and TALEN proteins played a pioneering role in engineering ATFs (Boch et al., 2009; Moscou and Bogdanove, 2009; Mahfouz et al., 2012; Perez-Pinera et al., 2013) but slow progress and difficulty in engineering failed to attract scientific community acceptance (Gupta et al., 2012). Currently, ATFs are easier and faster to generate and engineer in scientific laboratories using CRISPR-Cas-based platforms (Gilbert et al., 2013; Mali et al., 2013). CRISPR interference (CRISPRi) and CRISPR activation (CRISPRa) based ATFs (Bikard et al., 2013; Gilbert et al., 2013; Mali et al., 2013; Qi et al., 2013; Konermann et al., 2015; Zalatan et al., 2015; Chavez et al., 2016) have been extensively used in several plants including Arabidopsis, tobacco, wheat, maize, and rice (Lowder et al., 2015, 2017; Young et al., 2019). Thus, modulation of transcription in plants is entering an exciting era (Lowder et al., 2018). Currently, ATFs are attractive tools, especially for studying target gene regulation and expression and transcriptome manipulation in plants. Moreover, Lowder et al. (2018) developed CRISPR-Act2.0 transcriptional activation system, based on dCas9 approach, using synthetic biology tools to enhance and coordinate the transcription of several target genes simultaneously in rice.

Synthetic transcriptional regulators (CRISPR/dCas9-based ATFs) either modify or block plant TFs, followed by preinitiation of transcriptional complexes or by inducing changes in chromatin structure (Perez-Pinera et al., 2013). Several researchers have activated or repressed reporters or endogenous genes (Lowder et al., 2015, 2016, 2017, 2018; Li et al., 2017a, 2017b). Maximum transcriptional activation or repression is achieved when dCas9 activators or repressors are fused to promoter regions (Didovyk et al., 2016). Furthermore, CRISPR/dCas9-based ATFs, used to modulate transcription, guided by multiple sgRNAs can synergistically bind to a variety of different loci (Didovyk et al., 2016), mimicking the multiplexing approach with significant benefits. Several researchers have reported simultaneous activation and expression using the multiplexing approach as mentioned above (Perez-Pinera et al., 2013; Xing et al., 2014; Lowder et al., 2015; Ma et al., 2015; Didovyk et al., 2016; Li et al., 2017a, 2017b). Multiplexing CRISPR/dCas9-based ATFs offer significant advantages for exploring transcriptional networks (Perez-Pinera et al., 2013; Xing et al., 2014; Didovyk et al., 2016; Li et al., 2017a, 2017b) by facilitating the simultaneous expression of several gRNAs in different plants.

3.5 CRISPR/CAS-BASED CLIMATE-SMART CROPS WITH IMPROVED YIELD AND YIELD-RELATED TRAITS

CRISPR/Cas9 tools have been extensively used in model and economically important crop plants, including Arabidopsis, tobacco, wheat, sorghum, rice, barley, maize cotton, soybean, tomato, and potato, to counter various environmental stresses (Zhang et al., 2017; Li et al., 2018a, 2018b, 2018c; Lou et al., 2018; Shim et al., 2018). Genome editing provides new opportunities for the development of abiotic stress-tolerant crops, especially the multiple CRISPR/Cas9 tools that have converted conventional plant breeding to precision plant breeding (Feng et al., 2013; Jiang et al., 2013; Shan et al., 2013). It is well known that biotic (Savary et al., 2012; Rodriguez-Moreno et al., 2017) and abiotic (Bartels and Hussain, 2008; Hussain et al., 2011, 2012; Pandey et al., 2017; Mariani and Ferrante, 2017; Mushtaq et al., 2018) stresses can negatively affect gene expression in higher organisms at the spatial and temporal levels (Hrmova and Hussain, 2021). Nature provides several strategies to all the living organisms to respond to stress, comprising a cascade of complex molecular interactions (Hrmova and Hussain, 2021). Several research groups have attempted CRISPR/Cas-mediated mutagenesis in genes involved in salt tolerance (OsSOS1, OsMPK2, OsPDS, OsBADH2, OsRR22, OsNAC041), chilling tolerance (SlCBF1; OsAnn3), drought tolerance (TaDREB2, TaDREB3, SlMAPK3, ARGOS8, OsSAPK2, SlNPR1), and cadmium tolerance (OsNramp5), resulting in edited lines with improved tolerance to the respective stresses (Shan et al., 2013; Kim et al., 2017a, 2017b; Lou et al., 2017; Shi et al., 2017; Tang et al., 2017; Wang et al., 2017; Curtin et al., 2018; Bo et al., 2019; Li et al., 2019; Sadanandom et al., 2019; Zhang et al., 2019). One notable accomplishment of current gene-editing efforts is the generation of rice germplasm with herbicide-tolerant features that minimize the overall cost of rice cultivation (Shukla et al., 2009; D'Halluin et al., 2013).

The CRISPR/Cas9-based approaches have been successfully used to knockout genes such as OsAAP3, OsDEP1, TaDEP1, TaGW2, TaGASR7, OsGS3, OsGRF4, OsGn1a that negatively impact yield-related traits including tiller number, panicle size, grain weight, grain size, and grain number in cereal crop plants like wheat and rice. The results demonstrate that CRISPR/Cas9 is an efficient technology for improving crop yield (Li et al., 2016a, 2016b, 2017b; Xu et al., 2016; Liu et al., 2017; Ma et al., 2017; Lu et al., 2018; Zhang et al., 2018a, 2018b, 2018c). CRISPR/Cas tools have also been used extensively to improve plant yield, enhance nutritional properties, and increased biomass-related characters in several economically important crop plants (Qi et al., 2016; Zhang et al., 2016; Andersson et al., 2017; Kim et al., 2017a, 2017b; Sun et al., 2017; Huang et al., 2018; Li et al., 2018a, 2018b, 2018c; Okuzaki et al., 2018; Sánchez-León et al., 2018; Zhang et al., 2018a, 2018b, 2018c). The studies mentioned above comprehensively and conclusively demonstrated that CRISPR/Cas9-mediated modern breeding tools can successfully generate valuable mutations for improved crop yield and quality.

3.6 CONCLUSION AND FUTURE PERSPECTIVES

Crop productivity is a direct measure of the photosynthetic capacity of plants that is why it is regarded as the unique and foundation source of energy for life (Ambavaram et al., 2014). Furthermore, the actual photosynthetic efficiency of plants ranges from 0.1 to 8% due to the poor conversion of sunlight to chemical energy in plant leaves. Understanding the genetics of the photosynthetic process remains under investigation. Attempts (plant breeding and exploring natural variations) to improve crop yield by direct selection of plant photosynthetic efficiency have had limited success. The generation of transgenic plants for enhanced photosynthetic capacity is a possible strategy for overcoming inefficient photosynthesis. Transgenic plants carrying specific genes can potentially be developed with improved photosynthetic capacity, leading to higher growth rates.

Recent research reports have demonstrated that overexpression of miRNA408 in diverse plants, such as Arabidopsis, tobacco, and rice enhanced photosynthetic efficiency, growth and development, and high biomass yield (Pan et al., 2018; Song et al., 2018). Further, the miR408-carrying transgenic plants (Arabidopsis) showed tolerance to various stress conditions (Zhang et al., 2014a, 2014b; Ma et al., 2015). Therefore, it is logical to use miRNA408 alone or combined with other existing genes to develop super traits, such as increased photosynthetic efficiency in plants (Zhang and Li, 2013; Ma et al., 2015).

CRISPR/Cas-mediated genome editing opens a new dimension in plant research for phenomenal crop biotechnology success. The use of gene editing allows efficient, precise, and targeted mutagenesis for increasing yield and nutritional value in crops, especially cereal crops. Integrating these improvements, CRISPR/Cas gene editing will be a lucrative tool for developing new climate-smart stress-resilient crop plants with increased photosynthesis and yields, and tolerance to biotic and abiotic stresses, enhancing nutritional value for the ever-increasing global population.

CRISPR/Cas has immense potential for increasing agricultural production by increasing plant tolerance to environmental stresses and enhancing the nutritional value of plant-based products. There should not be any compromise on high-yielding crop varieties with improved nutritional value to meet the demand of the rapidly increasing human population around the globe.

KEYWORDS

- **photosynthesis**
- **environmental stress**
- **plant productivity**
- **genome editing**
- **CRISPR/Cas**

REFERENCES

Achard, P.; Herr, A.; Baulcombe, D. C.; Harberd, N. P. Modulation of Floral Development by A Gibberellin-Regulated MicroRNA. *Development* **2004,** *131*, 3357–3365.

Aharon, R.; Shahak, Y.; Wininger. S.; Bendov. R.; Kapulnik, Y.; Galili. G. Overexpression of a Plasma Membrane Aquaporin in Transgenic Tobacco Improves Plant Vigor under Favorable Growth Conditions But Not under Drought or Salt Stress. *Plant Cell* **2003,** *15*, 439–447.

Alexandratos, N.; Bruinsma, J. World Agriculture Towards 2030/2050: The 2012 Revision. ESA Working Paper No. 12-03; FAO: Rome,2012; p 154.

Alptekin, B.; Langridge, P.; Budak, H. Abiotic Stress miRNomes in Triticeae. *Funct. Integr. Genom.* **2017,** *17*, 145–170.

Alves, L.; Niemeier, S.; Hauenschild, A.; Rehsmeier, M.; Merkle, T. Comprehensive Prediction of Novel MicroRNA Targets in *Arabidopsis thaliana*. *Nucl.Acids Res.* **2009,** *37*, 4010–4021.

Ambavaram, M. M. R.; BAsu. S.; Krishnan. A.; Ramegowda. V.; Batlang. U.; Rahman. L.; Baisakh, N.; Periera, A. Coordinated Regulation of Photosynthesis in Rice Increases Yield and Tolerance to Environmental Stress. *Nat. Commun.* **2014,** *5*, 5302.

Andersson, M.; Turesson, H.; Nicolia, A.; Fält, A. S.; Samuelsson, M.; Hofvander, P. Efficient Targeted Multiallelic Mutagenesis in Tetraploid Potato (Solanum tuberosum) by Transient CRISPR-Cas9 Expression in Protoplasts. *Plant Cell Rep.* **2017,** *36*, 117–128.

Ashraf, M.; Foolad, M. R. Roles of Glycinebetaine and Proline in Improving Plant Abiotic Stress Resistance. *Environ. Exp. Bot.* **2007,** *59*, 206–216.

Ashraf, M.; Harris, P. J. C. Photosynthesis under Stressful Environments: An Overview. *Photosynthetica* **2013,** *51*, 163–190.

Baker, N. R. Chlorophyll Fluorescence: A Probe of Photosynthesis in Vivo. *Annu. Rev. Plant Biol.* **2008,** *59*, 89–113.

Baltes, N.; Voytas, D. Enabling Plant Synthetic Biology through Genome Engineering. *Trends Biotechnol.* **2015,** *33*, 120–131.

Bartels, D.; Hussain, S. S. Current Status and Implications of Engineering Drought Tolerance in Plants Using Transgenic Approaches. *CAB Rev: Perspect. Agric. Vet. Sci. Nutr. Nat. Resour*. **2008,** *3*, 020.

Bedell, V. M. et al. In Vivo Genome Editing Using a High Efficiency TALEN system. *Nature* **2012,** *491*, 114–118.

Berger, Y.; Harpaz-Saad, S.; Brand, A.; Melnik, H.; Sirding, N.; Alvarez, J. P.; Zinder, M.; Samach, A.; Eshed, Y.; Ori, N. The NAC-Domain Transcription Factor GOBLET Specifies Leaflet Boundaries in Compound Tomato Leaves. *Development* **2009,** *136*, 823–832.

Bikard, D.; Jiang, W.; Samai, P.; Hochschild, A.; Zhang, F.; Marrafinni, L. A. Programmable Repression and Activation of Bacterial Gene Expression Using an Engineered CRISPR-Cas System. *Nucl.Acids Res*. **2013,** *41*, 7429–7437.

Billis, K.; Billini, M.; Tripp, H. J.; Kyrpides, N. C.; Mavromatis, K. Comparative Transcriptomics between *Synechococcus* PCC 7942 and *Synechocystis* PCC 6803 Provide Insights into Mechanisms of Stress Acclimation. *PLoS One* **2014,** *9*, e109738.

Bo, W.; Zhaohui, Z.; Huanhuan, Z.; Xia, W.; Binglin, L.; Lijia, Y.; Xiangyan, H.; Deshui, Y.; Xuelian, Z.; Chunguo, W.; et al. Targeted Mutagenesis of NAC Transcription Factor Gene, OsNAC041, Leading to Salt Sensitivity in Rice. *Rice Sci*. **2019,** *26*, 98–108.

Boch, J.; Schotze, H.; Schornack, S.; Landgraf, A.; Hahn, S.; Kay, S.; Lahaye, T.; Nickstadt, A.; Bonas, U. Breaking the Code of DNA Binding Specificity of TAL-type III Effectors. *Science* **2009,** *326*, 1509–1512.

Bogdanove, A. J.; Voytas, D. F. TAL Effectors: Customizable Proteins for DNA Targeting. *Science* **2011,** *333*, 1843–1846.

Boyer, J. S. Plant Productivity and Environment. *Science*. **1982,** *218*, 443–448.

Brini, F.; Hanin, M.; Mezghani, I.; Berkowitz, G. A.; Masmoudi, K. Overexpression of Wheat Na+/H+antiporter TNHX1 and H+-pyrophosphatase TVP1 Improve Salt- and Drought-Stress Tolerance in *Arabidopsis thaliana* Plants. *J. Exp. Bot.* **2007,** *58*, 301–308.

Brooks, C.; Nekrasov, V.; Lippman, Z.; Van Eck, J. Efficient Gene Editing in Tomato in the First Generation Using the Clustered Regularly Interspaced Short Palindromic Repeats/ CRISPR-Associated9 System. *Plant Physiol*. **2014,** *166*, 1292–1297.

Buhtz, A.; Springer, F.; Chappell, L.; Baulcombe, D. C.; Kehr, J. Identification and Characterization of Small RNAs from the Phloem of *Brassica napus*. *Plant J*. **2008,** *53*, 739–749.

Bustos-Sanmamed, P.; Mao, G.; Deng, Y.; Elouet, M.; Khan, G. A.; Bazin, J.; Turner, M.; Subramanian, S.; Yu, O.; Crespi, M.; lelandais-Briere, C. Overexression of miR160 Affects Root Growth and Nitrogen-Fixing Nodule Number in *Medicago truncatula*. *Funct. Plant Biol*. **2013,** *40*, 1208–1220.

Cai, Y.; Chen, L.; Liu, X.; Sun, S.; Wu, C.; Jiang, B.; Han, T. et al. CRISPR/ Cas9-Mediated Genome Editing in Soybean Hairy Roots. *PLoS ONE*, **2015,** *10*, e0136064.

Capell, T.; Bassie, L.; Christou, P. Modulation of the Polyamine Biosynthetic Pathway in Transgenic Rice Confers Tolerance to Drought Stress. *Proc. Natl. Acad. Sci. USA* **2004,** *101*, 9909–9914.

Cermak, T.; Baltes, N. J.; Cegan, R.; Zhang, Y.; Voytas, D. F. High-Frequency, Precise Modification of the Tomato Genome. *Genome Biol*. **2015,** *16*, 232.

Chattopadhyay, S.; Ang, L. H.; Puente, P.; Deng, X. W.; Wei, N. Arabidopsis bZIP Protein HY5 Directly Interacts with Light-Responsive Promoters in Mediating Light Control of Gene Expression. *Plant Cell* **1998,** *10*, 673–683.

Chavez, A.; Tuttle, M.; Pruitt, B.; Ewen-Campen, B.; Chari, R.; Ter-Ovanesyan, D.; Haque, S.; Cecchi, R.; Kowal, E.; Buchthal, J.; Housden, B.; Perrimon, N.; Collins, J.; Church, G. Comparison of Cas9 Activators in Multiple Species. *Nat. Methods* **2016,** *13*, 563–567.

Chen, K.; Wang, Y.; Zhang, R.; Zhang, H.; Gao, C. CRISPR/Cas Genome Editing and Precision Plant Breeding in Agriculture. *Annu. Rev. Plant Biol.* **2019,** *70*, 667–697.

Chen, Y.; Zhou, B.; Li, J.; Tang, H.; Tang, J.; Yang, Z. Formation and Change of Chloroplast-Located Plant Metabolites in Response to Light Conditions. *Int. J. Mol. Sci.* **2018,** *19*, 654.

Chitwood, D. H.; Nogueira, F. T.; Howell, M. D.; Montgomery, T. A.; Carrington, J. C.; Timmermanns, M. C. Pattern Formation via Small RNA Mobility. *Genes Dev*. **2009,** *23*, 549–554.

Chuck, G.; Candela, H.; Hake, S. Big Impacts by Small RNAs in Plant Development. *Curr. Opin. Plant Biol.* **2009,** *12*, 81–86.

Comai, L.; Zhang, B. MicroRNAs: Key Gene Regulators with Versatile Functions. *Plant Mol. Biol.* **2012,** *80*, 1.

Cong, L. et al. Multiplex Genome Engineering Using CRISPR/Cas Systems. *Science* **2013,** *339*, 819–823.

Cortleven, A.; Nitschke, S.; Klaumunzer, M.; AbdElgawad, H.; Asard, H.; Grimm, B.; Riefler, M.; Schmulling, T. A Novel Protective Function for Cytokinin in the Light Stress Response in Mediated by the ARABIDOPSIS HISTIDINE KINASE2 and ARABIDOPSIS HISTIDINE KINASE3 Receptors. *Plant Physiol.* **2014,** *164*, 1470–1483.

Cuperus, J. T.; Fahlgren, N.; Carrington, J. C. Evolution and Functional Diversification of *MIRNA* Genes. *Plant Cell* **2011,** *23*, 431–442.

Curtin, S. J.; Xiong, Y.; Michno, J. M.; Campbell, B. W.; Stec, A. O.; Cermák, T.; Starker, C.; Voytas, D. F.; Eamens, A. L.; Stupar, R. M. CRISPR/Cas9 and TALENs Generate Heritable Mutations for Genes Involved in Small RNA Processing of *Glycine max* and *Medicago truncatula*. *Plant Biotechnol. J.* **2018,** *16*, 1125–1137.

D'Halluin, K.; Vanderstraeten, C.; Van Hulle, J.; Rosolowska, J.; Van Den Brande, I.; Pennewaert, A.; D'Hont, K.; Bossut, M.; Jantz, D.; Ruiter, R.; et al. Targeted Molecular Trait Stacking in Cotton through Targeted Double-Strand Break Induction. *Plant Biotechnol. J.* **2013,** *11*, 933–941.

Dann, M.; Leister, D. Enhancing (Crop) Plant Photosynthesis by Introducing Novel Genetic Diversity. *Phil. Trans. R. Soc. B* **2017,** *372*, 20160380.

Demmig-Adams, B.; Stewart, J. J.; Baker, C. R.; Adams, W. W. Optimization of Photosynthetic Productivity in Contrasting Environments by Regulons Controlling Plant Form and Function. *Int. J. Mol. Sci.* **2018,** *19*, 872.

Dhar, M. K.; Visha, P.; Sharma, R.; Kaul, S. Epigenetic Dynamics: Role of Epimarks and Underlying Machinery in Plants Exposed to Abiotic Stress. *Int. J. Genom.***2014,** 187146.

Didovyk, A.; Borek, B.; Tsimring, L.; Hasty, J. Transcriptional Regulation with CRISPR-Cas9: Principles, Advances, and Applications. *Curr. Opin. Biotechnol.* **2016,** *40*, 177–184.

Dreissig, S.; Schiml, S.; Schindele, P.; Weiss, O.; Rutten, T.; Schubert, V.; Gladilin, E.; et al. Live-Cell CRISPR Imaging in Plants Reveals Dynamic Telomere Movements. *Plant J.* **2017,** *91*, 565–573.

Driever, S. M.; Simkin, A. J.; Alotaibi, S.; Fisk, S. J.; Madgwick, P. J.; Sparks, C. A.; Jones, H. D.; Lawson, T.; Parry, M. A. J.; Raines, C. A. Increased sbpase Activity Improves Photosynthesis and Grain Yield in Wheat Grown in Greenhouse Conditions. *Phil. Trans. R. Soc. B* **2017,** *372*, 1730.

Du, H.; Zeng, X.; Zhao, M.; Cui, X.; Wang, Q.; Yang, H.; Cheng, H.; et al. Efficient Targeted Mutagenesis in Soybean by TALENs and CRISPR/Cas9. *J. Biotechnol.* **2016,** *217*, 90–97.

Dusenge, M. E.; Duarte, A. G.; Way, D. A. Plant Carbon Metabolism and Climate Change: Elevated CO_2 and Temperature Impacts on Photosynthesis, Photorespiration and Respiration. *New Phytol.* **2019,** *221*, 32–49.

Eberhard, S.; Finazzi, G.; Wollman, F. A. The Dynamics of Photosynthesis. *Annu. Rev. Genet.* **2008,** *42*, 463–515.

Efeoglu, B.; Terzioglu, S. Photosynthetic Responses of Two Wheat Varieties to High Temperature. *Eur. Asian J. Biosci.* **2009,** *3*, 97–106.

Eva, C.; Oszvald, M.; Tamas, L. Current and Possible Approaches for Improving Photosynthetic Efficiency. *Plant Sci.* **2019,** *280*, 433–440.

Fahad, S.; Bajwa, A. A.; Nazir, U.; et al. Crop Production under Drought and Heat Stress: Plant Responses and Management Options. *Front. Plant Sci.* **2017,** *8*, 1147.

Fan, W.; Zhang, M.; Zhang, H.; Zhang, P. Improved Tolerance to Various Abiotic Stresses in Transgenic Sweet Potato (*Ipomoea batatas*) Expressing Spinach Betaine Aldehyde Dehydrogenase. *PLoS ONE* **2012,** *7*, e37344.

Feng, C.; Yuan, J.; Wang, R.; Liu, Y.; Birchler, J.; Han, F. Efficient Targeted Genome Modification in Maize Using CRISPR/Cas9 System. *J. Genet. Genom.***2016,** *43*, 37–43.

Feng, Z. et al. Multigeneration Analysis Reveals the Inheritance, Specificity, and Patterns of CRISPR/Cas-Induced Gene Modifications in Arabidopsis. *Proc. Natl. Acad. Sci. USA* **2014,** *111*, 4632–4637.

Feng, Z.; Zhang, B.; Ding, W.; Liu, X.; Yang, D.-L.; Wei, P.; Cao, F.; Zhu, S.; Zhang, F.; Mao, Y. et al. Efficient Genome Editing in Plants Using a CRISPR/Cas System. *Cell Res.* **2013,** *23*, 1229–1232.

Ferdous, J.; Hussain, S. S.; Shi, B. Role of MicroRNAs in Plant Drought Tolerance. *Plant Biotechnol. J.* **2015,** *13*, 293–305.

Fernie, A. R.; Yan, J. De novo Domestication: An Alternative Route Toward New Crops for the Future. *Mol. Plant* **2019,** *12*, 615–631.

Foyer, C. H.; Neukermans, J.; Queval, G.; Noctor, G.; Harbinson, J. Photosynthetic Control of Electron Transport and the Regulation of Gene Expression. *J. Exp. Bot.* **2012,** *63*, 1637–1661.

Foyer, C. H.; Noctor, G. Redox Homeostasis and Antioxidant Signalling: A Metabolic Interface between Stress Perception and Physiological Responses. *Plant Cell Online* **2005,** *17*, 1866–1875.

Foyer, C. H.; Ruban, A. V.; Nixon, P. J. Photosynthesis Solutions to Enhance Productivity. *Phil. Trans. R. Soc. Lond. B Biol. Sci.* **2017,** *372*, 20160374.

Foyer, C. H.; Shigeoka, S. Understanding Oxidative Stress and Antioxidant Functions to Enhance Photosynthesis. *Plant Physiol.* **2011,** *155*, 93–100.

Fu, J.; Momcilović, I.; Clemente, T. E.; Nersesian, N.; Trick, H. N.; Ristic, Z. Heterologous Expression of a Plastid EF-Tu Reduces Protein Thermal Aggregation and Enhances CO_2 Fixation in Wheat (*Triticum aestivum*) Following Heat Stress. *Plant Mol. Biol.* **2008,** *68*, 277–288.

Georg, J.; Voss, B.; Scholz, I.; Mitschke, J.; Wilde, A.; Hess, W. R. Evidence for a Major Role of Antisense RNAs in Cyanobacterial Gene Regulation. *Mol. Syst. Biol.* **2009,** 5, 305.

Gierga, G.; Voss, B.; Hess, W. R. Non-Coding RNAs in Marine *Synechococcus* and Their Regulation under Environmentally Relevant Stress Conditions. *ISME J.* **2012,** *6*, 1544–1557.

Gilbert, L. A.; Larson, M. H.; Morsut, L.; Liu, Z.; Brar, G. A.; Torres, S. E.; Stern-Ginossar, N.; Brandman, O.; Whitehead, E. H.; Doudna, J. A.; Lim, W. A.; Weissman, J. S.; Qi, L. S. CRISPR-Mediated Modular RNA-Guided Regulation of Transcription in Eukaryotes. *Cell* **2013,** *154*, 442–451.

Griffiths-Jones, S.; Saini, H. K.; van Dongen, S.; Enright, A. J. miRBase: tools for microRNA Genomics. *Nucl.Acids Res.* **2008,** *36*, D154–D158.

Guo, Y.; Xu, Q.; Canzio, D.; Shou, J.; Li, J.; Gorkin, D. U. Jung, I.; et al. CRISPR Inversion of CTCF Sites Alters Genome Topology and Enhancer/Promoter Function. *Cell*, **2015,** *162*, 900–910.

Gupta, M.; Dekelver, R. C.; Palta, A.; Clifford, C.; Gopalan, S.; Miller, J. C.; Novak, S.; Desloover, D.; Gachotte, D.; Connell, J.; et al. Transcriptional Activation of Brassica Napus Beta-Ketocyl-ACP Synthase II with an Engineered Zinc Finger Protein Transcription Factor. *Plant Biotechnol. J.* **2012,** *10*, 783–791.

Gururani, M. A.; Mohanta, T. K.; Bae, H. Current Understanding of the Interplay between Phytohormones and Photosynthesis under Environmental Stress. *Int. J. Mole Sci.* **2015,** *16*, 19055–19085.

Haake, V.; Geiger, M.; Walch-Liu, P.; Engels, C.; Zrenner, R.; Stitt, M. Changes in Aldolase Activity in Wild-Type Potato Plants Are Important for Acclimation to Growth Irradiance and Carbon Dioxide Concentration, Because Plastid Aldolase Exerts Control Over the Ambient Rate of Photosynthesis across a Range of Growth Conditions. *Plant J.* **1999,** *17*, 479–489.

Haake, V.; Zrenner, R.; Sonnewald, U.; Stitt, M. A Moderate Decrease of Plastid Aldolase Activity Inhibits Photosynthesis, Alters the Levels of Sugars and Starch, and Inhibits Growth of Potato Plants. *Plant J.* **1998,** *14*, 147–157.

Harrison, E. P.; Olcer, H.; Lloyd, J. C.; Long, S. P.; Raines, C. A. Small Decreases in sbpase cause a linear decline in the apparent rubp regeneration rate, but do not affect Rubisco Carboxylation Capacity. *J. Exp. Bot.* **2001,** *52*, 1779–1784.

Harrison, E. P.; Willingham, N. M.; Lloyd, J. C.; Raines, C. A. Reduced Sedoheptulose-1,7-Bisphosphatase Levels in Transgenic Tobacco Led to Decreased Photosynthetic Capacity and Altered Carbohydrate Accumulation. *Planta* **1998,** *204*, 27–36.

He, C.; Yan, J.; Shen, G.; Fu, L.; Holaday, A. S.; Auld, D.; Blumwald, E.; Zhang, H. Expression of an Arabidopsis Vacuolar Sodium/Proton Antiporter Gene in Cotton Improves Photosynthetic Performance under Salt Conditions and Increases Fiber Yield in the Field. *Plant Cell Physiol.* **2005,** *46*, 1848–1854.

Hilton, I. B.; D'Ippolito, A. M.; Vockley, C. M.; Thakore, P. I.; Crawford, G. E.; Reddy, T. E.; Gersbach, C. A. Epigenome Editing by a CRISPR-Cas9-Based Acetyltransferase Activates Genes from Promoters and Enhancers. *Nat. Biotechnol.* **2015,** *33*, 510–517.

Hrmova, M.; Hussain, S. S. Plant Transcription Factors Involved in Drought and Associated Stresses. *Int. J. Mol. Sci.* **2021,** 22, 5662.

Hrmova, M.; Lopato, S.; Enhancing Abiotic Stress Tolerance in Plants by Modulating Properties of Stress Responsive Transcription Factors. In: *Genomics of Plant Genetic Resources. Part II: Crop Productivity, Food Security and Nutritional Quality*; Tuberosa, R., Granerm, A; Frisonm, E., Eds., Vol. 2; Springer: Dordrecht, 2014; pp 291–316.

Hsieh, L. C.; Lin, S. I.; Shih, A. C. C.; Chen, J. W.; Lin, W. Y.; Tseng, C. Y.; Lin, W. H.; Chiou, T. J. Uncovering Small RNA-Mediated Responses to Phosphate Deficiency in Arabidopsis by Deep Sequencing. *Plant Physiol.* **2009,** *151*, 2120–2132.

Huang, J.; Li, J.; Zhou, J.; Wang, L.; Yang, S.; Hurst, L. D.; Li, W. H.; Tian, D. Identifying a Large Number of High-Yield Genes in Rice by Pedigree Analysis, Whole-Genome Sequencing, and CRISPR-Cas9 Gene Knockout. *Proc. Natl. Acad. Sci. USA* **2018,** *115*, E7559–E7567.

Husbands, A. Y.; Chitwood, D. H.; Plavskin, Y.; Timmermanns, M. C. Signals and Pre-patterns: New Insights Into Organ Polarity in Plants. *Genes Dev*. **2009,** *23*, 1986–1997.

Hussain, S. S.; Kayani, M. A.; Amjad, M. Transcription Factors as Tools to Engineer Enhanced Drought Stress Tolerance in Plants. *Biotechnol. Prog*. **2011,** 27: 297–306.

Hussain, S. S.; Raza, H.; Afzal, I.; Kayani, M. A. Transgenic Plants for Abiotic Stress Tolerance: Current Status. *Arch. Agron. Soil Sci.* **2012,** *58*, 693–721.

Hussain, S. S.; Shi, B. Role of miRNAs in Abiotic and Biotic Stresses in Plants. In *Emerging Technologies and Management of Crop Stress Tolerance: Vol 1: Biological Techniques*; Ahmad, P., Rasool, S., Eds.; Elsevier, USA, 2014a; pp 181–207.

Hussain, S. S.; Siddique, K. H. M.; Lopato, S. Towards Integration of Bacterial Genomics in Plants for Enhanced Abiotic Stress Tolerance: Clues from Transgenics. *Adv. Environ. Res.* **2014b,** *33*, 65–122.

Hymus, G. J.; Cai, S.; Kohl, E. A.; Holtan, H. E.; Marion, C. M.; Tiwari, S.; Maszle, D. R.; Lundgren, M. R.; Hong, M. C.; Channa, N.; Loida, P.; Thompson, R.; Taylor, J. P.; Rice, E.; Repetti, P. P.; Ratcliffe, O. J.; Reuber, T. L.; Creelman, R. A. Application of HB17, an *Arabidopsis* Class II Homeodomain-Leucine Zipper Transcription Factor, to Regulate Chloroplast Number and Photosynthetic Capacity. *J. Exp. Bot.* **2013,** *64*, 4479–4490.

Ionescu, D.; Voss, B.; Oren, A.; Hess, W. R.; Muro-Pastor, A. M. Heterocyst-Specific Transcription of NsiR1, a Non-Coding RNA Encoded in a Tandem Array of Direct Repeats in Cyanobacteria. *J. Mol. Biol.* **2010,** *398*, 177–188.

Ito, Y.; Nishizawa-Yokoi, A.; Endo, M.; Mikami, M.; Toki, S. CRISPR/Cas9-Mediated Mutagenesis of the RIN Locus That Regulates Tomato Fruit Ripening. *Biochem. Biophys. Res. Commun.* **2015,** *467*, 76–82.

Jacobs, T. B.; LaFayette, P. R.; Schmitz, R. J.; Parrott, W. A. Targeted Genome Modifications in Soybean with CRISPR/Cas9. *BMC Biotechnol.* **2015,** *15*, 16.

Jagadeeswaran, G.; Saini, A.; Sunkar, R. Biotic and Abiotic Stress Down-Regulate miR398 Expression in Arabidopsis. *Planta* **2009,** *229*, 1009–1014.

Jiang, W.; Zhou, H.; Bi, H.; Fromm, M.; Yang, B.; Weeks, D. P. Demonstration of CRISPR/Cas9/sgRNA-Mediated Targeted Gene Modification in Arabidopsis, Tobacco, Sorghum and Rice. *Nucl.Acids Res*. **2013,** *41*, e188.

Jinek, M.; Chylinski, K.; Fonfara, I.; Hauer, M.; Doudna, J. A.; Charpentier, E. A Programmable Dual-RNA-Guided DNA Endonuclease in Adaptive Bacterial Immunity. *Science* **2012,** *337*, 816–821.

Jones-Rhoades, M. W.; Bartel, D. P.; Bartel, B. MicroRNAs and Their Regulatory Roles in Plants. *Annu. Rev. Plant Biol.* **2006,** *57*, 19–53.

Joshi, J.; Yan, Z.; Libault, M.; Jeong, D. H.; Park, S.; Green, P. J.; Sherrier, D. J.; Farmer, A.; May, G.; Meyers, B. C.; Xu, D.; Stacey, G. Prediction of Novel miRNAs and Associated Target Genes in *Glycine max*. *BMC Bioinf.* **2010,** *11*, S14.

Juarez, M. T.; Kui, J. S.; Thomas, J.; Heller, B. A.; Timmermans, M. C. MicroRNA-Mediated Repression of Rolled Leaf1 Specifies Maize Leaf Polarity. *Nature* **2004,** *824*, 84–88.

Kalaji, H. M.; Jajoo, A.; Oukarroum, A.; Brestic, M.; Zivcak, M.; Samborska, I. A. et al. Chlorophyll a Fluorescence as a Tool to Monitor Physiological Status of Plants under Abiotic Stress Conditions. *Acta Physiol. Plant.* **2016,** *38*, 102.

Kaneda, T. B. World Population Data Sheet. Population Reference Bureau, 2016 (www.prb.org), 22Kantar, M.; Lucas, S. J.; Budak, H. miRNA Expression Patterns of *Triticum dicoccoides* in Response to Shock Drought Stress. *Planta* **2011,** *233*, 471–484.

Kathuria, H.; Giri, J.; Tyagi, H.; Tyagi, A. K. Advances in Transgenic Rice Biotechnology. *Crit. Rev. Plant Sci.* **2007,** *26*, 65–103.

Kaur, R.; Yadav, P.; Sharma, A.; Kumar-Thukral, A.; Kumar, V.; Kaur Kohli, S. et al. Castasterone and Citric Acid Treatment Restores Photosynthetic Attributes in *Brassica juncea* L. under Cd (II) Toxicity. *Ecotoxicol. Environ. Saf.* **2017,** *145*, 466–475.

Khafagy, M. A.; Arafa, A. A.; El-Banna, M. F. Glycinebetaine and Ascorbic Acid Can Alleviate the Harmful Effects of NaCl Salinity in Sweet Pepper. *Aust. J. Crop Sci.* **2009,** *3*, 257–267.

Khan, G. A.; Declerck, M.; Sorin, C.; Hartmann, C.; Crespi, M.; Leladais-Briere, C. MicroRNAs as Regulators Root Development and Architecture. *Plant Mol. Biol.* **2011,** *77*, 47–58.

Khozaei, M.; Fisk, S.; Lawson, T.; Gibon, Y.; Sulpice, R.; Stitt, M.; Lefebvre, S. C.; Raines, C. A. Overexpression of Plastid Transketolase in Tobacco Results in a Thiamine Auxotrophic Phenotype. *Plant Cell* **2015,** *27*, 432–447.

Khraiwesh, B.; Zhu, J. K.; Zhu, J. Role of miRNAs and siRNAs in Biotic and Abiotic Stress Responses of Plants. *Biochem. Biophys. Acta* **2012,** *1819*, 137–148.

Kim, D.; Kim, D.; Alptekin, B.; Budak, H. CRISPR/Cas9 Genome Editing in Wheat. *Funct. Integr. Genom.* **2017a,** *18*, 31–41.

Kim, H.; Kim, S. T.; Ryu, J.; Kang, B. C.; Kim, J. S.; Kim, S. G. CRISPR/Cpf1-Mediated DNA-Free Plant Genome Editing. *Nat. Commun.* **2017b,** *8*, 14406.

Kim, J. I.; Kim, J. Y. New Era of Precision Plant Breeding Using Genome Editing. *Plant Biotechnol. Rep.* **2019,** *13*, 419–421.

Knott, G. J.; Doudna, J. A. CRISPR-Cas Guide the Future of Genetic Engineering. *Science* **2018,** *361*, 866–869.

Kobayashi, K.; Sasaki, D.; Noguchi, K.; Fujinuma, D.; Komatsu, H.; Kobayashi, M.; Sato, M, Toyooka, K.; Sugimoto, K.; Niyogi, K. K.; Wada, H.; Masuda, T. Photosynthesis of Root Chloroplasts Developed in Arabidopsis Lines Overexpressing *GOLDEN2-LIKE* Transcription Factors. *Plant Cell Physiol.* **2013,** *54*, 1365–1377.

Kohli, S. K.; Handa, N.; Sharma, A.; Gautam, V.; Arora, S.; Bhardwaj, R.; et al. Interaction of 24-Epibrassinolide and Salicylic Acid Regulates Pigment Contents, Antioxidative Defense Responses, and Gene Expression in *Brassica juncea* L. Seedlings under Pb Stress. *Environ. Sci. Pollut. Res.* **2018,** *25*, 15159–15173.

Konermann, S.; Brigham, M. D.; Trevino, A. E.; Joung, J.; Abudayyeh, O. O.; Barcena, C.; Hsu, P. D.; Habib, N.; Gootenberg, J. S.; Nishimasu, H.; Nureki, O.; Zhang, F. Genome-Scale Transcriptional Activation by an Engineered CRISPR-Cas9 Complex. *Nature* **2015,** *517*, 583–588.

Kopf, M.; Klahn, S.; Pade, N.; Weingärtner, C.; Hagemann, M. V. B.; Hess, W. R. Comparative Genome Analysis of the Closely Related *Synechocystis* Strains PCC 6714 and PCC 6803. *DNA Res.* **2014a,** *21*, 255–266.

Kopf, M.; Klahn, S.; Scholz, I.; Matthiessen, J. K.; Hess, W. R, Hagemann, M. V. B. Comparative Analysis of the Primary Transcriptome of *Synechocystis* sp. PCC 6803. *DNA Res.* **2014b,** *21*, 527–239.

Kopf, M.; Klahn, S.; Scholz, I.; Hess, W. R.; Hagemann, M. V. B. Variations in the Non-Coding Transcriptome as a Driver of Inter-Strain Divergence and Physiological Adaptation in Bacteria. *Sci. Rep.* **2015a,** *5*, 9560.

Kopf, M.; Moke, F.; Bauwe, H.; Hess, W. R.; Hagemann M. V. B. Expression Profiling of the Bloomforming Cyanobacterium *Nodularia spumigena* CCY9414 under High Light and Oxidative Stress Conditions. *ISME J.* **2015b,** *9*, 2142–2152.

Kossmann, J.; Muller-Rober, B.; Dyer, T. A.; Raines, C. A.; Sonnewald, U.; Willmitzer, L. Cloning and Expression Analysis of the Plastidic Fructose-1,6-bisphosphatase Coding Sequence from Potato: Circumstantial Evidence for the Import of Hexoses Into Chloroplasts. *Planta* **1992,** *188*, 7–12.

Kubis, A.; Bar-Even, A. Synthetic Biology Approaches for Improving Photosynthesis. *J. Exp. Bot.* **2019,** *70*, 1425–1433.

Kulcheski, F. R.; de Oliveira, L. F.; Molina, L. G.; Almerao, M. P.; Rodrigues, F. A.; Marcolino, J.; Barbosa, J. F.; Stolf-Moreira, R.; Nepomuceno, A. L.; Marcelino-Guimaraes, F. C.; Abdelnoor, R. V.; Nascimento, L. C.; Carazolle, M. F.; Pereira, G. A. G.; Margis, R. Identification of Novel Soybean MicroRNAs Involved in Abiotic and Biotic Stresses. *BMC Genom.* **2011,** *12*, 307.

Laporte, P.; Merchan, F.; Amor, B. B.; Wirth, S.; Crespi, M. Riboregulators in Plant Development. *Biochem. Soc. Trans.* **2007,** *35*, 1638–1642.

Lauter, N.; Kampani, A.; Carlson, S.; Goebel, M.; Moose, S. P. MicroRNA172 Down Regulates Glossy15 to Promote Vegetative Phase Change in Maize. *Proc. Natl. Acad. Sci. USA* **2005,** *102*, 9412–9417.

Lawson, T.; Bryant, B.; Lefebvre, S.; Lloyd, J. C.; Raines, C. A. Decreased sbpase Activity Alters Growth and Development in Transgenic Tobacco Plants. *Plant Cell Environ.* **2006,** *29*, 48–58.

Lee, Y. P, Kim, S. H.; Bang, J. W.; Lee, H. S, Kwak, S. S.; Kwon, S. Y. Enhanced Tolerance to Oxidative Stress in Transgenic Tobacco Plants Expressing Three Antioxidant Enzymes in Chloroplasts. *Plant Cell Rep.* **2007,** *26*, 591–598.

Lefebvre, S.; Lawson, T.; Fryer, M.; Zakhleniuk, O. V.; Lloyd, J. C.; Raines, C. A. Increased Sedoheptulose-1,7-Bisphosphatase Activity in Transgenic Tobacco Plants Stimulates Photosynthesis and Growth from an Early Stage in Development. *Plant Physiol.* **2005,** *138*, 451–460.

Lelandais-Briere, C.; Naya, L.; Sallet, E.; Calenge, F.; Frugier, F.; Hartmann, C.; Gouzy, J.; Crespi, M. Genome-Wide *Medicago truncatula* Small RNA Analysis Revealed Novel MicroRNAs and Isoforms Differentially Regulated in Roots and Nodules. *Plant Cell* **2009,** *21*, 2780–2796.

Lelandais-Briere, C.; Sorin, C.; Declerck, M.; Benslimane, A.; Crespi, M.; Hartmann, C. Small RBA Diversity in Plants and Its Impact in Development. *Curr. Genom.* **2010,** *11*, 14–23.

Li, T.; Liu, B.; Spalding, M. H.; Weeks, D. P.; Yang, B. High-Efficiency TALEN-Based Gene Editing Produces Disease-Resistant Rice. *Nat. Biotechnol.* **2012,** *30*, 390–392.

Li, A.; Jia, S.; Yobi, A.; Ge, Z.; Sato, S.; Zhang, C.; Angelovici, R.; Clemente, T. E.; Holding, D. R. Editing of an Alpha-Kafirin Gene Family Increases Digestibility and Protein Quality in Sorghum. *Plant Physiol.* **2018a,** *177*, 1425–1438.

Li, R.; Fu, D.; Zhu, B.; Luo, Y.; Zhu, H. CRISPR/Cas9-Mediated Mutagenesis of lncRNA1459 Alters Tomato Fruit Ripening. *Plant J.* **2018b,** *94*, 513–524.

Li, R.; Zhang, L.; Wang, L.; Chen, L.; Zhao, R.; Sheng, J.; Shen, L. Reduction of Tomato-Plant Chilling Tolerance by CRISPR-Cas9-Mediated SlCBF1 Mutagenesis. *J. Agric. Food Chem.* **2018c,** *66*, 9042–9051.

Li, X.; Wang, Y.; Chen, S.; Tian, H.; Fu, D.; Zhu, B.; Luo, Y.; Zhu, H. Lycopene Is Enriched in Tomato Fruit by CRISPR/Cas9-Mediated Multiplex Genome Editing. *Front. Plant Sci.* **2018d,** *9*, 559.

Li, H.; Deng, Y.; Wu, T.; Subramanian, S.; Yu, O. Misexpression of miR482, miR1512, and miR1515 Increases Soybean Nodulation. *Plant Physiol.* **2010,** *153*, 1759–1770.

Li, M.; Li, X.; Zhou, Z.; Wu, P.; Fang, M.; Pan, X.; Lin, Q.; Luo, W.; Wu, G.; Li, H. Reassessment of the Four Yield-Related Genes Gn1a, DEP1, GS3, and IPA1 in Rice Using a CRISPR/Cas9 System. *Front. Plant Sci.* **2016a,** *7*, 377.

Li, S.; Gao, F.; Xie, K.; Zeng, X.; Cao, Y.; Zeng, J.; He, Z.; Ren, Y.; Li, W.; Deng, Q. et al. The OsmiR396c-OsGRF4-OsGIF1 Regulatory Module Determines Grain Size and Yield in Rice. *Plant Biotechnol. J.* **2016b,** *14*, 2134–2146.

Li, Q.; Sapkota, M.; van der Knaap, E. Perspectives of CRISPR/Cas-Mediated Cis-Engineering in Horticulture: Unlocking the Neglected Potential for Crop Improvement. *Hort. Res.* **2020,** *7*, 36.

Li, R.; Liu, C.; Zhao, R.; Wang, L.; Chen, L.; Yu, W.; Zhang, S.; Sheng, J.; Shen, L. CRISPR/Cas9-Mediated SlNPR1 Mutagenesis Reduces Tomato Plant Drought Tolerance. *BMC Plant Biol.* **2019,** *19*, 38.

Li, X.; Zhou, W.; Ren, Y.; Tian, X.; Lv, T.; Wang, Z.; Fang, J.; Chu, C.; Yang, J.; Bu, Q. High-Efficiency Breeding of Early-Maturing Rice Cultivars via CRISPR/Cas9-Mediated Genome Editing. *J. Genet. Genom.* **2017a,** *44*, 175–178.

Li, Z.; Zhang, D.; Xiong, X.; Yan, B.; Xie, W.; Sheen, J.; Li, J. F. A Potent Cas9-Derived Gene Activator for Plant and Mammalian Cells. *Nat. Plants* **2017b,** *3*, 930.

Li, Z.; Liu, Z. B.; Xing, A.; Moon, B. P.; Koellhoffer, J. P.; Huang, L.; Ward, R. T. et al. Cas9-Guide RNA Directed Genome Editing in Soybean. *Plant Physiol.* **2015,** *169*, 960–970.

Liang, Z.; Chen, K.; Li, T.; Zhang, Y.; Wang, Y.; Zhao, Q.; Liu, J.; et al. Efficient DNA-Free Genome Editing of Bread Wheat Using CRISPR/Cas9 Ribonucleoprotein Complexes. *Nat. Publ.* Gr. **2017,** *8*, 1–5.

Liu, M.; Yu, H.; Zhao, G.; Huang, Q.; Lu, Y.; Ouyang, B. Profiling of Drought-Responsive MicroRNA and mRNA in Tomato Using High-Throughput Sequencing. *BMC Genom.* **2017a,** *18*, 481.

Liu, J.; Chen, J.; Zheng, X.; Wu, F.; Lin, Q.; Heng, Y.; Tian, P.; Cheng, Z. J.; Yu, X.; Zhou, K.; et al. GW5 Acts in the Brassinosteroid Signalling Pathway to Regulate Grain Width and Weight in Rice. *Nat. Plants* **2017b,** *3*, 17043.

Liu, Q.; Chen, Y. Q. Insights into the Mechanism of Plant Development: Interactions of miRNAs Pathway with Phytohormone Response. *Biochem. Biophys. Res. Commun.* **2009,** *384*, 1–5.

Long, S. P.; Marshall-Colon, A.; Zhu, X. G. Meeting the Global Food Demand of the Future by Engineering Crop Photosynthesis and Yield Potential. *Cell* **2015,** *161*, 56–66.

Lou, D.; Wang, H.; Liang, G.; Yu, D. OsSAPK2 Confers Abscisic Acid Sensitivity and Tolerance to Drought Stress in Rice. *Front. Plant Sci.* **2017,** *8*, 993.

Lou, D.; Wang, H.; Yu, D. The Sucrose Non-Fermenting-1-Related Protein Kinases SAPK1 and SAPK2 Function Collaboratively as Positive Regulators of Salt Stress Tolerance in Rice. *BMC Plant Biol.* **2018,** *18*, 203.

Lowder, L. G.; Zhang, D.; Baltes, N. J.; Paul, J. W.; Tang, X.; Zheng, X.; Voytas, D. F.; Hsieh, T. F.; Zhang, Y.; Qi, Y. A CRISPR/Cas9 Toolbox for Multiplexed Plant Genome Editing and Transcriptional Regulation. *Plant Physiol.* **2015,** *169*, 971–985.

Lowder, L. G.; Zhou, J.; Zhang, Y.; Malzahn, A.; Zhong, Z.; Hsieh, T. F.; Voytas, D. F.; et al. Robust Transcriptional Activation in Plants Using Multiplexed CRISPR-Act2.0 and mTALE-Act systems. *Mol. Plant*, **2017,** 11, 245–256.

Lowder, L. G.; Zhou, J.; Zhnag, Y.; Malzahn, A.; Zhong, Z.; Hseih, T. F.; Voytas, D. F.; Zhang, Y.; Qi, Y. Robust Transcriptional Activation in Plants Using Multiplexed CRISPR-Act2.0 and mTALE-Act Systems. *Mol. Plant* **2018,** *11*, 245–256.

Lowder, L.; Malzahn, A.; Qi, Y. Rapid Evolution of Manifold CRISPR Systems for Plant Genome Editing. *Front. Plant Sci.* **2016,** *7*, 1–112.

Lu, K.; Wu, B.; Wang, J.; Zhu,W.; Nie, H.; Qian, J.; Huang, W.; Fang, Z. Blocking Amino Acid Transporter OsAAP3 Improves Grain Yield by Promoting Outgrowth Buds and Increasing Tiller Number in Rice. *Plant Biotechnol. J.* **2018,** *16*, 1710–1722.

Ma, L.; Zhang, D.; Miao, Q.; Yang, J.; Xuan, Y.; Hu, Y. Essential Role of Sugar Transporter OsSWEET11 during the Early Stage of Rice Grain Filling. *Plant Cell Physiol.* **2017,** *58*, 863–873.

Ma, X.; Zhang, Q.; Zhu, Q.; Liu, W.; Chen, Y.; Qiu, Y.; Wang, B.; Yang, Z.; Li, H.; Lin, Y. et al. A Robust CRISPR/Cas9 System for Convenient, High Efficiency Multiplex Genome Editing in Monocot and Dicot Plants. *Mol. Plant* **2015,** *8*, 1274–1284.

Maayan, I.; Shaya, F.; Ratner, K.; Mani, Y.; Lavee, S.; Avidan, B.; Shahak, Y.; Ostersetzer-Biran, O. Photosynthetic Activity during Olive (*Olea europaea*) Leaf Development Correlates with Plastid Biogenesis and RuBisCO Levels. *Physiol. Plant.* **2008,** *134*, 547–558.

Mackelprang, R.; Lemaux, P. G. Genetic Engineering and Editing of Plants: An Analysis of New and Persisting Questions. *Annu. Rev. Plant Biol.* **2020**, *71*, 659–687.

Mahfouz, M. M.; Li, L.; Piatek, M.; Fang, X.; Mansour, H.; Bangarusamy, D. K. et al. Targeted Transcriptional Repression Using a Chimeric TALE-SRDX Repressor Protein. *Plant Mol. Biol.* **2012,** *78*, 311–321.

Mali, P.; Aach, J.; Stranges, P. B.; Esvelt, K. M.; Moosburner, M.; Kosuri, S.; Yang, L.; Church, G. M. CAS9 Transcriptional Activators for Target Specificity Screening and Paired Nickases for Cooperative Genome Engineering. *Nat. Biotechnol.* **2013,** *31*, 833–838.

Mariani, L.; Ferrante, A. Agronomic Management for Enhancing Plant Tolerance to Abiotic Stresses-Drought, Salinity, Hypoxia, and Lodging. *Horticulturae* **2017,** *3*, 52.

Mattana, M.; Biazzi, E.; Consonni, R.; Locatelli, F.; Vannini, C.; Provera, S.; Coraggio, I. Overexpression of *Osmyb4* Enhances Compatible Solute Accumulation and Increases Stress Tolerance of *Arabidopsis thaliana. Physiol. Plant.* **2005,** *125*, 212–223.

Maxwell, B. B.; Andersson, C. R.; Poole, D. S.; Kay, J. S.; Chory, J. HY5, Circadian Clock-Associated 1, and a Cis-Element, DET1 Dark Response Element, Mediate DET1 Regulation of Chlorophyll a/b-Binding Protein 2 Expression. *Plant Physiol.* **2003,** *133*, 1565–1577.

Meng, Y.; Huang, F.; Shi, Q.; Cao, J.; Chen, D.; Zhang, J.; Ni, J.; Wu, P.; Chen, M. Genome-Wide Survey of Rice microRNAs and microRNA-Target Pairs in the Root of a Novel Auxin–Resistant Mutant. *Planta* **2009,** *230*, 883–898.

Miao, J.; Guo, D.; Zhang, J.; Huang, Q.; Qin, G.; Zhang, X. et al. Targeted Mutagenesis in Rice Using CRISPR/Cas System. *Cell Res.* **2013,** *23*, 1233–1236.

Morales, F.; Ancin, M.; Fahket, D.; Gonzalez-Torralba, J.; Gpmez, A. L.; Seminario, A.; Soba, D.; Ben Mariem, S.; Garriga, M.; Aranjuelo, I. 2020. Photosynthetic Metabolism under Stressful Growth Conditions as a Bases for Crop Breeding and Yield Improvement. *Plants* **2020,** *9*, 88.

Moscou, M. J.; Bogdanove, A. J. Introduction of a Chimeric Chalcone Synthase Gene into Petunia Results in Reversible Co-suppression of Homologous Genes in Trans. *Plant Cell* **2009,** *2*, 279–289.

Mu, X.; Chen, Q.; Chen, F.; Yuan, L.; Mi, G. A RNA-seq Analysis of the Response of Photosynthetic System to Low Nitrogen Supply in Maize Leaf. *Int. J. Mol. Sci.* **2017,** *18*, 2624.

Mu, X.; Chen, Q.; Chen, F.; Yuan, L.; Mi, G. Within-Leaf Nitrogen Allocation in Adaptation to Low Nitrogen Supply in Maize during Grain-Filling Stage. *Front. Plant Sci.* **2016,** *7*, 699.

Muhammad, I.; Shalmani, A.; Ali, M.; Yang, Q. H.; Ahmad, H.; Li, F. B. Mechanisms Regulating the Dynamics of Photosynthesis Under Abiotic Stresses. *Front. Plant Sci.* **2021,** *11*, 615942.

Mushtaq, M.; Bhat, J. A.; Mir, Z. A.; Sakina, A.; Ali, S.; Singh, A. K.; Tyagi, A.; Salgotra, R. K.; Dar, A. A.; Bhat, R. CRISPR/Cas Approach: A New Way of Looking at Plant-Abiotic Interactions. *J. Plant. Physiol.* **2018,** *224–225*.

Nakano, T.; Suzuki, K.; Fujimura, T.; Shinshi, H. Genome-Wide Analysis of the ERF Gene Family in Arabidopsis and Rice. *Plant Physiol.* **2006,** *140*, 411–432.

Nguyen, H. C.; Lin, K. H.; Hsiung, T. C.; Huang, M. Y.; Yang, C. M.; Weng, J. H. et al. Biochemical and Physiological Characteristics of Photosynthesis in Plants of Two Calathea Species. *Int. J. Mol. Sci.* **2018,** *19*, 704.

Nikovics, K.; Blein, T.; Peaucelle, A.; Ishida, T.; Morin, H.; Aida, M.; Laufs, P. The Balance between the *MIR164A* and *CUC2* Genes Controls Leaf Margin Serration in Arabidopsis. *Plant Cell* **2006,** *18*, 2929–2945.

Nishihara, M.; Higuchi, A.; Watanabe, A.; Tasaki, K. Application of the CRISPR/Cas9 System for Modification of Flower Color in *Torenia fournieri*. *MBC Plant Biol.* **2018,** *18*, 331.

Nishiyama, Y.; Murata, N. Revised Scheme for the Mechanism of Photoinhibition and Its Application to Enhance the Abiotic Stress Tolerance of the Photosynthetic Machinery. *Appl. Microbiol. Biotechnol.* **2014,** *98*, 8777–8796.

Nouri, M. Z.; Moumeni, A.; Komatsu, S. Abiotic Stresses: Insight into Gene Regulation and Protein Expression in Photosynthetic Pathways of Plants. *Int. J. Mol. Sci.* **2015,** *16*, 20392–20416.

Nowicka, B. Target Genes for Plant Productivity Improvement. *J. Biotechnol.* **2019,** *298*, 21–34.

Nowicka, B.; Ciura, J.; Szymanska, R.; Kruk, J. Improving Photosynthesis, Plant Productivity and Abiotic Stress Tolerance-Current Trends and Future Perspectives. *J. Plant Physiol.* **2018,** *231*, 415–433.

Okuzaki, A.; Ogawa, T.; Koizuka, C.; Kaneko, K.; Inaba, M.; Imamura, J.; Koizuka, N. CRISPR/Cas9-Mediated Genome Editing of the Fatty Acid Desaturase 2 Gene in *Brassica napus*. *Plant Physiol. Biochem.* **2018,** *131*, 63–69.

Orr, D. J.; Pereira, A. M.; da Fonseca Pereira, P.; Pereira-Lima, I. A.; Zsogon, A.; Araujo, W. L. Engineering Photosynthesis: Progress and Perspectives. *F1000 Res.* **2017,** *6*, 1891.

Ort, D. R.; Merchant, S. S.; Alric, J.; Barkan, A.; Blankenship, R. E.; Bock, R.; Croce, R.; Hanson, M. R.; Hibberd, J. M.; Long, S. P. et al. Redesigning Photosynthesis to Sustainably Meet Global Food and Bioenergy Demand. *Proc. Natl. Acad. Sci. USA* **2015,** *112*, 8529–8536.

Ort, D. R.; Zhu, X. G.; Melis, A. Optimizing Antenna Size to Maximize Photosynthetic Efficiency. *Plant Physiol.* **2011,** *155*, 79–85.

Osakabe, Y.; Osakabe, K.; Shinozaki, S.; Tran L. S. P. Response of Plants to Water Stress. *Front. Plant Sci.* **2014,** *5*, 86.

Pan, J.; Huang, D.; Guo, Z.; Kuang, Z.; Zhang, H.; Xie, X.; Ma, Z.; Gao, S.; Lerdau, M. T.; Chu, C.; Li, L. Overexpression of MicroRNA408 Enhances Photosynthesis, Growth and Yield in Diverse Plants. *J. Integr. plant Biol.* **2018,** 60, 323–340.

Pandey, P.; Irulappan, V.; Bagavathiannan, M. V.; Senthil-Kumar, M. Impact of Combined Abiotic and Biotic Stresses on Plant Growth and Avenues for Crop Improvement by Exploiting Physio-morphological Traits. *Front. Plant. Sci.* **2017,** *8*, 537.

Pant, B. D.; Buhtz, A.; Kehr, J.; Scheible, W. R. MicroRNA399 Is a Long-Distance Signal for the Regulation of Plant Phosphate Homeostasis. *Plant J.* **2008,** *53*, 31–38.

Park, E. J.; Jeknić, Z.; Pino, M. T.; Murata, N.; Chen, T. H. Glycinebetaine Accumulation Is More Effective in Chloroplasts Than in the Cytosol for Protecting Transgenic Tomato Plants against Abiotic Stress. *Plant Cell Environ.* **2007,** *30*, 994–1005.

Paul, M. J.; Driscoll, S. P.; Andralojc, P. J.; Knight, J. S.; Gray, J. C.; Lawlor, D. W. Decrease of Phosphoribulokinase Activity by Antisense RNA in Transgenic Tobacco: Definition of the Light Environment under Which Phosphoribulokinase Is Not in Large Excess. *Planta* **2000,** *211*, 112–119.

Paunov, M.; Koleva, L.; Vassilev, A.; Vangronsveld, J.; Goltsev, V. Effects of Different Metals on Photosynthesis: Cadmium and Zinc Affect Chlorophyll Fluorescence in durum Wheat. *Int. J. Mol. Sci.* **2018,** *19*, 787.

Perez-Pinera, P.; Ousterout, D.; Brunger, J.; Farin, A.; Glass, K.; Guilak, F.; Crawford, G.; Hartemink, A.; Gersback, C. Synergistic and Tunable Human Gene Activation by Combinations of Synthetic Transcription Factors. *Nat. Methods* **2013,** *10*, 239–242.

Pfreundt, U.; Kopf, M.; Belkin, N.; Berman-Frank, I.; Hess, W. R. The Primary Transcriptome of the Marine diazotroph *Trichodesmium erythraeum* IMS101. *Sci. Rep.* **2014,** *4*, 6187.

Pilon, M. The Copper MicroRNAs. *New Phytol.* **2017,** *213*, 1030–1035.

Pinheiro, C.; Chaves, M. M. Photosynthesis and Drought: Can We Make Metabolic Connections from Available Data? *J. Exp. Bot.* **2011,** *62*, 869–882.

Qi, L. S.; Larson, M. H.; Gilbert, A. L.; Doudna, J. A.; Weissman, J. S.; Arkin, A. P.; Lim, W. A. Repurposing CRISPR as an RNA-Guided Platform for Sequence Specific Control of Gene Expression. *Cell* **2013,** *152*, 1173–1183.

Qi, W.; Zhu, T.; Tian, Z.; Li, C.; Zhang, W.; Song, R. High-Efficiency CRISPR/Cas9 Multiplex Gene Editing Using the Glycine tRNA-Processing System-Based Strategy in Maize. *BMC Biotechnol.* **2016,** *16*, 58.

Qi, Y. High Efficient Genome Modification by Designed Zinc Finger Nuclease. In *Advances in New Technology for Targeted Modifications of Plant Genomes*; Springer: New York, 2015; pp 39–53.

Reyon, D.; Tsai, S. Q.; Khayter, C.; Foden, J. A.; Sander, J. D.; Joung, J. K. FLASH Assembly of TALENs for High Throughput Genome Editing. *Nat. Biotechnol.* **2012,** *30*, 460–465.

Rodriguez-Moreno, L.; Song, Y.; Thomma, B. P. Transfer and Engineering of Immune Receptors to Improve Recognition Capacities in Crops. *Curr. Opin. Plant Biol.* **2017,** *38*, 42–49.

Rojas-Gonzalez, J. A.; Soto-Suarez, M.; Garcia-Diaz, A.; Romero-Puertas, M. C.; Sandalio, L. M.; Merida, A.; Thormahlen, I.; Geigenberger, P.; Serrato, A. J.; Sahrawy, M. Disruption of Both Chloroplastic and Cytosolic Fbpase Genes Results in a Dwarf Phenotype and Important Starch and Metabolite Changes in *arabidopsis thaliana*. *J. Exp. Bot.* **2015,** *66*, 2673–2689.

Rosenthal, D. M.; Locke, A. M.; Khozaei, M.; Raines, C. A.; Long, S. P.; Ort, D. R. Over-Expressing the c3 Photosynthesis Cycle Enzyme Sedoheptulose-1-7 Bisphosphatase Improves Photosynthetic Carbon Gain and Yield under Fully Open-Air CO_2 Fumigation (face). *BMC Plant Biol.* **2011,** *11*, 123.

Sadanandom, A.; Srivastava, A. K.; Zhang, C. Targeted Mutagenesis of the SUMO Protease, Overly Tolerant to Salt1 in Rice through CRISPR/Cas9-Mediated Genome Editing Reveals a Major Role of This SUMO Protease in Salt Tolerance. *BioRxiv* **2019,** 555706.

Sahrawy, M.; Avila, C.; Chueca, A.; Canovas, F. M.; Lopez-Gorge, J. Increased Sucrose Level and Altered Nitrogen Metabolism in *arabidopsis thaliana* Transgenic Plants Expressing Antisense Chloroplastic Fructose-1,6-bisphosphatase. *J. Exp. Bot.* **2004,** *55*, 2495–2503.

Saibo, N. J. M.; Lourenco, T.; Oliveira, M. M. Transcription Factors and Regulation of Photosynthetic and Related Metabolism under Environmental Stresses. *Ann. Bot.* **2009,** *103*, 609–623.

Sakamoto, A.; Murata, N. The Role of glycinebetaine in the Protection of Plants from Stress: Clues from Transgenic Plants. *Plant Cell Environ* **2002,** *25*, 163–171.

Sánchez-León, S.; Gil-Humanes, J.; Ozuna, C. V.; Giménez, M. J.; Sousa, C.; Voytas, D. F.; Barro, F. Low-gluten, nontransgenic wheat engineered with CRISPR/Cas9. *Plant Biotechnol. J.* **2018,** 16, 902–910.

Sanjana, N. E.; Cong, L.; Zhou, Y.; Cunniff, M. M.; Feng, G.; Zhang, F. A Transcription Activator-Like Effector Toolbox for Genome Engineering. *Nat. Protoc.* **2012,** *7*, 171–192.

Savary, S.; Ficke, A.; Aubertot, J.; Hollier, C. Crop Losses Due to Diseases and Their Implications for Global Food Production Losses and Food Security. *Springer Food Sec.* **2012,** *4*, 519–537.

Scheben, A.; Edwards, D. Genome Editors Take on Crops. *Science* **2017,** *355*, 1122–1123.

Scheben, A.; Wolter, F.; Batley, J.; Puchta, H.; Edwards, D. Towards CRISPR/Cas Crops-Bringing Together Genomics and Genome Editing. *New Phytol.* **2017,** *216*, 682–698.

Schiml, S.; Puchta, H. Revolutionizing Plant Biology: Multiple Ways of Genome Engineering by CRISPR/Cas. *Plant Methods* **2016,** *12*, 8.

Schmidt, C.; Pacher, M.; Puchta, H. Efficient Induction of Heritable Inversions in Plant Genome Using CRISPR/Cas System. *Plant J.* **2019,** *98*, 577–589.

Schreiber, A. W.; Shi, B. J.; Huang, C. Y.; Langridge, P.; Baumann, U. Discovery of Barley miRNAs through Deep Sequencing of Short Reads. *BMC Genom.* **2011,** *12*, 129.

Schwechheimer, C.; Zourelidou, M.; Bevan, M. W. Plant Transcription Factor Studies. *Annu. Rev. Plant Physiol. Plant Mol. Biol.* **1998,** *49*, 127–150.

Shan, Q.; Wang, Y.; Li, J.; Zhang, Y.; Chen, K.; Liang, Z.; Zhang, K.; Liu, J.; Xi, J. J.; Qiu, J. L. et al. Targeted Genome Modification of Crop Plants Using a CRISPR-Cas System. *Nat. Biotechnol.* **2013,** *31*, 686–688.

Shan, Q.; Zhang, Y.; Chen, K.; Zhang, K.; Gao, C. Creation of fragrant rice by targeted knockout of the *OsBADH2* gene using TALEN technology. *Plant Biotechnol. J.* **2015,** 13, 791–800.

Sharma, A.; Kumar, V.; Shahzad, B.; Ramakrishnan, M.; Sidhu, G. P. S.; Bali, A. S. et al. Photosynthetic Response of Plants under Different Abiotic Stresses: A Review. *J. Plant Growth Regul.* **2020,** *39*, 509–531.

Sharma, A.; Kumar, V.; Singh, R.; Thukral, A. K.; Bhardwaj, R. Effect of Seed Pre-soaking with 24-epibrassinolide on Growth and Photosynthetic Parameters of *Brassica juncea* L. in Imidacloprid Soil. *Ecotoxicol. Environ. Saf.* **2016a,** *133*, 195–201.

Sharma, A.; Thakur, S.; Kumar, V.; Kanwar, M. K.; Kesavan, A. K.; Thukral, A. K. et al. Pre-Sowing Seed Treatment with 24-epibrassinolide Ameliorates Pesticide Stress in *Brassica juncea* L. through the Modulation of Stress Markers. *Front. Plant Sci.* **2016b,** *7*, 1569.

Sharma, A.; Shahzad, B.; Kumar, V.; Kohli, S. K.; Sidhu, G. P. S.; Bali, A. S.; et al. Phytohormones Regulate Accumulation of Osmolytes under Abiotic Stress. *Biomolecules* **2019,** *9*, 285.

Shelake, R. M.; Pramanik, D.; Kim, J. Y. Exploration of Plant-Microbe Interactions for Sustainable Agriculture in CRISPR Era. *Microorganisms* **2019a,** *7*, 269.

Shelake, R. M.; Pramanik, D.; Kim J. Y. Evolution of Plant Mutagenesis Tools: A Shifting Paradigm from Random to Targeted Genome Editing. *Plant Biotechnol. Rep.* **2019b,** *13*, 423–445.

Sheng, X.; Sun, Z.; Wang, X.; Tan, Y.; Yu, D.; Yuan, G.; Yuan, D.; Duan, M. Improvement of Rice "Easy-to Shatter" Trait via CRISPR/Cas9-Mediated Mutagenesis of the *qSH1* Gene. *Front. Plant Sci.* **2020,** *11*, 619.

Shi, J.; Gao, H.; Wang, H.; Lafitte, H. R.; Archibald, R. L.; Yang, M.; Hakimi, S. M.; Mo, H.; Habben, J. E. ARGOS8 Variants Generated by CRISPR-Cas9 Improve Maize Grain Yield under Field Drought Stress Conditions. *Plant Biotechnol. J.* **2017,** *15*, 207–216.

Shim, J. S.; Oh, N.; Chung, P. J.; Kim, Y. S.; Choi, Y. D.; Kim, J. K. Overexpression of OsNAC14 Improves Drought Tolerance in Rice. *Front. Plant Sci.* **2018,** *9*, 310.

Shriram, V.; Kumar, V.; Devarumath, R. M.; Khare, T. S.; Wani, S. H. MicroRNAs as Potential Targets for Abiotic Stress Tolerance in Plants. *Front. Plant Sci.* **2016,** *7*, 817.

Shukla, V. K.; Doyon, Y.; Miller, J. C.; Dekelver, R. C.; Moehle, E. A.; Worden, S. E.; Mitchell, J. C.; Arnold, N. L.; Gopalan, S.; Meng, X.; et al. Precise Genome Modification in the Crop Species *Zea mays* Using Zinc-Finger Nucleases. *Nature* **2009,** *459*, 437–441.

Simkin, A. J.; López-Calcagno, P. E.; Davey, P. A.; Headland, L. R.; Lawson, T.; Timm, S.; Bauwe, H.; Raines, C. A. Simultaneous Stimulation of Sedoheptulose 1,7-bisphosphatase, Fructose 1,6-Bisphophate Aldolase and the Photorespiratory Glycine Decarboxylase h-Protein Increases CO_2 Assimilation, Vegetative Biomass and Seed Yield in Arabidopsis. *Plant Biotechnol. J.* **2017,** *15*, 805–816.

Simkin, A. J.; López-Calcagno, P. E.; Raines, C. A. Feeding the World: Improving Photosynthetic Efficiency for Sustainable Crop Production. *J. Exp. Bot.* **2019,** *70*, 1119–1140.

Simkin, A. J.; McAusland, L.; Headland, L. R.; Lawson, T.; Raines, C. A. Multigene Manipulation of Photosynthetic Carbon Assimilation Increases CO_2 Fixation and Biomass Yield in Tobacco. *J. Exp. Bot.* **2015,** *66*, 4075–4090.

Singh, K.; Foley, R. C.; Onate-Sanchez, L. Transcription Factors in Plant Defense and Stress Responses. *Curr. Opin. Plant Biol.* **2002,** *5*, 430–436.

Soares, C.; Branco-Neves, S.; de Sousa, A.; Azenha, M.; Cunha, A.; Pereira, R. et al. SiO_2 Nanomaterial as a Tool to Improve *Hordeum vulgare* L. Tolerance to Nano-NiO Stress. *Sci. Total Environ.* **2018,** *622–623*, 517–525.

Song, Z.; Zhang, L.; Wang, Y.; Li, H.; Li, S.; Zhao, H.; Zhang, H. Constitutive Expression of miR408 Improves Biomass and Seed Yield in Arabidopsis. *Front. Plant Sci.* **2018,** *8*, 2114.

Steglich, C.; Futschik, M. E.; Lindell, D.; Voss, B.; Chisholm, S. W.; Hess, W. R. The Challenge of Regulation in a Minimal Photoautotroph: Non-Coding RNAs in *Prochlorococcus*. *PLoS Genet.* **2008,** *4*, e1000173.

Su, J.; Hirji, R.; Zhang, L.; He, C.; Selvaraj, G.; Wu, R. Evaluation of the Stress-Inducible Production of Choline Oxidase in Transgenic Rice as a Strategy for Producing the Stress-Protectant Glycine Betaine. *J. Exp. Bot.* **2006,** *57*, 1129–1135.

Subramanian, S.; Fu, Y.; Sunkar, R.; Barbazuk, B. W.; Zhu, J. K.; Yu, O. Novel and Nodulation-Regulated microRNAs in Soybean Roots. *BMC Genom.* **2008,** *9*, 160.

Sun, G. MicroRNAs and Their Diverse Functions in Plants. *Plant Mol. Biol.* **2012,** *80*, 17–36.

Sun, Y.; Jiao, G.; Liu, Z.; Zhang, X.; Li, J.; Guo, X.; Du, W.; Du, J.; Francis, F.; Zhao, Y. et al. Generation of High-Amylose Rice through CRISPR/Cas9-Mediated Targeted Mutagenesis of Starch Branching Enzymes. *Front. Plant Sci.* **2017,** *8*, 298.

Sun, Y.; Zhang, X.; Wu, C.; He, Y.; Ma, Y.; Hou, H.; Guo, X. et al. Engineering Herbicide-Resistant Rice Plants through CRISPR/Cas9-Mediated Homologous Recombination of Acetolactate Synthase. *Mol. Plant*, **2016,** *9*, 628–631.

Sunkar, R.; Kapoor, A.; Zhu, J. K. Posttranscriptional Induction of Cu/Zn Superoxide Dismutase Genes in Arabidopsis Is Mediated by Downregulation of miR398 and Important for Oxidative Stress Tolerance. *Plant Cell* **2006,** *18*, 2051–2065.

Sunkar, R.; Zhu, J. K. Novel and Stress-Regulated MicroRNAs and Other Small RNAs from Arabidopsis. *Plant Cell* **2004,** *16*, 2001–2019.

Svitashev, S.; Young, J. K.; Schwartz, C.; Gao, H.; Falco, S. C.; Cigan, A. M. Targeted Mutagenesis, Precise Gene Editing and Site-Specific Gene Insertion in Maize Using Cas9 and Guide RNA. *Plant Physiol.* **2015,** *2*, 931–945.

Tang, L.; Mao, B.; Li, Y.; Lv, Q.; Zhang, L. P.; Chen, C. et al. Knockout of OsNramp5 Using the CRISPR/Cas9 System Produces Low Cd-Accumulating Indica Rice Without Compromising Yield. *Sci. Rep.* **2017,** *7*, 14438.

Todesco, M.; Rubio-Somoza, I.; Pazares, J.; Weigel, D. A Collection of Target Mimics for Comprehensive Analysis of MicroRNA Function in *Arabidopsis thaliana. PLoS Genet.* **2010,** *6*, 7.

Trindale, I.; Capitao, C.; Dalmay, T.; Fevereiro, M. P, Santos, D. M. miR398 and miR408 Are Up-regulated in Response to Water Deficit in *Medicago truncatula. Planta* **2009,** *231*, 705–716.

Turner,M.; Nizampatnam, N. R.; Baron, M.; Coppin, S.; Damodaran, S.; Adhikari, S.; Arunachalam. S.; Yu. O.; Subramaniun. S. Ectopic Expression of miR160 Results in Auxin Hypersensivity, Cytokinin Hyposensitivity, and Inhibition of Symbiotic Nodule Development in Soybean. *Plant Physiol.* **2013,** *162*, 2042–2055.

Uematsu, K.; Suzuki, N.; Iwamae, T.; Inui, M.; Yukawa, H. Increased Fructose 1,6-Bisphosphate Aldolase in Plastids Enhances Growth and Photosynthesis of Tobacco Plants. *J. Exp. Bot.* **2012,** *63*, 3001–3009.

United Nations, Department of Economic and Social Affairs, Population Division. World Population Prospects: The 2017 Revision, Key Findings and Advance Tables. Working Paper No. ESA/P/WP/248. 2017. https://esa.un.org/unpd/wpp/publications/Files/WPP2017_KeyFindings.pdf (accessed Dec 16, 2019).

Upadhyay, S.; Kumar, J.; Alok, A.; Tuli, R. RNA-Guided Genome Editing for Target Gene Mutations in Wheat. *Genet.Genes Genom.* **2013,** *3*, 2233–2238.

Usman, B.; Nawaz, G.; Zhao, N.; Liu, Y.; Li, R. Generation of High Yielding and Fragrant Rice (*Oryza sativa* L.) Lines by CRISPR/Cas9 Targeted Mutagenesis of Three Homeologs of Cytochrome P450 Gene Family and OsBADH2 and Transcriptome and Proteome Profiling of Revealed Changes Triggered by Mutations. *Plants* **2020,** *9*, 788.

Veillet, F.; Perrot, L.; Chauvin, L.; Kermarrec, M. P.; Guyon-Debast, A.; Chauvin, J. E.; Nogue, F. et al. Transgene-Free Genome Editing in Tomato and Potato Plants Using Agrobacterium-Mediated Delivery of a CRISPR/Cas9 Cytidine Base Editor. *Int. J. Mol. Sci.* **2019,** *20*, 402.

Voigt, K.; Sharma, C. M.; Mitschke, J.; Lambrecht, S. J.; Hagemann, M. V. B.; Hess, W. R.; Steglich, C. Comparative Transcriptomics of Two Environmentally Relevant Cyanobacteria Reveals Unexpected Transcriptome Diversity. *ISME J.* **2014,** *8*, 2056–2068.

Voss, B.; Georg, J.; Schön, V.; Ude, S.; Hess, W. R. Biocomputational Prediction of Non-Coding RNAs in Model Cyanobacteria. *BMC Genom.* **2009,** *10*, 123.

Wallace, J. G.; Rodgers-Melnick, E.; Buckler, E. S. On the Road to Breeding 4.0: Unraveling the Good, the Bad, and the Boring of Crop Quantitative Genomics. *Ann. Rev. Genet.* **2018,** *52*, 421–444.

Wang, C.; Li, Q. Z. Identification of Differentially Expressed miRNAs during the Development of Chinese Murine Mammary Gland. *J. Genet. Genom.* **2007,** *34*, 966–973.

Wang, H.; Wang, H.; Shao, H.; Tang, X. Recent Advances in Utilizing Transcription Factors to Improve Plant Abiotic Stress Tolerance by Transgenic Technology. *Front. Plant Sci.* **2016,** *7*, 1–13.

Wang, H.; Zhang, H.; Li, Z. Analysis of Gene Expression Profile Induced by Water Stress in Upland Rice (*Oryza sativa* L. var. IRAT109) Seedlings Using Subtractive Expressed Sequence Tags Library. *J. Integr. Plant Biol.* **2007,** *49*, 1455–1463.

Wang, J.; Zhou, J.; Zhang, B.; Vanitha, J.; Ramachandran, S.; Kiang, S. Y. Genome-Wide Expansion and Expression Divergence of the Basic Leucine Zipper Transcription Factors in Higher Plants with an Emphasis on Sorghum. *J. Integr. Plant Biol.* **2011,** *53*, 212–231.

Wang, L.; Chen, L.; Li, R.; Zhao, R.; Yang, M.; Sheng, J.; Shen, L. Reduced Drought Tolerance by CRISPR/Cas9-Mediated SlMAPK3 Mutagenesis in Tomato Plants. *J. Agric. Food Chem.* **2017,** *65*, 8674–8682.

Wang, L.; Sun, S.; Jin, J.; Fu, D.; Yang, X.; Weng, X.; Xu, C.; Li, X.; Xiao, J.; Zhang, Q. Coordinated Regulation of Vegetative and Reproductive Branching in Rice. *Proc. Natl. Acad. Sci. USA* **2015,** *112*, 15504–15509.

Wang, L.; Zhao, C. M.; Wang, Y. J.; Liu, J. Overexpression of Chloroplast-Localized Small Molecular Heat-Shock Protein Enhances Chilling Tolerance in Tomato Plant. *J. Plant Physiol. Mol. Biol.* **2005,** *31*, 167–174.

Wang, Y.; Cheng, X.; Shan, Q.; Zhang, Y.; Liu, J.; Gao, C.; Qiu, J. L. Simultaneous Editing of Three Homoeoalleles in Hexaploid Bread Wheat Confers Heritable Resistance to Powdery Mildew. *Nat. Biotechnol.* **2014,** *32*, 947–951.

Weber, A. P. M.; Bar-Even, A. Update: Improving the Efficiency of Photosynthetic Carbon Reactions. *Plant Physiol.* **2019,** *179*, 803–812.

Weeks, D. P.; Spalding, M. H.; Yang, B. Use of Designer Nucleases for Targeted Gene and Genome Editing in Plants. *Plant Biotechnol. J.* **2016,** *14*, 483–495.

Woo, J. W.; Kim, J.; Kwon, S.; Corvalan, C.; Cho, S. W.; Kim, H.; Kim, S. G.; et al. DNA-free Genome Editing in Plants with Preassembled CRISPR-Cas9 Ribonucleoproteins. *Nat. Biotechnol.* **2015,** *33*, 1162–1164.

Wu, C. Y.; Li, Q. Z.; Feng, Z. X. Non-coding RNA Identification Based on Topology Secondary Structure and Reading Frame in Organelle Genome Level. *Genomics* **2016,** *107*, 9–15.

Wu, Z.; Stone, J. D.; Storchova, H.; Sloan, D. B. High transcript abundance, RNA Editing and Small RNAs in Intergenic Regions within the Massive Mitochondrial Genome of the Angiosperm, *Silene noctiflora. BMC Genom.* **2015,** *16*, 938.

Xie, F. L.; Huang, S. Q.; Guo, K.; Xiang, A. L.; Xhu, Y. Y.; Nie, L.; Yang, Z. M. Computational Identification of Novel microRNA and Targets in *Brassica napus. FEBS Lett.* **2007,** *58*, 1464–1474.

Xing, X. L.; Dong, L.; Wang, Z. P.; Zhang, H. Y.; Han, C. Y.; Liu, B.; Wang, X. C.; Chen, Q. J. A CRISPR/Cas9 Toolkit for Multiplex Genome Editing in Plants. *BMC Plant Biol.* **2014,** *14*, 327.

Xu, R.; Li, H.; Qin, R.; Wang, L.; Li, L.; Wei, P. et al. Gene Targeting Using the Agrobacterium Tumefaciens Mediated CRISPR-Cas System in Rice. *Rice* **2014,** *7*, 5.

Xu, R.; Yang, Y.; Qin, R.; Li, H.; Qiu, C.; Li, L.; Wei, P.; Yang, J. Rapid Improvement of Grain Weight via Highly Efficient CRISPR/Cas9-Mediated Multiplex Genome Editing in Rice. *J. Genet. Genom.* **2016,** *43*, 529–532.

Xue, Y.; Acar, M. Live-Cell Imaging of Chromatin Condensation Dynamics by CRISPR. *Science* **2018,** *4*, 216–235.

Yadav, P.; Kaur, R.; Kanwar, M. K.; Sharma, A.; Verma, V.; Sirhindi, G. et al. Castasterone Confers Copper Stress Tolerance by Regulating Antioxidant Enzyme Responses, Antioxidants, and Amino Acid Balance in *B. juncea* Seedlings. *Ecotoxicol. Environ. Saf.* **2018,** *147*, 725–734.

Yang, C.; Li, D.; Mao, D.; Liu, X.; Ji, C.; Li, X.; Zhao, X.; Cheng, Z.; Chen, C.; Zhu, L. Overexpression of microRNA319 Impacts Leaf Morphogenesis and Leads to Enhanced Cold Tolerance in Rice (*Oryza sativa* L.). *Plant Cell Environ.* **2013,** *36*, 2207–2218.

Yang, X.; Liang, Z.; Lu, C. Genetic Engineering of the Biosynthesis of Glycinebetaine Enhances Photosynthesis against High Temperature Stress in Transgenic Tobacco Plants. *Plant Physiol.* **2005,** *138*, 2299–2309.

Yao, Y.; Guo, G.; Ni, Z.; Sunkar, R.; Du, J.; Zhu, J. K.; Sun, Q. Cloning and Characterization of microRNAs from Wheat (*Triticum aestivum L.*). *Genome Biol.* **2007,** *8*, R96.

Young, J. K.; Gosior, S. L.; Jones, S.; Wang, L.; Navarroa, P.; Vickroy, B.; Barrangou, R. The Repurposing of Type I-E CRISPR Cascade for Gene Activation in Plants. *Commun. Biol.* **2019,** *2*, 383.

Zalatan, J. G.; Lee, M. E.; Almeida, R.; Gilbert, A. L.; Whitehead, E. H.; Russa, M. L.; Tsai, J. C.; Weissman, J. S.; Dueber, J. E.; Qi, L. S.; Lim, W. A. Engineering Complex Synthetic Transcriptional Programs with CRISPR RNA Scaffolds. *Cell* **2015,** *160*, 339–350.

Zandalinas, S. I.; Mittler, R.; Balfagón, D.; et al. Plant Adaptations to the Combination of Drought and High Temperatures. *Physiol. Plant.* **2018,** *162*, 2–12.

Zezulin, A.; Musunuru, K. Turning up the Heat with Therapeutic Epigenome Editing. *Cell Stem Cell*, **2018,** *22*, 10–11.

Zhang, A.; Liu, Y.; Wang, F.; Li, T.; Chen, Z.; Kong, D.; Bi, J.; Zhang, F.; Luo, X.; Wang, J.; et al. Enhanced Rice Salinity Tolerance via CRISPR/Cas9-Targeted Mutagenesis of the OsRR22 Gene. *Mol. Breed.* **2019,** *39*, 47.

Zhang, H.; Li, L. SQUAMOSA Promoter Binding Protein-like7 Regulated microRNA408 Is Required for Vegetative Development in Arabidopsis. *Plant J.* **2013,** *74*, 98–109.

Zhang, H.; Mao, X.; Wang, C.; Jing, R. Overexpression of a Common Wheat Gene TaSnRK2. 8 Enhances Tolerance to Drought, Salt, and Low Temperature in Arabidopsis. *PLoS ONE* **2010,** *30*, e16041.

Zhang, H.; Zhao, X.; Li, J.; Cai, H.; Deng, X. W.; Li, L. MicroRNA408 Is Critical for the HY5-SPL7 Gene Network That Mediates the Coordinated Response to Light and Copper. *Plant Cell* **2014a,** *26*, 4933–4953.

Zhang, H.; Zhang, J.; Wei, P.; Zhang, B.; Gou, F.; Feng, Z.; Mao, Y.; et al. The CRISPR/Cas9 System Produces Specific and Homozygous Targeted Gene Editing in Rice in One Generation. *Plant Biotechnol. J.* **2014b,** *12*, 797–807.

Zhang, J. P.; Yu, Y.; Feng, Y. Z.; Zhou, Y. F.; Zhang, F.; Yang, Y. W.; Lei, M. Q.; Zhang, Y. C.; Chen, Y. Q. MiR408 Regulates Grain Yield and Photosynthesis via a Phytocyanin Protein. *Plant Physiol.* **2017,** *175*, 1175–1185.

Zhang, Y.; Li, D.; Zhang, D.; Zhao, X.; Cao, X.; Dong, L.; Liu, J.; Chen, K.; Zhang, H.; Gao, C. et al. Analysis of the Functions of TaGW2 Homoeologs in Wheat Grain Weight and Protein Content Traits. *Plant J.* **2018a,** *94*, 857–866.

Zhang, M.; Cao, Y.; Wang, Z.; Wang, Z. Q.; Shi, J.; Liang, X.; Song, W.; Chen, Q.; Lai, J.; Jiang, C. A Retrotransposon in an HKT1 Family Sodium Transporter Causes Variation of Leaf Na+Exclusion and Salt Tolerance in Maize. *New Phytol.* **2018b,** *217*, 1161–1176.

Zhang, J.; Zhang, H.; Botella, J. R.; Zhu, J. Generation of New Glutinous Rice by CRISPR/Cas9-Targeted Mutagenesis of the Waxy Gene in Elite Rice Varieties. *J. Integr. Plant Biol.* **2018c,** *60*, 369–375.

Zhang, X.; Wollenweber, B.; Jiang, D.; Liu, F.; Zhao, J. Water Deficit and Heat Shock Effects on Photosynthesis of a Transgenic *Arabidopsis thaliana* Constitutively Expressing ABP9, a bZIP Transcription Factor. *J. Exp. Bot.* **2008,** *59*, 839–848.

Zhang, X.; Zou, Z.; Gong, P.; Zhang, J.; Ziaf, K.; Li, H.; Xiao, F.; Ye, Z. Over-Expression of microRNA169 Confers Enhanced Drought Tolerance to Tomato. *Biotech. Lett.* **2011,** *33*, 403–409.

Zhang, Y.; Liang, Z.; Zong, Y.; Wang, Y.; Liu, J.; Chen, K.; Qiu, J. L.; Gao, C. Efficient and Transgene-Free Genome Editing in Wheat Through Transient Expression of CRISPR/Cas9 DNA or RNA. *Nat. Commun.* **2016,** *7*, 12617.

Zhao, C. Z.; Xia, H.; Frazier, T. P.; Yao, Y. Y.; Bi, Y. P.; Li, A. Q.; Li, M. J.; Li, C. S.; Zhang, B. H.; Wang, X. J. Deep Sequencing Identifies Novel and Conserved microRNAs in Peanuts (*Arachis hypogaea* L.). *BMC Plant Biol.* **2010,** *10*, 3.

Zhao, X. Y.; Hong, P.; Wu, J. Y.; Chen, X. B.; Ye, X. G.; Pan, Y. Y.; Wang, J.; Zhang, X. S. The Tae-miR408-Mediated Control of *TaTOC1* Genes Transcription Is Required for the Regulation of Heading Time in Wheat. *Plant Physiol.* **2016,** *170*, 1578–1594.

Zhao, Y.; Lin, S.; Qiu, Z.; Cao, D.; Wen, J.; Deng, X.; Wang, X.; Lin, J.; Li, X. MicroRNA857 is Involved in the Regulation of Secondary Growth of Vascular Tissues in *Arabidopsis*. *Plant Physiol.* **2015,** *169*, 2539–2552.

Zhou, C.; Li, C. A Novel R2R3-MYB Transcription Factor BpMYB106 of Birch (Betula platyphylla) Confers Increased Photosynthesis and Growth Rate through Up-regulating Photosynthetic Gene Expression. *Front. Plant Sci.* **2016,** *7*, 315.

Zhou, H.; Liu, B.; Weeks, D. P.; Spalding, M. H.; Yang, B. Large Chromosomal Deletions and Heritable Small Genetic Changes Induced by CRISPR/Cas9 in Rice. *Nucl. Acids Res.* **2014,** *42*, 10903–10914.

Zhou, J.; Peng, Z.; Long, J.; Sosso, D.; Liu, B.; Eom, J. S.; Huang, S. et al. Gene Targeting by the TAL Effector PthXo2 Reveals Cryptic Resistance Gene for Bacterial Blight of Rice. *Plant J.* **2015,** *82*, 632–643.

Zhou, L.; Liu, Y.; Liu, Z.; Kong, D.; Duan, M.; Luo, L. Genome-Wide Identification and Analysis of Drought-Responsive MicroRNAs in *Oryza sativa*. *J. Exp. Bot.* **2010,** *61*, 4157–68.

Zhou, M.; Li, D.; Li, Z.; Hu, Q.; Yang, C.; Zhu, L.; Luo, H. Constitutive Expression of a miR319 Gene Alters Plant Development and Enhances Salt and Drought Tolerance in Transgenic Creeping Bentgrass. *Plant Physiol.* **2013,** *161*, 1375–1391.

Zhu, J.; Song, N.; Sun, S.; Yang, W.; Zhao, H.; Song, W. et al. Efficiency and Inheritance of Targeted Mutagenesis in Maize Using CRISPR-Cas9. *J. Genet. Genomics*. **2016,** *43*, 25–36.

CHAPTER 4

Light-Harvesting Antenna Complex and Its Role in Environmental Stress

SUDIP CHOUDHURY*, SUNAYANA GOSWAMI, ANURADHA ROY CHOUDHURY, and SAURAV PAUL

Department of Chemistry, Assam University, Silchar 788011, India

**Corresponding author. E-mail: sudip.choudhury@aus.ac.in, sudipch1@gmail.com*

ABSTRACT

Photosynthesis is an extremely efficient photochemical process. In photosynthetic process, the light-harvesting antenna complex (LH or LHC), also known as antenna complex, is a very important integral component. The photosynthesis begins with photons absorption through the antenna molecules, resulting in the excitation of chlorophyll molecules. The LHC transfers the energy to a chlorophyll *a* molecule at the photosystem reaction center. However, the LHC gets reorganized in response to various environmental stress factors. For instance, under extreme high-light conditions, various plants and algae adopt "molecular strategies" to activate thermal dissipation of the excess energy for the protection of the photosynthetic system from getting damaged. Similarly, when exposed to iron deficiency, the LHC complex declines functional association with the photosystem I (PSI) core to protect the later. Many plants respond to drought conditions by downregulating the light-harvesting Chl-proteins, thereby reducing the light absorption by antennae complex and the electron transport rates of PSI and photosystem II. Initiating with a brief discussion on important concepts of

Photosynthesis and Respiratory Cycles during Environmental Stress Response in Plants. Aryadeep Roychoudhury (Ed.)

photochemistry, this chapter intends to visualize the mechanism of the LHCs and portray their response under environmental stress.

4.1 INTRODUCTION

Photosynthetic organisms are directly dependent on sunlight to harvest energy. In vascular plants and green algae, the energy from solar radiation is received and transferred to the reaction center by a series of protein–pigment complexes called light-harvesting complexes (LHCs) (Roach et al., 2014). The LHCs basically contain certain membrane-associated aggregates of proteins and photosensitive pigments (around 250–400 pigment molecules per complex) embedded in the chloroplasts of photosynthetic organisms. They facilitate absorption and channeling of light energy for photosynthetic reaction centers (Antenna Complex, Nature Portfolio). Mostly chlorophyll b (Chl *b*), xanthophylls, and carotenes are present as the photosensitive pigments present in LHC.

4.2 ONE SIP OF THE FUNDAMENTALS

The LHCs absorb solar energy in low-light conditions to provide necessary excitation energy for photosynthetic processes. However, under high-intensity irradiation, they switch to photoprotection mode by dissipating the excess energy as thermal radiation through nonphotochemical quenching (NPQ). The fundamental principles of photochemistry inherently control the underlying mechanism of these processes.

The excitation of the photosensitive molecules occurs by absorbing light radiations (photons) of suitable wavelength in the visible and near ultra-violet region of the electromagnetic spectrum. The absorbed light radiations should be of sufficient energy so that it can promote atoms or molecules from ground electronic state to an excited electronic state, which in turn involves the promotion of a ground electron from bonding molecular orbital to an excited antibonding molecular orbital.

4.2.1 *THE SPIN STATE*

Most molecules have paired spins in ground state. The quantity 2S+1 is called spin multiplicity, where S denotes the total electron spin. When the spins are paired in their ground state, the total spin S becomes 0 as upward spin of the electron cancels the downward spin of the electron. Hence 2S+1=1, and this state of the molecule is termed as singlet ground state. When the light of suitable wavelength has been absorbed by the molecule, the paired electrons in ground state goes into the excited state and the spin orientations of the two single electrons in ground and excited levels respectively may be either parallel or antiparallel. If the spins are parallel then, S=1 and 2S+1=3, the spin multiplicity of the molecule is 3 and the state is called triplet excited state. However, if the spins are antiparallel, S=0 and 2S+1=1, the spin multiplicity is 1 and the state is called singlet excited state.

4.2.2 CONSEQUENCES OF LIGHT ABSORPTION

The light absorbed by a molecule or a complex may undergo different processes as well (Fig. 4.1), for instance, it may show only a decrease in the intensity of absorbed light as governed by the Lambert–Beer law. However, it is also possible that the light absorbed may be re-emitted instantaneously in one or more steps and the phenomenon is termed as *fluorescence*. The chlorophylls (Chl*) are nice instances of fluorescent pigments. However, absorbed light may re-emit at a very slow rate; even after the removal of source light, it continues to emit light. The phenomenon is called as *phosphorescence*. The possible consequences of absorption of light can be visualized through the Jablonski diagram (Fig. 4.1).

It has found that a singlet excited state has higher energy than corresponding triplet excited state. Depending on the absorbed photon energy, the molecule in singlet ground state (S_0) may jump to any of its singlet excited state (S_1, S_2, S_3 ...) as shown in the Jablonski diagram. The molecule either in singlet excited state (S_1, S_2, S_3...) or triplet excited state (T_1 T_2, T_3 ...) is said to be activated and the activated molecule returns to the ground state by releasing energy through different processes as shown in the diagram.

In some transitions, the excited molecule from any higher energy level such as S_3, S_2, or T_3, T_2 return to the first excited state say S_1 or T_1. These transitions do not comprise any kind of emission of radiation and are thus termed as *nonradiative* transitions and the process involved is called *internal conversion (IC)*. The molecule may also dissipate energy in the form of heat by another type of nonradiative process called *intersystem crossing*. This involves the returning of molecule from one spin state to other spin state such as S_2 to T_2 or S_1 to T_1.

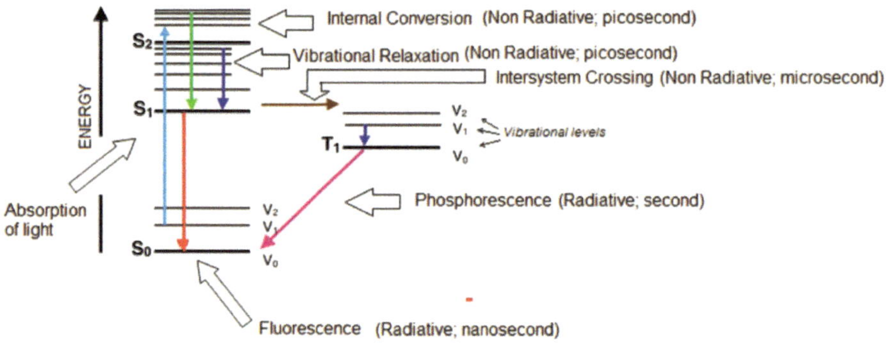

FIGURE 4.1 The Jablonski diagram.

However, the molecules can also return to their original ground state by dissipating energy as light of different wavelength. Transition from S_1 to S_0 is quite fast and is spectroscopically allowed, the transition termed as *fluorescence*, whereas transition from T_1 to S_0 is rather slow and forbidden and the emission of radiation in this transition is called *phosphorescence*.

4.3 ROLE OF ANTENNA COMPLEX IN PHOTOSYNTHESIS

Photosynthesis is seemingly the most efficient light-induced process allowing plants, algae, cyanobacteria, and anoxygenic photosynthetic bacteria to convert light energy into a chemical energy (Horton, 2004; Blankenship, 2014; Johnson, 2016). Photosynthesis may be divided into the "light" and "dark" reactions. In case of light reactions, water is splitted into oxygen, protons and electrons, using light, whereas in the dark reaction, the protons and electrons are used to reduce CO_2 to carbohydrate. The four key protein complexes embedded in the thylakoid membranes are photosystem II (PSII), photosystem I (PSI), cytochrome b_6f (Cyt b_6f), and ATP-synthase. Shoji and coworkers using quantum mechanics /molecular mechanics calculations provided an insight of the involvement of hydrogen bonding networks and proton wires during water oxidation by PSII (Shoji et al., 2015). The photon-energized electron transfer reactions of photosynthesis carried out by both PSII and PSI are surrounded by antenna system (AS) consisting of Chl*–protein complex (Eaton et al., 2012). In oxygenic organisms, the photosynthetic apparatus is present in the thylakoid membranes of chloroplasts, containing both PS I and PS II, operating in series. The AS surrounds each photosystem, consisting of a reaction-center (RC) complex. The function of both photosystem processes is that the PS II pulls out electrons and protons from water and passes the electrons to PS I. The photosynthetic apparatus of the nonoxygenic photosynthetic organism is very simple in the sense that they consist of only one RC (Hunter et al., 2009, Drews and Golecki, 1995). Table 4.1 enlists important molecular components of PSI and PSII.

TABLE 4.1 Important Molecular Components of PSI and PSII.

Component	Role	Component	Role
Plastocyanin	It acts as electron carrier to transfer electrons from cytochrome b_6f to the P700 cofactor of PSI	**Plastoquinone**	It acts as a mobile intra-thylakoid membrane electron carrier.
Ferredoxin	Ferredoxin is required for ferredoxin–$NADP^+$ reductase (FNR) enzyme to reduce $NADP^+$ to NADPH.	**Cytochrome b_6f**	It oxidizes plastoquinol to plastoquinone and also reduces plastocyanin, present in the lumen.
PsaA and PsaB	They are transmembrane proteins. They are involved in the binding of the cofactors P_{700}, Acc, A_0, A_1, and F_x	**Beta-Carotene**	It is used for quenching excess photoexcitation energy.
Phylloquinone	It is an electron acceptor; oxidizes A_1 and reduces F_x to pass the electron to F_b and F_a	**Beta-Carotene**	It is used for quenching excess photoexcitation energy.

As mentioned previously, a photosystem consists of numerous pigment–protein antenna complexes to increase the overall energy capture turnover. The initial component of photosynthetic antennae is pigment molecules that absorb light and transfer energy to the photosynthetic reaction centers, where the photochemical electron-transfer reactions occur. In the reaction center, Chl* or bacteriochlorophyll, many pigment molecules are there that collect the light. Naturally, three diverse classes of pigments are found, namely (bacterio)chlorophylls, phycobilins, and carotenoids, which fulfill the criteria for the antenna that are attached to wide collections of proteins (Sergej et al., 2004). Higher plants of large protein complexes carry chlorophyll a (Chl a), Chl b, and xanthophyll pigments, while some marine organisms like diatoms and brown algae have xanthophylls, Chl a, and chlorophyll c embedded in single light-harvesting antenna complex (LC or LHC). The common functions of antenna molecules are to absorb light energy, to control dissipation of excess excitation energy under light stress, and finally to balance flow of energy between PSI and PSII (Horton, 2004; Baghbanzadeh and Kassal, 2016; Yandu et al., 2021).

4.4 LIGHT HARVESTING AND PHOTOPROTECTION

4.4.1 MECHANISM OF LIGHT HARVESTING

The light-harvesting antenna is regarded as one of the major components of photosynthetic machinery. By means of the AS, the photosynthetic organisms harvest the photon quanta efficiently and channel the energy to the reaction centers, thereby enabling the captured light energy to convert into chemical energy with high quantum yield. The energy produced in this process is stored in the form of carbohydrates and other organic compounds. The PSII primary donor, known as P680 (P stands for "pigment") implicate to either of the two special Chl* excitonic dimer (also termed as "special pairs"). Though there are two entities in this "special pair," they function as a single unit and respond to excitation as a single molecule. The antenna molecules accumulate the excitation energy and transfer it to a "special pair" of redox-active Chl* molecules stationed in the RC. Upon excitation (Chl*) they can return to ground state (S_0) by releasing electrons to another one, thereby forming both oxidized "special pair" (Chl^+) and reduced acceptor (A^-). But, owing to the small spatial and spectral cross section, the special pair of Chl* can absorb only a limited amount of light. That is why large array of LHC complexes are there to increase the spatial and spectral cross section, to increase the absorption of energy by about two orders of magnitude (Johnson, 2016).

In photosynthesis, carotenoids serve as additional antenna pigments by capturing light energy in the blue–green region (nearly between 400 and 550 nm). The mechanistic pathway involved in light harvesting is briefly shown in the following equations,

$$\textit{Excitation}\ \text{Carotenoid} + h\nu \rightarrow {}^{1}\text{Carotenoid}^{*} \tag{4.1}$$

$$\textit{Singlet-singlet energy transfer:}\ {}^{1}\text{Carotenoid}^{*} \rightarrow {}^{1}\text{Car}^{*} \tag{4.2}$$

$$\textit{Finally,}\ {}^{1}\text{Car}^{*} \rightarrow {}^{1}\text{P680}^{*} \rightarrow \text{P680}^{+} + \text{Pheo}^{-} \tag{4.3}$$

By absorbing photon, the carotenoid molecule gets excited from a vibrational level in the electronic ground state (S_0) to different vibrational levels in the excited singlet state ($^1Car^*$) (here all electrons in the molecule are in spin-paired condition) (eq 4.1). Interpretation of carotenoid photophysics has been shown in terms of two low-lying excited singlet states, called S_2 and S_1 (Frank and Cogdell, 1996). In view of the selection rules, one photon is allowed to undergo transition from S_1 to S_2, which is internally converted again to S_1 by dissipating the excessive energy as heat (Kandori et al, 1994), and finally decay from S_1 to S_0 occurs. Carotenoids have the ability to donate energy from lowest lying both singlet excited state S_1 and S_2 to nearby Chl and then excite it to the singlet excited state ($^1Chl^*$), the so called singlet–singlet energy transfer (eq 4.2). The efficiency of energy transfer varies from organism to organism (Ritz et al., 2000; Macpherson et al., 2001). Carotenoids orient nearby Chl in unique configurations for capturing optimal singlet energy and migration of these efficient excitation energy with minimal energy losses (van Amerongen et al., 2001; Britton et al, 2008. The Chl later on transfers this excitation energy to the RC, which is utilized to induce a charge separation of $P680^+$cation and pheophytin anion ($Pheo^-$) pair (eq 4.3).

In case of oxygenic photosynthetic organisms, carotenoids can act as quencher. There the Bacteriochlorophyll ((B)Chl) acting as the sensitizers absorbs photon, goes to singlet excited state, and later on gets converted to triplet state. The long-lived triplet state of sensitizer molecules forms singlet oxygen from the ground triplet-state oxygen ($^3O_2$). Carotenoids deactivate (B)Chl triplet excited states (^3Chl) through a triplet−triplet energy-transfer reaction (Britton, 2008). Oxygen in the ground state (a triplet state) can have the ability to accept excitation energy efficiently from other molecules, which are in the excited triplet state. If the ground state of oxygen is in singlet state, it can oxidize pigments, proteins, and lipids and thus considered as lethal to living tissues (Foote et al., 1970).

4.4.2 PHOTOPROTECTION AND PHOTOINHIBITION

Under the influence of excessive light, carotenoids quench singlet excited state Chl (Horton et al., 1996). The presence of excessive light on a photosynthetic organism may be extremely harmful. Actually, excessive light absorption required for maximal CO_2 fixation will lead to the formation of a long lifetime triplet excited state Chl ($^3Chl^*$) (eq 4.4), which may react with molecular oxygen, thereby producing excited singlet oxygen ($^{-1}O_2^*$) (eq 4.8), formed in PSII RC (Knox and Dodge, 1985). Oxidation of lipids, proteins, and pigments may be caused due to photooxidative damage

(Formaggio et al., 2001), leading to photoinhibition of the photosynthetic machinery and also to photobleaching. At this stage, carotenoids can respond in two possible ways: one is either by quenching the triplet excited state of Chl (eq 4.5) (Schödel et al., 1999) or by scavenging the singlet excited state oxygen directly, thereby inducing interaction of triplet excited state of Chl with that of molecular oxygen (eq 4.6) (Farmilo and Wilkinson,1973). The photoprotection property of the carotenoid depends on the capacity of the triplet state (^{3}Carotenoid*), whose energy is quite low to effectively quench triplet state Chl and singlet state oxygen (Polívka and Sundström, 2004). In PSII RC, β-carotenes do not quench the triplet excited state of P680 (^{3}P680*) by triplet–triplet transfer as they are not in close contact with P680, instead scavenge the singlet excited state oxygen formed by the interaction of the triplet excited state of P680 with molecular oxygen. In both cases, triplet excited carotenoids are relaxed to their ground state by dissipation of heat energy (eq 4.7). Both the molecules must be positioned in close contact with each other for a well organized triplet–triplet transfer between Chl and carotenoid (Cogdell and Frank, 1987), which is maintained by the proteins, via coordinating pigments.

The mechanisms involved in carotenoid mediated photoprotection induced the antenna complex system is summarized below:

$$\text{Excitation Chl} + h\nu \rightarrow {}^1\text{Chl}^* \rightarrow {}^3\text{Chl}^* \qquad (4.4)$$

Triplet–triplet energy transfer: $^3\text{Chl}^* + \text{Carotenoid} \rightarrow \text{Chl} + {}^3\text{Carotenoid}^*$ (4.5)

Or

$$^1\text{O}_2^* + \text{Carotenoid} \rightarrow {}^3\text{Carotenoid}^* \qquad (4.6)$$

Thus either

Photoprotection $^3\text{Carotenoid}^* \rightarrow \text{Carotenoid} + \text{heat}$ (4.7)

Else, **Photoinhibition** $^3\text{Chl}^* + \text{O}_2 \rightarrow {}^1\text{O}_2^* + \text{Chl}$ (4.8)

4.5 RESPONSES OF ANTENNA COMPLEX UNDER STRESS

4.5.1 GENERAL STRESS SURVIVAL PROTOCOL OF THE ANTENNA COMPLEX

LHC provides principal solar energy and funnel that energy to the RC of the photosystems. As mentioned earlier, the LHC is situated in the thylakoid membrane of the chloroplast, thereby representing a trimeric structure with

Chl a, Chl b, along with carotenoid molecules. The chromophores present in the LHC complex are responsible for sunlight absorption (Sen et al., 2021; Suga and Shen, 2020). Environmental stresses such as irregular light and temperature, drought, salinity, and metal stress have adverse effects on physiological and biochemical processes in LHC. Photosynthetic organisms exhibit tolerance mechanisms under abiotic stress through acclimation mechanisms in LHC system (Sen, 2020; Ashraf and Harris, 2013). In various prototrophs, LHC system formed by the large multisubunit Chl*-carotenoid–protein complexes trap light energy and brings it to the photochemical reaction centers (Nelson and Junge, 2015; Qin et al., 2015; Nelson, 2013). The large ASs of all these organisms consist of basic units of polypeptides, which combine pigment molecules in a specific and organized manner. LHC complex apparatus develops cascade of biochemical and physiological responses that increase their possibility of surviving the stress (Xiong et al., 2018). Responses occurring in the LHC system under various environmental stress conditions are schematically presented in Figure 4.2. Table 4.2 represents some of the important responses of photosynthetic apparatus under various stress conditions.

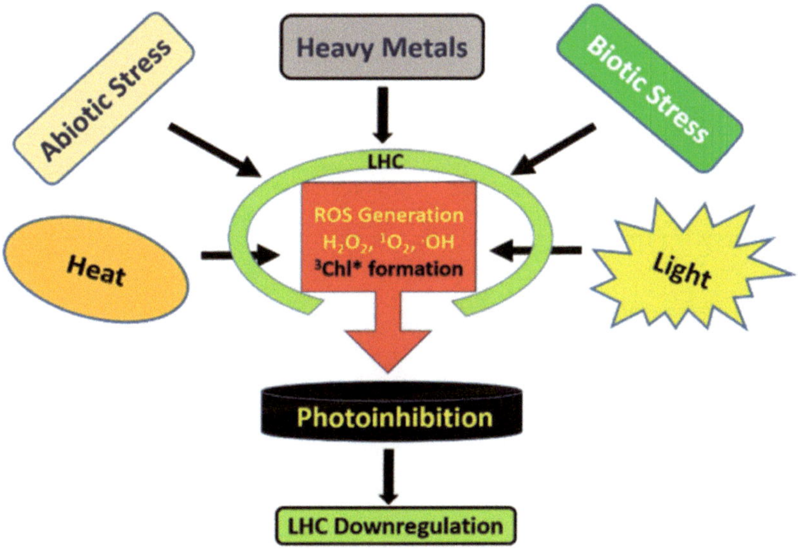

FIGURE 4.2 Schematic presentations of sequential events occurring in the LHC antenna system under environmental stress conditions.

TABLE 4.2 Some of the Responses of Photosynthetic Apparatus under Various Stress Conditions.

Stress Source	Stress type	Selected study organism	Response	References
Light	High Light	Green plants	Light-harvesting system is down-regulated to limit light energy transfer to the reaction center	Sen et al. (2021)
Water	Deficiency	Potato plants	Reduction of light absorption and photosystem activity	Li et al. (2017)
Fe	Deficiency	Cyanobacteria	Accumulation of IsiA protein	Lokstein et al. (2021)
Cu	Excess	*Rubrivivax gelatinosus*	Inactivation of bacteriochlorophyll a (B800)	Tambosi et al., 2018
Cu	Excess	*Ulva compressa*	Increased transcription of genes belonging to PSI, PSII etc. and proteins required to repair Photosystem II and to protect of Photosystem I.	Laporte et al. (2020)
Cd	Excess	*S. sclarea*	Increased accumulation of carotenoids.	Dobrikova et al. (2021a)
Ag	Excess	*Rubrivivax gelatinosus*	Inactivation of bacteriochlorophyll a (B800)	Tambosi et al. (2018)
Zn	Excess	*Salvia sclarea* L.	Stimulation of the PSI and PSII	Dobrikova et al. (2021b)

4.5.2 LIGHT-INDUCED STRESS

Though all photosynthetic prototroph organisms utilize light energy during photosynthesis, yet paradoxically, excess of light also causes damage to the photosynthetic apparatus (Sen et al., 2021; Guangxi, et al., 2020). Prototroph organisms have developed several defensive mechanisms against excess light. In green plants, the light-harvesting antennae proteins, which bind Chl a, Chl b, and carotenoids together form an extended LHC superfamily. The primary role of these LHC proteins is to assemble light energy but they are also known to have another crucial responsibility to protect the photosynthetic apparatus (Pascal et al., 2016; Guangxi, et al., 2020; Lishuan et al.,

2020). Under high light stress and when the excitation rate exceeds, light absorption may be detrimental to the photosynthetic apparatus. To reimburse the light stress, light-harvesting activity of the LHC system is downregulated and limit the amount of light energy transmitted to the reaction center (Yuval et al., 2017; Xiaodong, et al., 2019).

In addition to light absorption, LHC protein complex actively regulates the photosystem quantum efficiency and also catalyze the photoprotective reactions. Antenna complex with associated photosystem is susceptible to high light illumination resulting in photoinhibition. This situation causes photoinhibition, which induces the production of reactive oxygen species (ROS), which in turn inactivate the photochemical reaction center (Büchel, 2015; Chakib, et al., 2020; Sen et al., 2021). LHC antenna complex systems regulate photochemical effectiveness of photosystems and give enhanced level of photoprotection. Fluctuations of light intensity also cause changes of excitation on photosystem apparatus and results into photochemical quenching of Chl* molecules and increased ROS generation. Photoprotective LHC subunits play an important role in this regard by activation of ROS scavenging or by limiting their release. Antenna complex proteins take an important part in downregulation of Chl* through photoprotective mechanism of Excess Energy Dissipation (EED) which safely dissipates excitation energy in excess. LHC system helps in the reduction of excitation energy concentrate from the antenna into the photochemical reaction centers. The overall process decreases the accumulation of ROS produced during secondary reactions (Xiaodong, et al., 2019; Lokstein et al., 2021). In some mutant cases, it was found that the depletion of LHC proteins results into the depletion of EED machinery and increase in photosensitivity reactions. Xanthophylls bound to the LHC proteins protect the antenna complex against singlet oxygen (^{1}O2) formation, by quenching Chl* or directly scavenging singlet $^1O_2$ (Nicol et al., 2019; Ruban, 2018; Zhao et al., 2017; Moejes et al., 2017). High light intensity probably causes lowering of photosynthetic efficiency followed by the death of the photosynthetic cell, tissue, or organism. Further, NPQ also plays an important protective role. NPQ process causes dissipation of excess light energy in the form of heat. NPQ occurs in the photosynthetic membranes of photosynthetic organisms. Antenna complex proteins regulate the mechanisms involved in NPQ of Chl* fluorescence that can dissipate excessive light energy absorbed by Chl* as heat. Thus, NPQ process gets itself involved directly or indirectly into the LHC system physiology (Cao et al., 2020; Pinnola, 2019; Nicol et al., 2019; Ruban, 2018; Murata et al., 2007).

4.5.3 DROUGHT STRESS

Climatic inconsistency is responsible for more recurrent extreme heat and drought stress events. Heat and water stress have a distressing effect on photosystem. At the molecular level, organisms express or repress genes when they experience heat and drought stresses, commence a cascade of biochemical and physiological responses that increase their prospect of surviving the stress. Heat stress or drought provoke alteration in photosynthesis and other metabolic pathways and results into the imbalance of photosynthesis, along with exhibiting devastating impact on Chl* fluorescence and carbon assimilation processes. These occur due to the excessive production of ROS in the stressed plants. Drought stress-induced alterations in the expression pattern of light-harvesting antenna protein complex and reaction center proteins and result in reduced light absorption, which bring down the utilization of the absorbed energy thereby decrementing the photosystem activity (Jahan and Hasan, 2018; Dalal and Tripathy, 2018; Zhao et al., 2010; Tambussi et al., 2000).

Under water stress conditions, photosynthesis promotes ROS or ROS generation due to excess absorption of light, which ill-effects the photosynthetic apparatus and the symmetry of LHC system. Drought stress registers adverse effects on the thylakoid membrane stability, which in turn lowers the LHC antenna proteins activity. In higher photosynthetic organisms, both Chl a and Chl* bare bound to LHC complex proteins, and the availability of both Chl* is essential for the assembly and functioning of the LHC proteins. Generation of ROS in chloroplast due to water stress downregulate the Chl* biosynthesis process, lower the Chl a fluorescence and photosystem reaction center activity (Busch and Montgomery, 2015; Dalal and Tripathy, 2012; Tambussi et al., 2000).

Thus, in the situation of water deficiency, there is an overall reduction in the amount of Chl* biosynthesis, assembly of LHC proteins, and reaction center activity, which ultimately result in reduced light absorption, and reduced quantity of oxygen-evolving complexes leading to a decline in the use of the absorbed energy and eventually a steep drop in the activity of the photosystem (Dalal and Tripathy, 2018; Busch and Montgomery, 2015).

4.5.4 HEAVY METAL STRESS

The environmental contamination with different heavy metals is a worldwide concern that has resulted from anthropogenic activities. Heavy metal or metalloids are naturally occurring components found in the earth crust, soil, sediments, and water bodies, and are major persistent environmental pollutants. Indeed, some of the heavy metals such as copper (Cu), zinc (Zn), and iron (Fe) are essential in trace elements for various metabolic activities. However, excess of any kind of metal adversely affects the metabolic system. In photosynthetic apparatus, metal stress exerts their toxic effects by damaging the photosystem while generating ROS along with impairing the antenna complex pigment–protein activity. Metal ion interference causes reduction of photosynthetic enzymes and chloroplast membrane functioning, ultimately inhibiting the photosynthesis activity (Sen et al., 2021; Jahan and Hasan, 2018; Chandra and Kang, 2016; Gururani et al., 2015).

4.5.4.1 COPPER STRESS

Cu is one of the important micronutrients for plant organisms and plays a crucial role in regulating the photosynthetic electron transport process (Maksymicc et al., 1997). Obviously, excess of Cu in the chloroplast could be damaging for the photosynthetic reaction chain. Using FT-IR spectroscopic studies, Janik et al. (2010) observed LHCII aggregation in plants under exposure of excess Cu. In another work, Tambosi et al. (2018) observed that Cu exposure caused inactivation of the 800 nm absorbing bacteriochlorophyll a (B800) in *Rubrivivax gelatinosus*.

The marine alga *Ulva compressa* has a clever solution to excess copper situation. This alga is mainly found in the copper-polluted coastal areas in northern Chile. The high tolerance of the species arises from the elevating photosynthesis and respiration. An increase in the levels of transcripts for LHC II, PSII, Cyt b_6f, PSI, LHCI, and ATP synthase was observed under excess copper exposure. Simultaneously, an increase in the transcription of proteins involved in the repair of PSII and protection of PSI was observed, to sustain the elevated rate of photosynthesis (Laporte et al., 2020).

4.5.4.2 IRON STRESS

Iron is an important trace element. It is required for many vital life processes, including respiration and photosynthesis. Due to the key involvement of iron in the photosynthetic apparatus in thylakoid membranes, iron deficiency imposes a marked effect on the reaction center activity between two PSI and PSII), causing oxidative damage. The decline in photosystem activity under iron deficiency adversely impacts the electron transfer from PSII to PSI. LHC provides an adaptive mechanism in photosynthetic apparatus to protect against the iron deficiency. Iron stress reduces the efficiency of excitation energy transfer between LHC AS and reaction center. The structural remodeling changes of the antenna complex protein system diminish the oxidative stress to the thylakoid membrane under iron-deficient conditions through thermal dissipation of excess energy absorbed (Sen et al., 2021; Chen et al., 2005; Bibby et al., 2001; Green, 2003). Cyanobacteria addresses iron deficiency by accumulating an additional Chl a- and ß-carotene-binding protein called iron stress-induced protein A, IsiA, which is structurally related to the PSII core antenna complex CP43 (Lokstein et al., 2021)

4.5.4.3 CADMIUM STRESS

Cd is one of the most toxic contaminants that may cause harmful effects on photosynthetic apparatus. The most common toxicity of Cd is to cause oxidative stress through the assembly of ROS. In photosynthetic organisms, Cd stress is responded by modification of the functionality of photosynthetic membranes by affecting the Chl* biosynthesis, LHC, and the activity of both the photosystems. Cd stress-induced inhibition of photosystems is a result of Cd interference with electron transport on the PSI as well as a significant reduction of antenna pigment–protein complex. The light-harvesting antennae protein with associated Chl* and carotenoid system together form an extended signaling mechanism system under Cd stress and provides tolerance mechanisms to cope up with Cd toxicity.LHC subunits play an important role here by activation of ROS scavenging or by limiting their release (Dobrikova and Apostolova, 2019; Hasan et al., 2011). In a recent work, Dobrikova et al. (2021a) observed that *Salvia sclarea* plants take up a number of protective strategies like increased uptake of Fe (to protect photosynthetic electron transport system), increased accumulation of carotenoids (to counter the ROS produced due to Cd stress), etc. to enhance Cd tolerance and to protect the photosynthetic apparatus.

4.5.4.4 SILVER STRESS

Ag^{+} ions are extensively used in industry, agriculture, cosmetics, and health care. Although possessing remarkable benefits and applications, the toxic effect of Ag ions in biological system is still to be explored completely. Tambosi et al. (2018) demonstrated that Ag exposure severely affected the LHC in *Rubrivivax gelatinosus*. They further identified that the Agions specifically inactivate the 800-nm absorbing Chl a (B800) through severing of the interaction of B800 pigments with light-harvesting protein.

4.5.4.5 ZINC STRESS

Zinc is a cofactor of many enzymes and therefore is a required micronutrient for plant. The element has involvement in various essential biochemical processes of a plant including photosynthesis. Exposure to high concentrations of Zn ions causes phytotoxicity in plants. Interestingly many plants exhibit tolerance to zinc by inducing various protective mechanisms. *S. sclarea*, a medicinal plant known to be a good zinc accumulator can tolerate quite large amount zinc in the leaves. Dobrikova et al. (2021b) observed that during Zn surge, *S. sclarea* exhibits two major responses, (1) stimulation of the PSI and PSII activities by increased accumulation of Fe, Ca, and Mn ions, and (2) increase in total phenolics and anthocyanins for Zn detoxification and to defend against oxidative stress.

4.6 CONCLUSION

LHCs systems regulate light-controlled energy accumulation during photosynthesis process. Antenna complex is responsible for capturing light energy and delivering it to the photochemical reaction centers. Although collection of light energy is the primary role of the LHC, they also need to shoulder another crucial role of protecting the photosynthetic apparatus. LHC protein complex actively regulates the photosystem quantum efficiency and also catalyzes the photoprotective reactions. Antenna complex with associated photosystem is susceptible to high light illumination, resulting in photoinhibition. Environmental stress such as salinity, drought, ultraviolet, and excess light and temperature forces a severe reduction in photosynthetic activity and an increased generation of ROS causing oxidative damage. This chapter highlights the mechanism of downregulation of LHC complex protein system under different stress conditions.

DISCLAIMER

The authors declare no conflicts of interest.

KEYWORDS

- **antenna complex**
- **photosynthesis**
- **photochemistry**
- **stress-response**

REFERENCES

Albanese, P.; Manfredi, M.; Meneghesso, A.; Marengo, E.; Saracco, G.; Barber, J.; et al. Dynamic Reorganization of Photosystem II Supercomplexes in Response to Variations in Light Intensities. *Biochim. Biophys. Acta—Bioenerg.* **2016,** *1857*, 1651–1660.

Antenna complex, Nature Portfolio. https://www.nature.com/subjects/antenna-complex (accessed July 04, 2021).

Ashraf, M.; Harris, P. J. C. Photosynthesis under Stressful Environments: An Overview. *Photosynthetica* **2013,** *51*, 163–190.

Baghbanzadeh, S.; Kassal, I. Distinguishing the Roles of Energy Funnelling and Delocalization in Photosynthetic Light Harvesting. *Phys. Chem. Chem. Phys.* **2016,** *18*, 7459–7467.

Bibby, T. S.; Nield, J.; Barber, J. Iron Deficiency Induces the Formation of an Antenna Ring around Trimeric Photosystem I in Cyanobacteria. *Nature* **2001,** *412*, 743–745.

Blankenship, R. E. *Molecular Mechanisms of Photosynthesis*; Wiley- Blackwell: Hoboken, NJ, 2014.

Britton, G. Functions of Intact Carotenoids. In *Carotenoids, Vol. 4: Natural Functions*; Britton, G., Liaaen-Jensen, S., Pfander, H., Eds.; BirkhäuserVerlag AG: Basel, Switzerland, 2008.

Büchel, C. Evolution and Function of Light Harvesting Proteins. *J Plant Physiol.* **2015,** *172*, 62–75.

Busch, A. W. U.; Montgomery, B. L. Interdependence of Tetrapyrrole Metabolism, the Generation of Oxidative Stress and the Mitigative Oxidative Stress Response. *Redox Biol.* **2015,** *4*, 260–271.

Cao, P.; Pan, X.; Su, X.; Liu, Z.; Li, M. Assembly of Eukaryotic Photosystem II with Diverse Light-Harvesting Antennas. *Curr. Opin. Struct. Biol.* **2020,** *63*, 49–57.

Chandra, R.; Kang, H. Mixed Heavy Metal Stress on Photosynthesis, Transpiration Rate, and Chlorophyll Content in Poplar Hybrids. *Forest Sci. Technol.* **2016,** *12*, 55–61.

Chen, M.; Bibby, T. S.; Nield, J.; Larkum, A.; Barber, J. Iron Deficiency Induces a Chlorophyll d-Binding PCB Antenna System around Photosystem I in *Acaryochloris marina. Biochim. Biophys. Acta—Bioenerg.* **2005,** *1708*, 367–374.

Cogdell, R. J.; Frank, H. A. How Carotenoids Function in Photosynthetic Bacteria. *Biochim. Biophys. Acta—Rev. Bioenerg.* **1987,** *895*, 63–79.
Dalal, V. K.; Tripathy, B. C. Modulation of Chlorophyll Biosynthesis by Water Stress in Rice Seedlings during Chloroplast Biogenesis. *Plant Cell Environ.* **2012,** *35*, 1685–1703.
Dalal, V. K.; Tripathy, B. C. Water-Stress Induced Downsizing of Light-Harvesting Antenna Complex Protects Developing Rice Seedlings from Photo-Oxidative Damage. *Sci. Rep.* **2018,** *8*, 5955.
Djediat, C.; Feilke, K.; Brochard, A.; Caramelle, L.; Kim Tiam, S.; Sétif, P. et al. Light Stress in Green and Red Planktothrix Strains: The Orange Carotenoid Protein and Its Related Photoprotective Mechanism. *Biochim. Biophys. Acta—Bioenerg.* **2020,** *1861*, 148037.
Dobrikova, A. G.; Apostolova, E. L. Damage and Protection of the Photosynthetic Apparatus under Cadmium Stress. In *Cadmium Toxicity and Tolerance in Plants*; Elsevier, 2019.
Dobrikova, A. G.; Apostolova, E. L.; Hanć, A.; Yotsova, E.; Borisova, P.; Sperdouli, I.; Adamakis, I.-D. S.; Moustakas, M. Dadmium Toxicity in *Salvia sclarea* L.: An Integrative Response of Element Uptake, Oxidative Stress Markers, Leaf Structure and Photosynthesis. *Ecotoxicol. Environ.Saf.* **2021,** *209*, 111851.
Drews, G.; Golecki, J. R. Structure, Molecular Organization, and Biosynthesis of Membranes by Purple Bacteria. In *Anoxygenic Photosynthetic Bacteria: Advances in Photosynthesis and Respiration*; Blankenship, R. E., Madigan, M. T., Bauer, C. E., Eds.; Kluwer Academic Publishers: Dordrecht, The Netherlands, 1995.
Eaton-Rye, J. J.; Tripathy, B. C.; Sharkey, T. D., Eds. *Photosynthesis: Plastid Biology, Energy Conversion and Carbon Assimilation; Advances in Photosynthesis and Respiration*; Springer: Dordrecht, The Netherlands, 2012.
Farmilo, A.; Wilkinson, F. On The Mechanism of Quenching of Singlet Oxygen in Solution. *Photochem. Photobiol.* **1973,** *18*, 447–450.
Foote, C. S.; Chang, Y. C.; Denny, R. W. Chemistry of Singlet Oxygen. X. Carotenoid Quenching Parallels Biological Protection. *J. Am. Chem. Soc.* **1970,** *92*, 5216–5218.
Formaggio, E.; Cinque, G.; Bassi, R. Functional Architecture of the Major Light-Harvesting Complex from Higher Plants. *J. Mol. Biol.* **2001,** *314*, 1157–1166.
Frank, H. A.; Cogdell, R. J. Carotenoids in Photosynthesis. *Photochem. Photobiol.* **1996,** *63*, 257–264.
Green, B. R. *The Evolution of Light-harvesting Antennas*, 2003; pp 129–168.
Gururani, M. A.; Venkatesh, J.; Tran, L. S. P. Regulation of Photosynthesis during Abiotic Stress-Induced Photoinhibition. *Mol. Plant.* **2015,** *8*, 1304–1320.
Hasan, S. A.; Hayat, S.; Ahmad, A. Brassinosteroids Protect Photosynthetic Machinery against the Cadmium Induced Oxidative Stress in Two Tomato Cultivars. *Chemosphere* **2011,** *84*, 1446–1451.
Horton, P. Molecular Design of the Photosystem II Light-Harvesting Antenna: Photosynthesis and Photoprotection. *J. Exp. Bot.* **2004,** *56*, 365–373.
Horton, P.; Ruban, A. V.; Walters, R. G.; Regulation of Light Harvesting in Green Plants. *Annu.Rev. Plant Physiol. Plant Mol. Biol.* **1996,** *47* (1), 655–684.
Hunter, C. N.; Daldal, F.; Thurnauer, M. C.; Beatty, J. T., Eds. *The Purple Phototrophic Bacteria; Advances in Photosynthesis and Respiration*; Springer: Dordrecht, The Netherlands, 2009.
Jahan, M. S.; Hasan, M. M. Light-Harvesting Complexes Communicate Growth and Physiology of Plants. *Indian J. Plant Physiol.* **2018,** *23*, 1–6.
Johnson, M. P. Photosynthesis. *Essays Biochem.* **2016,** *60*, 255–273.

Kandori, H.; Sasabe, H.; Mimuro, M. Direct Determination of a Lifetime of the S2 State of Beta.-Carotene by Femtosecond Time-Resolved Fluorescence Spectroscopy. *J. Am. Chem. Soc.* **1994,** *116*, 2671–2672.

Knox, J. P.; Dodge, A. D. Singlet Oxygen and Plants. *Phytochemistry* **1985,** *24*, 889–896.

Laporte, D.; Rodríguez, F.; González, A. et al. Copper-Induced Concomitant Increases in Photosynthesis, Respiration, and C, N and S Assimilation Revealed by Transcriptomic Analyses in *Ulvacompressa* (Chlorophyta). *BMC Plant Biol.* **2020,** *20*, 25.

Li, J.; Cang, Z.; Jiao, F.; Bai, X.; Zhang, D.; Zhai, R.; Influence of Drought Stress on Photosynthetic Characteristics and Protective Enzymes of Potato at Seedling Stage. *J. Saudi Soc. Agric. Sci.* **2017,** *16* (1), 82–22.

Lokstein, H.; Renger, G.; Götze, J. Photosynthetic Light-Harvesting (Antenna) Complexes—Structures and Functions. *Molecules*. **2021,** *26*, 3378.

Lu, Y.; Gan, Q.; Iwai, M.; Alboresi, A.; Burlacot, A.; Dautermann, O. et al. Role of an Ancient Light-Harvesting Protein of PSI in Light Absorption and Photoprotection. *Nat. Commun.* **2021,** *12*, 679.

Macpherson, A. N.; Arellano, J. B.; Fraser, N. J.; Cogdell, R. J.; Gillbro, T. Efficient Energy Transfer from the Carotenoid S2 State in a Photosynthetic Light-Harvesting Complex. *Biophys J.* **2001,** *80*, 923–30.

Mazor, Y.; Borovikova, A.; Caspy, I.; Nelson, N. Structure of the Plant Photosystem I Supercomplex at 2.6 Å Resolution. *Nat Plants*. **2017,** *3*, 17014.

Moejes, F. W.; Matuszyńska, A.; Adhikari, K.; Bassi, R.; Cariti, F.; Cogne, G. et al. A Systems-Wide Understanding of Photosynthetic Acclimation in Algae and Higher Plants. *J. Exp. Bot.* **2017,** *68*, 2667–2681.

Murata, N.; Takahashi, S.; Nishiyama, Y.; Allakhverdiev, S. I. Photoinhibition of Photosystem II under Environmental Stress. *Biochim. Biophys. Acta—Bioenerg.* **2007,** *1767*, 414–421.

Nelson, N. Evolution of Photosystem I and the Control of Global Enthalpy in an Oxidizing World. *Photosynth. Res.* **2013,** *116*, 145–151.

Nelson, N.; Junge, W. Structure and Energy Transfer in Photosystems of Oxygenic Photosynthesis. *Annu. Rev. Biochem.* **2015,** *84*, 659–683.

Nicol, L.; Nawrocki, W. J.; Croce, R. Disentangling the Sites of Non-Photochemical Quenching in Vascular Plants. *Nat. Plants.* **2019,** *5*, 1177–1183.

Pi, X.; Tian, L.; Dai, H-E.; Qin, X.; Cheng, L.; Kuang, T. et al. Unique Organization of Photosystem I–Light-Harvesting Supercomplex Revealed by Cryo-EM from a Red Alga. *Proc. Natl. Acad. Sci.* **2018,** *115*, 4423–4428.

Pinnola, A. The Rise and Fall of Light-Harvesting Complex Stress-Related Proteins as Photoprotection Agents during Evolution. Griffiths, H., Ed. *J. Exp. Bot.* **2019,** *70*, 5527–5535.

Polívka, T.; Sundström, V.; Ultrafast Dynamics of Carotenoid Excited States−From Solution to Natural and Artificial Systems. *Chem. Rev.* **2004,** *104*, 2021–2072.

Qin, X.; Suga, M.; Kuang, T.; Shen, J-R. Structural Basis for Energy Transfer Pathways in the Plant PSI-LHCI Supercomplex. *Science*. **2015,** *348*, 989–995.

Ritz, T.; Damjanović, A.; Schulten, K.; Zhang, J.; Koyama, Y.; Efficient Light Harvesting through Carotenoids. *Photosynth.Res.* **2000,** *66*, 125–144.

Ruban, A. V. Light Harvesting Control in Plants. *FEBS Lett.* **2018,** *592*, 3030–3039.

Schödel, R.; Irrgang, K-D.; Voigt, J.; Renger, G. Quenching of Chlorophyll Fluorescence by Triplets in Solubilized Light-Harvesting Complex II (LHCII). *Biophys. J.* **1999,** *76*, 2238–2248.

Sen, S. Cyanobacterial Membrane Biology under Environmental Stresses with Particular Reference to Photosynthesis and Photomorphogenesis. In *Advances in Cyanobacterial Biology*; Elsevier, 2020.

Sen, S.; Mascoli, V.; Liguori, N.; Croce, R.; Visscher, L.; Understanding the Relation between Structural and Spectral Properties of Light-Harvesting Complex II. *J. Phys. Chem. A.* **2021,** *27* (125), 4313–4322.

Shoji, M.; Isobe, H.; Yamanaka, S.; Umena, Y.; Kawakami, K.; Kamiya, N.; Shen, J.-R.; Nakajima, T.; Yamaguchi, K.; Large-Scale QM/MM Calculations of Hydrogen Bonding Networks for Proton Transfer and Water Inlet Channels for Water Oxidation—Theoretical System Models of the Oxygen-Evolving Complex of Photosystem II. In *Advances in Quantum Chemistry*; Sabin, J. R., Brändas, E. J., Eds., Vol. 70; Academic Press, 2015; pp 325–413.

Su, X.; Ma, J.; Pan, X.; Zhao, X.; Chang, W.; Liu, Z. et al. Antenna Arrangement and Energy Transfer Pathways of a Green Algal Photosystem-I–LHCI Supercomplex. *Nat. Plants.* **2019,** *5*, 273–281.

Suga, M.; Shen, J-R. Structural Variations of Photosystem I-Antenna Supercomplex in Response to Adaptations to Different Light Environments. *Curr. Opin. Struct. Biol.* **2020,** *63*, 10–17.

Tambosi, R.; Liotenberg, S.; Bourbon, M. L.; Steunou, A. S.; Babot, M.; Durand, A.; Kebaili, N.; Ouchane, S. Silver and Copper Acute Effects on Membrane Proteins and Impact on Photosynthetic and Respiratory Complexes in Bacteria. *mBio.* **2018,** *9* (6), e01535-18.

Tambussi, E. A.; Bartoli, C. G.; Beltrano, J.; Guiamet, J. J.; Araus, J. L. Oxidative Damage to Thylakoid Proteins in Water-Stressed Leaves of Wheat (Triticumaestivum). *Physiol. Plant* **2000,** *108*, 398–404.

van Amerongen, H.; van Grondelle, R. Understanding the Energy Transfer Function of LHCII, the Major Light-Harvesting Complex of Green Plants. *J. Phys. Chem. B.* **2001,** *105*, 604–617.

Vasil'ev, S.; Bruce, D. Optimization and Evolution of Light Harvesting in Photosynthesis: The Role of Antenna Chlorophyll Conserved between Photosystem II and Photosystem I. *Plant Cell.* **2004,** *16*, 3059–3068.

Vinklárek, I. S.; Bornemann, T. L. V.; Lokstein, H.; Hofmann, E.; Alster, J.; Pšenčík, J. Temperature Dependence of Chlorophyll Triplet Quenching in Two Photosynthetic Light-Harvesting Complexes from Higher Plants and Dinoflagellates. *J. Phys. Chem. B.* **2018,** *122*, 8834–8845.

Wu, G.; Ma, L.; Sayre, R. T.; Lee, C-H. Identification of the Optimal Light Harvesting Antenna Size for High-Light Stress Mitigation in Plants. *Front. Plant Sci.* May 15, **2020a,** *11*.

Wu, L.; Zhang, L.; Tu, W.; Sun, R.; Li, F.; Lin, Y. et al. Photosynthetic Inner Antenna CP47 Plays Important Roles in Ephemeral Plants in Adapting to High Light Stress. *J. Plant Physiol.* **2020b,** *251*, 153189.

You, Z-Q.; Hsu, C-P.; AbInito Study on Triplet Excitation Energy Transfer in Photosynthetic Light-Harvesting Complexes. *J. Phys. Chem. A.* **2011,** *115*, 4092–4100.

Zhao, L.; Cheng, D.; Huang, X.; Chen, M.; Dall'Osto, L.; Xing, J. et al. A Light Harvesting Complex-Like Protein in Maintenance of Photosynthetic Components in Chlamydomonas. *Plant Physiol.* **2017,** *174*, 2419–2433.

Zhao, M.; Running, S. W. Drought-Induced Reduction in Global Terrestrial Net Primary Production from 2000 through 2009. *Science* **2010,** *329*, 940–943.

CHAPTER 5

Role of C3 Cycle Genes and Enzymes in Overcoming Environmental Stress

MONIKA BANSAL[1] and SHABIR H WANI[2*]

[1]*SIILAS campus, Jaipur national University, Jaipur 302025, Rajasthan, India*

[2]*Mountain Research Center for Field Crops, Khudwani, Sher-e-Kashmir University of Agricultural Sciences and Technology of Kashmir, Jammu and Kashmir, India*

[*]*Corresponding author. E-mail: shabirhwani@skuastkashmir.ac.in*

ABSTRACT

There are different environmental stresses like salinity, drought, flooding, and adverse temperatures all of which together with their abrupt changes, have a detrimental influence on the photosynthetic carbon fixation process in plants. By limiting photosynthesis, these stresses reduce agricultural productivity and cause yield losses. Photosynthesis is the product of a multistep and complex mechanism involving many biological pathways. Unfortunately, plants have only developed few stress resistance mechanisms to sustain photosynthesis efficiency so far. This chapter will offer a description of how abiotic stresses affect and regulate carbon fixation in C3 cycle during photosynthesis.

5.1 INTRODUCTION

Abiotic stress has a substantial impact on photosynthesis functioning by interfering with parts of photosynthetic machinery such as photosystem

Photosynthesis and Respiratory Cycles during Environmental Stress Response in Plants. Aryadeep Roychoudhury (Ed.)

I and II, electron transfer, carbon fixation, ATP production process, and stomatal conductance. Plant photosynthetic systems respond to stress in different ways, based on plant species, kind of stress, frequency and duration of occurrence, and a variety of other factors. These pathways are photosynthetic electron transport systems, where light energy is converted to ATP and NADPH. The Calvin Benson cycle is also called as carbon fixation cycle where CO_2 is converted into carbohydrates. Because the Calvin cycle is the first route of carbon fixation during photosynthesis, its regulation is critical for significantly increasing photosynthetic CO_2 fixation capability. The Calvin cycle occurring in chloroplasts is integral part of metabolism in algae and plants and is mediated by a sequence of enzymatic processes involving 11 enzymes in total.

Abiotic stressors reduce agricultural productivity and cause loss of yield by limiting photosynthesis (De Oliveira et al., 2013). Abiotic stresses have the potential to modify the ultrastructure of cellular organelles. The plants react to stress by coordinating the expression of nucleus and chloroplast genes. These genes have clear roles; they work together in the context of an intricately organized network of genes, regulatory elements, signaling factors and metabolites. Photosynthesis in plants is governed by a dynamic network of proteins. This research paper offers a description of how abiotic stresses affect gene regulation during photosynthesis, particularly in the C3 cycle. Abiotic stresses such as drought, heat, and intense light cause overreduction of the electron transport chain, which culminates in photooxidation (Foyer et al., 2012; Nishiyama and Murata, 2014). Plants have the capacity to withstand this stress by slowing down the rate of electron transport, resulting in the transfer of excess absorbed light into thermal energy. The dissipation of excitation energy received by chlorophyll as heat is referred to as nonphotochemical quenching (Nath et al., 2013; Spetea et al., 2014). Similary, complicated regulatory and signaling systems are involved where signals are transmitted to regulatory system via ion channels and secondary messengers (Cortleven et al., 2014; Tikkanen and Aro, 2014).

The expression of numerous stress-inducible genes is known to be under control by regulatory system components such as phytohormones, transcription factors, kinases, and phosphatases (Osakabe et al., 2014; Gururani et al., 2015). However, improving photosynthetic efficiency for increased plant yield needs to be accomplished (Ashraf and Harris, 2013; Ambavaram et al., 2014). Photosynthesis is a process which involves multiple steps which if damaged at any point, results in a drop in overall plant yield .

5.2 THE C3 CYCLE PATHWAY

Plants using the C3 pathway for carbon fixation may be found all over the world. They make up more than 95% of all plant species on the planet, particularly in wet and cold climates with low humidity. Carbon fixation takes place during the initial stage of the Calvin cycle. The enzyme that catalyzes basic reaction during photosynthesis is ribulose bisphosphate carboxylase (RuBisCO). The enzyme RuBisCO is the most abundant around the world. Rubisco is used by majority of plants in the Calvin–Benson cycle to stabilize atmospheric CO_2. Since first stable component of this cycle is a three-carbon compound, this phase is known as the C3 method. During C3 photosynthesis, mesophyll cells that contain RuBisCO enzyme maintain direct contact with the intercellular air space, which is linked with outside environment with the help of stomatal opening in the epidermis (Long et al., 2006). RuBisCO causes bottlenecks during the process of photosynthesis (Parry et al., 2007). The poor performance of RuBisCO carboxylase and specificity for CO_2 because of the greater O_2/CO_2 ratio at high temperatures are evident goals for improvement (Peterhansel and Maurino, 2011). So far, there has been minimal progress in enhancing the catalytic characteristics of RuBisCO for photorespiration reduction (Esquivel et al., 2013; Matsumura et al., 2005; Parry et al., 2007). This enzyme catalyses two processes, oxygenation and carboxylation (Portis and Parry, 2007). The oxygenation step guides the movement of carbon into the photorespiratory pathway, which can result in loss of 25% to 30% of the carbon. Although the bulk of the triose phosphate generated in the Calvin cycle is used to synthesize RuBP, one-sixth of the carbon departs the cycle to be used in the biosynthesis of a variety of chemicals. This proves that the Calvin cycle plays a critical role during carbon metabolism. Modification of this pathway has a possibility for increasing yield and carbon supply for secondary metabolite production. It has long been acknowledged that the catalytic properties of the enzyme RuBisCO play a major role in the limitation of carbon absorption in plants via the C3 cycle (Portis and Parry, 2007). Eight enzymes are involved in the regeneration of the RuBP. Antisense plants with decreased amounts of specific enzymes were produced and examined to find enzymes that have considerable regulation for carbon flux (Stitt and Schulze, 1994; Raines, 2003).

5.3 EFFECT OF ABIOTIC STRESS ON PHOTOSYNTHESIS

Abiotic stress causes major limitations for all living creatures, but plants face additional problems since they cannot move as freely as other organisms. Under extreme abiotic stress conditions, the structure of biological molecules like lipids, proteins, and nucleic acids is prone to degradation. Photosynthesis is a complicated process involving many biological pathways. Different environmental stresses like salt, drought, floods, light, temperatures, and their fast changes, show a negative effect on the photosynthetic carbon metabolism process in plants and has the potential to modify the structure of organelles, concentration of different pigments, and metabolite production.

5.3.1 TEMPERATURE STRESS AND PHOTOSYTHESIS

Photosynthesis is among the most delicate biological processes, especially when exposed to high temperature. Heat stress causes increase in the formation of reactive oxygen and nitrogen species, yet plants have the ability to activate antioxidant enzymes like catalase, peroxidase and superoxide mutase as well as nonenzymatic antioxidants like ascorbate. According to Gururani et al., (2015), reactive oxygen species (ROS) damage and impair function of the photosynthetic machinery, especially Photosystem II (PSII), results in photoinhibition due to imbalance in the photosynthetic redox signaling pathways and suppression of PSII repair.

In case of poplar, high temperatures have both transitory and permanent effects on photosynthesis and productivity (Song et al., 2014). It was further suggested that after 6 hours of exposure to heat, photosynthesis was able to recover, which in trees may be considered a turning point in terms of photosynthetic response. Genome-wide expression analysis resulted in total of 29,896 differentially expressed genes from which 15,670 were upregulated and 14,226 are downregulated on exposure to 6 hours of heat stress treatment. Majority of these genes were photosynthetic regulatory genes. Under high temperature, reduced Pn and stomatal conductance (Gs) was due to the partial closure of stomata, which resulted from CO_2 buildup reductions (Cui et al., 2006). High temperature stress causes photoinhibition that causes accumulation of reduced electron acceptors. Overproduction of ROS induces lipid peroxidation and damage to pigments, nucleic acids, and proteins (Wang et al., 2014). Chen et al., (2012) provide an evidence that heat stress-induced distortion in structure of thylakoid cause great loss in PSII activity. Increased temperature stress decreases stacking of grana near core complexes (Lan et al., 2014).

Heat stress reduces activity of RuBisCO enzyme, affecting the photosynthesis process (Anjana and Allakhverdiev, 2017). High temperature and drought greatly increase photorespiratory deficits (Ogren, 1984; Brooks and Farquhar, 1985). Because CO_2 solubility reduces with increasing temperature, lowering CO_2 availability at the RuBisCO catalytic site accelerates oxygenase process (Jordan and Ogren, 1984). Transgenic antisense lines with decreased RuBisCO protein levels showed that RuBisCO limitation on C3 carbon fixation is exacerbated under high intensity of temperature and light (Stitt and Schulze, 1994), but growth under increased CO_2 concentration may have a compensatory effect (Masle et al., 1993). Transgenic experiments indicated that RuBisCO does not restrict C3 carbon fixation during all conditions and enzymes from the regenerative phase enzymes also play a role in controlling photosynthetic rate. RuBisCO activase (RCA) proteins with molecular weights of 41and 47 kDa (isoform) have the ability for activating RuBisCO in various species; they also show physiological relevance in thermal sensitivity during heat stress (Crafts-Brandner and Salvucci, 2000). In rice (Wang et al., 2010) and spinach (Crafts-Brandner et al., 1997; Kim and Portis, 2006), α-isoform is more thermostable in comparison with β-isoform, which indicates that α-RCA isoform have major role in acclimation during mild heat stress.

5.3.2 DROUGHT STRESS AND PHOTOSYNTHESIS

Drought has been a crucial factor for restricting plant development and growth. In plants, drought affects physiological, molecular, and biochemical processes. Closure of stomata is the first response to drought, resulting in a drop in mesophyll CO_2 transport and a reduction in rate of photosynthesis (Chaves et al., 2003; Liu et al., 2010). Under water-stressed conditions, closure of stomata and metabolic impairment results in reduced photosynthetic rates (Arve et al., 2011; Ashraf and Harris, 2013; Gupta and Thind, 2015). After brief exposure to drought stress, photosynthesis is inhibited, resulting in growth inhibition, stomatal closure with subsequent reduction in transpiration, a drop in chlorophyll content, photosynthesis inhibition, and alterations in the functions of many essential proteins (Lawlor and Cornic, 2002) to deal with osmotic alterations in plants. Photosynthesis is among the primary processes affected by drought (Chaves, 1991) and salinity (Munns *et al.,* 2006). The effects may be direct, in the form of decrease in CO_2 availability caused by diffusion limitations from mesophyll cells and stomata (Flexas *et al.,* 2004) or from the changes in metabolism rate of photosynthesis (Lawlor and Cornic, 2002) or can arise as secondary effects like oxidative stress. Abiotic stressors decrease stomatal conductance, causing oxidative stress, which reduces the activity of RuBisCO and obstructs photosynthesis (Crafts-Brandner and Salvucci, 2004; Chaves et al., 2009; Kohli et al., 2017). Gene expression studies have provided information about the tolerant genotypes. In case of C3 plants, genes involved in Photosystem I (PSI) and PSII and the Calvin–Benson cycle, such as triose phosphate isomerase, fructose 1,6-bisphosphatase, RuBisCO small subunit, and RuBisCO activase, have been shown to be repressed in the tolerant cells types (Degenkolbe et al., 2009; Moumeni et al., 2011). During drought stress, numerous genes in photosynthetic pathways were downregulated in susceptible and tolerant genotypes. It is interpreted that plants avoid photooxidation and the production of free radicals, which have a detrimental impact on plant photosynthetic ability (Xu et al., 2010). The RuBisCO activity was reduced as a result of water deficit stress (Tezara et al., 1999). RuBisCO holoenzyme has been found to be reasonably stable under drought situations (Webber et al., 1994). However, investigations have revealed that the quantity of RuBisCO small subunit transcripts in Arabidopsis, rice, and tomato decreases rapidly (Bartholomew et al., 1991; Williams et al., 1994). Drought reduces RuBisCO activity in tobacco, largely due to tightly bound inhibitors (Parry et al., 2002). *Jatropha curcas* L. seedlings when exposed to drought indicated differential genes expression of PSI, PSII, and the Calvin cycle, genes encoding light-harvesting complex proteins, and genes coding for key enzymes during Calvin cycle. The genes coding for RuBisCO small subunit, phosphoglycerate kinase, and phosphoribulokinase were downregulated significantly (Zhang et al., 2015).

In the majority of photosynthetic species, the C3 cycle is the main mechanism for carbon assimilation. It is the single highest organic carbon flux in the biosphere, absorbs per year about 100 billion tons of carbon. Gene expression analyses revealed large information about tolerant C3 and C4 genotypes. Studies proved that during drought stress, genes play crucial functions in PSI and PSII and the Calvin–Benson cycle like triose phosphate isomerase, fructose-1,6-bisphosphatase, RuBisCO small subunit, and RuBisCO activase were repressed in the tolerant C3 species (Degenkolbe et al., 2009; Moumeni et al., 2011) and also in tolerant C4 maize plants (Hayano-Kanashiro et al., 2009). In response to drought stress, several common genes in photosynthetic pathways were downregulated in both tolerant and sensitive genotypes. As a result, plants avoid photooxidation and the production of free radicals like ROS, which have deleterious impacts on plant photosynthetic activity (Xu et al., 2010). The level and acitivity of RuBisCO are reduced when there is water shortage (Tezara et al., 1999). RuBisCO holoenzyme has been relatively stable during drought (Webber et al., 1994). Studies showed that tomato, rice, and Arabidopsis plants exhibited decrease in the number of transcript of RuBisCO small subunit (Williams et al., 1994; Vu et al., 1999). Similarly, drought stress reduces RuBisCO activity in tobacco, which is mostly due to tight binding of inhibitors (Parry et al., 2002)

5.3.3 SALINITY STRESS AND PHOTOSYNTHESIS

The concentration of salt in plant tissues is primarily increased by significant salt accumulation around the root zone. Primary effect of salt stress is water deficit with effect on leaf expansion rate in given plant species, irrelevant of the degree of tolerance. Physiological changes in leaves during salt stress include a restriction of CO_2 transport into the chloroplast and reduced carbon metabolism.

The expression of various photosynthesis-related genes may vary on exposure to salinity stress. The occurrence of salt in the soil solution limits root water intake, resulting in stunted plants with lower growth rates due to diminished cell expansion (Zhu, 2002; Parida and Das, 2005; Ranjit et al., 2016). Excessive salt ions around roots result in loss of volume and turgor, and reduction in plant growth (Passioura and Munns, 2000; Fricke and Peters, 2002; Liu et al., 2015). *Raphanus sativus* growth is reduced under salinity by 80% and 20%, respectively, due to stomatal closure (Marcelis and Hooijdonk, 1999). Similar results were reported in tomato and rice, cotton, and *Brassica campestris* subsp. *chinensis* (Mohammad et al., 1998; Rahman et al., 2016, Meloni et al., 2001, Memon et al., 2010) where there is reduction in shoot weight, height of plants, number of leaves per plant, root length and root surface when exposed with high salt concentration. Transcriptome analysis in tomato was performed under salt and drought stress, where the results suggested that during high salt level, hundreds of stress-responsive genes were induced (Chaves et al., 2009; Prins et al., 2011). Arabidopsis transcriptome during salt stress revealed that 1118 genes were upregulated by osmotic and ionic stresses (Sewelam et al., 2014).

5.4 GENETIC ENGINEERING APPROACHES FOR OVERCOMING ABIOTIC STRESS

It is crucial to combine traditional physiological approaches with functional genomics techniques, such as transcriptomic, metabolomic, and proteomics to understand regulatory principles and realize plant metabolism dynamics during stress conditions.This will help to study the dynamism by using this comprehensive analysis. The Calvin cycle is initiated by RuBisCO, catalyzing the carboxylation of the CO_2 acceptor molecule RuBP. Its catalytic property majorly determines carbon assimilation rate. Because of its importance in the Calvin cycle, considerable efforts have been made in recent years to improve the photosynthetic capability of this essential enzyme.

Sedoheptulose-1,7-bisphosphatase (SBPase) is related with carbon fixation during the Calvin cycle. The deviations in activity of SBPase affect photosynthesis, development, growth and stress tolerance in SBPase antisense transgenic tomato. In transgenic lines with enhanced SBPase activity, the photosynthetic rates was increased with an increase in the accumulation of sucrose and starch. Leaf area and total biomass were increased in SBPase sense plants, while in antisense plants, these parameters were reduced in comparison with the wild-type tomato. Under chilling stress, tomato plants with enhanced SBPase activity were more chilling tolerant as shown by less electrolyte leakage, increased photosynthetic capacity, and elevated RuBP regeneration rate. The enzyme sedoheptulose-1,7-bisphosphatase functions at the branch point where assimilated carbon may go to the regenerative phase to produce RuBP molecule in the cycle or exported from the cycle for sucrose or transitory starch biosynthesis. A large portion of the reductions in glyceraldehyde-3-P dehydrogenase, Fru-1,6-bisphosphatase, and phosphoribulokinase level do not limit photosynthetic capacity, which proves that these enzymes exist in excess and do not restrict carbon fixation through the Calvin cycle. On the contrary, even minor changes in SBPase activity leads to major reduction in photosynthesis of SBPase antisense transgenic tobacco lines.

In transgenic Arabidopsis and tobacco, the rate constant for the induction of net CO_2 assimilation was dictated by the rate of RuBisCO activation by RuBisCO activase (Mott et al., 1997; Hammond et al., 1998). Transgenic rice plants with different levels of RuBisCO activase have been found to have similar results (Fukayama et al., 2012; Yamori et al., 2012). The relaxation times, which is reciprocal of rate constant for net CO_2 assimilation (Woodrow and Mott, 1989) was shorter in rice that overexpress the β-RuBisCO activase from barley and maize (Fukayama et al., 2012).

The enzyme activity of sedoheptulose-1,7-bisphosphatase was enhanced in transgenic rice cultivar by overexpression of OsSbp cDNA from another cultivar 9311 of rice. Transgenic lines accumulated SBPase in chloroplasts, which resulted in enhanced stress tolerance. Overexpression of SBPase was found to be an efficient strategy for increasing salinity tolerance in rice. Results proved that during salt stress, SBPase regulates RuBisCO activation by causing regeneration of the acceptor molecule, ribulose-1,5-bisphosphate in stroma, and prevents the sequestration of RuBisCO activase to the thylakoid membrane and enhanced tolerance of photosynthesis to salinity (Feng et al., 2007a). In another study, genetic engineering of rice allowed accumulation of SBPase in chloroplasts, which resulted in increased tolerance to high temperature during seedling growth (Feng et al., 2007a, 2007b). Furthermore, CO_2 absorption in transgenic plants was higher in plants that were resistant to severe temperature stress compared with control plants.

The role of fructose-1,6-bisphosphatase and SBPase to the carbon flux during the Calvin cycle was examined in transgenic tobacco that express cyanobacterial FBPase-II in chloroplasts or *Chlamydomonas* SBPase in chloroplasts. The transgenic line TpF-11 showed more than 2.3-fold high FBPase activity and TpS-11 and TpS-10 plants showed 1.6- and 4.3-fold greater SBPase activity in chloroplasts in comparison with wild-type plants. The final dry matter accumulation was also about 1.3–1.5-fold higher when compared with the wild-type plants (Tamoi et al., 2006). In tobacco plants, overexpression of bifunctional SBPase resulted in enhanced rate of fixation of CO_2 (Miyagawa et al, 2001; Lefebvre et al., 2005). Besides this, overexpression of photosynthesis associated genes such as glycine betaine (Su et al., 2006; Park et al., 2007); betaine aldehyde dehydrogenase (Fan et al., 2012); OsSbp cDNA (Feng et al., 2007a, 2007b); plastidal protein synthesis elongation factor (Fu et al., 2008); chloroplast small heat shock protein (Wang et al., 2005); sucrose nonfermenting 1-related protein kinase 2 (Zhang et al., 2010); Na^+/H^+ antiporter (He et al., 2005; Brini, 2017); and aquaporins (Aharon et al., 2003) resulted in improvement of plant performance during abiotic stress. Transcription factors (TFs) function as master switches during environmental stress. Based on their binding domains, TFs may be divided into many groups. Indeed, gene regulatory element alteration is receiving a lot of scientific interest for genetic engineering purposes, and it has generated encouraging results during abiotic stress. Zhang et al., (2008) investigated the role of *ABP9* gene during different abiotic stresses like drought and heat. Transgenic Arabidopsis lines that express ABP9 showed a considerable increase in the photosynthetic capacity of plants under both drought and heat stress. This is due to dissipating excess light energy, elevating carbon-use efficiency, and increasing abscisic acid content and water use efficiency (Zhang et al., 2008a, 2008b). Similarly, HY5 (LONG HYPOCOTYL 5), a bZIP-type transcription factor, was involved with *CAB* gene expression and regulation by light (Chattopadhyay et al., 1998; Maxwell et al., 2003; Saibo et al., 2009). Moreover, besides regulating Chl a/b binding protein 2 (CAB2) expression (Maxwell et al., 2003) it regulates the expression of the gene for the RuBisCO small subunit (RbcS1A) (Chattopadhyay et al., 1998). Overexpression of OsMYB4 results in accumulation of glycine betaine in Arabidopsis and resulted in stress tolerance and helped to maintain photosynthesis (Mattana et al., 2005). Glycine betaine has the capacity to maintain the structure of RuBisCO when it exposed to high salinity (Sakamoto and Murata, 2002, Yang et al., 2005,). Thus, OsMYB4 TF has indirect influence on the regulation of photosynthetic genes under stress environment (Ashraf and Harris, 2013). Overexpression of BpMYB106, an R2R3MYB TF, upregulates numerous genes involved in photosynthesis and oxidative phosphorylation pathways (Zhou and Li, 2016). RNA-Seq profiling revealed that BpMYB106 activates the expression of photosynthesis related genes in transgenic birch by their interaction with the MYB2 element in promoters. As a result, all transgenic plants demonstrated greater leaf water usage efficiency, increased net photosynthetic rate, and improved growth rate (Zhou and Li, 2016).

5.5 CONCLUSIONS AND FUTURE PERSPECTIVES

Environmental stressors provide significant obstacles to plant growth and development, resulting in a significant reduction in photosynthetic effectiveness. Because many factors are responsible for controlling photosynthetic pathways, improving crop performance under stress has proven to be difficult. The level of stress tolerance is greatly reliant on the level of tolerance or susceptibility, which is regulated by nuclear genes and protein expression. Nonetheless, breakthroughs in genomic tools and functional research, as well as high-throughput sequencing technologies, have increased possibility of raising stress tolerant plants. Several key genes involved in plant response to environmental stresses have been identified as a result of these investigations, and many of these genes have already been demonstrated in model plants to be successful in introducing stress tolerance. Improved photosynthetic performance is a viable technique for increasing crop productivity, and it has the potential to reproduce the advantages of Green Revolution. Several worldwide research consortia have recently been established to improve the productivity of crops like wheat and rice. In fact, improving photosynthetic performance is a promising technique for addressing future yield associated issues in plants.

KEYWORDS

- **C3 cycle**
- **environmental stress**
- **photosynthesis**
- **RuBisCO**

REFERENCES

Aharon, R.; Shahak, Y.; Wininger, S. et al. Overexpression of a Plasma Membrane Aquaporin in Transgenic Tobacco Improves Plant Vigor under Favorable Growth Conditions But Not under Drought or Salt Stress. *Plant Cell* **2003,** *15*, 439–447.

Ambavaram, M. M. R.; Basu, S.; Krishnan, A. et al. Coordinated Regulation of Photosynthesis in Rice Increases Yield and Tolerance to Environmental Stress. *Nat. Commun.* **2014,** *5*, 5302.

Arve, L. E. L.; Torre, S. L.; Olsen, J. E. L.; Tanino, K. K. Stomatal Responses to Drought Stress and Air Humidity. In *Abiotic Stress in Plants—Mechanisms and Adaptations*; Shanker, A., Venkateswarlu, B., Eds.; IntechOpen, 2011; pp 267–280.

Ashraf, M.; Harris, P. J. C. Photosynthesis under Stressful Environments: An Overview. *Photosynthetica* **2013,** *51*, 163–190. DOI: 10.1007/s11099-013-0021-6.

Bartholomew, D. M.; Bartley, G. E.; Scolnik, P. A. Abscisic Acid Control of rbcS and Cab Transcription in Tomato Leaves. *Plant Physiol.* **1991,** *96*, 291–296.

Bray, E. A. Genes Commonly Regulated by Water-Deficit Stress in *Arabidopsis thaliana*. *J. Exp. Bot.* **2004,** *55*, 2331–2341.

Brini, F. Photosynthesis under Stressful Environmental Conditions: Existing Challenges. In *Environment and Photosynthesis: A Future Prospect*; Singh, S., Ed.; Studium Press (Pvt.), Limited a Joint venture of Studium Press: USA, 2017; pp 68–91.

Brooks, A.; Farquhar, G. D. Effect of Temperature on the CO_2/O_2 Specificity of Ribulose-1,5-Biphosphate Carboxylase/Oxygenase and the Rate of Respiration in the Light. *Planta* **1985,** *165*, 397–406.

Chattopadhyay, S.; Ang, L. H.; Puente, P. et al. Arabidopsis bZIP Protein HY5 Directly Interacts with Light-Responsive Promoters in Mediating Light Control of Gene Expression. *Plant Cell* **1998,** *10*, 673–683.

Chaves, M. M.; Flexas, J.; Pinheiro, C. Photosynthesis under Drought and Salt Stress: Regulation Mechanisms from Whole Plant to Cell. *Ann. Bot.* **2009,** *103*, 551–560.

Chen, W. R.; Zheng, J. S.; Li, Y. Q.; Guo, W. D. Effects of High Temperature on Photosynthesis, Chlorophyll Fluorescence, Chloroplast Ultrastructure, and Antioxidant Activities in Fingered Citron. *Russ. J. Plant Physiol.* **2012,** *59* (6), 732–740.

Cortleven, A.; Nitschke, S.; Klaumunzer, M. et al. A Novel Protective Function for Cytokinin in the Light Stress Response in Mediated by the ARABIDOPSIS HISTIDINE KINASE2 and ARABIDOPSIS HISTIDINE KINASE3 Receptors. *Plant Physiol.* **2014,** *164*, 1470–1483.

Crafts-Brandner, S. J.; Salvucci, M. E. Rubisco Activase Constrains the Photosynthetic Potential of Leaves at High Temperature and CO(2). *Proc. Natl. Acad. Sci. USA* **2000,** *97*, 13430–13435.

Cui, L.; Li, J.; Fan, Y. et al. High Temperature Effects on Photosynthesis, PSII Functionality and Antioxidant Activity of Two *Festuca arundinacea* Cultivars with Different Heat Susceptibility. *Bot. Stud.* **2006,** *47* (1), 61–69.

De Oliveira, A. B.; Mendes Alencar, N. L.; Gomes-Filho, E. Comparison between the Water and Salt Stress Effects on Plant Growth and Development, 2013.

Degenkolbe, T.; Do, P. T.; Zuther, E. et al. Expression Profiling of Rice Cultivars Differing in Their Tolerance to Long-Term Drought Stress. *Plant Mol. Biol.* **2009,** *69*, 133–153.

Dinneny, J. R.; Long, T. A.; Wang, J. Y.; Jung, J. W.; Mace, D.; Pointer, S.; Barron, C.; Brady, S. M.; Schiefelbein, J.; Benfey, P. N. Cell Identity Mediates the Response of Arabidopsis Roots to Abiotic Stress. *Science* **2008,** *320*, 942–945.

Esquivel, M. G.; Genkov, T.; Nogueira, A. S.; Salvucci, M. E.; Spreitzer, R. J. Substitutions at the Opening of the RuBisCO Central Solvent Channel Affect Holoenzyme Stability and CO_2/O_2 Specificity But Not Activation by RuBisCO Activase. *Photosynth. Res.* Dec **2013,** *118* (3), 209–218.

Fan, W.; Zhang, M.; Zhang, H.; Zhang, P. Improved Tolerance to Various Abiotic Stresses in Transgenic Sweet Potato (Ipomoea batatas) Expressing Spinach Betaine Aldehyde Dehydrogenase. *PLoS One* **2012,** *7*, e37344.

Feng, L.; Han, Y.; Liu, G.; An, B.; Yang, J.; Yang, G.; Li, Y.; Zhu, Y. Overexpression of Sedoheptulose-1,7-Bisphosphatase Enhances Photosynthesis and Growth under Salt Stress in Transgenic Rice Plants. *Funct. Plant Biol.* Sep **2007a,** *34* (9), 822–834.

Feng, L.; Wang, K.; Li, Y.; Tan, Y.; Kong, J.; Li, H.; Li, Y.; Zhu, Y. Overexpression of SBPase Enhances Photosynthesis Against High Temperature Stress in Transgenic Rice Plants. *Plant Cell Rep.* Sep **2007b,** *26* (9), 1635–1646.

Flexas, J.; Bota, J.; Loreto, F. et al. Diffusive and Metabolic Limitations to Photosynthesis under Drought and Salinity in C(3) Plants. *Plant Bio.* **2004,** *6*, 269–279.

Foyer, C. H.; Neukermans, J.; Queval, G. et al. Photosynthetic Control of Electron Transport and the Regulation of Gene Expression. *J. Exp. Bot.* **2012,** *63*, 1637–1661.

Fricke, W.; Peters, W. S. The Biophysics of Leaf Growth in Salt-Stressed Barley. A Study at the Cell Level. *Plant Physiol.* **2002,** *129*, 374–388.

Gorantla, M.; Babu, P. R.; Lachagari, V. B. R.; Reddy, A. M. M.; Wusirika, R.; Jeffrey, L.; Bennetzen, J. L.; Reddy, A. R. Identification of Stress-Responsive Genes in an Indica Rice (*Oryza sativa* L.) Using ESTs Generated from Drought-Stressed Seedlings. *J. Exp. Bot.* **2007,** *58*, 253–265.

Gupta, N.; Thind, S. Improving Photosynthetic Performance of Bread Wheat under Field Drought Stress by Foliar Applied Glycine Betaine. *J. Agric. Sci. Tech.* **2015,** *17*, 75–86.

Gururani, M. A.; Venkatesh, J.; Tran, L. S. P. Regulation of Photosynthesis during Abiotic Stress-Induced Photoinhibition. *Mol. Plant* **2015,** *8* (9), 1304–1320.

Hayano-Kanashiro, C.; Calderon-Vazquez, C.; Ibarra-Laclette, E. et al. Analysis of Gene Expression and Physiological Responses in Three Mexican Maize Landraces under Drought Stress and Recovery Irrigation. *PLoS One* **2009,** *4*, e7531.

Hussain, S. S.; Kayani, M. A.; Amjad, M. Transcription Factors as Tools to Engineer Enhanced Drought Stress Tolerance in Plants. *Biotechnol. Prog.* **2011,** *27*, 297–306.

Jordan, D. B.; Ogren, W. L. The CO_2/O_2 Specificity of Ribulose 1,5- Biphosphate Carboxylase/ Oxygenase. *Planta* **1984,** *161*, 308–313.

Lawlor, D. W.; Cornic, G. Photosynthetic Carbon Assimilation and Associated Metabolism in Relation to Water Deficits in Higher Plants. *Plant Cell Environ.* **2002,** *25*, 275–294.

Lefebvre, S.; Lawson, T.; Fryer, M.; Zakhleniuk, O. V.; Lloyd, J. C.; Raines, C. A. Increased Sedoheptulose-1,7-bisphosphatase Activity in Transgenic Tobacco Plants Stimulates Photosynthesis and Growth from an Early Stage in Development. *Plant Physiol.* **2005,** *138*, 451–460.

Long, S. P.; Zhu, X. G.; Naidu, S. L.; Ort, D. R. Can Improvement in Photosynthesis Increase Crop Yields? *Plant Cell Environ.* Mar **2006,** *29* (3), 315–330.

Marcelis, L. F. M.; Hooijdonk, J. V. Effect of Salinity on Growth, Water Use and Nutrient Use in Radish (*Raphanus sativus* L.). *Plant Soil* **1999,** *215*, 57–64.

Mattana, M.; Biazzi, E.; Consonni, R. et al. Overexpression of Osmyb4 Enhances Compatible Solute Accumulation and Increases Stress Tolerance of *Arabidopsis thaliana. Physiol. Plant.* **2005,** *125*, 212–223.

Maxwell, B. B.; Andersson, C. R.; Poole, D. S. et al. HY5, Circadian Clock-Associated 1, and a Cis-Element, DET1 Dark Response Element, Mediate DET1 Regulation of Chlorophyll *a/b*-Binding Protein 2 Expression. *Plant Physiol.* **2003,** *133*, 1565–1577.

Meloni, D. A.; Oliva, M. A.; Ruiz, H. A.; Martinez, C. A. Contribution of Proline and Inorganic Solutes to Osmotic Adjustment in Cotton under Salt Stress. *J. Plant Nutr.* **2001,** *24*, 599–612.

Memon, S. A.; Hou, X.; Wang, L. Morphological Analysis of Salt Stress Response of Pak Choi. *Electron. J. Environ. Agric. Food Chem.* **2010,** *9*, 248–254.

Miyagawa, Y.; Tamoi, M.; Shigeoka, S. Overexpression of a Cyanobacterial Fructose-1,6-/ sedoheptulose-1,7-bisphosphatase in Tobacco Enhances Photosynthesis and Growth. *Nat. Biotechnol.* **2001,** *19*, 965–969

Mohammad, M.; Shibli, R.; Ajlouni, M.; Nimri, L. Tomato Root and Shoot Responses to Salt Stress under Different Levels of Phosphorus Nutrition. *J. Plant Nutr.* **1998,** *21*, 1667–1680.

Moumeni, A.; Satoh, K.; Kondoh, H. et al. Comparative Analysis of Root Transcriptome Profiles of Two Pairs of Drought-Tolerant and Susceptible Rice Near-Isogenic Lines under Different Drought Stress. *BMC Plant Biol.* **2011,** *11*, 174.

Nath, K.; Jajoo, A.; Poudyal, R. S. et al. Towards a Critical Understanding of the Photosystem II Repair Mechanism and Its Regulation during Stress Conditions. *FEBS Lett.* **2013,** *587*, 3372–3381.

Nishiyama, Y.; and Murata, N. Revised Scheme for the Mechanism of Photoinhibition and Its Application to Enhance the Abiotic Stress Tolerance of the Photosynthetic Machinery. *Appl. Microbio. Biotechnol.* **2014,** *98*, 8777–8796.

Ogren, W. L. Photorespiration: Pathways, Regulation, and Modification. *Annu. Rev. Plant Physiol.* **1984,** *35*, 415–442.

Osakabe, Y.; Osakabe, K.; Shinozaki, S.; Tran, L. S. P. Response of Plants to Water Stress. *Front. Plant Sci.* **2014,** *5*, 86.

Parida, A. K.; Das, A. B. Salt Tolerance and Salinity Effects on Plants: A Review. *Ecotoxicol. Environ. Saf.* **2005,** *60*, 324–349.
Park, E. J.; Jeknić, Z.; Pino, M. T. et al. Glycinebetaine Accumulation Is More Effective in Chloroplasts Than in the Cytosol for Protecting Transgenic Tomato Plants Against Abiotic Stress. *Plant Cell Environ.* **2007,** *30*, 994–1005.
Parry, M. A. J.; Andralojc, P. J.; Khan, S. et al. Rubisco Activity: Effects of Drought Stress. *Ann. Bot.* **2002,** *89*, 833–839.
Passioura, J. B.; Munns, R. Rapid Environmental Changes That Affect Leaf Water Status Induce Transient Surges or Pauses in Leaf Expansion Rate. *Funct. Plant Biol.* **2000,** *27*, 941–948.
Portis, A. R., Jr.; Parry, M. A. J. Discoveries in Rubisco (Ribulose 1,5-Bisphosphate Carboxylase/Oxygenase): A Historical Perspective. *Photosynth. Res.* **2007,** *94*, 121–143.
Prins, A.; Mukubi, J. M.; Pellny, T. K. et al. Acclimation to High CO_2 in Maize Is Related to Water Status and Dependent on Leaf Rank. *Plant Cell Environ.* **2011,** *34*, 314–331
Rabello, A. R.; Guimarães, C. M.; Rangel, P.,H.,N.; Silva, F. R.; Seixas, D.; Souza, E.; Brasileiro, A. C. M.; Spehar, C. R.; Ferreira, M. E.; Mehta, A. Identification of Drought-Responsive Genes in Roots of Upland Rice (*Oryza sativa* L.) *BMC Genom.* **2008,** *9*, 485. DOI: 10.1186/1471-2164-9-485.
Rahman, M. A.; Moser, A.; Rotzer, T.; Pauleit, S. Within Canopy Temperature Differences and Cooling Ability of Tiliacordata Trees Grown in Urban Conditions. *Build. Environ.* **2017,** *114*, 118–128.
Raines, C. A. The Calvin Cycle Revisited. *Photosynth Res* **2003,** *75*, 1–10.
Ranjit, S. L.; Manish, P.; Penna, S. Early Osmotic, Antioxidant, Ionic, and Redox Responses to Salinity in Leaves and Roots of Indian Mustard (*Brassica juncea* L.). *Protoplasma* **2016,** *253*, 101–110.
Saibo, N. J. M.; Lourenco, T.; Oliveira, M. M. Transcription Factors and Regulation of Photosynthetic and Related Metabolism under Environmental Stresses. *Ann. Bot.* **2009,** *103*, 609–623.
Sakamoto, A.; Murata, N. The Role of Glycinebetaine in the Protection of Plants from Stress: Clues from Transgenic Plants. *Plant Cell Environ.* **2002,** *25*, 163–171.
Sewelam, N.; Oshima, Y.; Mitsuda, N.; Ohmetakagi, M. A Step Towards Understanding Plant Responses to Multiple Environmental Stresses: A Genome-Wide Study. *Plant Cell Environ.* **2014,** *37*, 2024–2035.
Sharkey, T. D. Estimating the Rate of Photorespiration in Leaves. *Physiol Plant* **1988,** *73*, 147–152.
Song, Y.; Chen, Q.; Ci, D. et al. Effects of High Temperature on Photosynthesis and Related Gene Expression in Poplar. *BMC Plant Biol.* **2014,** *14* (111), 1–20.
Spetea, C.; Rintamaki, E.; and Schoefs, B. (2014). Changing the Light Environment: Chloroplast signaling and response mechanisms. *Phil. Trans. R. Soc. B Biol. Sci.* **2014,** *369*, 20130220.
Stitt, M.; Schulze, D. Does Rubisco Control the Rate of Photosynthesis and Plant-Growth—An Exercise in Molecular Ecophysiology. *Plant Cell Environ.* **1994,** *17*, 465–487.
Su, J.; Hirji, R.; Zhang, L. et al. Evaluation of the Stress-Inducible Production of Choline Oxidase in Transgenic Rice as a Strategy for Producing the Stress-Protectant Glycine Betaine. *J. Exp. Bot.* **2006,** *57*, 1129–1135.
Tamoi, M.; Nagaoka, M.; Miyagawa, Y.; Shigeoka, S. Contribution of Fructose-1,6-Bisphosphatase and Sedoheptulose-1,7-Bisphosphatase to the Photosynthetic Rate and

Carbon Flow in the Calvin Cycle in Transgenic Plants. *Plant Cell Physiol.* Mar **2006,** *47* (3), 380–390.

Tezara, W.; Mitchell, V. J.; Driscoll, S. D.; Lawior, D. W. Water Stress Inhibits Plant Photosynthesis by Decreasing Coupling Factor and ATP. *Nature* **1999,** *401*, 914–917.

Tikkanen, M.; Aro, E. M. Integrative Regulatory Network of Plant Thylakoid Energy Transduction. *Trends Plant Sci.* **2014,** *19*, 10–17.

Trujillo, L.; Menendez, C.; Ochogavia, M. E. et al. Engineering Drought and Salt Tolerance in Plants Using SodERF3, a Novel Sugarcane Ethylene Responsive Factor. *Biotechnol. Aplicada* **2009,** *26*, 168–171.

Wang, H.; Zhang, H.; Li, Z. Analysis of Gene Expression Profile Induced by Water Stress in Upland Rice (*Oryza sativa* L. var. IRAT109) Seedlings Using Subtractive Expressed Sequence Tags Library. *J. Integr. Plant Biol.* **2007,** *49*, 1455–1463.

Webber, A. N.; Nie, G. Y.; Long, S. P. Acclimation of Photosynthetic Proteins to Rising Atmospheric CO_2. *Photosynth. Res.* **1994,** *39*, 413–425.

Williams, J.; Bulman, M.; Neill, S. Wilt-Induced ABA Biosynthesis, Gene Expression and Down-Regulation of RBCs Messenger-RNA Levels in *Arabidopsis thaliana. Physiol. Plant.* **1994,** *91*, 177–182.

Yang, X.; Liang, Z.; Lu, C. Genetic Engineering of the Biosynthesis of Glycinebetaine Enhances Photosynthesis against High Temperature Stress in Transgenic Tobacco Plants. *Plant Physiol.* **2005,** *138*, 2299–2309.

Zhang, C.; Zhang, L.; Zhang, S. et al. Global Analysis of Gene Expression Profiles in Physic Nut (*Jatropha curcas* L.) Seedlings Exposed to Drought Stress. *BMC Plant Biol.* **2015,** *15*, 17.

Zhang, X.; Wollenweber, B.; Jiang, D. et al. Water Deficit and Heat Shock Effects on Photosynthesis of a Transgenic *Arabidopsis thaliana* Constitutively Expressing ABP9, a bZIP Transcription Factor. *J. Exp. Bot.* **2008,** *59*, 839–848.

Zhou, J.; Wang, X.; Jiao, Y.; Qin, Y.; Liu, X.; He, K.; Chen, C.; Ma, L.; Wang, J.; Xiong, L. et al. Global Genome Expression Analysis of Rice in Response to Drought and High-Salinity Stresses in Shoot, Flag Leaf, and Panicle. *Plant Mol. Biol.* **2007,** *63*, 591–608.

Zhou, C.; Li, C. A Novel R2R3-MYB Transcription Factor BpMYB106 of Birch (*Betula platyphylla*) Confers Increased Photosynthesis and Growth Rate through Up-Regulating Photosynthetic Gene Expression. *Front. Plant Sci.* **2016,** 7, 315.

Zhu, J. K. Salt and Drought Stress Signal Transduction in Plants. *Annu. Rev. Plant Biol.* **2002,** *53*, 247–273.

CHAPTER 6

RuBisCo Activase as a Response to Environmental Stress in Plants

SANCHAYITA RAJKHOWA[1*], BISWAJIT BOSE[2], JYOTIRMOY SARMA[3], and ROOPA KUMARI[4]

[1]*Department of Chemistry, Jorhat Institute of Science & Technology, Jorhat, Assam, India*

[2]*Department of Botany, School of Basic Sciences, Shri Ramasami Memorial (SRM) University, Sikkim, East Sikkim, India*

[3]*Department of Chemistry, The Assam Kaziranga University, Jorhat, Assam, India*

[4]*Department of Chemistry, Ramjas College, University of Delhi, New Delhi, India*

[*]*Corresponding author. E-mail: sanchayita.rajkhowa@gmail.com*

ABSTRACT

Ribulose-1,5-bisphosphate carboxylase/oxygenase activase, RuBisCO (RCA) is a plant nuclear gene that encodes a chloroplast protein, which plays a crucial role in photosynthesis (Pn). RCA is known to facilitate the process of releasing sugar-phosphate inhibitors at RCA catalytic sites during CO_2 fixation. Most plant species exhibit two RCA isoforms, that is, shorter RCA-β and larger RCA-α. They are present either by alternative splicing from one gene or expressed from different genes. The process of RuBisCO activation by RCA isoforms has widely been studied in higher plants. The catalytic properties of RuBisCO are mostly regulated as a response to short-term

Photosynthesis and Respiratory Cycles during Environmental Stress Response in Plants. Aryadeep Roychoudhury (Ed.)

variations in the environment and limit the efficiency of Pn, although these regulations may not be optimally correlated with crop yield. Many of these environmental factors are closely related to decrease RuBisCO activation, which is caused by impairment in RCA functions. With the socioeconomic development, environment has undergone a drastic change over the years bearing the negative impacts of various anthropogenic activities. Environmental (abiotic) stress [e.g., heat, hypoxia, drought, pH, salinity, heavy metal (HM) pollution, elevated ozone (O_3) concentration, and ultraviolet-B irradiance] can strongly affect senescence and decrease RuBisCO activity related to downregulation of the enzyme's activation state by either de-carbamylation or binding of inhibitory sugar phosphates in response to high temperature or drought. On the other hand, the harmful effects of HM stress (like cadmium), low temperature, O_3 concentration, and ultraviolet-B (UV-B) stress on the activity of RuBisCO are linked with variations in the enzyme concentration.

6.1 INTRODUCTION

The key enzyme in C_3 carbon fixation and the most abundant protein present in plants, RuBisCO [ribulose-1,5-bisphosphate (RuBP) carboxylase/oxygenase] involves carbon assimilation through the carboxylation of RuBP. RuBisCO must be activated by carbamylation with CO_2 and Mg^{2+} before binding to its substrate RuBP (Miziorko and Lorimer, 1983). Several factors such as pH, variation in stromal Mg^{2+} concentrations, binding of sugar phosphates to the catalytic sites chiefly regulate the function of RuBisCO (Badger and Lorimer, 1981; Seemann et al., 1990). RuBisCO activase (RCA) is basically an adenosine triphosphate (ATP)-dependent enzyme, which activates RuBisCO during daytime by dissociating these sugars from the catalytic sites (Portis, 1995; Spreitzer and Salvucci, 2002). Carboxy arabinitol 1-P, a phosphorylated sugar performs a special role by inhibiting RuBisCO activity at night (Moore and Seemann, 1994). RCA interacts with RuBisCO large subunit (LSU), that is, the catalytic site of the protein making a conformational change and thereby making the active site of the enzyme free for further reaction. However, there is a lack of information on the exact reaction mechanism. Nevertheless, studies revealed that most of the species (e.g., wheat and rice), contain two isoforms of RCA: a longer redox-sensitive α-isoform of 46–48 kDa and a shorter redox insensitive β-isoform of 41–43 kDa (Zhang and Portis, 1999). On the other hand, several other species like tobacco and maize do not have the longer isoform (Salvucci et al., 1987). Any significant variation in the redox condition and ADP/ATP ratio of the chloroplast regulates the RCA activity that consequently controls the RuBisCO activation process and CO_2 assimilation (Carmo-Silva and Salvucci, 2013; Mott and Woodrow, 2000; Scales et al., 2014). Although RuBisCO has a major role in CO_2 assimilation, the complex reaction mechanism is the reason for its inefficacy making a compromise with the photosynthetic productivity. For example, apart from CO_2 assimilation, RuBisCO also catalyzes wasteful side reactions and consumes energy during the process of photorespiration. RuBisCO has a lower turnover number as compared with other enzymes involved in Calvin cycle; thus, a sufficient amount of RuBisCO can only ensure satisfactory photosynthetic rates. Even though RuBisCO and other photorespiratory enzymes account for up to 25% leaf nitrogen and 50% of leaf soluble protein, RuBisCO activity is still limited (Parry et al., 2013).

RuBisCO is hexadecameric in structure, also known as Form I RuBisCO, that consists of eight chloroplast-encoded LSUs being arranged as four dimers and eight nuclear-encoded small subunits. Each LSU contains an N-terminal domain and comparatively larger C-terminal, that is, an alpha/beta-barrel domain. The larger domain contributes most of the active residues to interact with the substrates or their analogs, while N-terminal domain is responsible for the remaining residues within a single dimer (Parry et al., 2008). RuBisCO has evolved ~3 billion years ago in an environment of high CO_2 and very low O_2 (Whitney et al., 2011). Since then, RuBisCO has progressively developed to its present form with the gradual change of the earth's atmosphere by a complex reaction mechanism.

RCA was discovered in 1985 in *Arabidopsis* mutant that displayed no RuBisCO activity in light even at normal CO_2 concentration (Salvucci et al., 1985). It was later detected in a wide range of species with the help of antibodies (Salvucci et al. 1987). RCA, being present mostly in higher plants, has two subunits of 41–46 kDa, although a few species are known to possess a single polypeptide chain (Salvucci et al. 1987). The larger form differs from the shorter polypeptide by containing additional 37 amino acids (approximately) at the C-terminal. The enzyme is encoded by one or more nuclear genes, produced in cytosol (precursor) and forms chloroplast to result in a mature protein (Qian and Rodermel, 1993). Both the proteins can be produced from a single transcript by alternative splicing resulting into two separate mRNAs, for example in *Arabidopsis* (Werneke et al., 1989).

RuBisCO activity is organ specific as well as highly dependent on environmental conditions. It is mainly regulated to control flux via Calvin cycle or photosynthetic carbon reduction cycle in response to any kind of fluctuations (mostly temporary/short term) in the environment (Geiger and Servaites, 1994). The amount of RuBisCO protein, being the key factor in potential RuBisCO activity, is evaluated from the relative rate of biosynthesis and degradation. All these processes are controlled by gene expression (Krapp et al., 1993; Sheen, 1990), polypeptide synthesis, mRNA stability, active holoenzyme assembly, posttranslational modification, and various other factors that affect the protein degradation (Desimone et al., 1996; Eckardt and Pell, 1995; Mehta et al., 1992;). RuBisCO activity must match with the existing rate of RuBP utilization for its regeneration. However, RuBisCO activity regulation does not only depend on the accessibility of substrate, as excess RuBisCO than that of optimum requirement during photosynthesis (Pn) is deactivated (Mott et al., 1984; Sage et al., 1990a, 1990b). Activity of RuBisCO (E) in vivo is regulated by either of the two ways (i) carbamylation of lysine residue at catalytic site of the enzyme and subsequently stabilization of the carbamate produced by Mg^{2+} ion, and thus forms a ternary complex ($E.CO_2.Mg^{2+}$) that is catalytically active, and (ii) via strong binding of inhibitors (I) having low molecular weight as shown in Fig. 1. It is noteworthy that the CO_2 involved in carbamylation is different from that which reacts with acceptor molecule, that is, RuBP, during catalysis.

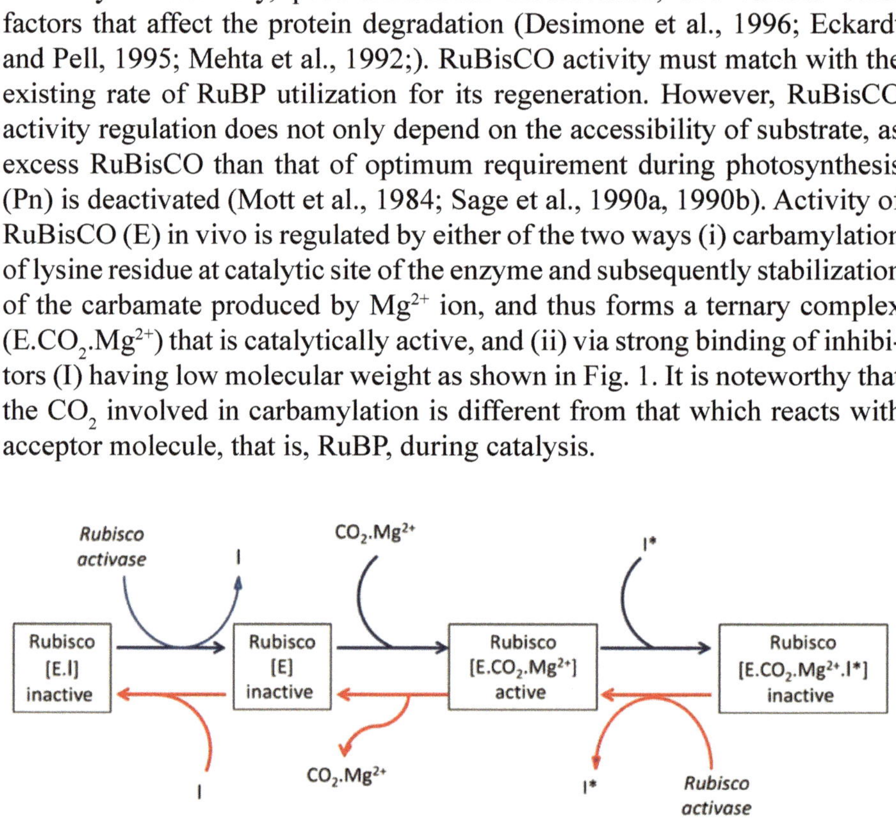

FIGURE 6.1 Regulation of RuBisCO catalytic activity pathways.

Inhibitors can bind to RuBisCO (E) either before (forming E.I) or after carbamylation ($E.CO_2.Mg^{2+}.I$) that blocks the active sites of the enzyme, and hence prevents further substrate binding and/or carbamylation. To remove the strongly bound inhibitors from the catalytic site of Rubisco (carbamylated or decarbamylated forms), it requires RCA along with the hydrolysis of ATP. This will ensure that the active sites of RuBisCO are free from any inhibitors and hence ready to be carbamylated and/or to spontaneously participate in the process of catalysis.

Extreme environmental conditions also termed environmental stress such as heat, hypoxia, drought, pH, salinity; heavy metal (HM) pollution, elevated ozone (O_3) concentration, and ultraviolet-B irradiance affect the enzyme and its activity at various levels. Till now no work has covered all kinds of environment stress in a systematic way. Therefore, this chapter will discuss those abiotic stress factors in details.

6.2 BIOTIC AND ABIOTIC STRESS

Environmental stress (biotic/abiotic) displays deteriorating effects on the overall plant growth as well as the yield although the impacts are studied in laboratories under controlled conditions. The natural environment, however, differs from the artificially created environment in laboratories in terms of simultaneous exposure to more than one stress condition, which could be a combination of abiotic and/or biotic stress like heat and drought, cold and drought, heat and salinity, drought and high light intensity, or any abiotic stress in combination with pathogenic infection. Studies have shown that plants respond to these combined stresses in a unique and different way, which cannot be predicted from the data if each single stress had affected individually (Rejeb et al., 2014). Suzuki et al. demonstrated cumulative effects of various stresses in a stress matrix (as shown in Fig. 2), their positive and negative impacts on plant growth, productivity, physiology, and yield in details (Suzuki et al., 2014). In addition, a simultaneous occurrence of various stresses causes higher complexity in plant responses and each response may sometimes be opposing and inhibiting one another.

The easily purified, obtainable, and abundant RuBisCO promises to be an ideal candidate for model protein to study various effects of stress during photosynthetic metabolism. However, cellular resources of RuBisCO protein turnover, that is, synthesis, maintenance, and degradation remain uncertain. Nevertheless, the extent of this turnover could be predicted if the abundance of RuBisCO is considered in a stress-induced oxidizing environment (Feller et al., 2008). In presence of light, heat, and cold stress, active oxygen species get augmented in chloroplast, which consequently intensifies the oxidative damage caused to thylakoid-bound stromal proteins. Stress-induced oxidative modification in certain RuBisCO residues leads to degradation of the enzyme (Marin-Navarro and Moreno, 2003; Moreno et al., 2008). Besides, the exact mechanism of RuBisCO degradation is still obscure, but the process is expected to proceed through the association of chloroplast membranes with oxidatively-damaged RuBisCO (Mehta et al., 1992). Oxidative stress is known to influence the RuBisCO abundance by reducing transcript levels in smaller subunits, while increasing the ethylene content (Glick et al., 1995). It is evident that RuBisCO is regulated by those factors that influence regulation of other cytosolic enzymes and chloroplast. For example, partial reduction in RuBisCO LSU transcript in *Chlamydomonas* by active oxygen species with simultaneous drop in stromal signals occurs under high-light stress (Irihimovitch and Shapira, 2000). Moreover, reduction in RuBisCO small subunit (RbcS) transcript along with an increase in cytosolic glutamine synthetase expression is observed in drought condition (Bauer et al., 1997). These are a few instances that illustrate the influence of oxidative stress on expression of enzymes of primary metabolism.

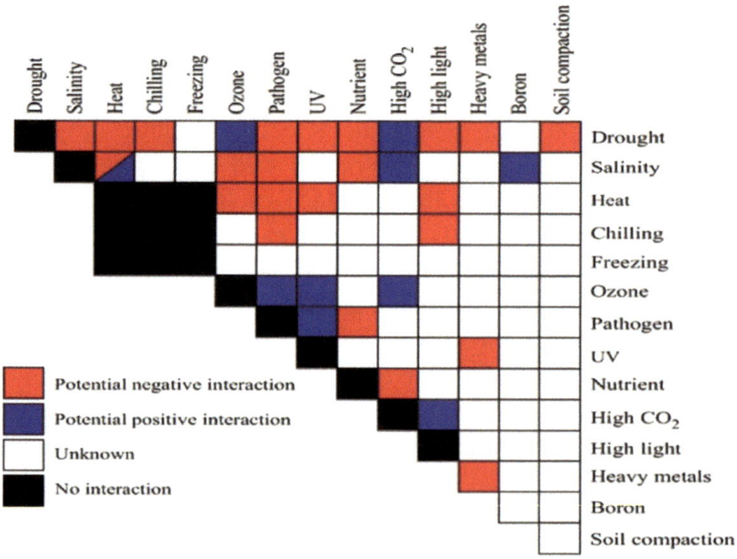

FIGURE 6.2 Matrix representation of simultaneous biotic and abiotic environmental stresses on crops grown in the field known as stress matrix (Reprinted from Suzuki et al., 2014; Open access).

Irving and Robinson (2006) suggested a plausible RuBisCO. Rubisco turnover model based on their study on cereal leaves and this theoretical model is found to be in accordance with previous experimental data (Friedrich and Huffaker, 1980; Suzuki et al., 2001; Mae et al., 1983). According to this dynamic model, concentration of RuBisCO can be controlled by changing biosynthesis, where degradation is considered to follow a first-order reaction and the quantity of RuBisCO at a particular time, *t* can be evaluated from the below equation

$$\begin{aligned}&[\text{Rubisco}]_t = [\text{Rubisco}]_0\, e^{-kt}\\&\text{Where } k - \text{The first-order rate constant for degradation}\\&[\text{Rubisco}]_0 - \text{Initial concentration of Rubisco}\end{aligned} \tag{4.1}$$

Feller et al. (1998) investigated the effect of high temperature on RuBisCO activity and RCA d in vivo with cotton and wheat leaves (Feller et al., 1998). Another study on CO_2 assimilation and RuBisCO activity above thermal optimum revealed that RuBisCO and Pn inhibition due to elevated temperature is a consequence of temperature sensitiveness of RCA (Cen and Sage, 2005). Apart from abiotic stress, plants are under constant threat from pathogens (bacteria, viruses, fungi, and nematodes) as well as attack by several herbivore pests (Atkinson and Urwin, 2012). The climatic conditions may influence the range/numbers of pathogens and pests (Bale et al., 2002).

6.3 DEPENDENCE OF RUBISCO ACTIVATION ON THE AVAILABILITY OF CO_2

At optimal temperature and sufficient sunlight, the rate of net CO_2 assimilation in C_3 plants is limited by RuBisCO if the partial pressure of CO_2 goes below the ambient level, that is, 370 μbar, whereas one of the RuBP regeneration processes limits CO_2 assimilation above 370 μbar (von Caemmerer, 2000). A number of studies made on numerous species have analyzed the CO_2 response during Pn by plotting CO_2 assimilation vs. intercellular CO_2 partial pressure and found that the RuBisCO activity is limited even at high CO_2 level at optimum temperature (Sage et al., 1990a, 1990b; Sage and Sharkey, 1987; Salvucci and Crafts-Brandner, 2004; von Caemmerer and Farquhar, 1981). In general, a maximum of ~90% RuBisCO activation is observed in most of the C3 species (Sage et al., 1990a, 1990b). Variation in the availability of CO_2 in the atmosphere is directly correlated to the ratio of R_{Rcap}/R_{Ccap} where R_{Rcap} is the RuBP regeneration capacity and R_{Ccap} represents capacity of RuBisCO to consume RuBP. This ratio indicates the energy supply to the leaf cells against its demand (Cen and Sage, 2005). The ratio of R_{Rcap}/R_{Ccap} is a useful index and it has two conditions as follows

(i) If $R_{Rcap}/R_{Ccap} > 1$ then abundant available ATP, less power is required for maximum activase activity and thus highest activation of RuBisCO (Sage et al., 1990).
(ii) If $R_{Rcap}/R_{Ccap} < 1$ then deactivation of RuBisCO occurs (Sage et al., 1990).

For example, RuBisCO activation decreases in *Phaseolus vulgaris* and *Chenopodium album* if kept under shade where the ratio $R_{Rcap}/R_{Ccap} < 1$ (Sage et al., 1990). Interestingly, there is an increase in this ratio and reactivation of RuBisCO in both species if the concentration of surrounding CO_2 is being reduced (Sage et al., 1990). If an increase in temperature lowers the R_{Rcap} as compared with R_{Ccap}, it can be concluded that the RuBisCO activation state is sensitive to CO_2 level and lowering of CO_2 availability reactivates RuBisCO even at elevated temperatures. Studies on *Nicotiana tabacum* (tobacco) and *Gossypium hirsutum* (cotton) also support this conclusion (Crafts-Brandner and Salvucci, 2000). On the other hand, reduction in activation state of RuBisCO due to heat impairment of RCA at elevated temperature will have either of the two consequences, (i) little effect by the reduction of CO_2 availability, or (ii) the activation state of RuBisCO gets declined further. Cen and Sage (2005) in their study have showed how the activity and activation state of RuBisCO vary in response to atmospheric CO_2 in *Ipomoea batatas* (sweet potato) grown under greenhouse conditions. Their study revealed that there is a significant decrease in activation state of RuBisCO on doubling CO_2 availability above 370 mbar partial pressure, while the same increase of CO_2 availability is reduced to one-half. At temperature below 16°C, the activation state of RuBisCO enhances with lowering of the partial pressure of CO_2 (Cen and Sage, 2005).

An extensive study on the effects of $[CO_2]$ on RuBisCO activity inferred the followings:

1. When atmospheric carbon dioxide concentration, that is, $[CO_2]$ >200% (as compared with controlled treatment), there is a decrease in RuBisCO content that drives the RuBisCO activity when plants are exposed to high level of CO_2. Photosynthetic acclimation at elevated $[CO_2]$ depends upon the ability of the species to adjust the "source-sink" equilibrium (Ainsworth et al., 2004). According to this "source-sink" hypothesis, elevated $[CO_2]$ limits the increasing sink strength, resulting in decrease in RuBisCO content and eventually the photosynthetic rates that compensate for activity of the source and capacity of the sink (Aranjuelo et al., 2013). Under the condition of

elevated [CO_2], enhancement of RuBisCO binding inhibitors leads to the protection of RuBisCO minimizing its proteolytic activity (Kane et al., 1998; Pérez et al., 2011).

2. On the contrary, plants fumigated with [CO_2] <200% (as compared with corresponding ambient treatment) resulted in reduction of RuBisCO content and its activation state that suggests a pivotal role of RuBisCO inhibitors during decline of caboxylation process (Galmés et al., 2013).

6.4 EFFECTS OF O_3 ON RUBISCO ACTIVITY

Plant physiological and biochemical processes are greatly influenced by O_3, an abundant air pollutant, and therefore, an overall decline of Pn with subsequent increase in respiration together with chlorosis leads to a drop in plant productivity (Kangasjärvi et al., 1997; Farage and Long, 1995). O_3 concentration in atmosphere reaches to its highest value around middle of the day during summer (Lorenzini et al., 1994). Plants receive O_3 through stomata and all plants respond to O_3 stress differently. This may be due to the fact that all plant species have different stomatal conductance, which create differences in O_3 sensitivity of the plants (Iyer et al., 2013)

RuBisCO, being susceptible to O_3, its activity and transcript levels are reduced during O_3 exposure (Pell et al., 1994). Pelloux et al. (2001) studied the O_3 stress on *Pinus halepensis* (Aleppo pine) and found that a prolonged exposure induced a slight reduction in plant dry weight (Pelloux et al., 2001). Their results are in accordance with previous studies on *Pinus pinea* (Italian stone pine) and *Pinus ponderosa* (bull pine), where the total biomass reduction was 16% and 35%, respectively (Andersen et al., 1997; Edwards et al., 1990). In the case of Aleppo pine, a reduced root/shoot ratio indicated a decrease in carbon allocation into the roots as a result of compensation of increased demand of carbon in detoxifying processes involved in leaves (Matyssek et al., 1995). In rural areas, O_3 present during night time can also significantly reduce the plant biomass, particularly if the O_3 hinders the guard cell from opening in the dark (McCurdy, 1994; Winner et al., 1989; Torsethaugen et al., 1999). A decrease in the total chlorophyll content, that is, chlorosis signifies impairment of Pn due to increased O_3 stress (Wallin et al., 1990). At molecular and biochemical level, there is a 40% reduction in overall RuBisCO activity in the Aleppo pine needles under O_3 stress, which can be correlated to both a reduction in the RuBisCO large subunits (RbcL) protein and *RbcL* transcripts quantities (Pelloux et al., 2001). Decrease in LSU protein quantity reflects on the minimal requirement for the RuBisCO activation under O_3 stress that could regulate the RCA synthesis and degradation characterizing the aging and maturation of leaves. The O_3-induced RuBisCO reduction thus offers premature senescence of plants.

A number of studies carried out by several groups shed light on the effects of O_3 on RuBisCO. There is 50–80% drop-in RuBisCO activity in rice leaves when treated with 0.12 $\mu L\ L^{-1}$ O_3 for 2 h (Nakamura and Saka, 1978). A similar decrease of 36–80% in RuBisCO activity without foliar symptoms is observed when alfalfa, also called lucerne, a perennial flowering plant is kept under 0.25 $\mu L\ L^{-1}$ O_3 for 2 h (Pell and Pearson, 1983). It has also been reported that RuBisCO activity decreased in flag leaves of field-grown wheat when exposed to 0.1 $\mu L\ L^{-1}$ O_3 for 8 h per day for 2 months (Lehnherr et al., 1987).

In another study, an increase in O_3 proposes a pattern that is moderately independent of [O_3] (Intensity of O_3 stress). Hence, this pattern can be summed up as a comparative decrease in the preliminary and overall activities up to around 70% of the quantities estimated under the controlled conditions, which can be defined by a comparable decrease in RuBisCO concentration (Galmés et al., 2013). This result is as per the noticeable decrease in the amounts of *RbcS* and *RbcL* mRNA transcripts, and also LSU polypeptides in plants exposed to O_3 (Dizengremel, 2001). Likewise, O_3 effect results in enhanced oxidizing stress in chloroplast, presumably changing the form of RuBisCO into an unstable structure that might be easily broken down by proteases (Pell et al., 1997). Despite the fact that the concentration of RCA reduces on the exposure of O_3, the carbamylation process of RuBisCO in O_3-treated plants has not been shown to change significantly. This may be due to a steady or rising [RuBisCO]/[RCA] ratio. Contrary to this general trend, O_3 induces phytotoxic effects on RuBisCO activity as well as concentration of RCA depends on various factors such as the species or the genotypes of the species, duration of exposure, or the age of the leaf where immature leaves are less sensitive to O_3 than the matured ones (Pell et al., 1997; Lütz et al., 2000; Wittig et al., 2007).

6.5 PATTERNS OF THE C_3 PHOTOSYNTHETIC RESPONSE TO TEMPERATURE

Temperature is one of the major determinants in controlling the growth and development of a plant and each plant species requires an optimum temperature for its normal growth (Szymańska et al., 2017). Global climate change is a prime factor that is responsible for temperature rise (daily/seasonal/annual mean temperature). This rise in global temperature is well experienced by plants and reflected in their growth. Pn is considered as a temperature-sensitive physiological process. Most of the plant species show remarkable ability to adjust their efficiency in terms of Pn to varied temperature conditions (Yamori et al., 2012). Temperature stress (heat and cold) can alter cellular homeostasis by unfavorably affecting the normal physiological and/or biochemical processes such as respiration, Pn, water balance, and membrane stability (Awasthi et al., 2015). Thus, the inherent capacities for thermal acclimation of Pn vary among the plants having various photosynthetic routes such as C_3, C_4, and Crassulacean acid metabolism plants (Yamori et al., 2014). Out of them, C_4 plants were found to have a higher ability for photosynthetic acclimation to elevated temperature than the C_3 plants (Osborne et al., 2008). The C_3 plants address around 85% of all higher plant species and C_3 Pn was reported to show a temperature optimum between 20°C and 35°C (Sage et al., 2008). Species acclimated to cooler temperatures were found exhibit lower temperature optima, while species adapted to the high temperature showed thermal threshold exceeding 35°C at elevated CO_2 concentration (Sage and Kubien, 2007). Thermal response of Pn was found to depend on the particular temperature response of CO_2 diffusion and biochemical constraints modulating the Pn rate. It was found that at elevated temperature, the relative rates of both light and dark respiration increased as compared with the ability for RuBisCO carboxylation in vivo (Sage et al., 2008). In the C_3 plants, Pn was found to be restricted by the RuBisCO efficiency, regeneration of RuBP, and/or regeneration of Pi (Farquhar et al., 1980). The rate of CO_2 diffusion through stomatal and mesophyll conductance, the amount and activation of RuBisCO are also known to seriously affect the RuBP carboxylation and thereby net photosynthetic rate (Yamori et al., 2014). At limiting concentrations of CO_2, RuBP was found to be the pivotal parameter for carboxylation with saturation of RuBP, which is known as major limiting step of Pn (Yamori et al., 2014). On the contrary, rate of overall Pn and RuBP saturation was reported to be limited by RuBP regeneration, which is again regulated by either the chloroplast electron transport system to generate ATP and NADPH or the activity of various Calvin cycle enzymes involved in RuBP regeneration such as SBPase, etc. (Raines, 2006; Yamori et al., 2011a, 2011b).

6.5.1 *LIMITATIONS BY RCA AT ELEVATED TEMPERATURE*

Investigating the rate-limiting step of Pn at different temperature scenarios ends up in basic understanding about the temperature-responsive mechanism of Pn and productivity to the various temperature. Global increase of temperature associated with CO_2 levels since the beginning of industrial revolution directly affects the total carbon balance and thereby alters the photosynthetic carbon fixation (Jensen, 2000). High temperature is considered as one of the important abiotic stresses that affects the overall plant growth-development-productivity worldwide. With a rise in temperature (up to 50 °C) the rate of RuBisCO carboxylation increases, while the solubility of CO_2 RuBisCO specificity for CO_2 decreases (Salvucci et al., 2001). These situations favor RuBP oxygenation by RuBisCO over carboxylation in combination with light, and dark respiration lowers the potential of CO_2 fixation at high temperature (Kobza and Edwards, 1987; Salvucci, 2008). The limitation of RuBisCO is considered as one of the major limitations for Pn due to the limited efficiency of completely activated RuBisCO to utilize RuBP. On the other hand, a RCA limitation is considered as the decreased capability of RuBisCO to utilize RuBP by the deactivation of catalytic site to such an extent that RuBP consumption-ability falls under the regeneration capacity of RuBP (Sage et al., 2008). Cessation of photosynthetic activity above the optimum temperature is therefore considered as a result of deactivation of RuBisCO and thereby RCA activity (Yamori et al., 2006; Galmés et al., 2013). Moderate heat stress above temperature optimum was also reported to hinder electron transport chain, RuBP generation, and ATP synthesis (Carmo-Silva and Salvucci, 2011; Schrader et al., 2004; Yamori et al., 2008). Pn above the extreme optimum temperature is reported to be gradually restricted due to inhibition of electron transport components of the photosynthetic process (Salvucci and Crafts-Brandner, 2004). Similar experimental evidences on elevated temperature-induced RuBisCO deactivation have been reported in *Arabidopsis*, spinach, wheat, cotton, tobacco, pea, and rice (Kobza and Edwards, 1987; Crafts-Brandner and Law, 2000; Law and Crafts-Brandner, 1999; Crafts-Brandner and Salvucci, 2000; Sharkey et al., 2001; Haldimann and Feller, 2005; Kim and Portis, 2005; Salvucci et al., 2006; Yamori et al., 2006; Makino and Sage, 2007).

6.6 RUBISCO ACTIVITY UNDER DROUGHT CONDITION

Drought is considered as one of the emerging global concerns, which affect plant growth, development, and productivity due to global climate warming, particularly in arid and semiarid regions (Parry et al., 2002; Zhou et al., 2007). Drought stress imparts many biochemical, physiological, metabolic, and molecular responses in plants, where Pn being the primary physiological concern, also hampers the crop yield (Zhou et al., 2007). Stomatal closure in response to the drought restricts the entry of atmospheric CO_2 to the leaf mesophylls and thereby limits CO_2 assimilation and water loss (Parry et al., 2002). Stomatal limitation was found to be responsible for declining the rate of Pn at the onset of drought, followed by series of metabolic and biochemical events such as limited photophosphorylation, RuBP regeneration, and RuBisCO activity (Gunasekera and Berkowitz, 1993; Tezara et al., 1999; Cornic, 2000; Parry et al., 2002). Impaired ATP synthesis and ATP availability due to the reduced electron transport rate were reported to be major metabolic responses to drought stress, which in turn restrict RuBP regeneration, thus limiting the rate of atmospheric CO_2 fixation in plants (Galmés et al., 2007; Tezara et al., 1999; Flexas et al., 2004; Singh et al., 2014). The activation of intracellular RuBisCO with atmospheric CO_2 resulted due to carbamylation of a lysyl residue in the active site of the enzyme, and the concentration of tight-binding inhibitors (Galme´s et al., 2011). It was also observed that a certain number of the RuBisCO catalytic sites are occupied by the strongly bound inhibitors under severe drought conditions, which decrease the catalytic sites available for carboxylation of RuBisCO (Galme´s et al., 2011, 2013). Holaday et al., (1992) showed that concentration variation of RuBisCO has been witnessed under water stress; however, the RuBisCO concentration was found to be sufficient without limiting the photosynthetic process (Pickersgill, 1986; Warren and Adams, 2004). Increase in strongly bound RuBisCO inhibitors was previously accounted in many of the plant species under moderate to severe water deficit conditions and thus RuBisCO regulation can be attributed to two factors: (i) the activation state, and (ii) carbamylation of RuBisCO (Parry et al., 2002; Carmo-Silva et al., 2010). However, under extreme drought conditions, RuBisCO becomes inactivated due to the decarbamylation of its catalytic sites, which is probably due to the lower CO_2 availability and the degree of stress response of the various plant species was found to be regulated by degree of CO_2 starvation and decarbamylation of the RuBisCO catalytic sites (Galme´s et al., 2011, 2013). The RuBisCO holoenzyme was found to be stable having a half-life of several days even under the drought stress, indicating that RuBisCO limited the photosynthetic process if the drought stress continued for longer periods (Webber et al., 1994; Flexas et al., 2006). The impact of drought on RuBisCO activity depends on its intensity and plant species. Several investigations have reported a remarkable reduction in RuBisCO activity whereas a few other studies have found little or no inhibition of its activity particularly in vivo (Pelloux et al., 2001; Parry et al., 2002; Bota et al., 2003; Zhou et al., 2007).

6.7 EFFECTS OF SALINITY AND pH ON RUBISCO ACTIVITY

Salt stress is one of the major threats, which adversely affect the crop quality and productivity around the globe. Increased saline conditions in agricultural lands adversely affect plant metabolism, enzymatic dysfunction, nutrient imbalance, and osmotic potential (Xu et al., 2016). Excessive intracellular accumulation of various salt concentrations (Cl^- and, especially, Na^+) in plants imparts ionic imbalance and osmotic stress, which in turn hamper normal physiological functions like Pn, mineral homeostasis, etc. A higher level of salt concentration is known to affect net Pn by irreversibly damaging the photosynthetic apparatus and thereby causing reduced plant growth and crop yield (He et al., 2014). Damage induced by salt stress to the photosynthetic method is reflected by carboxylation/oxygenation efficiency and RuBisCO enzyme activity, modulation of various plastidial proteins, function of thylakoid membrane, ATP generation, and the rate of electron transport system (Chaves et al., 2009; Lee et al., 2014; Taj and Challabathula, 2021). Several studies have indicated that limitation in the content or activity of RuBisCO is one of the major factors in salt stress-related downregulation of photosynthetic activity in plants (Parry et al., 2002; James et al., 2006; He et al., 2013). Reduced rate of net Pn is also found to be associated with lower intracellular CO_2 concentration, which in turn makes less CO_2 available to bind with RuBisCO enzyme and thereby enhancing the photorespiration rate (Igamberdiev, 2015). The lower availability of CO_2 was due to the restricted CO_2 diffusion in the leaves caused by reduced stomatal and internal conductance. Mae et al. (1983) showed a rapid degradation of RuBisCO and translocation of nitrogen in developing organs during leaf senescence. Salinity-induced reduction of RuBisCO activity was found to be observed in salt-sensitive soybean, which was correlated with the reduction in RuBisCO large subunit contents (Lu et al., 2009). Similar type of findings on effect of salt stress in relation with reduction of RuBisCO content and inhibition of RCA activity was observed in spinach, tobaco, ginger, *Prunus*, sugar beet, etc (Delfine et al., 1998; Wang et al., 2017; Lee et al., 2014; Wang et al., 2021; Yin et al., 2021). Further, ultrastructural changes of the chloroplast such as the development of various plastoglobuli, disintegrated thylakoid membranes, and vesicles of RuBisCO or RuBisCO-containing bodies (RCB) were found to be reported by several workers with the degradation of RuBisCO during salt stress in plants (Krupinska, 2006; Prins et al., 2008). Formation of chloroplast protrusion body (CP) is considered as one of the unique ultrastructural changes under salinity stress (Paramonova et al., 2004). At higher salt concentration (150 mM NaCl), reduced RuBisCO content and activity were associated with excess production of CPs and RCBs, which significantly hampered net Pn rate (He et al., 2014). However, the physiological role of CPs has not been elucidated in detail till date.

Soil acidification also has adverse effects on plant growth, development, and crop productivity, all over the world, especially in arid and semiarid regions (Ryu and Cho, 2015). The soil acidification has become a concern due to the improper applications of various chemical fertilizers, poor irrigation practises, industrial pollution, and seawater intrusion in the coastal areas (Gaudio et al., 2015; Ding et al., 2020). The catalysis and activation of RuBisCO are greatly dependent on pH and alkalinization was found to increase the RuBP carboxylase equilibrium activation state at air level CO_2 in vitro (Lorimer et al., 1976). Mott and Berry (1986) showed that the rate of activation by RuBP carboxylase is reduced at low pH in the presence of RUBP and a high steady-state activation of RuBP carboxylase was evident at pH 8.6 and above in the presence of RuBP at air level CO_2. Low pH of soil induces severe oxidative stress through enhanced production of reactive oxygen species and thereby affects the net CO_2 assimilation during Pn, RuBisCO activity, quantum yield of photosystem II, and electron transport system (Zhang et al., 2014). Long et al. (2017) based on experimental evidences reported reduced leaf CO_2 assimilation, stomatal conductance, transpiration, and RuBisCO activity at low pH (pH 2.5) as compared to pH 3–6 in two *Citrus* species (C. *sinensis* and C. *grandis*) (Long et al., 2017). However, higher pH had insignificant effects on seedling growth and development of both the *Citrus* species. On the contrary, low pH was found to improve the net Pn by reducing the absorption of sodium (Na^+) ions, and increased the mineral contents like potassium (K), magnesium (Mg), iron (Fe), and zinc (Zn) in the leaves of ginger plants under the influence of salt stress (Yin et al., 2020).

6.8 EFFECTS OF HMS

Most of the toxic metals belong to transition elements of the periodic table having atomic masses >20 with specific gravity >5 g cm^{-3} (Rascio and Navari-Izzo, 2011). In terms of biology, "HMs" refer to those metals and/or metalloids that are toxic to plant and animal kingdom even at trace amounts (Rascio and Navari-Izzo, 2011). However, metals like copper (Cu), Fe, Zn, manganese (Mn), molybdenum (Mo), chromium (Cr), selenium (Se), and nickel (Ni) present at low concentration are regarded as essential for the growth and metabolism of living beings; while higher concentrations of the same proved to be toxic. Although the concentration of metals in soil can be lesser than 1 mg kg^{-1} up to as high as 1×10^5 mg kg^{-1}, rise in various human activities adds up to their concentration in the environment (Ashfaque et al., 2016). HMs like arsenic (As), cadmium (Cd), Cr, Cu, lead (Pb), mercury (Hg), Ni, Se, and Zn produced from metal industries, mining, burning of fossil fuels, thermal power stations, automobiles, animal rearing, agrochemicals, waste disposal, etc. are regarded as environment pollutants (Dubey, 2010). Agency for Toxic Substances and Disease Registry (ATSDR) declared As, Pb, Cd, and Hg as the most toxic elements due to their toxicity, availability, potential exposure, and frequency of occurrence (ATSDR, 2019). The metals present in soil can enter the plants through the roots (Pourrut et al., 2011). Reports on foliar HM uptake by crops cultivated around metal-recycling industries are found as another major route to enter the plants (Uzu et al., 2010; Schreck et al., 2013). This transfer of HMs from soil to roots can be (i) indirect that involves adsorption of metal ions on to the roots followed by bonding of uronic acid to the root through carboxylic groups, or (ii) direct via polysaccharide mucilage of rhizoderm cell surface (Pourrut et al., 2011). In addition, metal uptake completes via several other steps such as desorption of metal from soil, transportation of metals to the roots, metal uptake by roots, and finally translocation of these metals to other parts of the plants (Saifullah et al., 2009). Once HMs enter into plants, there are various toxic effects as shown in Fig. 3. Bioavailability of a metal in soil depends on its amount that is easily accessible or made accessible over a period of time to any organism from the environment (Pauget et al., 2012). Most of the HMs, particularly Cu and Pb are comparatively less soluble and thus readily available for plants (Punamiya et al., 2010; Sammut et al., 2010; Vega et al., 2010). Although total HM concentration in soil is high, strong bonding of these metals with soil components makes only a small fraction of HMs phytoavailable. Among all the HMs, Cd is readily available to plants as it primarily exists in soil solution or solid phase (Verbruggen et al., 2009).

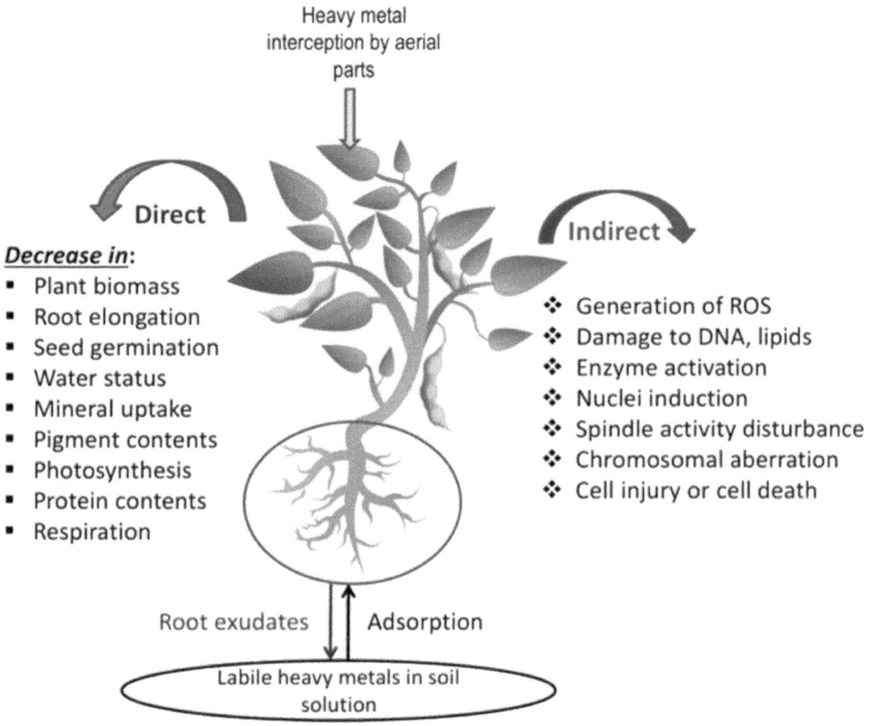

FIGURE 6.3 Reduced crop production upon HM uptake by plant roots and their direct and indirect toxic impacts due to ROS (Adapted from Shahid et al., 2015).

Effects of HMs on RuBisCO and then on Pn are well documented in literature, particularly focusing on the toxicity of Cd (Chaffei et al., 2004; Krantev et al., 2008; Ying et al., 2010). Reports suggest that the Cd concentration ([Cd]) in the media (soil/air/water) plays a vital role in regulation of RuBisCO activity. Galmés et al. (2013) statistically analyzed the RuBisCO activity under different [Cd] concentrations and observed that its initial activity remains almost same at [Cd] < 50 µM, whereas the same gets reduced upon exposed to [Cd] >50 µM (Galmés et al., 2013). Their study did not find any concrete correlation of RuBisCO activity with its concentration difference, although several workers linked these variations with change in RuBisCO concentration on comparing controlled and uncontrolled plants (Pietrini et al., 2003; Chaffei et al., 2004; Ying et al., 2010). Different mechanisms have been proposed to justify the effect of Cd on RuBisCO activity. Binding of bivalent cations excluding Mg^{2+} shifts the RuBisCO activity toward RuBP oxygenation (Christeller and Laing, 1979; Robison et al., 1979). Alternatively, Cd^{2+} is likely to reduce the carboxylase activity by replacing the Mg^{2+} ions in the catalytic sites of the enzyme (Dias et al., 2013). Another mechanism suggested a decrease in RuBisCO carboxylase activity resulting in minimum chlorophyll synthesis under [Cd] stress (Pietrini et al., 2003). Irreversible dissociation of the RuBisCO LSU and SSU occurred under high [Cd] that eventually led to complete inhibition of RCA (Malik et al., 1992). Later, it has been shown that binding of thiol residues of the enzyme with Cd^{2+} makes RuBisCO inactive (Portis, 2003) followed by the formation of several active oxygen species, for example, hydrogen peroxide (H_2O_2), hydroxyl radical (•OH), and superoxide anion ($O_2^{\bullet-}$) along with lipid peroxides, damaging the biopolymer and cell membrane by causing oxidative stress (Romero-Puertas et al., 2004; Heyno et al., 2008). Cu is primarily absorbed by the roots causing physiological imbalances in the plants (Påhlsson, 1989). If Cu is absorbed excessively, decrease in biomass due to chlorosis and hindrance of electron transport in Pn takes place (Quartacci et al., 2000). Similarly, zinc (Zn) also interferes several crucial physiological processes like plant growth inhibition, toxic lipid peroxide formation, etc (Panda et al., 2003). Arsenic affects the stomatal conductance, restricts CO_2 conacentration, and thus inhibits RuBisCO activity (Silva, 2008). Parys et al. (1998) suggested that RuBisCO activity is potentially decreased by 20% upon treatment with 5 mmol dm^{-3} of Pb $(NO_3)_2$ after 24 h (Parys et al., 1998). The effect of HMs on various plants depends on their concentration in soil. Different species has different level of sensitivities towards each HM. A recent qualitative analysis of RuBisCO under HM stress showed that both large and small subunits of RuBisCO in *Spirulina platensis* are adversely affected under high Ni concentration (Singh and Singh, 2020).

6.9 CONCLUSION

Plants respond to various stressors depending on their stage of development and the environmental criteria to which they are exposed to or treated. Global climate change is an inevitable phenomenon that creates a variety of abiotic stress conditions affecting the plant life cycle by imposing physiological, morphological, biochemical, and molecular changes. Plant metabolism gets interrupted by the influence of abiotic stress: single or combination of stressors. These abiotic stressors hinder the photosynthetic activities, plant growth and development, and their yield. Stomatal closure is the primary reason in leaf Pn reduction, while photosynthetic mechanism remains unchanged, thus making the leaf responsive toward vapor pressure variation. High CO_2 and O_3 atmosphere concentration influence photosynthetic methods by changing the capacity of Pn as well as stomatal conductance, which eventually affect the ability of the plants to water uptake and sequester carbon, and finally alter the hydrological and carbon cycles. Reduced photosynthetic process in plants can be linked to a number of factors like:

- Change in optimum temperature that is necessary for Pn.
- Decrease in CO_2 supply due to stomatal closure.
- Dehydration of cell membrane in drought condition.
- Effect of salinity.
- HM toxicity.
- Change in RuBisCO activity due to alteration in cytoplasmic structure.

RuBisCO, being the most important enzyme for Pn, degrades by the influence of abiotic stress. For instance, stomatal closure restricts the process of synthesis by lowering CO_2 availability for assimilation by the enzyme. Moreover, rise in leaf/canopy temperature due to inability of evaporative cooling imposes metabolic restrictions on Pn via RuBisCO inactivation. In terms of future perspectives, RuBisCO degradation, study of proteolytic enzyme, and investigation of photosynthetic process at subcellular level under various abiotic stress must be systematically elucidated. Nevertheless, plants adapt to environmental conditions and stress to survive even in adverse circumstances. However, human activity must not cross the threshold beyond which plants fail to evolve themselves to survive.

KEYWORDS

- **RuBisCO activase**
- **photosynthesis (Pn)**
- **CO_2**
- **drought**
- **heavy metals (HMs)**
- **ozone (O_3)**
- **salinity**
- **temperature**
- **water stress**

REFERENCES

Ainsworth, E. A.; Rogers, A.; Nelson, R.; Long, S. P. Testing the "Source–Sink" Hypothesis of Down-Regulation of Photosynthesis in Elevated [CO_2] in the Field with Single Gene Substitutions in Glycine Max. *Agric. For. Meteorol.* **2004,** *122*, 85–94.

Andersen, C. P.; Wilson, R.; Plocher, M.; Hogsett, W. E. Carry-Over Effects of Ozone on Root Growth and Carbohydrate Concentrations of Ponderosa Pine Seedlings. *Tree Physiol.* **1997,** *17* (12), 805–811.

Aranjuelo, I.; Sanz-Sáez, Á.; Jauregui, I.; Irigoyen, J. J.; Araus, J. L.; Sánchez-Díaz, M.; Erice, G. Harvest Index, a Parameter Conditioning Responsiveness of Wheat Plants to Elevated CO_2. *J. Exp. Bot.* **2013,** *64* (7), 1879–1892.

Ashfaque, F.; Inam, A.; Sahay, S.; Iqbal; S.Influence of Heavy Metal Toxicity on Plant Growth, Metabolism and Its Alleviation by Phytoremediation-a Promising Technology. *J. Agric. Ecol. Int.* **2016,** *6* (2); 1–19.

Atkinson, N. J.; Urwin, P. E. The Interaction of Plant Biotic and Abiotic Stresses: From Genes to the Field. *J. Exp. Bot.* **2012,** *63*, 3523–3543.

ATSDR. ATSDR, Agency for Toxic Substances and Disease Registry. ATSDR's Substance Priority List. 2019. https://www.atsdr.cdc.gov/spl/index.html (accessed May 22, 2021).

Awasthi, R.; Bhandari, K.; Nayyar, H. Temperature Stress and Redox Homeostasis in Agricultural Crops. *Front. Environ. Sci.* **2015,** *3*, 11. https://doi.org/10.3389/fenvs.2015.00011

Badger, M. R.; Lorimer, G. H. Interaction of Sugar Phosphates with the Catalytic Site of Ribulose-1, 5-Bisphosphate Carboxylase. *Biochemistry* **1981,** *20* (8), 2219–2225.

Bale, J. S.; Masters, G. J.; Hodkinson, I. D.; Awmack, C.; Bezemer, T. M.; Brown, V. K.; Butterfield, J.; Buse, A. ; Coulson, J. C.; Farrar, J. Herbivory in Global Limate Change Research: Direct Effects of Rising Temperature on Insect Herbivores. *Global Change Bio.* **2002,** *8*, 1–16.

Bauer, D.; Biehler, K.; Fock, H.; Carrayol, E.; Hirel, B.; Migge, A.; Becker, T. W. A Role for Cytosolic Glutamine Synthetase in the Remobilization of Leaf Nitrogen during Water Stress in Tomato. *Physiol. Plant.* **1997,** *99*, 241–248.

Carmo-Silva, A. E.; Salvucci, M. E. The Regulatory Properties of Rubisco Activase Differ Among Species and Affect Photosynthetic Induction during Light Transitions. *Plant Physiol.* **2013,** *161*, 1645–1655.

Carmo-Silva, A. E.; Keys, A. J.; Andralojc, P. J.; Powers, S. J.; Arrabaça, M. C.; Parry, M. A. Rubisco Activities, Properties, and Regulation in Three Different C_4 Grasses under Drought. *J. Exp. Bot.* **2010,** *61*, 2355–2366.

Cen, Y. P.; Sage, R. F. The Regulation of Rubisco Activity in Response to Variation in Temperature and Atmospheric CO_2 Partial Pressure in Sweet Potato. *Plant Physiol.* **2005,** *139* (2), 979–990.

Chaffei, C.; Pageau, K.; Suzuki, A.; Gouia, H.; Ghorbel, M. H.; Masclaux-Daubresse, C. Cadmium Toxicity Induced Changes in Nitrogen Management in *Lycopersicon esculentum* Leading to a Metabolic Safeguard through an Amino Acid Storage Strategy. *Plant Cell Physiol.* **2004,** *45*, 1681–1693.

Chaves, M. M.; Flexas, J.; Pinheiro, C. Photosynthesis under Drought and Salt Stress: Regulation Mechanisms from Whole Plant to Cell. *Ann. Bot.* **2009,** *103*, 551–560.

Christeller, J. T.; Laing, W. A. Effects of Manganese Ions and Magnesium Ions on the Activity of Soya-Bean Ribulose Bisphosphate Carboxylase/Oxygenase. *Biochem J.* **1979,** *183* (3), 747–750.

Cornic, G. Drought Stress Inhibits Photosynthesis by Decreasing Stomatal Aperture—Not by Affecting ATP Synthesis. *Trends Plant Sci.* **2000,** *5*, 187–188.

Crafts-Brandner, S. J.; Law, R. D. Effect of Heat Stress on the Inhibition and Recovery of the Ribulose-1,5-bisphosphate Carboxylase/Oxygenase Activation State. *Planta* **2000,** *212* (1), 67–74.

Crafts-Brandner, S. J.; Salvucci, M. E. Rubisco Activase Constrains the Photosynthetic Potential of Leaves at High Temperature and CO_2. *Proc. Natl. Acad. Sci. USA* **2000,** *97*, 13430–13435.

Crafts-Brandner, S. J.; Salvucci, M. E. Analyzing the Impact of High Temperature and CO_2 on Net Photosynthesis: Biochemical Mechanisms, Models and Genomics. *Field Crops Res.* **2004,** *90*, 75–85.

Delfine, S.; Alvino, A.; Zacchini, M.; Loreto, F. Consequences of Salt Stress on Conductance to CO_2 Diffusion, Rubisco Characteristics and Anatomy of Spinach Leaves. *Funct. Plant Biol.* **1998,** *25*, 395–402.

Desimone, M.; Henke, A.; Wagner, E. Oxidative Stress Induces Partial Degradation of the Large Subunit of Ribulose-1,5-bisphosphate Carboxylase/Oxygenase in Isolated Chloroplasts of Barley. *Plant Physiol.* **1996,** *111*, 789–796.

Dias, M. C.; Monteiro, C.; Moutinho-Pereira, J.; Correia, C.; Gonçalves, B.; Santos, C. Cadmium Toxicity Affects Photosynthesis and Plant Growth at Different Levels. *Acta Physiol. Plant.* **2013,** *35* (4), 1281–1289.

Dizengremel, P. Effects of Ozone on the Carbon Metabolism of Forest Trees. *Plant Physiol. Biochem.* **2001,** *39*, 729–742.

Dubey, R. S. Metal Toxicity, Oxidative Stress and Antioxidative Defense System in Plants. In *Reactive Oxygen Species and Antioxidants in Higher Plants*; Gupta, S. D., Ed.; CRC Press: Boca Raton, 2010; pp 177–203.

Eckardt, N.; Pell, E. Oxidative Modification of Rubisco from Potato Foliage in Response to Ozone. *Plant Physiol. Biochem.* **1995,** *33*, 273–282.

Edwards, N. T.; Taylor, G. E.; Adams, M. B.; Simmons, G. L.; Kelly, J. M. Ozone, Acidic Rain and Soil Magnesium Effects on Growth and Foliar Pigments of Pinus taeda L. *Tree Physiol.* **1990,** *6* (1), 95–104.

Farage, P. K.; Long, S. P. An in Vivo Analysis of Photosynthesis during Short-Term O_3 Exposure in Three Contrasting Species. *Photosynth. Res.* **1995,** *43* (1), 11–18.

Farquhar, G. D.; Von Caemmerer, S.; Berry, J. A. A Biochemical Model of Photosynthetic CO_2 Assimilation in Leaves of C_3 Species. *Planta*, **1980,** *149*, 78–90.

Feller, U.; Anders, I.; Mae, T. Rubiscolytics: Fate of Rubisco after Its Enzymatic Function in a Cell Is Terminated. *J. Exp. Bot.* **2008,** *59* (7), 1615–1624.

Feller, U.; Crafts-Brandner, S. J.; Salvucci, M. E. Moderately High Temperatures Inhibit Ribulose-1,5-Bisphosphate Carboxylase/Oxygenase (Rubisco) Activase-Mediated Activation of Rubisco. *Plant Physiol.* **1998,** *116*, 539–546.

Flexas, J.; Bota, J.; Loreto, F.; Cornic, G.; Sharkey, T. D. Diffusive and Metabolic Limitations to Photosynthesis under Drought and Salinity in C3 Plants. *Plant Biol. (Stuttg).* **2004,** *6*, 269–279.

Flexas, J.; Ribas-Carbó, M.; Bota, J.; Galmés, J;, Henkle, M.; Martínez-Cañellas, S.; Medrano, H. Decreased Rubisco Activity during Water Stress Is Induced by Stomatal Closure, Not by Decreased Relative Water Content. *New Phytol.* **2006,** *172*, 73–82.

Friedrich, J. W.; Huffaker, R. C. Photosynthesis, Leaf Resistances, and Ribulose-1,5-bisphosphate Carboxylase Degradation in Senescing Barley Leaves. *Plant Physiol.* **1980,** *65*, 1103–1107.

Galmés, J.; Abadía, A.; Cifre, J.; Medrano, H.; Flexas, J. Photoprotection Processes under Water Stress and Recovery in Mediterranean Plants with Different Growth Forms and Leaf Habits. *Physiol. Plant.* **2007,** *130*, 495–510.

Galmés, J.; Aranjuelo, I.; Medrano, H.; Flexas, J. Variation in Rubisco Content and Activity under Variable Climatic Factors. *Photosynth. Res.* **2013,** *117* (1), 73–90.

Galmés, J.; Ribas-Carbó, M.; Medrano, H.; Flexas, J. Rubisco Activity in Mediterranean Species Is Regulated by the Chloroplastic CO_2 Concentration under Water Stress. *J. Exp. Bot.* **2011,** *62*, 653–665.

Geiger, D.; Servaites, J. C. Dynamics of Self-Regulation of Photosynthetic Carbon Metabolism. *Plant Physiol. Biochem.* **1994,** *32*, 173–183.

Glick, R. E.; Schlagnhaufer, C. D.; Arteca, R. N.; Pell, E. J. Ozone-Induced Ethylene Emission Accelerates the Loss of Ribulose-1,5-Bisphosphate Carboxylase/Oxygenase and Nuclear-Encoded mRNAs in Senescing Potato Leaves. *Plant Physiol.* **1995,** *109*, 891–898.

Gunasekera, D.; Berkowitz, G. A. Use of Transgenic Plants with Rubisco Antisense DNA to Evaluate the Rate Limitation of Photosynthesis under Water Stress. *Plant Physiol.* **1993,** *103*, 629–635.

Haldimann, P.; Feller, U. Growth at Moderately Elevated Temperature Alters the Physiological Response of the Photosynthetic Apparatus to Heat Stress in Pea (*Pisum sativum* L.) Leaves. *Plant Cell Environ.* **2005,** *28*, 302–317.

He, Y.; Yu, C.; Zhou, L.; Chen, Y.; Liu, A.; Jin, J.; Hong, J.; Qi, Y.; Jiang, D. Rubisco Decrease Is Involved in Chloroplast Protrusion and Rubisco-Containing Body Formation in Soybean (Glycine max.) under Salt Stress. *Plant Physiol. Biochem.* **2014,** *74*, 118–124.

Holaday, A. S.; Ritchie, S. W.; Nguyen, H. T. Effects of Water Deficit on Gas-Exchange Parameters and Ribulose 1, 5-bisphosphate Carboxylase Activation in Wheat. *Environ. Exp. Bot.* **1992,** *32*, 403–410.

Igamberdiev, A. U. Control of Rubisco Function via Homeostatic Equilibration of CO_2 Supply. *Front. Plant Sci.* **2015,** *6*, 106. DOI: 10.3389/fpls.2015.00106.

Irihimovitch, V.; Shapira, M. Glutathione Redox Potential Modulated by Reactive Oxygen Species Regulates Translation of Rubisco Large Subunit in the Chloroplast. *J. Bio. Chem.* **2000,** *275*, 16289–16295.

Irving, L. J.; Robinson, D. A Dynamic Model of Rubisco Turnover in Cereal Leaves. *New Phytol.* **2006,** *169*, 493–504.

Iyer, N. J.; Tang, Y.; Mahalingam, R. Physiological, Biochemical and Molecular Responses to a Combination of Drought and Ozone in *Medicago truncatula. Plant Cell Environ.* **2013,** *36* (3), 706–720.

James, R. A.; Munns, R.; von Caemmerer, S.; Trejo, C.; Miller, C.; Condon, T. A. G. Photosynthetic Capacity Is Related to the Cellular and Subcellular Partitioning of Na^+, K^+ and Cl^- in Salt-Affected Barley and Durum Wheat. *Plant Cell Environ.* **2006,** *29*, 2185–2197.

Jensen, R. G. Activation of Rubisco Regulates Photosynthesis at High Temperature and CO_2. *Proc. Natl. Acad. Sci.* **2000,** *97*, 12937–12938.

Kane, H. J.; Wilkin, J. M.; Portis, A. R. J.; Andrews, T. J. Potent Inhibition of Ribulose-Bisphosphate Carboxylase by an Oxidized Impurity in Ribulose-1, 5-bisphosphate. *Plant Physiol.* **1998,** *117* (3), 1059–1069.

Kangasjärvi, J.; Talvinen, J.; Utriainen, M.; Karjalainen, R. Plant Defence Systems Induced by Ozone. *Plant, Cell Environ.* **1997,** *17* (7), 783–794.

Kim, K.; Portis, A. R. Temperature Dependence of Photosynthesis in *Arabidopsis* Plants with Modifications in Rubisco Activase and Membrane Fluidity. *Plant Cell Physiol.* **2005,** *46*, 522–530.

Kobza, J.; Edwards, G. E. Influences of Leaf Temperature on Photosynthetic Carbon Metabolism in Wheat. *Plant Physiol.* **1987,** *83*, 69–74.

Krantev, A.; Yordanova, R.; Janda, T.; Szalai, G.; Popova, L. Treatment with Salicylic Acid Decreases the Effect of Cadmium on Photosynthesis in Maize Plants. *J. Plant Physiol.* **2008,** *165*, 920–931.

Krapp, A.; Hoffman, B.; Schafer, C.; Stitt, M. Regulation of the Expression of rbcS and Other Photosynthetic Genes by Carbohydrates: A Mechanism for Sink Regulation of Photosynthesis. *Plant J.* **1993,** *3*, 817–828.

Krupinska, K. Fate and Activities of Plastids DURING Leaf Senescence. In *The Structure and Function of Plastids*; Wise, R. R., Hoober, J. K., Eds., Vol. 23; Springer: Dordrecht, 2006; pp 433–449.

Law, R. D.; Crafts-Brandner, S. J. Inhibition and Acclimation of Photosynthesis to Heat Stress Is Closely Correlated with Activation of Ribulose-1,5-Bisphosphate Carboxylase/Oxygenase. *Plant Physiol.* **1999,** *120*, 173–181.

Lee, S. Y.; Damodaran, P. N.; Roh, K. S. Influence of Salicylic Acid on Rubisco and Rubisco Activase in Tobacco Plant Grown under Sodium Chloride in Vitro. *Saudi J. Biol. Sci.* **2014,** *21*, 417–426.

Lehnherr, B.; Grandjean, A.; Mächler, F.; Fuhrer, J. The Effect of Ozone in Ambient Air on Ribulose Bisphosphate Carboxylase/Oxygenase Activity Decreases Photosynthesis and Grain Yield in Wheat. *J. Plant Physiol.* **1987,** *130* (2–3), 189–200.

Long, A.; Zhang, J.; Yang, L. T.; Ye, X.; Lai, N. W.; Tan, L. L.; Lin, D.; Chen, L. S. Effects of Low pH on Photosynthesis, Related Physiological Parameters, and Nutrient Profiles of Citrus. *Front. Plant Sci.* **2017,** *8*, 185.

Lorenzini, G.; Nali, C.; Panicucci, A. Surface Ozone in Pisa (Italy): A Six-Year Study. *Atmos. Environ.* **1994,** *28* (19), 3155–3164.

Lorimer, G. H.; Badger, M. R.; Andrews, T. J. The Activation of Ribulose-1, 5-Bisphosphate Carboxylase by Carbon Dioxide and Magnesium Ions. Equilibria, Kinetics, a Suggested Mechanism, and Physiological Implications. *Biochemistry* **1976,** *15*, 529–536.

Lu, K. X.; Cao, B. H.; Feng, X. P.; He, Y.; Jiang, D. A. Photosynthetic Response of Salt-Tolerant and Sensitive Soybean Varieties. *Photosynthetica* **2009,** *47*, 381–387.

Lütz, C.; Anegg, S.; Gerant, D.; Alaoui-Sossé, B.; Gérard, J.; Dizengremel, P. Beech Trees Exposed to High CO_2 and to Simulated Summer Ozone Levels: Effects on Photosynthesis, Chloroplast Components and Leaf Enzyme Activity. *Physiol. Plantarum* **2000,** *109* (3), 252–259.

Sage, R. F. A Model Describing the Regulation of Ribulose-1,5-Bisphosphate Carboxylase, Electron-Transport, and Triose Phosphate Use in Response to Light-Intensity and CO_2 in C_3 Plants. *Plant Physiol.* **1990,** *94*, 1728–1734.

Sage, R. F.; Sharkey, T. D. The Effect of Temperature on the Occurrence of O_2 and CO_2 Insensitive Photosynthesis in Field-Grown Plants. *Plant Physiol.* **1987,** *84*, 658–664.

Sage, R. F.; Sharkey, T. D.; Pearcy, R. W. The Effect of Leaf Nitrogen and Temperature on the CO_2 Response of Photosynthesis in the C_3 Dicot Chenopodium Album L. *Aust. J. Plant Physiol.* **1990a,** *17*, 135–148.
Sage, R. F.; Sharkey, T. D.; Seemann, J. R. Regulation of Ribulose-1,5-Bisphosphate Carboxylase Activity in Response to Light Intensity and CO_2 in the C_3 Annuals *Chenopodium album* L. and *Phaseolus vulgaris* L. *Plant Physiol.* **1990b,** *94*, 1735–1742.
Sage, R. F.; Kubien, D. S. The Temperature Response of $C_3$and C_4 Photosynthesis. *Plant Cell Environ.* **2007,** *30*, 1086–1107.
Sage, R. F.; Way, D. A.; Kubien, D. S. Rubisco, Rubisco Activase, and Global Climate Change. *J. Exp. Bot.* **2008,** *59*, 1581–1595.
Saifullah, M. E.; Qadir, M.; de Caritat, P.; Tack, F. M. G.; Du Laing, G.; Zia, M. H. EDTA-Assisted Pb Phytoextraction. *Chemosphere* **2009,** *74*, 1279–1291.
Salvucci, M. E.; Crafts-Brandner, S. J. Inhibition of Photosynthesis by Heat Stress: The Activation State of Rubisco as a Limiting Factor in Photosynthesis. *Physiol. Plant* **2004,** *120*, 179–186.
Salvucci, M. E.; Portis, A. R. J.; Ogren, W. L. A Soluble Chloroplast Protein Catalyses Ribulose Bisphosphate Carboxylase/Oxygenase in Vivo. *Photosyn. Res.* **1985,** *7*, 193–201.
Salvucci, M. E.; Werneke, J.; Ogren, W.; Portis; A. R. J. Purification and Species Distribution of Rubisco Activase. *Plant Physiol.* **1987,** *84*, 930–936.
Salvucci, M. E. Association of Rubisco Activase with Chaperonin-60β: A Possible Mechanism for Protecting Photosynthesis during Heat Stress. *J. Exp. Bot.* **2008,** *59*, 1923–1933.
Sammut, M. L.; Noack, Y.; Rose, J.; Hazemann, J. L.; Proux, O.; Depoux, M.; Ziebel, A.; Fiani, E. Speciation of Cd and Pb in Dust Emitted from Sinter Plant. *Chemosphere* **2010,** *78*, 445–450.
Scales, J. C.; Parry, M. A.; Salvucci, M. E. A Non-Radioactive Method for Measuring Rubisco Activase Activity in the Presence of Variable ATP: ADP Ratios, Including Modifications for Measuring the Activity and Activation State of Rubisco. *Photosynth. Res.* **2014,** *119*, 355–365.
Schrader, S. M.; Wise, R. R.; Wacholtz, W. F.; Ort, D. R.; Sharkey, T. D. Thylakoid Membrane Responses to Moderately High Leaf Temperature in Pima Cotton. *Plant Cell Environ.* **2004,** *27*, 725–735.
Seemann, J. R.; Kobza, J.; d MOORE, B. R. A. N. D. O. N. Metabolism of 2-carboxyarabinitol 1-Phosphate and Regulation of Ribulose-1, 5-Bisphosphate Carboxylase Activity. *Photosynth. Res.* **1990,** *23* (2), 119–130.
Shahid, M.; Khalid, S.; Abbas, G.; Shahid, N.; Nadeem, M.; Sabir, M.; Aslam, M.; Dumat, C. Heavy Metal Stress and Crop Productivity. In *Crop Production and Global Environmental Issues*; Hakeem, K., Ed.; Springer: Cham, 2015; pp 1–25.
Sharkey, T. D.; Badger, M. R.; von Caemmerer, S.; Andrews, T. J. Increased Heat Sensitivity of Photosynthesis in Tobacco Plants with Reduced Rubisco Activase. *Photosyn. Res.* **2001,** *67*, 147–156.
Sheen, J. Metabolic Repression to Transcription in Higher Plants. *The Plant Cell* **1990,** *2*, 1027–1038.
Singh, A. D.; Singh, G. P. Biopigments and Rubisco Expression under Heavy Metal Stress in Spirulina Platensis. *Eco. Environ. Cons. 26 (November Suppl. Issue)* **2020,** S351–S356.
Singh, J.; Pandey, P.; James, D.; Chandrasekhar, K.; Achary, V. M. M.; Kaul, T. et al. Enhancing C_3 Photosynthesis: An Outlook on Feasible Interventions for Crop Improvement. *Plant Biotechnol. J.* **2014,** *12*, 1217–1230.

Spreitzer, R. J.; Salvucci, M. E. Rubisco: Structure, Regulatory Interactions, and Possibilities for a Better Enzyme. *Annu. Rev. Plant Biol.* **2002,** *53*, 449–475.

Suzuki, N.; Rivero, R. M.; Shulaev, V.; Blumwald, E.; Mittler, R. Abiotic and Biotic Stress Combinations. *New Phytol.* **2014,** *203*, 32–43.

Suzuki, Y.; Makino, A.; Mae, T. Changes in the Turnover of Rubisco and Levels of mRNAs of rbcL and rbcS in Rice Leaves from Emergence to Senescence. *Plant Cell Environ.* **2001,** *24*, 1353–1360.

Szymańska, R.; Ślesak, I.; Orzechowska, A.; Kruk, J. Physiological and Biochemical Responses to High Light and Temperature Stress in PLANTS. *Environ. Exp. Bot.* **2017,** *139*, 165–177.

Taj, Z.; Challabathula, D. Protection of Photosynthesis by Halotolerant Staphylococcus Sciuri ET101 in Tomato (*Lycoperiscon esculentum*) and Rice (*Oryza sativa*) Plants during Salinity Stress: Possible Interplay between Carboxylation and Oxygenation in Stress Mitigation. *Front. Microbiol.* **2021,** *11*, 3232.

Tezara, W.; Mitchell, W. J.; Driscoll, S. D.; Lawlor, D. W. Water Stress Inhibits Plant Photosynthesis by Decreasing Coupling Factor and ATP. *Nature* **1999,** *401*, 914–917.

Torsethaugen, G.; Pell, E. J.; Assmann, S. M. Ozone Inhibits Guard Cell K^+ Channels Implicated in Stomatal Opening. *Proc. Natl. Acad. Sci.* **1999,** *96* (23), 13577–13582.

Vega, F. A.; Andrade, M. L.; Covelo, E. F. Influence of Soil Properties on the Sorption and Retention of Cadmium, Copper and Lead, Separately and Together, by 20 Soil Horizons: Comparison of Linear Regression and Tree Regression Analyses. *J. Hazard. Mater* **2010,** *174*, 522–533.

Verbruggen, N.; Hermans, C.; Schat. H. Molecular Mechanisms of Metal Hyperaccumulation in Plants. *New Phytol.* **2009,** *181*, 759–776.

von Caemmerer, S. *Biochemical Models of Leaf Photosynthesis;* CSIRO Publishing: Collingwood, Australia, 2000.

von Caemmerer, S.; Farquhar, G. D. Some Relationships between the Biochemistry of Photosynthesis and the Gas-Exchange of Leaves. *Plants,* **1981,** *153*, 376–387.

Wallin, G.; Skärby, L.; Selldén, G. Long-Term Exposure of Norway Spruce, Picea Abies (L.) Karst., to Ozone in Open-Top Chambers: I. Effects on the Capacity of Net Photosynthesis, Dark Respiration and Leaf Conductance of Shoots of Different Ages. *New Phytol.* **1990,** *115* (2), 335–344.

Wang, X.; Song, Z.; Ti, Y.; Liu, Y.; Li, Q. Physiological Response and Transcriptome Analysis of *Prunus mume* to Early Salt Stress. *J. Plant Biochem. Biotechnol.* **2021,** 1–13.

Wang, Y.; Stevanato, P.; Yu, L.; Zhao, H.; Sun, X.; Sun, F.; Li, J.; Geng, G. The Physiological and Metabolic Changes in Sugar Beet Seedlings under Different Levels of Salt Stress. *J. Plant Res.* **2017,** *130*, 1079–1093.

Warren, C. R.; Adams, M. A. Evergreen Trees Do Not Maximize Instantaneous Photosynthesis. *Trends Plant Sci.* **2004,** *9*, 270–274.

Webber, A. N.; Nie, G. Y.; Long, S. P. Acclimation of Photosynthetic Proteins to Rising Atmospheric CO_2. *Photosyn. Res.* 1994. 39, 413–425.

Werneke, J. M.; Chatfield, J. M.; Ogren, W. L. Alternative Splicing mRNA Generates the Two Ribulose Bisphosphate Carboxylase/Oxygenase Activase in Spinach and Arabidopsis. *Plant Cell* **1989,** *1*, 815–825.

Whitney, S. M.; Houtz, R. L.; Alonso, H. Advancing Our Understanding and Capacity to Engineer Nature's CO_2-Sequestering Enzyme, Rubisco. *Plant Physiol.* **2011,** *155*, 27–35.

Winner, W. E.; Lefohn, S.; Cotter, I. S.; Greitner, C. S.; Nellessen, J.; Mcevoy, L. R.; Olson, R. L.; Atkinson, C. J.; Moore, L. D. Plant-Responses to Elevational Gradients of O_3 Exposures in Virginia. *Proc. Natl. Acad. Sci. USA* **1989,** *86*, 8828–8832.

Wittig, V. E.; Ainsworth, E. A.; Long, S. P. To What Extent Do Current and Projected Increases in Surface Ozone Affect Photosynthesis and Stomatal Conductance of Trees? A Meta-Analytic Review of the Last 3 Decades of Experiments. *Plant Cell Environ.* **2007,** *30* (9), 1150–1162.

Xu, C.; Tang, X.; Shao, H.; Wang, H. Salinity Tolerance Mechanism of Economic Halophytes from Physiological to Molecular Hierarchy for Improving Food Quality. *Curr. Genom.***2016,** *17*, 207–214.

Yamori, W.; Hikosaka, K.; Way, D. A. Temperature Response of Photosynthesis in C3, C4, and CAM Plants: Temperature Acclimation and Temperature Adaptation. *Photosyn. Res.* **2014,** *119*, 101–117.

Yamori, W.; Nagai, T.; Makino, A. The Rate-Limiting Step for CO_2 Assimilation At Different Temperatures Is Influenced by the Leaf Nitrogen Content in Several C_3 Crop Species. *Plant Cell Environ.* **2011a,** *34*, 764–777.

Yamori, W.; Sakata, N.; Suzuki, Y.; Shikanai, T.; Makino, A. Cyclic Electron Flow around Photosystem I via ChloroplastNAD(P)H Dehydrogenase (NDH) Complex Performs a Significant Physiological Role during Photosynthesis and Plant Growth at Low Temperature in Rice. *Plant J.* **2011b,** *68*, 966–976.

Yamori, W.; Suzuki, K.; Noguchi, K.; Nakai, M.; Terashima, I. Effects of Rubisco Kinetics and Rubisco Activation State on the Temperature Dependence of the Photosynthetic Rate in Spinach Leaves from Contrasting Growth Temperatures. *Plant Cell Environ.* **2006,** *29*, 1659–1670.

Yin, F.; Zhang, S.; Cao, B.; Xu, K. Low pH Alleviated Salinity Stress of Ginger Seedlings by Enhancing Photosynthesis, Fluorescence, and Mineral Element Contents. *Peer J.* **2021,** *9*, 10832.

Ying, R. R.; Qiu, R. L.; Tang, Y. T.; Hu, P. J.; Qiu, H.; Chen, H. R.; Shi, T. H. Cadmium Tolerance of Carbon Assimilation Enzymes and Chloroplast in Zn/Cd Hyperaccumulator Picris Divaricata. *J. Plant Physiol.* **2010,** *167*, 81–87.

Zhang, C. P.; Meng, P.; Li, J. Z.; Wan, X. C. Interactive Effects of Soil Acidification and Phosphorus Deficiency on Photosynthetic Characteristics and Growth in *Juglans regia* Seedlings. *Chin. J. Plant Ecol.* **2014,** *38*, 1345–1355.

Zhang, N; Portis, A. J. Mechanism of Light Regulation of Rubisco: A Specific Role for the Larger Rubisco Activase isoform Involving Reductive Activation by Thioredoxin-f. *Proc. Natl. Acad. Sci. USA* **1999,** *96*, 9438–9443.

Zhou,Y.; Lam, H. M.; Zhang, J. Inhibition of Photosynthesis and Energy Dissipation Induced by Water and High Light Stresses in Rice. *J. Exp. Bot.* **2007,** *58*, 1207–1217.

CHAPTER 7

Photosynthetic Response of Crop Plants Under UV Stress

T. T. DHANYA THOMAS, K. P. ASWATHI RAJ, M. S. AMRITHA, and JOS T. PUTHUR*

Plant Physiology and Biochemistry Division, Department of Botany, University of Calicut, C.U. Campus P.O., Kerala, 673635, India

**Corresponding author. E-mail: jtputhur@yahoo.com*

ABSTRACT

Sunlight provides the necessary energy for plant growth and development. Photosynthetic organisms totally depend on sunlight for their source of energy. Ultraviolet (UV) irradiations cause stress to plants. UV radiation above the threshold causes reduction in plant biomass, decreases the integrity of membranes with degradation of enzymes, damages the photosystem proteins, and augments the ROS production. UV radiation directly degrades the photosynthetic pigments and causes damage to the light-harvesting systems in plants. The partial breakdown of thylakoid membrane causes the reduction of photosystem II (PSII) activity during UV stress conditions. The activity of PSII was more reduced under UV stress and it was correlated with the extensive accumulation of reactive oxygen species (ROS) in thylakoid membranes, which hinders the oxygen evolution at the oxygen-evolving complex of PSII. Moreover, UV stress also reduces the leaf photosynthetic rate (*Pn*), stomatal conductance (*gs*), ribulose-1,5-bisphosphate carboxylase/oxygenase content, protein content, nitrogen concentration, and alter photosynthesis-related gene expressions. A high level of UV radiation leads

Photosynthesis and Respiratory Cycles during Environmental Stress Response in Plants. Aryadeep Roychoudhury (Ed.)

to a stress effect in plants, while low levels of UV radiation bring priming effect. UV priming significantly influences the photosynthetic activity, which increases the photosynthetic pigments, photosystem activities, *Pn*, *gs*, etc. This chapter provides an overview of different ways and means by which UV radiation influences the photosynthetic capacity of plants.

7.1 INTRODUCTION

Plants grow in varying environmental conditions, which sometimes turn adverse or create stressful condition for plant growth. These stressful situations are UV radiations, cold, heat, salinity, heavy metal, drought, and nutrient deficiency, which unfavorably affect biomass production and yield and can create food insecurity throughout the world (Ashraf et al., 2018; He et al., 2018). In plants, stress is an external condition that negatively affects its growth, development, and productivity. Different stresses prompt various plant responses such as reduced growth rate, abnormal cellular metabolism, differential gene expression, reduced crop yield, etc. Plants are experienced with various stress conditions that make them to adapt toward specific stress conditions (Gull et al., 2019). Stress may cause cell damage, abnormal metabolism that affects the homeostatic equilibrium, leading to various changes in physiological and chemical parameters (Fathi and Tari, 2016). Abiotic stresses are the major reasons for the reductions in plant growth and food production, which decreases the average yield of crops by 65–87% (León-Chan et al., 2017).

Photosynthesis is a successive redox reaction process and this multistep process takes place while the light-harvesting complexes (LHCs) absorb light energy and transfer it into the photosystem reaction centers through excitations (Baker, 2008). In photosynthesis, CO_2 and water are converted to complex organic molecules like carbohydrates. Photosynthesis converts around 200 billion tons of CO_2 to complex organic molecules and generates around 140 billion tons of oxygen to the atmosphere annually. Solar energy is converted into chemical energy, so that the photosynthesis acts as the prime energy conversion process and is the unavoidable part of the global food chain. Almost every living organism utilizes the complex organic compounds resulting from photosynthesis as the major energy source (Johnson, 2017).

Photosynthesis is the major mechanisms affected by abiotic stresses, which includes reduction in CO_2 diffusion, photosystem II (PSII), and RuBisCO activity. Abiotic stresses can also cause a decrease in the chlorophyll biosynthesis, photosystem performance, electron transport mechanisms, and gas exchange parameters. Various structural and functional components of the photosynthetic apparatus are sensitive to abiotic stresses, of which PSII and RuBisCO are the more vulnerable components (Gururani et al., 2015, Kohli et al., 2017, Singh and Thakur, 2018, Sharma et al., 2019).

The solar radiation spectrum is composed of ultraviolet (UV), visible, and infrared radiations. UV radiations falling on Earth surface have both beneficial and harmful effects on plant growth. Elevated level of UV radiation reaching the earth surface is a major threat of the present era (Choudhary and Agrawal, 2014). UV radiation is one of the main abiotic stresses, which has many harmful impacts on normal plant functions. This may cause various biochemical, physiological, metabolic, and molecular changes (Negrão et al., 2017). UV exposure creates damages to the biomolecules and has detrimental consequence on growth and productivity, resulting in considerable loss in the agricultural sector. Among the various physiological processes, altered photosynthesis is the most critical cause for impaired plant growth and crop reduction. This chapter elaborates on photosynthetic responses of crop plants under UV radiation (Fig. 7.1).

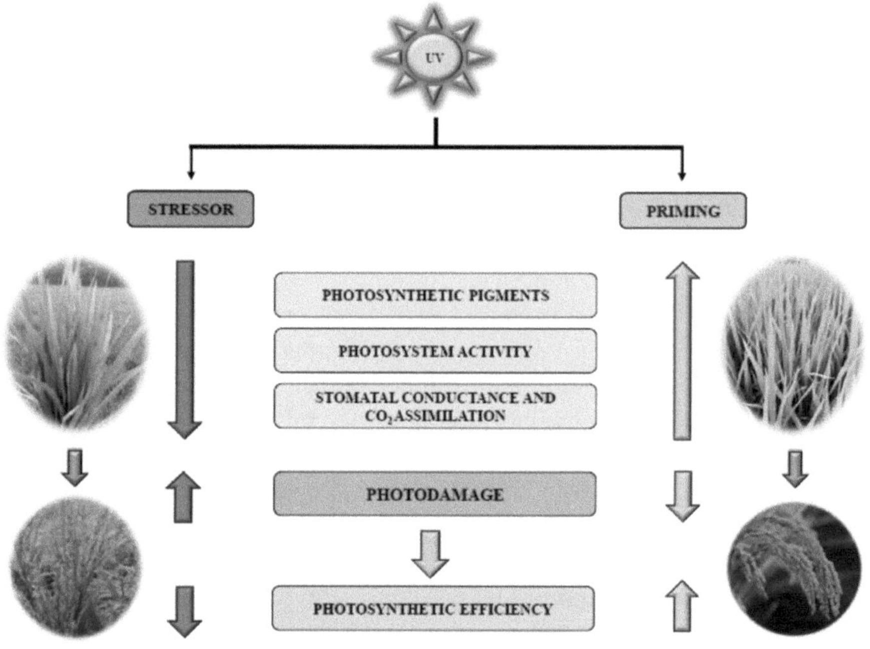

FIGURE 7.1 Impact of UV radiation as a stressor on plants and ameliorative effects of the same as a priming.

7.2 UV RADIATION AS A STRESSOR

In plants, UV radiation acts as a stress factor and the extent of stress is dependent on the wavelength, exposure as well as irradiance time, and on the morphological, genetic as well as protective disposition of species or cultivar of the plant. In UV radiation, UV-C radiation is the most harmful one, with wavelength shorter than 280 nm. It is not present in ambient solar radiation, but is emitted from artificial sources of UV. UV-B radiation is a major environmental stress, because of the stratospheric ozone depletion by chlorofluorocarbons (Tevini, 2004).

UV spectrum is classified into three zones: UV-A (315–400 nm), UV-B (280–315 nm), and UV-C (200–280 nm). The stratospheric ozone depletion in the Antarctic and Arctic areas has been observed since 1974 and 1990, respectively. The major consequence of stratospheric ozone depletion is the major reason for augmentation in the UV radiation reaching the Earth. The ozone layer destruction has been much more intense at high latitudes, mainly in Antarctica, where ozone concentrations have been reduced by 40–50% as related to the status of 1980 and changes have also occurred in the area over Ecuador (3–6%), where UV radiation is intense in nature (UNEP, 2002). UV-B radiation influences plant physiological as well as morphological parameters and negatively impacts its development and proliferation (Jansen et al., 2010). UV-B radiation directly harms the plants through photo modification or photosensitization of proteins, nucleotides, pigments, and lipids (Jordan, 2002). Not only physiological and morphological damages, UV-B can also initiate signal transduction pathways and bring about changes in gene expression that can negatively and positively impact plant growth (Hollósy, 2002; Brosché and Strid, 2003).

Augmented UV radiation may cause direct or indirect effects on plants. Direct effects such as DNA damages by the formation of pyrimidine–pyrimidine photoproducts or cyclobutane dimers, reduce the photosynthetic activity or RuBisCO activity, as well as alter the stomatal function. Direct effects of UV on DNA and phytohormones may cause various indirect morphogenetic changes. Indirect effects include decrease in the growth of leaf and stem of many crop plants like wheat, rice, maize, soybean, and pea, decreased leaf number and branches per plant, decreased fresh and dry mass, altered flowering and fruit ripening time, curling of cotyledons, etc (Tevini, 2004).

UV radiation leads to decreases in plant biomass, reduction of membrane integrity, degradation of enzymes, damages the photosystem I (PSI) and PSII proteins, augments the reactive oxygen species (ROS) production, and decreases the phytohormones and morphogenesis (Kataria et al., 2014; Robson et al., 2015; Köhler et al., 2017; Faseela and Puthur, 2018). Plants have inherent systems to encounter the UV stress conditions, of which the production of secondary metabolites (e.g., epicuticular wax, flavonoids, and anthocyanins) is considered as a major one (Qaderi et al., 2010). UV can stimulate several stress responses, which include physiological and photomorphogenic responses. Mechanisms to encounter UV irradiations in plants are augmented leaf thickness, modification in cuticle thickness, and augmented UV protective pigment production (Shaukat et al., 2013).

Previous studies have reported that UV radiation showed drastic effects on the morphological, biochemical, physiological, and molecular characters of many crop plants (Agrawal and Rathore, 2007; Mishra et al., 2008; Kataria et al., 2014). In response to high UV-B doses, variety of ROS is produced, which oxidizes biomolecules such as phospholipids in the plasma membranes, proteins, and nucleic acids, leading to severe damages to the plant cells (Hideg et al., 2013; Köhler et al., 2017). The significant impact of UV radiation at various levels was reported by various researchers, which includes visual symptoms (e.g., tissue necrosis and chlorosis), distortion in chloroplast ultrastructure and anatomy of leaf (e.g., changes in epidermal thickness and palisade mesophyll layers), photosynthetic pigments, lipid peroxidation, stomatal functioning, photosynthesis, transpiration, and respiration (Kakani et al., 2003; He et al., 2006; Jansen et al., 2010). The UV stress-induced decrease in the photosynthetic efficiency in crop plants is detailed in Table 7.1.

TABLE 7.1 UV Stress Significantly Reduces the Photosynthetic Efficiency in Different Crop Species.

S. no	Crop	UV radiation	Effects on photosynthesis	References
1	Tartary buckwheat	UV-B	Decrease of chlorophyll and carotenoid content	Yao et al. (2006)
2	Barley	UV-B	Decrease of chlorophyll and carotenoid content	Cicek et al. (2012)
3	Wheat and mung bean	UV-B	Decrease of chlorophyll and carotenoid content	Agrawal and Rathore (2007); Yan et al. (2016)
4	Cowpea	UV-B	Decrease of chlorophyll and carotenoid content	Mishra et al. (2008)
5	*Coriandrum sativum*	UV-B	Decrease of chlorophyll and carotenoid content	Kumar and Pandey (2017)

TABLE 7.1 *(Continued)*

S. no	Crop	UV radiation	Effects on photosynthesis	References
6	Rice	UV-B	Reduced chlorophyll and carotenoid content	Faseela and Puthur (2018); Thomas and Puthur (2019, 2020)
7	Cluster bean	UV-A and UV-B	Reduces the photosynthetic pigment content	Joshi et al. (2007)
8	Barley	UV-A and UV-B	Reduces the photosynthetic pigment content	Štroch et al. (2015)
9	*Capsicum longum*	UV-A UV-C	Reduces the chl a, chl b and carotenoid content	Sarghein et al. (2008)
10	Soybean and maize	UV-B	Reduced the total chlorophyll content	Shen et al. (2015)
11	Rice	UV-B	Photosynthetic efficiency, foliar gas-exchange parameters (Pn, gs, and Ci) were reduced	Thomas et al. (2019)
12	Rice	UV-B	Photosynthetic efficiency, foliar gas-exchange parameters (Pn, gs, and Ci) were reduced	Faseela and Puthur (2018)
13	*Pisum sativum*	UV-B	Plastoquinone-binding site of PSII complexes damaged decreased CO_2 assimilation rate	Nogués and Baker (1995)
14	Mung bean	UV-B	Increased the degradation of RuBisCO.	He et al. (2004)
15	Rice	UV-B	Reduced PSI, PSII, and mitochondrial activity	Thomas and Puthur (2019)

7.3 EFFECT OF UV RADIATION ON PHOTOSYNTHESIS

7.3.1 PHOTOSYNTHETIC PIGMENTS

UV radiation has an immense effect on the ratio and the content of pigments of the LHCs of the photosystems (Kosobryukhov et al., 2015). UV radiation influences the photosynthetic pigments, either by inhibition of pigment biosynthesis or by influencing the enzymes included in the chlorophyll (Chl) biosynthetic pathways. UV radiation reduces the Chl *a* content than Chl *b* and thus causes a selective damage of Chl *a* or its precursors (Kataria et al., 2014). Chlorophyll and carotenoid ratio in leaves can be used as a potential indicator of photooxidative damages induced by UV-B exposure (Salama et al., 2011). Carotenoids save the chlorophyll from photo-oxidative damages, so that decreases in carotenoids might enhance the deleterious effects by UV-B radiation on chlorophyll pigments (Mishra et al., 2008; Kataria et al., 2014). UV radiations differently affect the chlorophyll and carotenoid pigments. Usually, carotenoids are less affected by UV radiation than chlorophyll pigments (Yao et al., 2006).

A significant decrease of chlorophyll and carotenoids was seen in tartary buckwheat (Yao et al., 2006), barley (Cicek et al 2012), wheat (Agrawal and Rathore, 2007; Yan et al., 2016), cowpea (Mishra et al., 2008), *Coriandrum sativum* (Kumar and Pandey, 2017), and rice (Faseela and Puthur, 2018; Thomas and Puthur 2019, 2020) due to the UV-B exposure. The Chl *a*, Chl *b*, β-carotene, and lutein content were decreased during UV-B stress condition (Piccini et al 2020). The ratio between Chl *a* and Chl *b* was reduced with an increases of UV-B radiation, as seen in oats and soybeans (Kosobryukhov et al., 2015). UV-A and UV-B radiation reduces the photosynthetic pigment content in cluster bean (Joshi et al., 2007) and barley (Štroch et al., 2015). UV-C exposure reduces the Chl *a*, Chl *b,* and carotenoid content in *Capsicum longum* (Sarghein et al., 2008)

7.3.2 PHOTOSYSTEMS AND PHOTODAMAGES

UV radiation affects the structural and functional units of photosynthesis called as photosystems. They are located in the thylakoid membrane of chloroplast. Based on the wavelength of light at which the maximum performance undertaken, photosystems are termed as PSI and PSII, which have absorbance peaks at 700 and 680 nm, respectively (Nelson and Yocum, 2014). PSII is a multiprotein-pigment complex, an important component in the light reaction of photosynthesis. It catalyzes the electron transfer from water to primary and secondary plastoquinone acceptors. Exposure of plants to high quantity of light, more than that of their requirement, may harm the photosystems. Excessive light energy leads to photodamage of photosystems due to the excessive production of ROS in the chloroplast.

Studies suggest that the extent of photodamage is associated with the wavelength of light spectrum and is highly critical in UV region (Takahashi and Badger, 2010). The main target of the UV radiation is PSII, especially it causes damage to D1 protein and leads to photoinhibition. UV-B irradiation in pea plant downregulates the activity of genes such as *Lhcb* and *psbA* involved in photosynthesis (Mackerness et al., 1998). Similar findings were also noted in pea plants, where the exposure of plants to UV irradiations significantly reduced the activity of PSII, ATP synthase, and RuBisCO (Strid et al., 1990). Moreover, higher dosages of UV-B seem to reduce the activity of PSII in rice seedlings with respect to the control seedlings (Faseela and Puthur, 2018; Thomas and Puthur, 2019). Photosystem stability is a measure of the stress tolerance capacity of a plant. Depending on the intensity of the light and spectral properties, the relative excitation pressure on photosystems changes. Inappropriate sharing of light energy between PSI and PSII results in photoinhibition (Bellafiore et al., 2005).

PHOTOSYSTEM II PROTEIN 33 (PSB33) has vital role in stabilizing PSII, which is involved in the distribution of excitation energy between photosystems, and regulates the process of photoinhibition (Fristedt et al., 2015). It is located in the nonappressed region of thylakoids. It plays important role in maintaining the stability of D1 protein of PSII and the knock-out mutant plants showed increased degradation of D1 protein, which leads to photosystem damage (Fristedt et al., 2017). Mutant *Arabidopsis* plant lacking PSB33, on exposure to UV-A radiation showed reduced transcript level of D1 and D2 proteins of PSII (Nilsson et al., 2020). This in turn reduced the PSII pool in chloroplast. As a result of lower abundance of PSII, the process of photosynthesis gets impaired and results in reduced biomass (Nilsson et al., 2020). Similarly, exposure to high dose of UV-B radiation in pea plant reduces the PSII maximum quantum efficiency. Carbon assimilation was also reduced on further exposure to UV-B, which ultimately leads to photodamage (Nogués and Baker, 1995).

7.3.3 RUBISCO ACTIVITY

Ribulose-1,5-bisphosphate carboxylase-oxygenase (RuBisCO) is the key enzyme involved in the fixation of CO_2. Under UV stress, the activity of RuBisCO is highly affected. Five days of UV-B exposure to *Arabidopsis thaliana,* showed significant decrease in the RuBisCO protein. Larger and smaller subunits of RuBisCO were reduced in plants exposed to 7 days of UV irradiation (Rao et al., 1995). Reduced activity of RuBisCO resulted in reduced carbon assimilation. Long term of UV-B exposure reduced the content and activity of RuBisCO (Strid et al., 1990). Decrease in the mRNA transcript of both larger and smaller subunits of RuBisCO under UV stress led to a sudden fall in the rate of CO_2 assimilation and thereby the rate of photosynthesis (Nogués and Baker, 1995). Studies in *Oryza sativa* revealed that the activity of RuBisCO under UV-B radiation is dependent on the maturation stage of leaves. At the early leaf stage, the level of rbcS and rbcL were reduced, affecting the synthesis process of RuBisCO and an increased degradation of RuBisCO was noted at the time of the maturation of leaves (Takeuchi et al., 2002). Exposure of pea plant to UV irradiations showed a drastic reduction in the mRNA transcript level of RuBisCO subunits, which in turn reduced the content of total soluble proteins (Jordan et al., 1992).

Accumulation of ROS is shown to inhibit the activity of RuBisCO. In *Phaseolus radiatus*, increased level of H_2O_2 under UV-B stress resulted in the proteolytic activity of RuBisCO and subsequent reduction in the amount of RuBisCO. Scavenging of H_2O_2 by the exogenous application of ascorbic acid reduced the rate of RuBisCO degradation under UV-B stress condition (He et al., 2004). Maintenance of the average amount and activity of RuBisCO is essential for the proper functioning of the photosynthetic machinery.

7.3.4 CHLOROPHYLL A FLUORESCENCE

Negative impacts of UV on photosynthesis and photosystems can be studied by utilizing chlorophyll *a* (Chl *a*) fluorescence parameters. Chlorophyll *a* (chl *a*) fluorescence is a measurement that has been used to study the function and performance of photosynthetic machinery of plants and this technique also enables to identify the stress factor that limits plant growth. It also helps for measuring the progressions in the photosynthetic mechanism (Kalaji et al., 2018). Chl *a* fluorescence can be utilized as a perfect tool to test the direct correlation between UV-B stress and photosynthesis (Faseela and Puthur, 2018).

Exposure to UV-B irradiation results in a substantial increase in the dissipation of excitation energy, which can be measured by Chl *a* fluorescence emission (Olsson et al., 2000; Ripoll et al., 2016). However, UV-B stress significantly reduces the maximum fluorescence (Fm), minimum fluorescence, variable fluorescence (Fv)/Fm, Fv, and photochemical efficiency of PSII (ΦPSII) in plants (Hazrati et al., 2016; Reyes et al., 2018). An experiment was conducted by excluding UV-B radiation in *Salix arctica* leaves during the developing season where they get maximal radiation. The leaves got around 40% of the UV-B radiation as they were covered with Teflon and Mylar foil. Decrease in the UV-B irradiance resulted in an increase in Chl *a* fluorescence parameter. Active PSII reaction centers (RC/CSm) as well as electron transport rate (ETM/CSm) and all performance indexes such as performance index on absorption basis, PICSm, and PICSo were found to be elevated in reduced dosages of UV-B, which clearly indicates the negative effects of UV-B on plants (Albert et al., 2005). The effects of UV-B irradiation on PSII were studied in leaves of *Chenopodium album*. The study revealed that the efficiency of PSII was negatively correlated with the increase in intensity and time in the radiation of UV-B. In another study conducted, rice seedlings were subjected to different dosages of UV-B radiations and various parameters such as trapping flux (TR), absorption flux (ABS), electron transport flux (ET), specific membrane model, and dissipated energy (DI) per PSII reaction center (RC) was studied. Exposure to UV-B irradiation significantly decreased ABS/RC, which further proved the ameliorative effect of UV-B on photochemistry (Faseela and Puthur 2018).

Study conducted in *Arabidopsis* leaves states that UV-B irradiation negatively affects the maximum quantum yield of PSII as well as the gene expressions associated with photosynthesis (Sztatelman et al., 2015). Analysis of energy trapping revealed that the UV-B radiation-induced damage occurs initially at the acceptor side, and later at the donor side of PSII; this once again proved the negative correlation between UV-B radiation and Chl *a* fluorescence (van Rensen et al., 2007). Decrease in Chl *a* fluorescence is directly correlated with the reduction in the electron transport activity and it can also be correlated with the increased peroxidation of thylakoid lipids upon UV-B exposure in plants. Under UV-B exposure, the lipids that are required for photosynthetic electron transport activity gets degraded due to heavy exposure of UV-B radiation (Zhu et al., 2008). A study conducted in wheat leaves indicated the decrease in photosynthetic electron transport activity in isolated chloroplasts, showing an increase in the oxidative degradation of polyunsaturated fatty acyl residues of the thylakoid lipids, which decreases the chlorophyll activity and thereby directly reduces the photosynthetic efficiency of the same (Mishra and Singhal 1992).

7.4 GAS EXCHANGE

Gas exchange estimations are nondestructive, nonobtrusive procedures to study photosynthetic carbon assimilation rate in plants. Measurements are fast and can be utilized to analyze the modified photosynthetic rates (*Pn*) under various environmental stress conditions (Souza et al., 2004; Massacci et al., 2008; Wang et al., 2016). UV-B radiations in higher dosages are indirectly correlated with plant performance. The photosynthetic functions can be influenced by UV-B radiation at various levels, especially the leaf morphology that diminishes light capture (Jansen, 2002). *Pn*, transpiration rate (*E*), and the stomatal behavior, generally fluctuates when plants are exposed to UV-B light. It also makes changes in stomatal capacity that limit the accessibility of CO_2, photosynthetic pigments, and photosynthetic genes and enzymes involved in various carbon-fixing pathways (Strid and Porra, 1992).

Platycerium bifurcatum and *Olea europaea* exposed to UV-B showed reduced *Pn* value and stomatal conductance (*gs*) (Sanusi et al., 2011; Koubouris et al., 2015). Lowered *gs* generally reduces the diffusion of ambient CO_2 into mesophyll, which in turn result in the reduction of *Pn* (Larcher 2003). UV-B induced limitations on stomatal closure and nonstomatal (including damage to photosynthetic apparatus) parameters generally lead to reduction of *Pn* in plants (Jiang et al., 2006; Zervoudakis et al., 2012). A study conducted in grapevines with elevated levels of UV-B radiation for three consecutive seasons also showed similar results. Results indicated that limitation in gas exchange due to UV-B radiation directly limited photosynthesis and *gs* of grapevines. Gas exchange was reduced due to the reduction in the leaf area, which further diminished the net carbon fixation and development of the plant. The UV-B induced leaf thickness and stomatal closure caused hindrance in gas exchange (Berli et al., 2013). *Helianthus annus* subjected to UV-B radiation also showed negative impacts of the same on gas exchange.

UV-B radiation decreased various parameters such as photosynthesis, *gs*, and transpiration. Intercellular CO_2 concentrations were decreased in UV-B irradiated plants as a result of diminished *gs* (Cechin et al., 2008). However, in *Populus deltoids* Bartr. ex Marsh (eastern cottonwood), a quickly developing and shade-bigoted angiosperm, exposure to UV-B radiation upgraded the photosynthesis and gas exchange (Bassman et al., 2001). Consequent exposure to UV-B damages the photosynthetic machinery, imposes an additional nonstomatal limitation to various metabolic process (Osmond, 1994; Zlatev et al., 2012). Even low UV-B levels impair the *gs*, net *Pn*, and transpiration rate as a result of reduction in the time taken for stomatal closure due to a loss of turgor pressure along with ion leakage from the guard cells (Yao et al., 2008). Exposure to UV-B rays can induce stomatal closure by directly inhibiting potassium ion (K^+) accumulation and it indirectly affects the *gs* by reducing pore size by inducing some changes in the cell wall elasticity as well as cytoskeleton of the guard cells (Wright and Murphy 1982; Nogues et al., 1999). The study conducted in different rice varieties exposed to UV-B showed that photosynthetic efficiency of seedlings was exceptionally affected. UV stress treatments induced reduction in various foliar gas-exchange parameters such as *Pn*, *gs*, and *Ci*. The profoundly diminished *gs* adversely impacted *Ci* as well as *Pn*; the reduction in *gs* induced further reduction in *Ci* and *Pi* parameters of gas exchange (Thomas et al., 2019).

7.5 REPAIR MECHANISMS OF PHOTODAMAGES

UV radiations cause photoinhibition and the extent of photoinhibition depends on the rate of damage and the concomitant repair mechanism activated within plants. Filtering of UV wavelengths, repairing of UV-induced damage and ROS scavenging, and cyclic electron flow around PSI are the major mechanisms adopted by the plants to avoid the damages due to UV radiations (Takahashi and Badger, 2011). Recovery from photodamage largely depends on PSII repair. It is a complex cycle involving the lateral movement of the monomerized PSII from grana to stromal lamellae, their partial disassembly, proteolytic cleavage of D1 protein, re-synthesis of D1 protein, and assembly of PSII followed by back migration to grana (Bergo et al., 2003; Takahashi and Badger, 2011). It has been shown that irradiation with UV-B hinders the process of PSII repair cycle because the damaged PSII does not move from grana to the stromal lamellae (Barbato et al., 1992).

Plants produce antioxidant enzymes to scavenge the ROS generated during UV stress to mitigate the photodamages induced; furthermore, several secondary metabolites are also involved in the process of UV protection. Especially flavonoids and other phenolics are involved in the detoxification of ROS and preventing lipid peroxidation and membrane damage (Agati and Tattini, 2010; Thomas and Puthur, 2020). It was found that plants tolerate UV stress through the production of carotenoids, epicuticular wax, antioxidants, and secondary metabolites (Thomas and Puthur, 2020). In rice seedlings subjected to UV stress, it was found that an increase in the quantity of UV absorbing compounds such as flavonoids, anthocyanin and increase in the activity of phenylalanine ammonia lyase and cuticular wax content in leaves occurred (Thomas and Puthur, 2020). Under UV stress, tomato UVR8 (SlUVR8) promotes elongation of hypocotyl region in seedlings and accumulation of anthocyanin. Expression of photolyase was also promoted by S1UVR8. This revealed the role of SlUR8 in protection against UV-B stress (Liu et al., 2020). UV-B upregulated the transcription of SlMYB12, involved in biosynthetic pathway of anthocyanins. Similarly, expression of CHS1 and CHS2 encoding key enzymes in biosynthesis of flavonoids were also induced under UV-B.

DNA repair pathways play a vital role in photodamage repair. Maintenance of DNA methylation at the photodamaged sites is accomplished through various DNA repair pathways. UV-induced DNA lesions activate Direct Repair, Global Genome Repair (GGR), and small RNA-mediated GGR, and this prevent the excessive alterations in DNA methylation landscape, which are essential to maintain genome integrity at the damaged sites (Graindorge et al., 2019). Maintenance of rapid repair pathways reduces the harmful effect of exposure to UV radiation.

7.6 UV PRIMING

Plants under the exposure of high UV radiation are causing stress effects, whereas low levels of the UV radiation have positive effects, i.e., priming effect. Several reports have discussed about the positive effects of low-level UV irradiation on plant growth (Hideg et al., 2013; Bornman et al., 2015, Thomas and Puthur 2019; Thomas et al., 2020; Sen et al., 2020). The seeds that are exposed to UV radiation at low levels are a secure way to enhance productivity of crop plants and yield as well as enhances tolerance to different abiotic stresses (Delibaltova and Ivanova, 2006; Thomas and Puthur, 2017). Plants have been found to enhance the photosynthetic efficiency and activate antioxidant systems on low radiation of UV-B and UV-C treatments. The UV priming enhances stress adoption in plants, which include different biochemical, morphological, and physiological variations that beneficially confer the ability of the plants to tolerate various abiotic stresses (Mazza et al., 2013; Bussotti et al., 2014; Thomas and Puthur, 2017). The UV priming increases the chlorophyll synthesis and *Pn* in crop plants (Badridze et al., 2015, 2016; Thomas and Puthur, 2019; Sen et al., 2020). UV priming induced enhancement in photosynthetic features of various plants is detailed in Table 7.2.

TABLE 7.2 UV Priming of Seed/Seedlings of Different Crop Species Enhances Various Photosynthetic Features.

S. no	Crop	UV radiation	Effects on photosynthesis	References
1	Kidney bean	UV-B	Increased the content of carotenoids, anthocyanins	Kacharava et al. (2009)
2	Cabbage	UV-B	Increased the content of carotenoids, anthocyanins	Kacharava et al. (2009)
3	Beet	UV-B	Increased the content of carotenoids, anthocyanins	Kacharava et al. (2009)
4	*Vigna mungo*	UV-B	Enhanced anthocyanin content	Shaukat et al. (2013)
5	Maize	UV-B	Enhanced anthocyanin content	Rudnóy et al. (2015)
6	Wheat	UV	Increased chlorophyll synthesis	Badridze et al. (2016)
7	Mung bean	UV-C	Chlorophyll content was enhanced	Siddiqui et al. (2011)
8	Groundnut	UV-C	Chlorophyll content was enhanced	Siddiqui et al. (2011)
9	Wheat	UV-B	Enhanced stomatal conductance, intercellular CO_2 concentration and transpiration rate	Li et al. (2010)
10	Cowpea	UV-B	Caroteniod content was enhanced	Mishra et al. (2008)
11	Rice	UV-B	Chlorophyll content was increased	Xu and Qiu (2007)
12	Peanut	UV	Enhanced total chlorophyll content and carotenoid content	Siddiqui et al. (2011)
13	Mung bean	UV	Enhanced total chlorophyll content and carotenoid content	Hamid and Jawaid (2011)
14	Mash bean	UV	Enhanced total chlorophyll content and carotenoid content	Shaukat et al. (2013)
15	Barley	UV-B	Enhanced chlorophyll and carotenoid content; RC/CSm, ABS/CSm, TR/CSm, and ET/CSm were increased and DI/CSm was reduced; *Pn*, *gs*, and *Ci* were augmented	Singh et al. (2019)

TABLE 7.2 *(Continued)*

S. no	Crop	UV radiation	Effects on photosynthesis	References
16	Oats	UV-B	Enhanced chlorophyll and carotenoid content; RC/CSm, ABS/CSm, TR/CSm, and ET/CSm were increased and DI/CSm was reduced; *Pn, gs,* and *Ci* were augmented	Singh et al. (2019)
17	Rice	UV-B	Reduction in the PSI and PSII damages	Sen et al. (2020)
18	Rice	UV-B	PSI and PSII activity was enhanced; RC/CSm, ABS/CSm, TR/CSm, and ET/CSm were increased and DI/CSm was reduced	Thomas and Puthur (2019)
19	Rice	UV-B	Increased PSI and PSII activities; *Pn, gs,* and *Ci* were augmented	Thomas et al. (2020)

The augmentation in photosynthetic pigments is a major physiological response seen in plants as an effect of UV priming. Seeds when exposed with UV-B radiation (760 m/Wcm^{-2} for 90 min), the photosynthetic pigment content was augmented by 35% in cabbage and in varieties of kidney bean, while in beet, the enhancement was up to 45%. The low irradiations of UV (460 and 760 m/Wcm^{-2} for 30 min) enhanced the carotenoid synthesis in beet and cabbage; up to 62% in red cabbage, 40% in red beet, in white cabbage up to 20%, and 28% in white beet. The content of anthocyanins and carotenoid were amplified in kidney bean seeds up to 12% and 37% with UV exposure (90 min) (Kacharava et al., 2009). The augmented carotenoid content points to a significant stress tolerance mechanisms adapted in plants (Jaleel et al., 2009).

Upon irradiation of seeds with UV, anthocyanin and carotenoid contents were found to be elevated in red cabbage and in white beet. Likewise, when seeds of *Vigna mungo* were irradiated with UV-B for 40 min, the content of anthocyanin was enhanced up to threefold in the seedlings (Shaukat et al., 2013). The anthocyanin content was augmented because of the irradiation with UV-B (350 m/Wcm^{-2}) in maize (Rudnóy et al., 2015). The UV irradiation stimulated the chlorophyll synthesis in wheat. The seeds of wheat exposed to UV radiation showed prominent enhancement of photosynthesis in three different varieties such as black dika, macha, and durum wheat (Badridze et al., 2016). Chlorophyll content was enhanced when the seeds of groundnut and mungbean were irradiated by UV-C for various time intervals. In groundnut, the content of chlorophyll increased upon 20 min exposure, while in the latter, 15 min irradiation could bring about the same level of changes (Siddiqui et al., 2011). In field condition, seeds of wheat were irradiated to UV-B radiation (5 kJ/m^{-2}) and it was shown that the *gs*, intercellular CO_2 concentration, and transpiration rate were prominently augmented (Li et al., 2010). While cowpea seedlings were exposed to UV-B irradiation (0.4 W/m^{-2}) for various time intervals such as 15, 30, 45, and 60 min, it was found that the level of careteniods was enhanced (Mishra et al., 2008). While the seedlings of rice under field conditions were exposed to UV-B (9.7 kJ/m^{-2}) radiation, it stimulated photosynthesis and enhanced the concentration of chlorophyll up to 35% than control. UV-B exposed rice seedlings also showed increases in tolerance to photoinhibition (Xu and Qiu, 2007).

UV radiations have been reported to enhance total chlorophyll and carotenoid contents in peanut (Siddiqui et al., 2011), mung bean (Hamid and Jawaid, 2011; Siddiqui et al., 2011), and mash bean (Shaukat et al., 2013). UV-B priming alters the phenomenological energy flux ratios of Chl *a* fluorescence parameters, such as RC/CSm, ABS/CSm, TR/CSm, and ET/CSm that were increased and at the same time, DI/CSm was seen to be reduced in barley and oats (Singh et al., 2019) and also in rice (Thomas and Puthur, 2019). *Pn*, *gs*, and *Ci* were also augmented in barley and oats seeds exposed to UV-B priming. Total chlorophyll and carotenoids were significantly enhanced with UV-B exposure of barley and oats (Singh et al., 2019). The seed primed with UV-B (4 kJ/m^{-2}) radiation reduces the PSI and PSII damages in tolerant (Vaisakh) and in sensitive (Neeraja) varieties of rice when treated with NaCl and PEG stress conditions (Sen et al., 2020). PSI and PSII activities as well as *Pn*, *gs*, and *Ci* were also augmented in rice seeds radiated with UV-B priming (Thomas et al., 2020).

7.7 CONCLUSION

UV radiation effects on photosynthesis explore the molecular and physiological responses of crop plant toward UV radiation. UV irradiation is a major environmental aspect, which damages the photosynthetic apparatus and ultimately reduces the plant productivity. The major targets of UV radiation in plant are the nucleic acid molecules, Calvin–Benson cycle enzymes, and the photosystems. PSII is more vulnerable to UV radiation and within PSII, oxygen-evolving complex, LHCII, and D1/D2 proteins are greatly affected. UV-induced ROS generation also leads to the reduction of photosynthetic pigments, photosynthetic efficiency, and RuBisCO activity, which finally results in reduced crop yield. This chapter explains the effect of UV radiations on photosynthetic pigments, photosystems, RuBisCO activity, Chl *a* fluorescence parameters, and gas exchange parameters. It also details the photodamage and the repair mechanisms in plants under UV radiation. UV radiation mainly acts as a stressor for plants, while low levels of the same bring about positive influence in plants. UV priming significantly increases the photosynthetic efficiency in various crop plants. It beneficially increases the photosynthetic pigments, photosytem activities, Chl *a* fluorescence parameters, and gas exchange parameters, especially under stress conditions.

KEYWORDS

- **UV radiation**
- **stressor**
- **photosynthesis**
- **UV priming**
- **photosystem**
- **pigments**
- **Chl *a* fluorescence**

REFERENCES

Agati, G.; Tattini, M. Multiple Functional Roles of Flavonoids in Photoprotection. *New Phytol.* **2010,** *186*, 786–793.

Agrawal, S. B.; Rathore, D. Changes in Oxidative Stress Defense System in Wheat (*Triticum aestivum* L.) and Mung Bean (*Vigna radiata* L.) Cultivars Grown with and Without Mineral Nutrients and Irradiated by Supplemental Ultraviolet-B. *Environ. Exp. Bot.* **2007,** *59*, 21–33.

Ashraf, M. A.; Akbar, A.; Askari, S. H.; Iqbal, M.; Rasheed, R.; Hussain, I. Recent Advances in Abiotic Stress Tolerance of Plants through Chemical Priming: An Overview. In *Advances in Seed Priming*; Springer: Singapore, 2018; pp 51–79.

Badridze, G.; Kacharava, N.; Chkhubianishvili, E.; Rapava, L.; Kikvidze, M.; Chanishvili, S.; Shakarishvili, N.; Mazanishvili, L.; Chigladze, L. Effect of UV Radiation and Artificial Acid Rain on Productivity of Wheat. *Russ. J. Ecol.* **2016,** *47*, 158–166.

Badridze, G.; Kacharava, N.; Chkhubianishvili, E.; Rapava, L.; Kikvidze, M.; Chanishvili, S. H.; Chigladze, L. Influence of Ultraviolet Irradiation and Acid Precipitations on the Content of Antioxidants in Wheat Leaves. *Appl. Ecol. Environ. Res.* **2015,** *13*, 993–1013.

Baker, N. R. Chlorophyll Fluorescence: A Probe of Photosynthesis In Vivo. *Annu. Rev. Plant Biol.* **2008,** *59*, 89–113.

Barbato, R.; Friso, G.; Rigoni, F.; Dalla Vecchia, F.; Giacometti, G. M. Structural Changes and Lateral Redistribution of Photosystem II during Donor Side Photoinhibition of Thylakoids. *J. Cell Biol.* **1992,** *119*, 325–335.

Bellafiore, S.; Barneche, F.; Peltier, G.; Rochaix, J. D. State Transitions and Light Adaptation Require Chloroplast Thylakoid Protein Kinase STN7. *Nature* **2005,** *433*, 892–895.

Bergo, E.; Segalla, A.; Giacometti, G. M.; Tarantino, D.; Soave, C.; Andreucci, F.; Barbato, R. Role of Visible Light in the Recovery of Photosystem II Structure and Function from Ultraviolet-B Stress in Higher Plants. *J. Exp Bot.* **2003,** *54*, 1665–1673.

Bornman, J. F.; Barnes, P. W.; Robinson, S. A.; Ballare, C. L.; Flint, S. D.; Caldwell, M. M. Solar Ultraviolet Radiation and Ozone Depletion-Driven Climate Change: Effects on Terrestrial Ecosystems. *Photochem. Photobiol. Sci.* **2015,** *14*, 88–107.

Brosché, M.; Strid, Å. Molecular Events Following Perception of Ultraviolet-B Radiation by Plants. *Physiol. Plant.* **2003,** *117*, 1–10.

Bussotti, F.; Ferrini, F.; Pollastrini, M.; Fini, A. The Challenge of Mediterranean Sclerophyllous Vegetation under Climate Change: From Acclimation to Adaptation. *Environ. Exp. Bot.* **2014,** *103*, 80–98.

Choudhary, K. K.; Agrawal, S. B. Ultraviolet-B Induced Changes in Morphological, Physiological and Biochemical Parameters of Two Cultivars of Pea (*Pisum sativum* L.). *Ecotoxicol. Environ. Saf.* **2014,** *100*, 178–187.

Cicek, N.; Fedina, I.; Cakirlar, H.; Velitchkova, M.; Georgieva, K. The Role of Short-Term High Temperature Pretreatment on the UV-B Tolerance of Barley Cultivars. *Turk J. Agric. For.* **2012,** *36*, 153–165.

Delibaltova, V.; Ivanova, R. Impact of the Pre-Sowing Irradiation of Seeds by Helium-Neon Laser on the Dynamics of Development of Some Cotton Varieties. *J. Environ. Prot. Ecol.* **2006,** *7*, 909–917.

Faseela, P.; Puthur, J. T. The Imprints of the High Light and UV-B Stresses in *Oryza sativa* L. 'Kanchana' Seedlings Are Differentially Modulated. *J. Photochem. Photobiol. B.* **2018,** *178*, 551–559.

Fathi, A.; Tari, D. B. Effect of Drought Stress and Its Mechanism in Plants. *Int. J. Life Sci.* **2016,** 10, 1–6.

Fristedt, R.; Herdean, A.; Blaby-Haas, C. E.; Mamedov, F.; Merchant, S. S.; Last, R. L.; Lundin, B. Photosystem II Protein33, a Protein Conserved in the Plastid Lineage, Is Associated with the Chloroplast Thylakoid Membrane and Provides Stability to Photosystem II Supercomplexes in *Arabidopsis*. *Plant Physiol.* **2015,** *167*, 481–492.

Fristedt, R.; Trotta, A.; Suorsa, M.; Nilsson, A. K.; Croce, R.; Aro, E. M.; Lundin, B. PSB33 Sustains Photosystem II D1 Protein under Fluctuating Light Conditions. *J. Exp. Bot.* **2017,** *68*, 4281–4293.

Graindorge, S.; Cognat, V.; Mutterer, J.; Molinier, J. Photodamage Repair Pathways Contribute to the Accurate Maintenance of the DNA Methylome Landscape Upon UV Exposure. *PLoS Genet.* **2019,** *15*, 476–482.

Gull, A.; Lone, A. A.; Wani, N. U. Biotic and Abiotic Stresses in Plants. In *Abiotic and Biotic Stress in Plants*; Intech Open, 2019; pp 7–29.

Gururani, M. A.; Venkatesh, J.; Tran, L. S. P. Regulation of Photosynthesis during Abiotic Stress-Induced Photoinhibition. *Mol. Plant.* **2015,** *8*, 1304–1320.
Hamid, N.; Jawaid, F. Influence of Seed Pre-Treatment by UV-A and UV-C Radiation on Germination and Growth of Mung Beans. *Pak. J. Chem.* **2011,** *1*, 164–167.
He, J. M.; Liu, Z. H.; Xu, H.; She, X. P.; and Huang, C. The Involvement of Hydrogen Peroxide in UV-B-Inhibited Pollen Germination and Tube Growth of *Paeonia suffruticosa* and *Paulownia tomentosa* In Vitro. *Plant Growth Regul.* **2006,** *49*, 199–208.
He, J. M.; She, X. P.; Meng, Z. N.; Zhao, W. M. Reduction of Rubisco Amount by UV-B Radiation Is Related to Increased H_2O_2 Content in Leaves of Mung Bean Seedlings. *J. Plant Physiol. Mol. Biol.* **2004,** *30*, 291–296.
He, M.; He, C. Q.; Ding N. Z. Abiotic Stresses: General Defenses of Land Plants and Chances for Engineering Multistress Tolerance. *Front. Plant Sci.* **2018,** *9*, 1771–1786.
Hideg, É.; Jansen, M. A.; Strid, Å. UV-B Exposure, ROS, and Stress: Inseparable Companions or Loosely Linked Associates? *Trends Plant Sci.* **2013,** *18*, 107–115.
Hollósy, F. Effects of Ultraviolet Radiation on Plant Cells. *Micron.* **2002,** *33*, 179–197.
Jaleel, C. A.; Manivannan, P. A.; Wahid, A.; Farooq, M.; Al-Juburi, H. J.; Somasundaram, R. A.; Panneerselvam, R. Drought Stress in Plants: A Review on Morphological Characteristics and Pigments Composition. *Int. J. Agric. Biol.* **2009,** *11*, 100–115.
Jansen, M. A.; Martret, B. L.; Koornneef, M. Variations in Constitutive and Inducible UV-B Tolerance; Dissecting Photosystem II Protection in *Arabidopsis thaliana* Accessions. *Physiol. Plant.* **2010,** *138*, 22–34.
Johnson, M. P. Correction: Photosynthesis. *Essays Biochem.* **2017,** *61*, 429–441.
Jordan, B. R.; He, J.; Chow, W. S.; Anderson, J. M. Changes in mRNA Levels and Polypeptide Subunits of Ribulose 1, 5-Bisphosphate Carboxylase in Response to Supplementary Ultraviolet-B Radiation. *Plant Cell Environ.* **1992,** *15*, 91–98.
Jordan, B.R. Molecular Response of Plant Cells to UV-B Stress. *Funct. Plant Biol.* **2002,** *29*, 909–916.
Joshi, P.N.; Ramaswamy, N. K.; Iyer, R. K.; Nair, J. S.; Pradhan, M. K.; Gartia, S.; Biswal, B.; Biswal, U. C. Partial Protection of Photosynthetic Apparatus from UV-B-Induced Damage by UV-A Radiation. *Environ. Exp. Bot.* **2007,** *59*, 166–172.
Kacharava, N.; Chanishvili, S.; Badridze, G.; Chkhubianishvili, E.; Janukashvili, N. Effect of Seed Irradiation on the Content of Antioxidants in Leaves of Kidney Bean, Cabbage and Beet Cultivars. *Aust. J. Crop Sci.* **2009,** *3*, 137–149.
Kakani, V.G.; Reddy, K. R.; Zhao, D.; Sailaja, K. Field Crop Responses to Ultraviolet-B Radiation: A Review. *Agric. For. Meteorol.* **2003,** *120*, 191–218.
Kataria, S.; Jajoo, A.; Guruprasad, K. N. Impact of Increasing Ultraviolet-B (UV-B) Radiation on Photosynthetic Processes. *J. Photochem. Photobiol. B.* **2014,** *137*, 55–66.
Köhler, H.; Contreras, R. A.; Pizarro, M.; Cortés-Antíquera, R.; Zúñiga, G. E. Antioxidant Responses Induced by UV-B Radiation in *Deschampsia antarctica* Desv. *Front. Plant Sci.* **2017,** *8*, 921–943.
Kosobryukhov, A. A.; Lyubimov, V. Y.; Kreslavski, V. D. Adaptive Mechanisms of Photosynthetic Apparatus to UV Radiation. In *Stress Responses in Plants*; 2015; pp 59–78.
Kumar, G.; Pandey, A. Effect of UV-B radiation on Chromosomal Organisation and Biochemical Constituents of *Coriandrum sativum* L. *Jordan J. Biol. Sci.* **2017,** *10.*
León-Chan, R. G.; López-Meyer, M.; Osuna-Enciso, T.; Sañudo-Barajas, J. A.; Heredia, J. B.; León-Félix, J. Low Temperature and Ultraviolet-B Radiation Affect Chlorophyll Content

and Induce the Accumulation of UV-B-Absorbing and Antioxidant Compounds in Bell Pepper (*Capsicum annuum*) Plants. *Environ. Exp. Bot.* **2017,** *139*, 143–151.

Li, Z.; Lu, G. Y.; Zhang, X. K.; Zou, C. S.; Cheng, Y.; Zheng, P. Y. Improving Drought Tolerance of Germinating Seeds by Exogenous Application of Gibberellic Acid (GA_3) in Rapeseed (*Brassica napus* L.). *Seed Sci. Technol.* **2010,** *38*, 432–440.

Liu, X.; Zhang, Q.; Yang, G.; Zhang, C.; Dong, H.; Liu, Y., et al. Pivotal Roles of Tomato Photoreceptor SlUVR8 in Seedling Development and UV-B Stress Tolerance. *Biochem. Biophys. Res. Commun.* **2020,** *522*, 177–183.

Mackerness, S.; Surplus, S. L.; Jordan, B. R.; Thomas, B. Effects of Supplementary Ultraviolet-B Radiation on Photosynthetic Transcripts at Different Stages of Leaf Development and Light Levels in Pea (*Pisum sativum* L.): Role of Active Oxygen Species and Antioxidant Enzymes. *Photochem. Photobiol.* **1998,** *68*, 88–96.

Mazza, C. A.; Giménez, P. I.; Kantolic, A. G.; Ballaré, C. L. Beneficial Effects of Solar UV-B Radiation on Soybean Yield Mediated by Reduced Insect Herbivory under Field Conditions. *Physiol. Plant.* **2013,** *147*, 307–315.

Mishra, V.; Srivastava, G.; Prasad, S. M.; Abraham, G. Growth, Photosynthetic Pigments and Photosynthetic Activity during Seedling Stage of Cowpea (*Vigna unguiculata*) in Response to UV-B and Dimethoate. *Pestic. Biochem. Phys.* **2008,** *92*, 30–37.

Negrão, S.; Schmöckel, S. M.; Tester, M. Evaluating Physiological Responses of Plants to Salinity Stress. *Ann. Bot.* **2017,** *119*, 1–11.

Nelson, N.; Yocum, C. F. Structure and Function of Photosystems I and II. *Annu. Rev. Plant Biol.* **2014,** *57*, 521–565.

Nilsson, A. K.; Pěnčík, A.; Johansson, O. N.; Bånkestad, D.; Fristedt, R.; Suorsa, M. et. PSB33 Protein Sustains Photosystem II in Plant Chloroplasts under UV-A Light. *J. Exp. Bot.* **2020,** *71*, 7210–7223.

Nogués, S.; Baker, N. R. Evaluation of the Role of Damage to Photosystem II in the Inhibition of CO_2 Assimilation in Pea Leaves on Exposure to UV-B radiation. *Plant Cell Environ.* **1995,** *18*, 781–787.

Piccini, C.; Cai, G.; Dias, M. C.; Romi, M.; Longo, R.; Cantini, C. UV-B Radiation Affects Photosynthesis-Related Processes of Two Italian *Olea europaea* (L.) Varieties Differently. *Plants* **2020,** *12,* 1712–1724.

Qaderi, M. M.; Basraon, N. K.; Chinnappa, C. C.; Reid, D. M. Combined Effects of Temperature, Ultraviolet-B Radiation, and Watering Regime on Growth and Physiological Processes in Canola (*Brassica napus*) Seedlings. *Int. J. Plant Sci.* **2010,** *171*, 466–481.

Rao, M. V.; Paliyath, G.; Ormrod, D. P. Differential Response of Photosynthetic Pigments, Rubisco Activity and Rubisco Protein of *Arabidopsis thaliana* Exposed to UV-B and Ozone. *Photochem. Photobiol.* **1995,** *62*, 727–735.

Robson, T. M.; Klem, K.; Urban, O.; Jansen, M. A. Re-Interpreting Plant Morphological Responses to UV-B Radiation. *Plant Cell Environ.* **2015,** *38*, 856–866.

Rudnóy, S.; Majláth, I.; Pál, M.; Páldi, K.; Rácz, I.; Janda, T. Interactions of S-Methylmethionine and UV-B Can Modify the Defence Mechanisms Induced in Maize. *Acta Physiol. Plant.* **2015,** *37*, 148–156.

Salama, H. M.; Al Watban, A. A.; Al-Fughom, A. T. Effect of Ultraviolet Radiation on Chlorophyll, Carotenoid, Protein and Proline Contents of Some Annual Desert Plants. *Saudi J. Biol. Sci.* **2011,** *18*, 79–86.

Sarghein, S. H.; Carapetian, J.; Khara, J. Effects of UV-Radiation on Photosynthetic Pigments and UV Absorbing Compounds in *Capsicum longum* (L.). *Int. J. Bot.* **2008**.

Sen, A.; Challabathula, D.; Puthur, J. T. UV-B Priming of *Oryza sativa* Seeds Augments the Innate Tolerance Potential in a Tolerant Variety More Effectively toward NaCl and PEG Stressors. *J. Plant Growth Regul.* **2020,** *30*, 1–5.

Sharma, A.; Kumar, V.; Shahzad, B.; Ramakrishnan, M.; Sidhu, G. P.; Bali, A. S; Handa, N.; Kapoor, D.; Yadav, P.; Khanna, K.; Bakshi, P. Photosynthetic Response of Plants under Different Abiotic Stresses: A Review. *J. Plant Growth Regul.* **2019,** *19*, 1–23

Shaukat, S. S.; Farooq, M. A.; Siddiqui, M. F.; Zaidi, S. A. Effect of Enhanced UV-B Radiation on Germination, Seedling Growth and Biochemical Responses of *Vigna mungo* (L.) Hepper. *Pak. J. Bot.* **2013,** *45*, 779–785.

Siddiqui, A., Dawar, S. H.; Zaki, M. J.; Hamid, N. Role of Ultraviolet (UV-C) radiation in the Control of Root Infecting Fungi on Groundnut and Mung Bean. *Pak. J. Bot.* **2011,** *43*, 2221–2224.

Singh, J.; Thakur, J. K. Photosynthesis and Abiotic Stress in Plants. In *Biotic and Abiotic Stress Tolerance in Plants*; 2018; 27–46.

Singh, S.; Rai, K.; Agrawal, S. B.; Agrawal, M. Comparative Assessment of UV-B Priming on Vegetative and Reproductive Stages of Oat and Barley. *Proc. Natl. Acad. Sci. India Sect. B* **2019,** *25*, 1–9.

Strid, Å.; Chow, W. S.; Anderson, J. M. Effects of Supplementary Ultraviolet-B Radiation on Photosynthesis in *Pisum sativum. Biochim. Biophys. Acta, Bioenerg.* **1990,** *1020*, 260–268.

Štroch, M.; Materová, Z.; Vrábl, D.; Karlický, V.; Šigut, L.; Nezval, J.; Špunda, V. Protective Effect of UV-A Radiation during Acclimation of the Photosynthetic Apparatus to UV-B Treatment. *Plant Physiol. Biochem.* **2015,** *96*, 90–96.

Takahashi, S.; Badger, M. R. Photoprotection in Plants: A New Light on Photosystem II Damage. *Trends Plant Sci.* **2011,** *16*, 53–60.

Takahashi, S.; Milward, S. E.; Yamori, W.; Evans, J. R.; Hillier, W.; Badger, M. R. The Solar Action Spectrum of Photosystem II Damage. *Plant Physiol.* **2010,** *153*, 988–993.

Takeuchi, A.; Yamaguchi, T.; Hidema, J.; Strid, A.; Kumagai, T. Changes in Synthesis and Degradation of Rubisco and LHCII with Leaf Age in Rice (*Oryza sativa* L.) Growing under Supplementary UV-B Radiation. *Plant Cell Environ.* **2002,** *25*, 695–706.

Tevini, M. Plant Responses to Ultraviolet Radiation Stress. In *Chlorophyll a Fluorescence*; Springer: Dordrecht, 2004; pp 605–621.

Thomas, D. T.; Challabathula, D.; Puthur, J. T. UV-B Priming of *Oryza sativa* var. Kanchana Seedlings Augments Its Antioxidative Potential and Gene Expression of Stress-Response Proteins under Various Abiotic Stresses. *3 Biotech.* **2019,** *9*, 375–389.

Thomas, D. T.; Puthur, J. T. Amplification of Abiotic Stress Tolerance Potential in Rice Seedlings with a Low Dose of UV-B Seed Priming. *Funct. Plant Biol.* **2019,** *46*, 455–466.

Thomas, D.T.; Puthur, J. T. UV Radiation Priming: A Means of Amplifying the Inherent Potential for Abiotic Stress Tolerance in Crop Plants. *Environ. Exp. Bot.* **2017,** *138*, 57–66.

Thomas, T. D.; Dinakar, C.; Puthur, J. T. Effect of UV-B Priming on the Abiotic Stress Tolerance of Stress-Sensitive Rice Seedlings: Priming Imprints and Cross-Tolerance. *Plant Physiol. Biochem.* **2020,** *147*, 21–30.

Thomas, T. T. D.; Puthur, J. T. UV-B Priming Enhances Specific Secondary Metabolites in *Oryza sativa* (L.) Empowering to Encounter Diverse Abiotic Stresses. *Plant Growth Regul.* **2020,** *92*, 169–180.

UNEP. *Environmental Effects of Ozone Depletion 2002 Assessment: High Level Ozone*; New York, 2002; pp 65–84.

Xu, K.; Qiu, B. S. Responses of Superhigh-Yield Hybrid Rice Liangyoupeijiu to Enhancement of Ultraviolet-B Radiation. *Plant Sci.* **2007,** *172*, 139–149.

Yan, F.; Liu, Y.; Sheng, H.; Wang, Y.; Kang, H.; Zeng, J. Salicylic Acid and Nitric Oxide Increase Photosynthesis and Antioxidant Defense in Wheat under UV-B Stress. *Biol. Plant.* **2016,** *60*, 686–694.

Yao, Y.; Xuan, Z.; Li, Y.; He, Y.; Korpelainen, H.; Li, C. Effects of Ultraviolet-B Radiation on Crop Growth, Development, Yield and Leaf Pigment Concentration of Tartary Buckwheat (*Fagopyrum tataricum*) under Field Conditions. *Eur. J. Agron.* **2006,** *25*, 215–222.

CHAPTER 8

C4 and CAM Plants with Better Resilience to Environmental Stresses

WASIFA HAFIZ SHAH[1], SEERAT SALEEM[1], NAVEED UL MUSHTAQ[1], AADIL RASOOL[1], INAYATULLAH TAHIR[2], and REIAZ UL REHMAN[1*]

[1]*Department of Bioresources, School of Biological Sciences, University of Kashmir, Srinagar 190006, India*

[2]*Department of Botany, School of Biological Sciences, University of Kashmir, Srinagar 190006, India*

[*]*Corresponding author. E-mail: rreaizbiores@gmail.com*

ABSTRACT

C3, C4, and crassulacean acid metabolism (CAM) are the three types of photosynthetic pathways in plants. These differ structurally, functionally, and also in terms of response to various environmental stresses. C4 plants are said to have emerged in somewhat arid regions, characterized by high temperatures in addition to water stress. C4 plants have two types of cells arranged in a specialized Kranz-type anatomy wherein mesophyll cell chloroplasts perform the light-dependent reactions, and the bundle sheath cells perform the light independent reaction. C4 plants show high water-use efficiency (WUE), as they can perform photosynthesis even at low intercellular CO_2 concentrations. Besides, the efficient use of RuBisCO suppresses photorespiration, thereby improving the nitrogen-use efficiency (NUE). In addition to C4 metabolism, efficient antioxidant defense system also contributes to stress tolerance. CAM plants have adapted to severe water deficit in semi-arid and arid regions of the world, including deserts. This involves a special type of photosynthetic adaptation that allows CO_2 uptake at night when the

Photosynthesis and Respiratory Cycles during Environmental Stress Response in Plants. Aryadeep Roychoudhury (Ed.)

stomata are open and its fixation during the day when the stomata are closed. These characteristics allow CAM plants to use 80% less water than C_3 plants in order to produce similar amount of biomass. High temperature and salinity tolerance is beneficial in fulfilling the agricultural demands during global climate change scenarios. Many stress-adaptive responses involve abscisic acid (ABA)-dependent and ABA-independent pathways which are involved in the stress signaling of CAM plants. Stress adaptation in C4 and CAM plants also includes accumulation of osmoprotectants, enhanced antioxidant system, and development of specialized structures. This chapter explains the numerous adaptive mechanisms of C4 and CAM plants toward various environmental stresses.

8.1 INTRODUCTION

Food supply to the ever-increasing population has become challenging due to transformation of cultivable land into waste land. Besides, urbanization and industrialization are also responsible for the degradation of agricultural land (Ahanger et al., 2017). Plants face both abiotic and biotic environmental stresses which impact the crop productivity and reduce average yield. In plants, the abiotic stresses such as drought, salinity, freezing, heat, and heavy metals lead to osmotic stress, resulting in the imbalances in cellular ion-distribution and homeostasis (Kumar, 2013). Being sessile in nature, plants have developed various defense mechanisms to overcome these stresses. Plants receive the stress stimuli and through various signal transduction pathways, they generate appropriate cellular responses (Gull et al., 2019). The key metabolic process, that is, photosynthesis is adversely affected by stress. Ribulose-1,5-bisphoshate carboxylase/oxygenase (RuBisCO) enzyme activity determines the efficiency of photosynthesis in a plant (Doubnerová and Ryšlavá, 2011). RuBisCO has a low affinity for CO_2 and hence has a low potential to differentiate between molecular CO_2 and O_2. This leads to photorespiration, where RuBP reacts with O_2 and CO_2 is released back into the atmosphere. Photorespiration increases with increasing temperature, whereas the suppressed photorespiration is an important feature of C4 photosynthesis (Ghannoum, 2009). The three forms of photosynthesis known in plants are C3, C4, and crassulacean acid metabolism (CAM). C3 photosynthesis involves the direct fixation of CO_2 by carboxylase activity of RuBisCO into five carbon sugar to produce a 3-carbon compound (3-phosphoglyceric acid). Photosynthesis in C4 and CAM plants involves CO_2 fixation into 3-carbon compound by phosphoenolpyruvate carboxylase (PEPC) which leads to the production of 4-carbon compound and its carboxylation to release CO_2 in the vicinity of RuBisCO, thereby increasing the CO_2 concentration and decreasing the chances of oxygenase reaction (Abogadallah, 2010). Eighty-five percent of the higher plants comprise by C3 species. Five percent are C4 species and CAM account for the remaining 10%. C3, C4, and CAM plants differ greatly in photosynthetic characteristics. Plant reactions to various stresses differ among the various plant species, stage of development and intensity of stress. C3 and C4 plants are structurally and functionally different from each other as well as in their climatic requirements. C3 plants have a temperate origin, while C4 plants have evolved in tropical and arid environments. C4 plants are better acclimated to water stress, and have superior water use efficiency than C3 plants (Nayyar and Gupta, 2006). Closing of stomata is one of the main reactions to salt and drought stress in plants. In comparison to C3 plants, C4 plants are able to utilize very low concentrations of CO_2 and hence assimilate CO_2 even during stomatal closure. It is also reported that C4 plants are equipped with better antioxidant defense system under oxidative stress (Stepien and Klobus, 2005). Climate change (rainfall or temperature) and increasing CO_2 in the atmosphere have led to the expansion of CAM plants in drylands. CAM plants are known for their diurnal CO_2 uptake, high water use efficiency, high water storage in tissues, and high drought resistance. *Cylindropuntia imbricata* (CAM species) and *Boutelouaeriopoda* (C_4 grass) when treated with CO_2 enrichment in drought conditions showed that CAM plant had a dominance over C4 grass and responded positively to stress conditions (Yu et al., 2019). C3, C4, and CAM plants use both morphological and metabolic mechanisms in order to adapt to drought conditions by accumulating various specialized metabolites. They accumulate various specialized metabolites with antioxidant activities, mainly comprised of alkaloids, terpenes, phenols, flavonoids, anthocyanins, and polyols (Zahedi et al., 2019). In higher plants, C4 and CAM are the two vital carbon concentrating mechanisms (CCM). Although it is rare to find the occurrence of both the CCM in a single leaf, it is present in *Portulaca* spp. CAM is different from C4 in its flexibility of action where many facultative plants practice CAM under stress and opt C3 mechanism in feasible conditions (Ferrari et al., 2020).

8.2 C4 PLANTS: PHYSIOLOGY AND CHARACTERISTICS

C4 occurs in 19 families, mostly present in hot and high-light environment (Ferrari et al., 2020). C4 photosynthesis occurs among the forage grasses at lower latitudes and a few major crops such as maize and sugarcane. Maize is the most widely produced grain in the world, while sugarcane is at the top of the list of bioenergy crop. In warm environment, C4 photosynthesis boosts plant productivity and the yields are 50% more than the C3 crops (Sage, 2014). Approximately, 25% of the total terrestrial photosynthesis is contributed by C4 plants and 5% of the total vascular plants are C4 plants. C4 photosynthesis is mostly seen in grasses and sedges, whereas lesser dicot species are known (Westhoff and Gowik, 2010). Among the 15,000 monocotyledonous plants, only 6000 plants perform C4 photosynthesis and comparatively lesser number of dicotyledons (1600 out of the 300, 000) perform C4 photosynthesis (Ehleringer and Cerling, 2002). The evolution of C4 plants roots back to early Oligocene about 30 million years ago and on the other hand, C4 dicots emerged less than 20 million years ago. C4 photosynthesis has a polyphyletic origin which led to an increased vein density and enhancement of bundle sheath cell layer giving rise to the basic "Kranz anatomy" (Westhoff and Gowik, 2010). The C4 syndrome is the carbon fixation during the day by PEPC and its refixation by RuBisCO. PEPC and RuBisCO are segregated into two different compartments within the leaf, that is, mesophyll and bundle-sheath which are anatomically and biochemically separate. This process elevates the concentration of CO_2 more than that of O_2 in the vicinity of RuBisCO, thereby increasing the carboxylation reactions catalyzed by the enzyme and lowering the photorespiration rate. The saturation of RuBisCO with its CO_2 substrate leads to an increased rate of CO_2 assimilation, improved photosynthetic and NUE in high temperature and low CO_2 conditions (Christin and Osborne, 2014). Organic acids have an important role to play in numerous metabolic pathways as the four carbon dicarboxylic acid is the first product of carbon fixation in C4 plants which signifies that organic acids are essential photosynthetic intermediates. In C4 photosynthesis, OAA (oxaloacetate) is the short-lived product of initial step of CO_2 fixation in the leaf mesophyll cells. OAA is converted to malate and aspartate which are decarboxylated in the bundle sheath cells releasing CO_2 that is used for making carbohydrates (Ludwig, 2016). The RuBisCO carboxylation reaction in C3 plants is strongly limited when the CO_2 levels in the atmosphere fall below 100 Pa. However, this limitation in C4 plants occurs when the atmospheric CO_2 levels are below 20 Pa. It is suggested that C4 photosynthesis has evolved in C3 species that were relocated from their ancestral niche (moist tropical forests) to drier woodland or savannah habitats (Osborne and Sack, 2012). Compared to C3 plants (rice, wheat, potato, etc.), C4 plants like millets, maize, sorghum, sugarcane have an approximate

50% higher photosynthetic potential. The initial carbon fixation involves the production of OAA from CO_2 and phosphoenolpyruvate (PEP) which is catalyzed by PEPC. OAA is further metabolized into malate which then diffuses into the bundle sheath cell for decarboxylation to increase the concentration of CO_2 around RuBisCO. Eventually, the regeneration of the substrate, PEP takes place in mesophyll cell by pyruvate orthophosphate dikinase (PPDK) (Wang et al., 2012). The morphological specialization known as "Kranz anatomy" spatially separates the C4 carboxylation pathways, comprising of peripheral mesophyll cells (MC) and an inner ring of bundle sheath (BS) (occasionally mestome sheath cells) that surround the vascular bundle. Kranz anatomy reduces the interveinal distance and increases the BS: MC area (Griffiths et al., 2013). As compared to the C3 plants, the number of mesophyll cells is less in C4 leaves. C4 leaf has a distinct appearance because of the increased number and size of chloroplast in the BS cells. The veins are dark green colored with light green mesophyll tissue because of a smaller number of chloroplasts present in mesophyll cells. Intermediate species between C3 and C4 that use C2 photosynthesis are recently termed as "C2" species that use two carbon glycine molecules to transfer CO_2 into the bundle sheath cells (Stata et al., 2014). C4 grass species such as elephant grass (*Pennisetum purpureum*), foxtail millet (*Setaria italica*), and switchgrass (*Panicum virgatum*) are ideal energy crops as they are highly efficient in the conversion of light into biomass energy (approx. 3.7% as compared to 2.4% in C3 plants), have a high water use efficiency (WUE) and leaf level NUE, radiation use efficiency, can grow well in marginal lands and are tolerant to salinity, drought and water-logging conditions. Compared to C3 species, these plants are capable of generating large quantities of biomass, particularly in high temperature and resource-limited regions. The high WUE of C4 plants is because these plants can fix CO_2 with less open stomata owing to the high affinity of PEP carboxylase for bicarbonate (Byrt et al., 2011).

C4 plants are divided into three biochemical subtypes: NADP malic enzyme-type (NADP-ME), NAD malic enzyme-type (NAD-ME) and phosphoenolpyruvate carboxykinase C4 photosynthesis (PEP-CK) (Fig. 8.1). This grouping is based upon the various steps involved during the concentration of CO_2, the transported metabolites, the site of decarboxylation reactions and the kind of enzymes used. NADP-ME type occurs in the chloroplast, whereas NAD-ME takes place in the mitochondria. These two processes utilize at least 5 ATP and 2 NADPH to fix each CO_2 (Turkan et al., 2018). CO_2 fixation by PEPC produces C4 acids which are translocated to bundle sheath cells (BSCs) where decarboxylation takes place which increases the CO_2 concentration around RuBisCO in order to reduce photorespiration. The form of C4 acid (malate) translocated to BSCs varies between the species. In case of NADP-ME type, oxaloacetate is reduced to malate, while as in NAD and PEPCK types, OAA is converted to aspartate and transported to BSCs (Furbank, 2011).

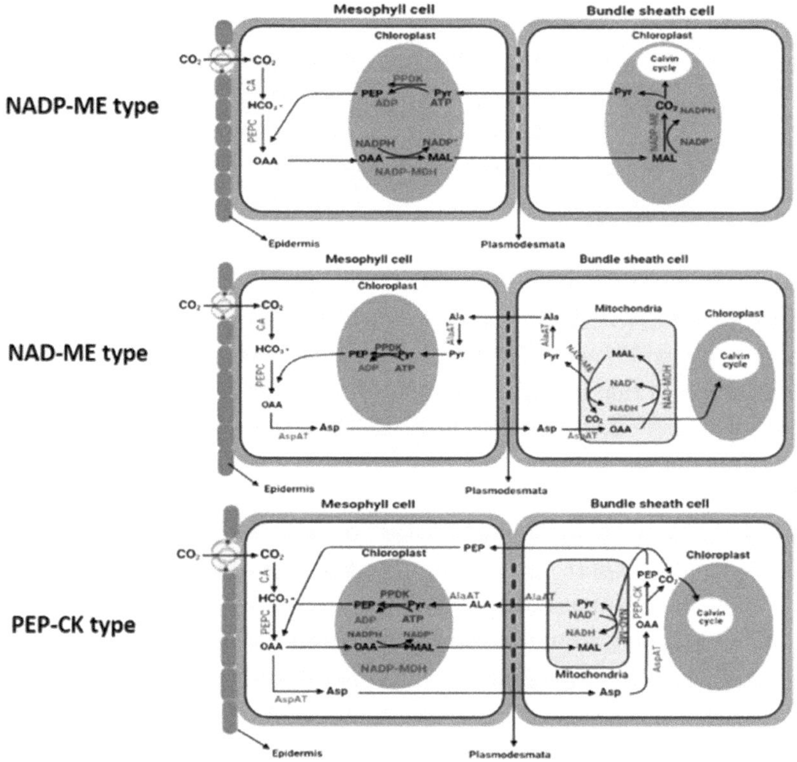

FIGURE 8.1 Systematic outline of C4 pathway in (A) NADP-ME type species, (B) NAD-ME species and (C) PEP-CK type species. HCO_3^-, bicarbonate; CA, carbonic anhydrase; PEPC, phosphoenolpyruvate carboxylase; OAA, oxaloacetate; PEP, phosphoenolpyruvate; MAL, malate; Pyr, pyruvate; ALA, Alanine; Asp, Aspartate; PPDK, pyruvate orthophosphate dikinase; NADP-MDH, NADP-malate dehydrogenase; NADP-ME, NADP dependent malic enzyme; NAD-ME, NAD dependent malic enzyme; PEP-CK, PEP carboxykinase; Asp-AT, Aspartate aminotransferase; Ala-AT, Alanine aminotransferase.

The various crops that belong to NADP-ME C4 subtype are maize (*Zea mays*), sugarcane (*Saccharum* spp.) and sorghum (*Sorghum bicolor*). A few examples of NAD-ME subtype are switchgrass (*P. virgatum* L.), pearl millet [*Pennisetum glaucum* (L.) R. Br] and amaranth (Amaranthaceae).

Geographic surveys have shown that drier areas show the presence of more NAD-ME subtypes, while increasing annual precipitation increases the percentage of NADP-ME subtypes. NADP-ME subtype plants (except Aristidoideae tribe) are known for their greater photosynthetic NUE when the nitrogen supply is low. "Kranz anatomy" varies in C4 subtypes. In NADP-ME subtype, there is the absence of a cellular layer in the middle of BS cells and the vascular bundle, with suberin deposition in the BS cell wall. NAD-ME vasculature is surrounded by an outer BS and an interior nonphotosynthetic mestome sheath with suberin deposits (Rao and Dixon, 2016). The first two common steps in all subtypes are the conversion of CO_2 into bicarbonate (HCO_3^-) by carbonic anhydrase (CA), followed by PEP carboxylation in the cytosol of mesophyll cells. The C4 subtype determines the fate of resulting C4 dicarboxylic acid oxaloacetate (OAA) (Khoshravesh et al., 2020). The various subtypes of C4 photosynthesis are explained below.

8.2.1 NADP-ME C4 SUBTYPE

NADP-ME subtype involves the import of OAA into the MC chloroplast which is then reduced to malate (MAL) by NADP-MDH (malate dehydrogenase). This malate diffuses into the BS cells along its concentration gradient and is taken up by the chloroplasts. Malate further undergoes oxidative decarboxylation by the NADP-dependent malic enzyme (NADP-ME), yielding one molecule of each of CO_2, NADPH, and pyruvate (Pyr). The CO_2 assimilation by RuBisCO yields two molecules of 3-phosphoglyceric acid (3-PGA) that can enter the Calvin cycle either in MCs or BSCs. This requires 3-PGA and triose phosphates (TP) shuttling between BSCs and MCs. Pyruvate exits from the BSCs, enters the MCs and is taken up by the MC chloroplasts and is transformed into PEP by PPDK. PEP exits from the chloroplast by PEP/phosphate translocator (PPT) and can enter a new cycle of CO_2 shuttle (Weber and Caemmerer, 2010). Malate decarboxylation releases NADPH in BS chloroplasts; hence, BS photosynthetic electron system does not need to produce NADPH but only ATP.

8.2.2 NAD-ME C4 SUBTYPE

NAD-ME involves the conversion of OAA into aspartate (Asp), by aspartate aminotransferase (Asp-AT) in the MC cytosol. This aspartate enters BSCs and is changed back into OAA in the mitochondria by mitochondrial Asp-AT. Mitochondrial NADP-MDH reduces OAA to malate which is further oxidatively decarboxylated by mitochondrial NAD-ME, producing NADH, pyruvate, and CO_2. Pyruvate exits from the mitochondria and CO_2 enters the Calvin cycle in the chloroplasts of BSCs. Next follows the pyruvate conversion to Alanine (Ala) in BSCs by alanine aminotransferase (Ala-AT). After entering MCs, alanine is deaminated to pyruvate which then regenerates the carbon acceptor, PEP (Yin and Struik, 2018).

8.2.3 PEP-CK C4 SUBTYPE

PEP-CK subtype involves multiple transamination reactions and is somewhat more complicated than the other subtypes. When this enzyme is used as a primary decarboxylase, both aspartate (Asp) and malate are produced from OAA. In the MCs, cytosolic aminotransferase (Asp-AT) produces aspartate and OAA which are then translocated to MC chloroplast. NADP-MDH converts OAA into malate (MAL), while Asp is translocated to BSCs where it is changed back to OAA by cytosolic Asp-AT. Further OAA is decarboxylated by PEP-CK in an ATP requiring reaction and CO_2 released enters the Calvin cycle in the chloroplasts of BSCs. PEP-CK activity generates PEP which is diffused back to MCs and is ultimately used by PEPC. Malate formed in MC chloroplasts is transported to BSC mitochondria for decarboxylation by NAD-ME and this result in a profit of NADH which is utilized for producing ATP that drives the PEP-CK reaction. This malate shuttle yields pyruvate that is transformed to alanine in BSCs that consumes the amino group associated with the import of Asp to the BSCs. Thus, malate shuttle maintains the ammonia homoeostasis between BSCs and MCs (Ludwig, 2016).

It has been reported that NADP-ME subtype C4 plants such as sugarcane, maize, sorghum, etc., can show flexibility in the context of existence, of more than one type of C4 acid decarboxylation pathway in a single plant, which is determined by the environmental conditions. The enzymatic activity and gene expression analysis suggested that in addition to the NADP-ME pathway, sugarcane also used PEP-CK pathway as an adaptation to water stress (Cacefo et al., 2019). Nonphotosynthetic forms of PEPC, NAD-ME, NADP-ME, and PPDK perform a variety of functions in plants. These have a role in nitrogen assimilation, pH regulation, and detoxification, and provide NADPH to withstand various biotic and abiotic stresses, help in lipid membrane regeneration, regulate stomatal opening and osmotic pressure and synthesize osmotic adjustment metabolites such as malate (Doubnerová and Ryšlavá, 2011). *NADP-ME* gene expression is induced by cold stress that leads to the production of NADP-ME protein that helps to maintain the liquidity of membrane in cold stress and increase cold tolerance (Hui et al., 2010).

8.3 C4 RESILIENCE TO VARIOUS ENVIRONMENTAL STRESSES

Low precipitation, high temperature, and salinity limit the plant growth and are the basis of C4 distribution in different environments. As the plants are sessile and cannot move from unfavorable to favorable conditions, they should be able to withstand the challenging conditions, in order to survive (Kumar et al., 2017). As per different scholarly estimates, stress influences most of the arable land in the world and only 10% is free from it. Stress can cause abnormal metabolism in the plants, hampered growth and even death. C4 plants are halophytes that can tolerate high salinity levels and adverse drought conditions. In order to cope with the environmental stresses, these plants utilize various efficient mechanisms such as increased photosynthetic rates, enhanced antioxidant defense system, high gene expression, accumulation of osmoprotectants, and elimination or uptake of selective ions (Yadav et al., 2021). Responses of C4 plants toward various stresses are discussed below.

8.3.1 SALINITY STRESS

Global food security is under threat due to soil salinity as the salt affected land accounts for 20% of the global cultivated area and 50% of the total irrigated lands of the world. Salinity leads to osmotic stress, limited water uptake, and unrestrained uptake of Na^+ and Cl^- ions. All these factors ultimately affect plant growth and development. Plant responses to salinity stress include various metabolic processes and gene networks that result in adaptation for survival. Stress results in increased reactive oxygen species (ROS) generation in plants that result in oxidative damages to the living cells (Hossain and Dietz, 2016). Salt stress strongly induces photorespiration which is one of the major sources of ROS. Owing to their CO_2 concentrating mechanism, C4 plants have lesser rates of photorespiration (3.5–6% of photosynthesis) in different environmental conditions, as compared to C3 plants where photorespiration rates are 20–30% of photosynthesis (Abogadallah, 2010). Salinity stress resulted in lipid peroxidation, decreased CO_2 assimilation, and increased antioxidant activity in both wheat (C3) and maize (C4). Superoxide dismutase (SOD) and glutathione reductase (GR) activities were higher in maize leaves as compared to wheat with better CO_2 assimilation. It was concluded that C4 plants are better equipped in resistance mechanisms against oxidative stress caused by salinity and drought (Stepien and Klobus, 2005). It is reported that C4 plants inoculated with arbuscular mycorrhizal fungi (AMF), *Funneliformis mosseae* maintained high cytosolic K^+/Na^+ ratio during salt stress. High K^+/Na^+ ratio increased plant salt tolerance by Na exclusion (Chandrasekaran et al., 2016). Plants with C4 photosynthesis inhabiting regions with salinity and limited water supply have an advantage over C3 plants as the ratio of CO_2 assimilation to water transpiration in C4 plants is twice that of C3 plants with WUE. Switch grass (C4) when compared with giant reed (C3) showed higher tolerance to water and salinity stresses due to high photosynthetic rates and it was concluded that switchgrass has higher tolerance toward stresses like water stress, salinity, cold temperature, and continuous darkness (Sánchez et al., 2016). Salt tolerant *Miscanthus sinensis* Anderss JM0119 can resist long term salinity exposure due to higher photosynthetic rate, transpiration rate, chlorophyll content, PSII operating efficiency, coefficient of photochemical quenching, activities of PEPC and PPDK and expression of *PEPC* and *PPDK* genes with relatively less activity of NADP-MDH and NADP-ME under salinity (Sun et al., 2021). The adaptation responses of maize to salinity stress involve an increase PPDK, PEPC, NADP-MDH, and NAD-MDH activities with a decrease of NADP-ME activity (Omoto et al., 2012). It was reported that elevated CO_2 increases the salt resistance in C3 species, *Chenopodium quinoa* and also in the C4 species, *Atriplex nummularia.* Even at elevated CO_2 concentration, the salt resistance of *A. nummularia* was higher than that of *C. quinoa* (Geissler et al., 2015). Plastid terminal oxidase (PTOX) protein is located at the stroma lamellae in chloroplasts and is involved in alternative electron transport. It was found that *Spartina alterniflora* could tolerate much higher concentrations of salt than *S. italica,* showing three to four-fold increase in PTOX implying its photoprotective role during salt stress (Essemine et al., 2020). C4 *Suaeda* species accumulated a variety of metabolites in various abiotic stresses under ambient CO_2 conditions. These included 14 amino acids, 20 sugars, 11 sugar acids, 6 fatty acids, 9 different compounds, and 1 flavonoid (kaempferol). Among the various amino acids accumulated under drought and salt stress are proline, serine, threonine, aspartic acid, glutamic acid, glutamine, and tryptophan. The various sugars accumulated are galactose, turanose, rhamnose, melibiose, cellobiose, sucrose, lactose, galactinol, lactose, galactinol, and ribono-1,4-lactone and sugar acids such as ribonic acid, galacturonic acid, glyceric acid, and threonic acid (Yadav et al., 2021). C4 plant, *Portulaca oleracea* L. is highly tolerant to saline stress. Increase in *PC5S* gene expression and proline content in leaves is reported to be an important adaptation of this plant in saline conditions (Sdouga et al., 2019).

8.3.2 DROUGHT STRESS

Prolonged period of water shortage varying from days to years, over a particular region is termed as drought. Drought is a significant threat to the food security in the current era of climate change. Drought stress affects photosynthesis by decreased CO_2 diffusion and metabolic constraints (Siddique et al., 2016). Among the various crop performance limiting factors, drought stress is a serious threat for crop production especially in the warm and dry regions of the world. One of the important traits that is related to better crop yield is drought tolerance. Plant hormone ABA (abscisic acid) plays a significant role in the plant life cycle and various plant adaptations to environmental stresses. Besides serving as a signalling molecule, ABA also has a role in the transport of reserves in drought stress (Fathi and Tari, 2016). Drought affects the plants physiologically, morphologically as well as at molecular levels including impaired carbon assimilation and increased ROS production. C4 photosynthesis is highly responsive to drought stress as the carbon concentration mechanism leads to the suppression of photorespiration which decreases leaf transpiration and stomatal conductance (Siddique et al., 2016). High temperature and drought stress implicated on sunflower (C3) and maize (C4) resulted in a negative impact on photosynthesis, showing more adversity in sunflower than in maize. Maize showed no reduction in stomatal conductance and a five to six fold greater WUE. Drought tolerant varieties of both the species showed an increase in root growth which enhances the water uptake (Killi et al., 2017). Fixation of CO_2 in the MSCs of C4 plants is aided by PEPC to form OAA. Drought causes an increase in PEPC activity and upregulation of NADP-ME. Severe drought stress leads to membrane damage which requires membrane regeneration. NADP-ME aids in providing NADH and pyruvate for fatty acid synthesis. NADP-ME also provides NADPH which is required for the antioxidant defense systems such as ascorbate-glutathione cycle (Rao and Chaitanya, 2016). Sorghum is reportedly more drought tolerant than maize which is attributed to its deep root system that is able to draw water from soil depths. Comparing the two, drought reportedly impacted maize severely in terms of shoot fresh and dry masses, reduction in leaf relative water content, inhibited gas exchange, and fluctuation in WUE. It was concluded that aquaporin genes *PIP1;5* in leaves and *PIP2;3* in roots were highly responsive to drought and may have an important role in imparting drought tolerance and high WUE to sorghum plants (Hasan et al., 2017). Drought-tolerant, C4 hybrid bermudagrass [*Cynodon dactylon* (L.) Pers. × *C. transvaalensis* Burtt Davy, cv. Tifway] when exposed to drought conditions showed lower electrolyte leakage (EL) and higher relative water content (RWC). The drought responsive genes identified were functionally arranged into the categories of metabolism, stress defense, osmoregulation, membrane system, signaling, protein synthesis and degradation. Around 36 upregulated genes included those for cuticle wax formation (provides protection against dehydration); dehydration protective proteins (dehydrins, HVA22-like protein) and oxidative stress defense (superoxide dismutase, dehydroascorbate reductase, 2-Cys peroxiredoxins), and for stress signaling (EREBP-4 like protein and WRKY transcription factor). It was concluded that such adaptations help the plant with drought tolerance (Zhou et al., 2014). A comparison of ROS formation and antioxidant enzymes between *Cleome gynandra* (C4) and *Cleome spinosa* (C3) under drought stress revealed an increase in antioxidant enzymes, peroxidase (POX), catalase (CAT), ascorbate peroxidase (APX) and GR in both the species. *Cleome spinosa* showed remarkable increase in malondialdehyde (MDA) and H_2O_2 than *Cleome gynandra.* It was concluded that *Cleome gynandra* has an efficient antioxidant system to survive the ROS generation under drought stress (Uzilday et al., 2012). Millets are C4 crops and can efficiently utilize little water available in semi-arid regions, thus preventing photorespiration. Pearl millet variety adjusted the flowering time according to the rainfall pattern where main panicle development coincides with the period of rain. The various morphological adaptations in drought tolerant millets are decreased shoot length, increased root length (little millet), and increased leaf tensile strength (tef millet). Physiological adaptations include osmotic adjustment of the cytoplasm by accumulating compatible solutes such as proline, free amino acids, sugars, organic acids, and glycine betaine. Increased activity of antioxidant ROS scavenging enzymes during drought stress reduces damaging effects of drought (Tadele, 2016).

8.3.3 TEMPERATURE

Both high and low soil temperature has a negative effect on the plant photosynthetic processes. Thermal stress is caused by the direct effect of temperature and results in homeostatic imbalance in an organism. Photosynthesis is the most temperature sensitive physiological process. In addition to this, thermal stress causes instability of proteins, membranes, RNA and cytoskeletal structures (Júnior et al., 2018). C3 photosynthesis is more productive at lower temperatures, while C4 photosynthesis is more efficient at elevated temperatures. Oxygenase reaction increases with increasing temperature and hence above 25°C, C3 plants shift to photorespiratory metabolism. Comparatively in such conditions, C4 plants maintain high NUE and WUE (Yadav and Mishra, 2019). At low temperature, photosynthesis is limited due to low sucrose synthesis. On the other hand, high temperature leads to increased mitochondrial respiration, resulting in faster degradation of reserves (Sánchez et al., 2016). *Panicum maximum* Jacq. when grown at elevated temperature and increased atmosphere CO_2 concentration showed different anatomical and physiological adjustments in order to increase the leaf biomass production. These include increased adaxial cuticle thickness, decreased leaf thickness, starch content, reduced vascular bundle size and bulliform cells. Elevated CO_2 affected the epidermis anatomy and leaf gas exchange while increased temperature mainly affected the leaf structure. These adjustments provide protection against the elevated temperature and enhance the WUE and carbon fixation, for better functioning in these conditions (Habermann et al., 2019). Maize plants colonized with AMF showed high photosynthetic activity under high temperature stress (Mathur et al., 2018). Pearl millet tolerance to high temperature is probably better than any other cereal (Prasad et al., 2020). Foxtail millet variety tolerant to high soil temperature accumulated protective metabolites (serine, threonine, valine, malate, isomaltose, maltose, raffinose and itaconate) with efficient utilization and assimilation of nitrogen and carbon (Aidoo et al., 2016). When the temperature exceeds 30°C, proso millet prepares for drought conditions by cessation of vegetative growth, flowering and maintaining a shorter stem height (Habiyaremye et al., 2017). C4 plants are generally maladapted to cold due to the reason that they have originated in warm climates. At cooler temperatures, C3 photosynthesis is generally superior to C4 photosynthesis. Irrespective of this, some C4 species have evolved mechanisms for the sustenance in cold environments, for example, *Miscanthus* spp., *P. virgatum* (switch grass), *Spartina pectinata* (prairie cordgrass) which are cold tolerant bioenergy crops. These form mass of rhizomes overwinter which give rise to new springtime growth. In addition to rhizomes, *S. pectinata* forms dormant shoots at the soil surface. C4 crops in cold climates start leaf growth early in spring in order to close the canopy before the start of long photoperiods of late spring. The chilling tolerance in *Miscanthus* enables the plant to maintain protein synthesis at cool temperature (10°C–15°C) and produce photoprotection by using zeaxanthin and carotenoids in the repairs of photosystem in high light conditions. Among the three, cordgrass has the highest ability to tolerate frost and spring chill (Sage et al., 2015). Owing to its NAD-ME C4 photosynthesis, switchgrass showed a higher photosynthesis rate and hence higher tolerance than giant reed (C3) after various stress treatments (water stress, salinity, and cold temperature) (Sánchez et al., 2016). In addition to other modifications, chilling tolerance in maize can be engineered by increasing the RuBisCO content. Increased RuBisCO content resulted in increased CO_2 assimilation, and improved photochemical quenching. The results indicated that increased RuBisCO content helped the maize plants to cope with chilling stress and recovery (Smith et al., 2020).

8.4 CAM PLANTS

Plants in arid, hot and high light conditions face photosynthetic obstacles because water loss restricts stomatal opening that reduces CO_2 supply for photosynthesis. C4 and CAM photosynthesis are the two solutions for such conditions. Unlike C4 evolution which involves expression loss of multiple genes, CAM species require gain of expression. This makes the CAM possibly sliding back to C3. Many of the CAM species retain the C3 photosynthetic ability, while the C4 cycle is obligatory and nonreversible (Bräutigam et al., 2017). CAM is found across 36 families in over 400 genera that outnumbers C4 species and has evolved multiple times in vascular plants. It is mostly characterized in families like Crassulaceae, Bromeliaceae, Orchidaceae and Agavaceae. CAM crops such as *Agave* spp., *Manfreda* spp., *Polianthes* spp., *Prochnyanthes mexicana* (Zucc.) Rose, *Aloe vera* (L.) Burm. f., *A. comosus*, *Hylocereus* spp., *Opuntia ficus-indica* (L.) Mill., *Stenocereus* spp., *Vanilla planifolia* Jacks. ex Andrews are all used for food, bioenergy, fiber and animal feed (Yang et al., 2015). CAM plants are known for their WUE and drought resistance. CAM plants account for 10% of the total higher plant species. CAM is a convergent physiological syndrome that has independently evolved multiple times across the terrestrial plants. CAM plants can withstand multiple stresses and inhabit diverse environment, ranging from deserts to cloud forests (Males and Griffiths, 2017).

The typical stem and leaf succulents of CAM families are Cactaceae (1500 species) and Agavaceae (300 species). These families mostly consist of CAM plants that inhabit deserts. The highest CAM species are found in the family Orchidaceae (9500) while Bromeliaceae consists of 1250 CAM species. Orchidaceae, Bromeliaceae and Clusiaceae species are mostly found in tropical forests, while the genus *Isoetes* includes submerged freshwater CAM species (Lüttge, 2010). CAM plants are typical succulents with succulent leaves and/or stems and succulence exhibits numerous large cells with enlarged vacuoles occupying 90% of the cell volume. Vacuoles accommodate the nocturnal storage of malic acid and increases water storage (Borland et al., 2018). CAM expression in a plant is determined by ontogenesis and environmental signals. In C3 and CAM plants, the photosynthetic activity takes place in the leaf MCs. CAM expression is not only species dependent but also varies with different organs and tissues. The two opposite leaves in *Clusia minor* (a C3-CAM intermediate species) can follow two distinct photosynthetic pathways, that is, C3 and CAM. Compartmentalization of C3-CAM pathways within a single leaf has also been reported in *C. minor*, where leaf lamina expresses C3 photosynthesis and veins follow CAM pathway. The midrib chloroplasts of a C3-CAM intermediate plant, *Mesembryanthemum crystallinum* revealed that C3 leaf midrib chloroplasts could contribute to the CO_2 supply for carboxylation, while in CAM plants, these were involved in H_2O_2 signaling and malate supply to various metabolic processes (Kuźniak et al., 2016).

Depending on various degrees of photosynthesis, CAM plants are grouped into obligate CAM, facultative CAM, CAM cycling and CAM idling. Plants that perform CAM irrespective of climatic conditions are termed as obligate (Pereira and Cushman, 2019). Under various stresses such as drought, salinity, temperature, nutrient availability, some plants which usually employ C3 or C4 photosynthesis switch to CAM pathway. This is known as facultative CAM and this phenomenon is reversible as the plants shifts back to C3 or C4 phenotype on the removal of stress. Facultative CAM has been reported in Aizoaceae (*Mesembryanthemum crystallinum* L.), Bromeliaceae [*Guzmania monostachia* (L.), Crassulaceae (*Sedum mite* Gilib.; *Sedum album* L.; *Sedum telephium* L.), Montiaceae [*Calandrinia (Parakeelya) polyandra* Benth.], Piperaceae (*Peperomia scandens* Ruiz and Pav.), Portulacaceae (*P. oleracea* L.), Talinaceae [*Talinum triangulare* (Jacq.)], [*Clusia cylindrica* (Hammel)] (Winter and Holtum, 2014). CAM cycling is a very weak CAM type which involves CO_2 uptake during the day and slight nocturnal accumulation of C4 acids. CAM idling is distinguished by nonatmospheric CO_2 fixation, involving the stomatal closure for 24 h and refixing of respiratory CO_2 (Pereira and Cushman, 2019). Unlike C4 plants, CO_2 fixation in CAM plants takes place in a single cell and various steps involved are performed in the same cell. These steps are separated temporally, and this minimizes transpiration water loss (Hartwell, 2018). CAM photosynthesis is a water saving method of carbon fixation where CO_2 uptake by the MCs and its fixation by RuBisCO are temporally separated. This involves the opening of stomata at night when the CO_2 is fixed followed by its vacuolar storage in the form of carboxylic acids (malate, citrate, and isocitrate). The following day, stomata remain shut to reduce the water loss and the stored CO_2 is refixed by RuBisCO in the chloroplast. Even though this cycle is energetically expensive, it efficiently saves precious water (Töpfer et al., 2020) (Fig. 8.2).

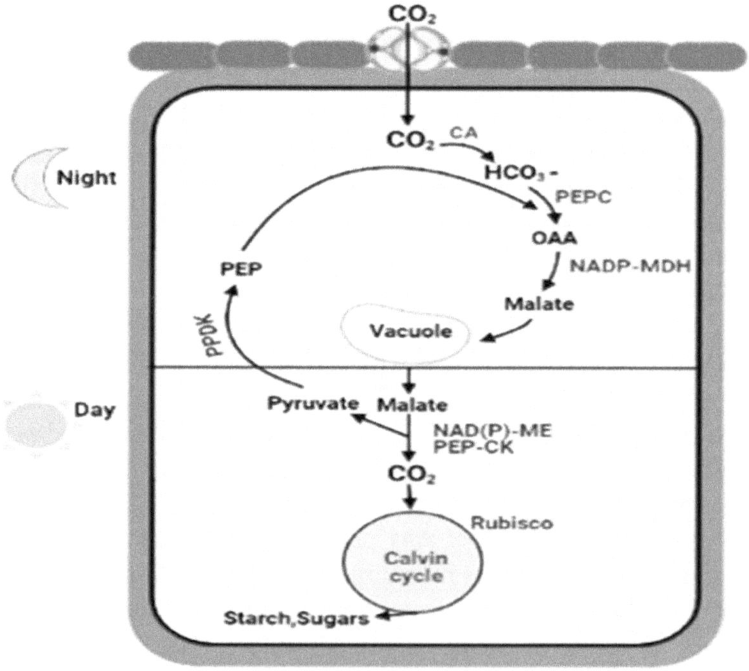

FIGURE 8.2 Systematic representation of CAM pathway. HCO_3^-, bicarbonate; CA, carbonic anhydrase; PEPC, phosphoenolpyruvate carboxylase; OAA, oxaloacetate; NADP-MDH, NADP-malate dehydrogenase; NADP-ME, NADP dependent malic enzyme; NAD-ME, NAD dependent malic enzyme; PEP-CK, PEP carboxykinase; PPDK, pyruvate orthophosphate dikinase; PEP, phosphoenolpyruvate.

8.5 CAM PLANTS: PHYSIOLOGY AND CHARACTERISTICS

Vascular plants with CAM metabolic pathway have improved WUE. These plants take up CO_2 at night when the evapotranspiration rates are low. At night, the stomata are open, and CO_2 is taken up with the C3 substrate, PEP, with the help of enzyme PEPC. Glycolytic breakdown of carbohydrate accumulated the previous day provides the PEP. The carbohydrate source can be vacuolar sugars, chloroplast starch or both. The subsequent breakdown of final 4-C product (malic acid) by malic enzyme (ME) or PEP carboxykinase yields CO_2, that is fixed the following day when the stomata is closed. This is considered as a water conserving strategy (Ceusters et al., 2016).

The stomatal pores in CAM leaves open at night and primary CO_2 fixation is catalyzed by PEPC. PEPC generates OAA (4 carbon product) using PEP and bicarbonate as substrates. NADP-MDH reduces OAA to malate which accumulates in the vacuoles of chloroplast containing MCs as malic acid, throughout the night. Approaching dawn, accumulated malic acid is discharged from the vacuole into the cytosol and depending on the families, it is decarboxylated by either NAD-ME/NADP-ME/PEP-CK, releasing pyruvate and CO_2. Pyruvate is transformed into PEP and CO_2 enters the Calvin cycle. Internal partial pressure, built up by the CO_2, mediates stomatal closure and is refixed by RuBisCO. The temporal control of CAM pathway is strictly maintained by a circadian clock (Hartwell, 2018). The high WUE in CAM plants can be 3–20-fold and 2–10-fold that of C_3 and C_4 plants, respectively. This high WUE is attributed to the various patterns of CO_2 fixation and stomatal movement in four sequential phases. Phase I constitutes the period of stomatal opening at night for primary atmospheric fixation of CO_2 by PEPC. This step takes place at the lower night temperature. Dark to dawn transition initiates the phase II which includes the shift from PEPC-mediated C4 carboxylation to RuBisCO-mediated CO_2 assimilation. In phase III, stomata are closed at higher temperatures during the daytime. CO_2 released from malate decarboxylation is refixed by RuBisCO. Phase IV is initiated with the end of light period and exhausted malate reserves. Here, the stomata are re-opened for CO_2 uptake. CAM plants reduce photorespiration by concentrating CO_2 in the proximity of RuBisCO during the daytime. This result in increased biochemical efficiency of CAM plants as compared to C3 plants (Yang et al., 2019). In order to optimize carbon assimilation and minimize futile cycling, carboxylation and decarboxylation reactions of CAM have to be supported by strict temporal control. The structural hallmarks of CAM include succulent leaves, shoots with low surface area: volume ratio and high-water storage capacitance. Chlorenchyma tissue is composed of densely packed, thin walled, highly vacuolated cells that provide maximum storage for malic acid. CAM plants are distinguished by comparatively low stomatal densities on their shoot surfaces and low maximal stomatal conductance. The transition from C3 through C3-CAM to strongly expressed CAM in families such as Orchidaceae is accompanied with increasing succulence (Winter et al., 2015). Transporters play important roles in CAM pathway, such as sugar transporters in vacuoles of pineapple reportedly have a role in maintaining carbohydrate partitioning. Proton-motive force that is required for malic acid accumulation in vacuoles is provided by vacuolar H^+-ATPase. Antiporters Na^+/H^+ are involved in N^+ compartmentalization in response to plant adaptation to stress and might also be involved in CAM metabolism. Aquaporins reportedly maintain water balance during CAM (Zhang et al., 2014).

It was reported that the stomatal responses in CAM plants is due to hypersensitivity of stomata to ABA. In addition to this, ABA also plays a role in biochemical transition from C3 to CAM metabolism, for example, *Kalanchoe blossfeldiana*, *Portula cariaafra*, and *Ananas comosus* (Negin and Moshelion, 2016).

8.6 CAM RESPONSE TO VARIOUS ENVIRONMENTAL STRESSES

8.6.1 DROUGHT STRESS

CAM plants can live under water deficit conditions. These plants minimize water loss by performing CO_2 assimilation and fixation at the night time, minimizing evaporative water losses. During the day time, the stomata are closed, and these plants perform C3 carboxylation reactions. CAM plants utilize 80% less water to produce similar amount of biomass as compared to C3 plants. CAM plants like *Agave salmiana, Opuntia ficus-indica*, and *Stenocereus queretaroensis* can tolerate subzero temperatures. Stress adaptive responses in CAM plants include ABA-dependent and ABA-independent pathways (Amin et al., 2019). It was reported that water stress in CAM plants, *Mesembryanthemum crystallinum* and *Kalanchoe fedtschenkoi* induced various transcription factor (TF) families such as AP2/ERF, MYB, WRKY, NAC, NF-Y, and bZIP. These TFs might play a role in regulation of water stress responses and CAM in *M. crystallinum* and *K. fedtschenkoi*. Using the biotechnological approach, the appropriate CAM TFs generated abiotic stress tolerance in *Arabidopsis thaliana* (Amin et al., 2019). Stress memory is the better response to a given stress by organisms that have been previously exposed to same stimulus, as compared to those which have not been previously exposed to the stress. During severe drought, CAM plants maintain a minimal metabolically active state and on re-watering, these are able to recover quickly. It was reported that CAM plant, *Aptenia cordifolia* showed drought stress memory and the phytohormones like gibberellins and ABA are involved in stress responses toward repeated drought periods. Enhanced accumulation of ABA in double-stressed plants is known to prevent photo-oxidative damages (Soriano et al., 2015). In drought stressed facultative, invasive CAM plant, *Aptenia cordifolia*, β-carotene and tocopherols played important role in redox regulation by physical and chemical scavenging of ROS (Marijuan et al., 2017). Photosynthetic organs of CAM plants (tissues, leaves, and stems) are highly succulent. The highly vacuolated cells maintain succulence. This is important for nocturnal storage of organic acids and differentiated water storage tissues such as hydrenchyma tissues. This is a general adaptation to water limited habitats (Lüttge, 2010). At cellular level, succulents avoid drought by a physiological trait known as hydraulic capacitance (C). Succulent cells or tissues have high values of C, that is, they can take up or loose large quantities of water for a given change in C (Borland et al., 2018). Along with succulence, epiphytic CAM species from Orchidaceae and the Bromeliaceae have specialized anatomical water storing structures called pseudobulbs. These are enlarged succulent internodal stem segments. These are composed of chlorenchyma cells and aerenchyma that are tightly packed and have aerial roots. In drought conditions, the roots can perform weak CAM and are enveloped by a spongy epidermis (velamen) that accumulates water and prevents its loss. In the family Bromeliaceae, the basal portion of leaves overlap to form tank-like structures called epidermal trichomes, that store water and nutrients absorbed. *Tillandsia*, another CAM species has dense epidermal trichomes that enhance water capture and absorption. CAM plants like *O. ficus-indica* have thick cuticles to prevent water loss. Many CAM plants, in addition to thick cuticle, possess intracuticular and epicuticular wax that prevents transpirational water loss from the leaf. Limited available water is optimized by CAM species (within Aizoaceae and Crassulaceae) using shallow rectifier like roots. Shallow rooting enhances water uptake and storage in the course of wet period and seems helpful in survival across dry season (Niechayev et al., 2019). *M. crystallinum is drought and salinity tolerant and shows accumulation of proline and stimulation of secondary metabolites such as* phenolics, carotenoids, soluble sugars, and other antioxidants such as ascorbic acid *in abiotic stress conditions to cope with oxidative stress (He et al., 2020).* Transcriptomic sequencing of *Agave sisalana* showed 3095 differentially expressed unigenes (DEG) between well-irrigated and drought-stressed leaf samples. The stress responsive genes were seen involved in various pathways such as hormonal responses, stress stimuli response, wax biosynthesis, and ROS metabolism. Heat shock protein unigenes showed upregulation as they are involved in the restoration of cellular homeostasis and depreciation of cellular injuries. Drought stress induced the expression of enzymatic antioxidants (catalase, ascorbate peroxidase, and peroxidase), nonenzymatic (glutathione) scavenging molecules and wax biosynthesis enzymes. Wax accumulation on plant cuticle prevents water loss and provides the plant with hydrophobic protection under osmotic stress (Sarwar et al., 2019).

8.6.2 SALINITY STRESS

Ice plant (M. *crystallinum*) with CAM photosynthesis showed a reduction in oxidative stress under salt stress. Higher oxidative stress requires higher SOD activities. The SOD activities were lower in wild type as compared to the CAM deficient mutant. It was concluded that lack of CAM in the mutant resulted in increased oxidative stress and CAM is more efficient in bringing down the oxidative stress as compared to C3 and C4 metabolism (Abogadallah, 2010). M. *crystallinum* undergoes a dramatic change in metabolism under salt stress. Salt stress induces transformation from C3 to CAM photosynthesis in M. *crystallinum* that enhances stress tolerance (Jouyban, 2012). After 6 days of salt treatment, transition from C3 metabolism to CAM initiated in *M. crystallinum* that resulted in increased WUE and stress tolerance. The various morphological and physiological changes that took place were increased CO_2 assimilation at night, inversed stomatal movements, and increased succulence (Guan et al., 2020). *M. crystallinum* is extremely salt tolerant and its adaptation to salinity includes development of epidermal bladder cells. These are nonglandular, nonsecreting, single cell trichomes lining the leaves, stems and flower buds. In unstressed conditions, these remain compressed but expand in response to stress and comprise 25% of the total aerial volume. When fully developed, these lead to osmotic adjustment by accumulating Na^+ and compatible solutes (mainly ononitol and pinitol) (Oh et al., 2015). *Hylotelephium erythrostictum*, a facultative CAM plant, shows high salt resistance. Quantitative proteomic analysis of leaves under salt stress revealed upregulation of proteins involved in glycolysis/gluconeogenesis and pyruvate metabolism. This advocates the initiation of CAM pathway to cope with stress. SAPK1 protein involved in signal transduction and reduced sensitivity toward salt stress was upregulated; hence, providing tolerance against stress (Zhao et al., 2020). Bromeliads, orchids of the tropical regions like *Salinas* avoid the intimate salt contact by growing epiphytically or without forming absorptive roots (Lüttge, 2010). *Aloe vera* plants tolerate salinity by increasing K mineral uptake and maintaining a higher K/Na ratio and lower Na/K ratio. The increase in proline content and PEPC activity on increasing salinity suggests their possible role in adaptation by influencing water retention and providing metabolic cellular protection (Amador et al., 2014). Salinity stress at high light intensity in *M. crystallinum*, showed a 6.4- and 17.7-fold increase in activities of NAD- and NADP-ME, respectively (Miszalski et al., 2001). NADP-ME is known to be a critical enzyme in CAM pathway. It showed 4–10 times increased activity under salt stress and might play an important role in switching from C3 to C4 pathway in facultative CAM species and resisting salt stress (Hui et al., 2010).

8.6.3 HEAT/COLD STRESS

Acclimation of photosynthesis to temperature fluctuations is different among plants following different photosynthetic pathways (C3, C4, and CAM). Day and night photosynthesis is differentially acclimated by CAM plants, because CAM plants are found in deserts. Day and night temperature vary considerably in deserts, CAM plants have a threshold of tolerating temperature up to 70°C and higher light intensity (> 1000 $\mu mol/m^2/s$) than C3 plants (Yamori et al., 2014). Stomata are shut during the day and opened at night when the temperature is lower, and humidity is higher as compared to the day which enhances WUE of the plant (Borland et al., 2018). CAM plants are less predominantly present in cooler climates of higher altitudes or latitudes. Some *Opuntia* species are found in freezing temperatures down to −50°C. Ice formation and dehydration is prevented by the various tolerance mechanisms involving partial water loss and cryoprotectant (sugars and sugar alcohols) accumulation. Extracellular ice nucleation leads to freezing. Intracellular water is discharged into the apoplastic spaces forming ice crystals and leading to dehydration. Cacti mucilage (highly branched polysaccharide) may be involved in delaying water loss from the cell to the extracellular ice crystals, which might be less harmful for the cells. CAM plants have thick cuticle which is made up of polymers that have a higher specific heat than the other polymers. This implies that in order to raise the plant surface temperature, a higher amount of heat is required (Niechayev et al., 2019). CAM plants change the angle of their leaves, or even loose all of their leaves that help in reducing the radiation incident on plants that in turn reduces internal temperature (Gilman and Edwards, 2020).

8.7 CONCLUSION AND FUTURE PERSPECTIVES

Climate change has led to agricultural stress conditions such as drought, salinity, increased global temperature and atmospheric CO_2. These conditions have led to land deterioration and decrease in agricultural productivity. One of the most effective ways of mitigating the adverse effects of climatic change on agriculture is the use of stress-resilient crops. Stress tolerance mechanisms interfere with cellular functions and lead to adaptation against various stresses. In order to facilitate food and fuel requirements of ever-growing population, biotechnological and breeding efforts need to be accelerated. CAM and C4 have evolved from common C3 ancestor and C4 photosynthesis can boost plant productivity in warmer regions where C3 photosynthetic pathway is insufficient because of less available water, high temperature and limited nutrient supply. Exploiting C4 photosynthesis can help in fulfilling the future agricultural demands by engineering the C4 pathway into the major C3 crops such as rice. CAM plants are recognized for their ability to withstand harsh climates. These can even sustain climatic conditions that are unbearable for C3 and C4 species. CAM plants possess immense agricultural values and potential to meet the food, feed and biofuel requirements of the future. Not many but a few CAM species and their products have received global attention; rest has remained undervalued. However, recent progress made in genomic resources has made us realize that some CAM species will help in facilitating genetic improvement via genome editing approaches and molecular-assisted breeding. These are agriculturally attractive commodities that need to be expanded in order to sustain climate change. Detailed understanding of plant response to various environmental stresses is also required in addition to the identification of upstream regulators that result in acclimation response. This can be used for the enhancement of tolerance in different plants toward various environmental stresses (Table 8.1).

TABLE 8.1 Comparison between C4 and CAM Plants.

	C4	CAM
Definition	These plants fix CO_2 during the day and produce a stable 4 carbon compound, OAA in the MCs followed by photosynthesis (C3 pathway) in the BSCs	CAM involves CO_2 fixation into OAA (C4 pathway) at night and storage of products until the next day, followed by photosynthesis (C3 pathway). It involves only MCs
Plant type	Mesophytes	Xerophytes
Habitat	Tropical or semi-tropical and dry climates	Desert and arid climates
Kranz anatomy	Present	Absent
Cells involved	MCs and BSCs	Only MCs
Stomatal movements	Stomata open during the day time and closed at night	Inverted stomata cycle (open at night and closed during the day
Worldwide species	5% of higher plant species	10% of higher plant species
Species	Sugarcane, millets, maize, etc.	Pineapple, Aloe, Cacti, etc.
Drought adaptive mechanisms	Increased root length, decreased shoot length, accumulation of compatible solutes, enhanced antioxidant system and adjusted flowering time; Up regulation of NADP-ME and PEPC activity; High WUE	Succulent leaves/ tissues/ stem for water storage. Special water storage structures pseudobulbs, epidermal trichomes. Presence of thick cuticle and epicuticular wax and high WUE Transformation of facultative CAM plants from C3 metabolism to CAM
Adaptive mechanisms in response to increased salinity	Accumulation of osmoprotectants (amino acids, sugars, fatty acids, flavonoids); High cytosolic K^+/Na^+ concentration; High photosynthetic rate, and chlorophyll content; Increased PEPC, PPDK activities; Enhanced antioxidant enzyme activities and increased proline content	Development of epidermal bladder cells that accumulate Na^+ and other compatible solutes; Increased uptake of K mineral to maintain K/ Na ratio; Increased proline content and PEPC activity; Transformation of facultative CAM plants from C3 metabolism to CAM

TABLE 8.1 *(Continued)*

	C4	CAM
Adaptive mechanisms to high/freezing temperatures	Increased adaxial cuticle thickness in leaves; decreased leaf thickness and starch content and reduced vascular bundle size and bulliform cells in high temperatures; accumulation of protective metabolites; cessation of vegetative growth, short stem height; Cold tolerance in some C4 plants involve formation of rhizomes and dormant shoots at the soil surface	Closed stomata during the day, thick cuticle, change in angle of leaves, loss of leaves that reduce internal temperature contribute toward the high temperature adaption; Cold tolerance in *Opuntia* involves cryoprotectant accumulation and delayed water loss from the cell

KEYWORDS

- **Kranz anatomy**
- **photorespiration**
- **nitrogen-use efficiency**
- **water use efficiency**
- **abiotic stress**
- **adaptations**

REFERENCES

Abogadallah, G. M. Insights into the Significance of Antioxidative Defense under Salt Stress. *Plant Signal. Behav.* **2010,** *5* (4), 369–374.

Ahanger, M. A.; Akram, N. A.; Ashraf, M.; Alyemeni, M. N.; Wijaya, L.; Ahmad, P. Plant Responses to Environmental Stresses—From Gene to Biotechnology. *AoB Plants* **2017,** *9* (4). https://doi.org/10.1093/aobpla/plx025

Aidoo, M. K.; Bdolach, E.; Fait, A.; Lazarovitch, N.; Rachmilevitch, S. Tolerance to High Soil Temperature in Foxtail Millet (*Setaria italica* L.) Is Related to Shoot and Root Growth and Metabolism. *Plant Physiol. Biochem.* **2016,** *106*, 73–81.

Amin, A. B.; Rathnayake, K. N.; Yim, W. C.; Garcia, T. M.; Wone, B.; Cushman, J. C.; Wone, B. W. Crassulacean Acid Metabolism Abiotic Stress-Responsive Transcription Factors: A Potential Genetic Engineering Approach for Improving Crop Tolerance to Abiotic Stress. *Front. Plant Sci.* **2019,** *10*, 129.

Borland, A. M.; Leverett, A.; Hurtado-Castano, N.; Hu, R.; Yang, X. Functional Anatomical Traits of the Photosynthetic Organs of Plants with Crassulacean Acid Metabolism. In *The Leaf: A Platform for Performing Photosynthesis*; Adams, W. W., Terashima, I., Eds.; Springer: Cham, Switzerland, 2018; pp 281–305.

Bräutigam, A.; Schlüter, U.; Eisenhut, M.; Gowik, U. On the Evolutionary Origin of CAM Photosynthesis. *Plant Physiol.* **2017,** *174* (2), 473–477.

Byrt, C. S.; Grof, C. P.; Furbank, R. T. C4 Plants as Biofuel Feedstocks: Optimising Biomass Production and Feedstock Quality from a Lignocellulosic Perspective Free Access. *J. Integr. Plant Biol.* **2011,** *53* (2), 120–135.

Cacefo, V.; Ribas, A. F.; Zilliani, R. R.; Neris, D. M.; Domingues, D. S.; Moro, A. L.; Vieira, L. G. E. Decarboxylation Mechanisms of C4 Photosynthesis in *Saccharum* spp.: Increased PEPCK Activity under Water-Limiting Conditions. *BMC Plant Biol.* **2019,** *19* (1), 1–14.

Ceusters, N.; Van den Ende, W.; Ceusters, J. Exploration of Sweet Immunity to Enhance Abiotic Stress Tolerance in Plants: Lessons from CAM. In *Progress in Botany*; Cánovas, F., Lüttge U., Matyssek, R., Eds., Vol. 78; Springer: Cham, Switzerland, 2016; pp. 145–166.

Chandrasekaran, M.; Kim, K.; Krishnamoorthy, R.; Walitang, D.; Sundaram, S.; Joe, M. M.; Selvakumar, G.; Hu, S.; Oh, S.; Sa, T. Mycorrhizal Symbiotic Efficiency on C3 and C4 Plants under Salinity Stress–a Meta-Analysis. *Front. Microbiol.* **2016,** *7*, 1246.

Christin, P. A.; Osborne, C. P. The Evolutionary Ecology of C4 Plants. *New Phytol.* **2014,** *204* (4), 765–781.

Doubnerová, V.; Ryšlavá, H. What Can Enzymes of C4 Photosynthesis do for C3 Plants under Stress? *Plant Sci.* **2011,** *180* (4), 575–583.

Ehleringer, J. R.; Cerling, T. E. C_3 and C_4 Photosynthesis. In *Encyclopedia of Global Environmental Change – Volume 2, the Earth System: Biological and Ecological Dimensions of Global Environmental Change*; Mooney, H. A., Canadell, J. G., Eds.; John Wiley & Sons, 2002; pp 186–190.

Essemine, J.; Lyu, M. J. A.; Qu, M.; Perveen, S.; Khan, N.; Song, Q.; Chen, G.; Zhu, X. G. Contrasting Responses of Plastid Terminal Oxidase Activity under Salt Stress in Two C4 Species with Different Salt Tolerance. *Front. Plant Sci.* **2020,** *11*,1009.

Fathi, A.; Tari, D. B. Effect of Drought Stress and Its Mechanism in Plants. *Int. J. Life Sci.* **2016,** *10* (1), 1–6.

Ferrari, R. C.; Bittencourt, P. P.; Rodrigues, M. A.; Moreno-Villena, J. J.; Alves, F. R.; Gastaldi, V. D.; Boxall, S. F.; Dever, L. V.; Demarco, D.; Andrade, S. C.; Edwards, E. J.; Hartwell, J.; Freschi, L. C4 and Crassulacean Acid Metabolism Within a Single Leaf: Deciphering Key Components Behind a Rare Photosynthetic Adaptation. *New Phytol.* **2020,** *225* (4), 1699–1714.

Furbank, R. T. Evolution of the C4 Photosynthetic Mechanism: Are There Really Three C4 Acid Decarboxylation Types? *J. Exp. Bot.* **2011,** *62* (9), 3103–3108.

Fleta-Soriano, E.; Pintó-Marijuan, M.; Munné-Bosch, S. Evidence of Drought Stress Memory in the Facultative CAM, *Aptenia cordifolia*: Possible Role of Phytohormones. *PLoS One* **2015,** *10* (8), e0135391.

Geissler, N.; Hussin, S.; El-Far, M. M.; Koyro, H. W. Elevated Atmospheric CO_2 Concentration Leads to Different Salt Resistance Mechanisms in a C3 (*Chenopodium quinoa*) and a C4 (*Atriplex nummularia*) Halophyte. *Environ. Exp. Bot.* **2015,** *118*, 67–77.

Ghannoum, O. C4 Photosynthesis and Water Stress. *Ann. Bot.* **2009,** *103* (4), 635–644.

Gilman, I. S.; Edwards, E. J. Crassulacean Acid Metabolism. *Curr. Biol.* **2020,** *30* (2), R57–R62.

Griffiths, H.; Weller, G.; Toy, L. F.; Dennis, R. J. You're So Vein: Bundle Sheath Physiology, Phylogeny and Evolution in C3 and C4 Plants. *Plant Cell Environ.* **2013,** *36* (2), 249–261.

Guan, Q.; Tan, B.; Kelley, T. M.; Tian, J.; Chen, S. Physiological Changes in *Mesembryanthemum crystallinum* during the C3 to CAM Transition Induced by Salt Stress. *Front. Plant Sci.* **2020,** *11*, 283.

Gull, A.; Lone, A. A.; Wani, N. U. I. Biotic and Abiotic Stresses in Plants. In *Abiotic and Biotic Stress in Plants*; de Oliveira, A. B., Ed.; IntechOpen: London, 2019; pp 1–19.

Habermann, E.; San Martin, J. A. B.; Contin, D. R.; Bossan, V. P.; Barboza, A.; Braga, M. R.; Groppo, M.; Martinez, C. A. Increasing Atmospheric CO_2 and Canopy Temperature Induces Anatomical and Physiological Changes in Leaves of the C4 Forage Species *Panicum maximum. PLoS One* **2019,** *14* (2), e0212506.

Habiyaremye, C.; Matanguihan, J. B.; D'Alpoim Guedes, J.; Ganjyal, G. M.; Whiteman, M. R.; Kidwell, K. K.; Murphy, K. M. Proso Millet (*Panicum miliaceum* L.) and Its Potential for Cultivation in the Pacific Northwest, US: A Review. *Front. Plant Sci.* **2017,** *7*, 1961.

Hartwell, J. The Circadian Clock in CAM Plants. In *Annual Plant Reviews Online*; Hall, A., McWatters, H., Eds.; Oxford: Blackwell, 2018; pp 211–236.

Hasan, S. A.; Rabei, S. H.; Nada, R. M.; Abogadallah, G. M. Water Use Efficiency in the Drought-Stressed Sorghum and Maize in Relation to Expression of Aquaporin Genes. *Biol. gia Plant.* **2017,** *61* (1), 127–137.

He, J.; Chua, E. L.; Qin, L. Drought Does Not Induce Crassulacean Acid Metabolism (CAM) But Regulates Photosynthesis and Enhances Nutritional Quality of *Mesembryanthemum crystallinum. Plos One* **2020,** *15* (3), e0229897.

Hossain, M. S.; Dietz, K. J. Tuning of Redox Regulatory Mechanisms, Reactive Oxygen Species and Redox Homeostasis under Salinity Stress. *Front. Plant Sci.* **2016,** *7*, 548.

Zeng-Hui, L.; Zheng-Bin, Z.; Li-Ye, C.; Hong-Bo, S. The Corresponding Relationship between Roles of NADP-Malicenzymes and Abiotic Stress in Plants. *Emirates J. Food Agric.* **2010,** *22* (4), 239–249.

Jouyban, Z. The Effects of Salt Stress on Plant Growth. *Tech. J. Eng. Appl. Sci.* **2012,** *2* (1), 7–10.

Júnior, R. D. S. N.; do Amaral, G. C.; Pezzopane, J. E. M.; Toledo, J. V.; Xavier, T. M. T. Ecophysiology of C3 and C4 Plants in Terms of Responses to Extreme Soil Temperatures. *Theor. Exp. Plant Physiol.* **2018,** *30* (3), 261–274.

Khoshravesh, R.; Stata, M.; Adachi, S.; Sage, T. L.; Sage, R. F. Evolutionary Convergence of C4 Photosynthesis: A Case Study in the Nyctaginaceae. *Front. Plant Sci.* **2020,** *11*, 578739.

Killi, D.; Bussotti, F.; Raschi, A.; Haworth, M. Adaptation to High Temperature Mitigates the Impact of Water Deficit during Combined Heat and Drought Stress in C3 Sunflower and C4 Maize Varieties with Contrasting Drought Tolerance. *Physiol. Plant.* **2017,** *159* (2), 130–147.

Kumar, M. Crop Plants and Abiotic Stresses. *J. Biomol. Res. Therap.* **2013,** *3* (1), 1000e125.

Kumar, V.; Sharma, A.; Soni, J. K.; Pawar, N. Physiological Response of C3, C4 and CAM Plants in Changeable Climate. *Pharma Innov. J.* **2017,** *6* (9), 70–79.

Kuźniak, E.; Kornas, A.; Kaźmierczak, A.; Rozpądek, P.; Nosek, M.; Kocurek, M.; Zellnig, G.; Müller, M.; Miszalski, Z. Photosynthesis-Related Characteristics of the Midrib and the Interveinal Lamina in Leaves of the C3–CAM Intermediate Plant *Mesembryanthemum crystallinum. Ann. Bot.* **2016,** *117* (7), 1141–1151.

Ludwig, M. The Roles of Organic Acids in C4 Photosynthesis. *Front. Plant Sci.* **2016,** *7*, 647.

Lüttge, U. Ability of Crassulacean Acid Metabolism Plants to Overcome Interacting Stresses in Tropical Environments. *AoB Plants* **2010,** *2010*, plq005.

Males, J.; Griffiths, H. Stomatal Biology of CAM Plants. *Plant Physiol.* **2017,** *174* (2), 550–560.

Mathur, S.; Sharma, M. P.; Jajoo, A. Improved Photosynthetic Efficacy of Maize (*Zea mays*) Plants with Arbuscular Mycorrhizal Fungi (AMF) under High Temperature Stress. *J. Photochem. Photobiol. B* **2018,** *180*, 149–154.

Miszalski, Z.; Niewiadomska, E.; Slesak, I.; Lüttge, U.; Kluge, M.; Ratajczak, R. The Effect of Irradiance on Carboxylating/Decarboxylating Enzymes and Fumarase Activities in *Mesembryanthemum crystallinum* L. Exposed to Salinity Stress. *Plant Biol.* **2001,** *3* (1), 17–23.

Murillo-Amador, B.; Córdoba-Matson, M. V.; Villegas-Espinoza, J. A.; Hernández-Montiel, L. G.; Troyo-Diéguez, E.; García-Hernández, J. L. Mineral Content and Biochemical Variables of *Aloe vera* L. under Salt Stress. *PLoS One* **2014,** *9* (4), e94870.

Nayyar, H.; Gupta, D. Differential Sensitivity of C3 and C4 Plants to Water Deficit Stress: Association with Oxidative Stress and Antioxidants. *Environ. Exp. Bot.* **2006,** *58* (1–3), 106–113.

Negin, B.; Moshelion, M. The Evolution of the Role of ABA in the Regulation of Water-Use Efficiency: From Biochemical Mechanisms to Stomatal Conductance. *Plant Sci.* **2016,** *251*, 82–89.

Niechayev, N. A.; Pereira, P. N.; Cushman, J. C. Understanding Trait Diversity Associated with Crassulacean Acid Metabolism (CAM). *Curr. Opin. Plant Biol.* **2019,** *49*, 74–85.

Oh, D. H.; Barkla, B. J.; Vera-Estrella, R.; Pantoja, O.; Lee, S. Y.; Bohnert, H. J.; Dassanayake, M. Cell Type-Specific Responses to Salinity–the Epidermal Bladder Cell Transcriptome of *Mesembryanthemum crystallinum*. *New Phytol.* **2015,** *207* (3), 627–644.

Omoto, E.; Taniguchi, M.; Miyake, H. Adaptation Responses in C4 Photosynthesis of Maize under Salinity. *J. Plant Physiol.* **2012,** *169* (5), 469–477.

Osborne, C. P.; Sack, L. Evolution of C4 Plants: A New Hypothesis for an Interaction of CO_2 and Water Relations Mediated by Plant Hydraulics. *Phil. Trans. R. Soc. B* **2012,** *367* (1588), 583–600.

Pereira, P. N.; Cushman, J. C. Exploring the Relationship between Crassulacean Acid Metabolism (CAM) and Mineral Nutrition with a Special Focus on Nitrogen. *Int. J. Mol. Sci.* **2019,** *20* (18), 4363.

Pintó-Marijuan, M.; Cotado, A.; Fleta-Soriano, E.; Munné-Bosch, S. Drought Stress Memory in the Photosynthetic Mechanisms of an Invasive CAM Species, *Aptenia cordifolia*. *Photosynthesis Res.* **2017,** *131* (3), 241–253.

Prasad, P. V. V.; Djanaguiraman, M.; Stewart, Z. P.; Ciampitti, I. A. Agroclimatology of Maize, Sorghum, and Pearl Millet. In *Agroclimatology: Linking Agriculture to Climate*; Hatfield, J. L., Sivakumar, M. V. K., Prueger, J. H., Eds.; American Society of Agronomy: Madison, WI, 2020; pp 201–241.

Rao, D. E.; Chaitanya, K. V. Photosynthesis and Antioxidative Defense Mechanisms in Deciphering Drought Stress Tolerance of Crop Plants. *Biol. Plant.* **2016,** *60* (2), 201–218.

Rao, X.; Dixon, R. A. The Differences between NAD-ME and NADP-ME Subtypes of C4 Photosynthesis: More Than Decarboxylating Enzymes. *Front. Plant Sci.* **2016,** *7*, 1525.

Sage, R. F. Photosynthetic Efficiency and Carbon Concentration in Terrestrial Plants: The C4 and CAM Solutions. *J. Exp. Bot.* **2014,** *65* (13), 3323–3325.

Sage, R. F.; de Melo Peixoto, M.; Friesen, P.; Deen, B. C4 Bioenergy Crops for Cool Climates, with Special Emphasis on Perennial C4 Grasses. *J. Exp. Bot.* **2015,** *66* (14), 4195–4212.

Salesse-Smith, C. E.; Sharwood, R. E.; Busch, F. A.; Stern, D. B. Increased Rubisco Content in Maize Mitigates Chilling Stress and Speeds Recovery. *Plant Biotechnol. J.* **2020,** *18* (6), 1409–1420.

Sánchez, E.; Gil, S.; Azcón-Bieto, J.; Nogués, S. The Response of *Arundo donax* L.(C3) and *Panicum virgatum* (C4) to Different Stresses. *Biomass Bioenergy* **2016,** *85*, 335–345.

Sarwar, M. B.; Ahmad, Z.; Rashid, B.; Hassan, S.; Gregersen, P. L.; Leyva, M. D. L. O.; Asp, T.; Husnain, T. De Novo Assembly of *Agave sisalana* Transcriptome in Response to Drought Stress Provides Insight into the Tolerance Mechanisms. *Sci. Rep.* **2019,** *9* (1), 1–14.

Sdouga, D.; Amor, F. B.; Ghribi, S.; Kabtni, S.; Tebini, M.; Branca, F.; Farah, N. T.; Marghali, S. An Insight from Tolerance to Salinity Stress in Halophyte *Portulaca oleracea* L.: Physio-Morphological, Biochemical and Molecular Responses. *Ecotoxicol. Environ. Saf.* **2019,** *172*, 45–52.

Siddique, Z.; Jan, S.; Imadi, S. R.; Gul, A.; Ahmad, P. Drought Stress and Photosynthesis in Plants. *Water Stress Crop Plants: Sustain. Approach* **2016,** *2*, 1–11.

Stata, M.; Sage, T. L.; Rennie, T. D.; Khoshravesh, R.; Sultmanis, S.; Khaikin, Y.; Ludwig, M.; Sage, R. F. Mesophyll Cells of C4 Plants Have Fewer Chloroplasts Than Those of Closely Related C3 Plants. *Plant Cell Environ.* **2014,** *37* (11), 2587–2600.

Stepien, P.; Klobus, G. Antioxidant Defense in the Leaves of C3 and C4 Plants under Salinity Stress. *Physiol. Plant.* **2005,** *125* (1), 31–40.

Sun, Q.; Yamada, T.; Han, Y.; Takano, T. Influence of Salt Stress on C4 Photosynthesis in *Miscanthus sinensis* Anderss. *Plant Biol.* **2021,** *23* (1), 44–56.

Tadele, Z. Drought adaptation in millets. In *Abiotic and Biotic Stress in Plants: Recent Advances and Future Perspectives*; Shanker, A., Shanker, C., Eds.; InTech: Rijeka, Croatia, 2016; pp 639–662.

Töpfer, N.; Braam, T.; Shameer, S.; Ratcliffe, R. G.; Sweetlove, L. J. Alternative Crassulacean Acid Metabolism Modes Provide Environment-Specific Water-Saving Benefits in a Leaf Metabolic Model. *Plant Cell* **2020,** *32* (12), 3689–3705.

Turkan, I.; Uzilday, B.; Dietz, K. J.; Bräutigam, A.; Ozgur, R. Reactive Oxygen Species and Redox Regulation in Mesophyll and Bundle Sheath Cells of C4 Plants. *J. Exp. Bot.* **2018,** *69* (14), 3321–3331.

Uzilday, B.; Turkan, I.; Sekmen, A. H.; Ozgur, R. E. N. G.; Karakaya, H. C. Comparison of ROS Formation and Antioxidant Enzymes in *Cleome gynandra* (C4) and *Cleome spinosa* (C3) under Drought Stress. *Plant Sci.* **2012,** *182*, 59–70.

Wang, C.; Guo, L.; Li, Y.; Wang, Z. Systematic Comparison of C3 and C4 Plants Based on Metabolic Network Analysis. *BMC Syst. Biol.* **2012,** *6*, S9.

Weber, A. P.; von Caemmerer, S. Plastid Transport and Metabolism of C3 and C4 Plants—Comparative Analysis and Possible Biotechnological Exploitation. *Curr. Opin. Plant Biol.* **2010,** *13* (3), 256–264.

Westhoff, P.; Gowik, U. Evolution of C4 Photosynthesis—Looking for the Master Switch. *Plant Physiol.* **2010,** *154* (2), 598–601.

Winter, K.; Holtum, J. A. Facultative Crassulacean Acid Metabolism (CAM) Plants: Powerful Tools for Unravelling the Functional Elements of CAM Photosynthesis. *J. Exp. Bot.* **2014,** *65* (13), 3425–3441.

Winter, K.; Holtum, J. A.; Smith, J. A. C. Crassulacean Acid Metabolism: A Continuous or Discrete Trait? *New Phytol.* **2015,** *208* (1), 73–78.

Yadav, S.; Mishra, A. Introgression of C4 Pathway Gene (s) in C3 Plants to Improve Photosynthetic Carbon Assimilation for Crop Improvement: A Biotechnological Approach. In *Photosynthesis, Productivity and Environmental Stress*; Ahmad, P., Ahanger, M. A., Alyemeni, M. N., Alam, P., Eds.; John Wiley: Blackwell, OK, 2019; pp 267–281.

Yadav, S.; Elansary, H. O.; Mattar, M. A.; M Elhindi, K.; A Alotaibi, M.; Mishra, A. Differential Accumulation of Metabolites in *Suaeda* Species Provides New Insights into Abiotic Stress Tolerance in C4-Halophytic Species in Elevated CO2 Conditions. *Agronomy* **2021,** *11* (1), 131.

Yamori, W.; Hikosaka, K.; Way, D. A. Temperature Response of Photosynthesis in C3, C4, and CAM Plants: Temperature Acclimation and Temperature Adaptation. *Photosynthesis Res.* **2014,** *119* (1), 101–117.

Yang, X.; Cushman, J. C.; Borland, A. M.; Edwards, E. J.; Wullschleger, S. D.; Tuskan, G. A.; Owen, N. A.; Griffiths. H.; Smith, J. A. et al. A Roadmap for Research on Crassulacean Acid Metabolism (CAM) to Enhance Sustainable Food and Bioenergy Production in a Hotter, Drier World. *New Phytol.* **2015,** *207* (3), 491–504.

Yang, X.; Liu, D.; Tschaplinski, T. J.; Tuskan, G. A. Comparative Genomics Can Provide New Insights into the Evolutionary Mechanisms and Gene Function in CAM plants. *J. Exp. Bot.* **2019,** *70* (22), 6539–6547.

Yin, X.; Struik, P. C. The Energy Budget in C4 Photosynthesis: Insights from a Cell-Type-Specific Electron Transport Model. *New Phytol.* **2018,** *218* (3), 986–998.

Yu, K.; D'Odorico, P.; Collins, S. L.; Carr, D.; Porporato, A.; Anderegg, W. R.; Gilhooly, W. P.; Wang, L.; Bhattachan, A.; Bartlett, M.; Hartzell, S.; Yin, J.; He, Y.; Li, W.; Tatlhego, M.; Fuentes, J. D. The Competitive Advantage of a Constitutive CAM Species Over a C4 Grass Species under Drought and CO_2 Enrichment. *Ecosphere* **2019,** *10* (5), e02721.

Zahedi, S. M.; Karimi, M.; Venditti, A. Plants Adapted to Arid Areas: Specialized Metabolites. *Nat. Prod. Res.* **2019,** *82*, 1–18.

Zhang, J.; Liu, J.; Ming, R. Genomic Analyses of the CAM Plant Pineapple. *J. Exp. Bot.* **2014,** *65* (13), 3395–3404.

Zhao, X.; Chen, Z.; Leng, P.; Hu, Z. iTRAQ-Based Quantitative Proteomic Analysis of the Response of *Hylotelephiumerythrostictum* Leaves to Salt Stress. *Sci. Horti.* **2020,** *264*, 109190.

Zhou, P.; An, Y.; Wang, Z.; Du, H.; Huang, B. Characterization of Gene Expression Associated with Drought Avoidance and Tolerance Traits in a Perennial Grass Species. *PLoS One* **2014,** *9* (8), e103611.

CHAPTER 9

Genetic Engineering of C4 Pathway in C3 Plants to Improve Stress Tolerance

L. JEYANTHI REBECCA[1*] and EIJI HIRASAWA[2]

[1]*Department of Agricultural Biotechnology, Bharath Institute of Higher Education and Research, Selaiyur, Chennai 600073, India*

[2]*Department of Biology and Geosciences, Osaka City University, Osaka, Japan*

[*]*Corresponding author. E-mail: jeyanthi.bt@bharathuniv.ac.in*

ABSTRACT

C_4 plants are considered to be more versatile when compared with C_3 plants in terms of their productivity and innate ability to withstand abiotic stress. This is because they exhibit reduced photorespiration owing to the Kranz anatomical structure of their leaf and increased water use and nitrogen assimilation efficiencies. Efforts have been made by researchers in transferring genes from C_4 plants that are responsible for carbon dioxide assimilation and photosynthesis, leading to elevated carbon dioxide levels. Overexpression of the C_4 photosynthetic genes, namely, phosphoenol pyruvate carboxylase, pyruvate, orthophosphate dikinase, and *NADP-ME* can increase the photosynthetic efficiency in C_3 plants like *Arabidopsis*, tobacco, rice, wheat, and potato. These genes also increased the resistance of the C_3 plants to abiotic stress. The elevated carbon dioxide levels resulted in increased crop productivity and thereby yield. The C_4 photosynthetic genes are also involved in reducing global warming and are implicated in the global climate change. This chapter highlights the importance of C_4, C_3, and Crassulacean

Photosynthesis and Respiratory Cycles during Environmental Stress Response in Plants. Aryadeep Roychoudhury (Ed.)

acid metabolism plants, their advantages over each other and the strategies to develop transgenic C_3 plants harbouring C_4 genes and having an increased photosynthetic efficiency and carbon assimilation, thereby increasing the productivity to meet global needs. This chapter also aims to highlight the importance of genetic engineering of C_3 plants in the aspect of how it might impact the global climate change.

9.1 INTRODUCTION

Carbon fixation in plants is the conversion of inorganic carbon namely carbon dioxide into organic compounds by way of photosynthesis. During photosynthesis, using light energy, plants convert water, carbon dioxide and minerals into oxygen and energy rich organic compounds namely carbohydrates such as sugars (Beadle and Long, 1985; Normile, 2009). This energy in the form of sugars is produced by a series of light independent reactions called the Calvin cycle. In this cycle, carbon is fixed in organic compounds such as ribulose-5-phosphate (RuBP). These organic compounds are later reduced to glucose by using ATP and NADPH generated during the light dependent reactions (Beale and Long, 1995). There is another phenomenon called cellular respiration, which is opposite to photosynthesis. In cellular respiration, oxygen and glucose is converted to water and carbon dioxide. Photosynthesis requires energy in the form of ATP, whereas cellular respiration releases energy. There are three different classes of plants, based on the metabolic pathways for carbon fixation adopted during photosynthesis. They are C_3, C_4, and Crassulacean acid metabolism (CAM) plants (Fig. 9.1).

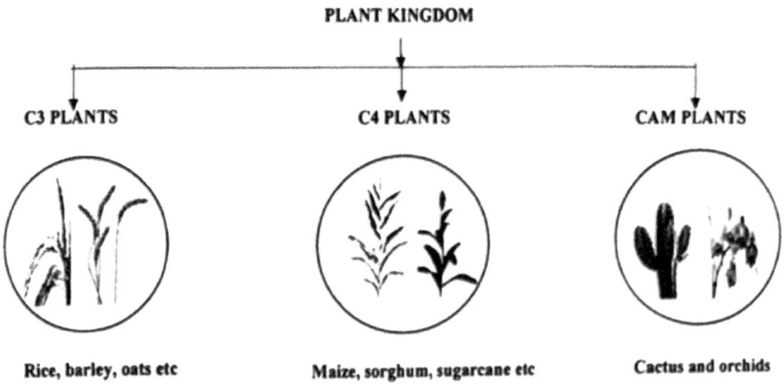

FIGURE 9.1 Classification of plants based on carbon fixation.

9.1.1 CARBON FIXATION IN PLANTS

The carbon cycle is a biogeochemical cycle involving the exchange of carbon among the living organisms, soil, atmosphere, and water bodies. It is a key indicator for the sustained life on earth. Carbon is the fourth most abundant element on earth, most of which being stored in the rocks. When the carbon in the atmosphere is increased, it results in warmer temperatures leading to global warming (Berry and Björkman, 1980). Plants fix carbon into 3-phosphoglyerate (PGA), a three-carbon compound through a cycle called the Calvin cycle named after Melvin Calvin.

9.1.2 C_3 CARBON FIXATION PATHWAY

C_3 plants are those where the carbon fixation occurs by way of conversion of carbon dioxide and RuBP (a 5-C sugar) into two molecules of PGA, which contains three carbon atoms by the following reaction catalysed by ribulose-1, 5-biphosphate carboxylase (RuBisCO):

$$CO_2 + H_2O + RuBP \rightarrow 2(3\text{-phosphogycerate})$$

This reaction is common in C_4 and CAM plants also and it is the first step of the Calvin–Benson cycle. It takes place in the chloroplast of the mesophyll cells. The C_3 plants grow in areas having fairly enough sunlight, moderate temperature, and abundant water (Sage and Pearcy, 2000). The required carbon dioxide concentration is 200 ppm or higher. These plants constitute more than 95% of the plants on earth and include the main food crops like rice, wheat, barley, oats, potato, tobacco, beetroot, soybean, etc. The enzyme RuBisCO that catalyzes the first step in carbon fixation, incorporates more amount of oxygen into RuBP with increase in temperature, resulting in photorespiration and leading to a net loss of carbon and nitrogen from the plants, limiting their growth (Edwards, 1999). In this reaction, RuBP is carboxylated. RuBisCO catalyzes both carboxylation and oxygenation reactions. The oxygenation pathway leads to photorespiration, which results in around 30% loss in the total carbon fixed. RuBisCO is the most abundant enzyme on earth. C_3 plants cannot grow under hot climatic conditions (Beale et al., 1999). C_3 plants have evolved in the Mesozoic and Paleozoic eras much earlier than the C_4 plants. Moreover, C_3 plants transpire more losing about 97% of water content and they close their stomata in dry areas to reduce water loss, which in turn reduces the amount of CO_2 entering the leaves (Sage, 2002). Hence, the CO_2 concentration in leaves is minimal leading to photorespiration as the CO_2: O_2 ratio is less.

9.1.3 C_4 CARBON FIXATION PATHWAY

The C_4 carbon fixation is also called Hatch–Slack pathway, wherein 4-C molecules are produced first. The C_4 plants include maize, sorghum, and sugarcane that have higher photosynthetic efficiency when compared with C_3 plants (Hatch, 1987). Nearly, 50% of the grass species are C_4 plants. The process of photorespiration is reduced in C_4 plants when compared with C_3 plants. The C_4 plants concentrate CO_2 around RuBisCO leading to an environment containing an increased CO_2 and reduced O_2 content (Beale and Long, 1995; Beale et al., 1999). The leaves of C_4 plants have two cells called mesophyll cells and bundle-sheath cells. The enzyme phosphoenol pyruvate (PEP) carboxylase helps in fixing CO_2 in the mesophyll cells, which in turn reacts with the three carbon PEP to form the four carbon oxaloacetic acid. The oxaloacetic acid is either reduced to malate or transaminated to aspartate (Poorter and Navas, 2003). These compounds enter the bundle sheath cells where they are decarboxylated, thereby increasing the CO_2 content surrounding RuBisCO to an extent much higher than that with the surrounding atmosphere and decreasing the chances of photorespiration. The RuBisCO produces pyruvate and phosphoglycerate in the subsequent reactions. The pyruvate and half of the phosphoglycerate diffuses back into the mesophyll cells where the phosphoglycerate is reduced chemically and once again it diffuses back into the bundle sheath. This cycle is called the reductive pentose phosphate cycle (Kapralov and Filatov, 2007).

The additional reactions leading to carbon fixation in C_4 plants when compared with C_3 plants results in the utilization of more energy in the form of ATP in C_4 plants, leading to the synthesis of PEP. The C_3 plants do not require much energy to produce PEP and so they can be grown more efficiently where the temperature is low or in shady places or where photorespiration is minimal (Edwards, 1999; Edwards and Andreo, 1992; Sage, 2002). There are two merits in C_4 plants resulting due to the accumulation of CO_2 around RuBisCO. Firstly, concentrating CO_2 helps to overcome the reduction of the solubility of gas with high temperature leading to increased rate of photosynthesis at increased temperatures (Sage, 2002; Dengler and Taylor, 2000). Secondly, the CO_2 concentrating mechanism tends to maintain high gradients of CO_2 concentration around the stomatal pores resulting in the suppression of oxygenation reaction by RuBisCO and the resultant process of photorespiration, thereby increasing the photosynthetic efficiency. Hence, C_4 plants have lower stomatal conductance, reduced water loss, and higher water use efficiency.

The C_4 plants are said to have evolved during the Oligocene period either due to elevated temperature or due to water deficit, which are the key factors responsible for higher rate of photorespiration. C_4 plants are said to have evolved many times independently owing to their wide and diverse distribution. This type of evolution is called convergent evolution (Sage, 2002). The plants belonging to the genus *Flaveria* have C_3–C_4 intermediates, which indicates that there are many independent evolutions that has occurred leading to a wholesome C_4 plant. There are many ways to predict the presence of C_3–C_4 intermediates. One way is the presence of rudimentary bundle sheath cells in C_3 plants that have very less chloroplasts and do not actively take part in photosynthesis. The other way to predict is by the detection of bundle sheath genes in C_3–C_4 intermediates.

9.1.3.1 THE KRANZ ANATOMY

The photosynthetic efficiency of C_4 plants is higher than C_3 plants. This is because the leaves of C_4 plants possess a distinctive leaf anatomy called the Kranz (meaning wreath or ring) anatomy (Hatch, 1987). In this type of anatomy, there are two morphologically and functionally distinct types of photosynthetic cells namely the bundle sheath cells and the mesophyll cells. The bundle sheath cells surround the vascular tissues and the mesophyll cells surround the bundle sheath cells just below the leaf epidermis (Hatch and Burnell, 1990). This wreath like arrangement gives rise to its name Kranz in German. This structure provides the outline for the compartmentalization or separation for two separate biochemical reactions, leading to CO_2 assimilation to occur. The chloroplasts are more abundant in bundle sheath cells than the mesophyll cells.

9.1.4 CAM CARBON FIXATION PATHWAY

CAM photosynthesis is a carbon fixation pathway that is found in certain plants and is a type of adaptation by plants to arid environments. The name Crassulacean was derived as this method of carbon fixation was first discovered in plants belonging to the family Crassulaceae. In CAM plants, the stomata in leaves remain shut during daytime to reduce evapotranspiration and opens in the night time to allow CO_2 to enter and later to diffuse inside the mesophyll cells. The CO_2 that had entered is stored as a four-carbon malate in vacuoles at night by the PEP reaction as in C_4 plants (Bonner and Bonner, 1948). They are stored in vacuoles at night as the Calvin cycle cannot function without NADPH and ATP that are produced during light-dependant reactions and which do not occur during the night. The concentrated CO_2 around the enzyme RuBisCO increases the photosynthetic efficiency of CAM plants as in the case of C_4 plants. During the day, the malate is transported to the chloroplasts where it is again converted back to CO_2 to be used for photosynthesis. The plants belonging to Cactaceae and pineapple plant follows the CAM pathway.

9.2 FACTORS AFFECTING CARBON FIXATION IN PLANTS

Carbon fixation in plants and other organisms that photosynthesise are affected by the following factors, namely, CO_2 and O_2 concentrations, light intensity, temperature, and nutrients. The internal factors are the leaf anatomy, leaf phyllotaxy, mesophyll cells, concentration of internal oxygen, and chlorophyll content. The external factors constitute firstly the availability of sunlight that includes the intensity, quality, and duration of light. Secondly, the level of CO_2 that is a limiting factor decides the rate of photosynthesis. Thirdly, it is water that has control over photosynthesis when there is drought or heat stress. Finally, it is the atmospheric temperature that has a direct impact on the performance of C_3 and C_4 plants (Gibbs et al., 1961).

9.3 EVOLUTION OF C_3 AND C_4 PLANTS

Carbon dioxide fixation in all plants occurs with the help of RuBisCO, which leads to the production of 3-phosphoglycerate through the carboxylation of ribulose-1,5-biphosphate. However, if RuBisCO adds oxygen to ribulose-1,5-biphosphate, it results in the production of two molecules one each of 3-phosphoglycerate and 2-phosphoglycolate. The latter at higher concentrations is toxic to the plants and inhibits several enzymes (Anderson, 1971). Photorespiration, the other pathway helps in scavenging the phosphoglycolate and this pathway is energy demanding and also leads to net loss of CO_2. This occurs under elevated temperature leading to an average of 40% reduction in photosynthesis (Ehleringer and Monson, 1993). The enzyme RuBisCO evolved more than three billion years ago when atmospheric CO_2 levels were higher than O_2 levels. Later, when atmospheric O_2 levels increased considerably, RuBisCO began to acquire oxygenase activity. Moreover, plants devised newer adaptations to cope up with the loss in net photosynthesis by way of cellular and molecular adaptations leading to the evolution of C_4 photosynthesis. This has resulted in the concentration of CO_2 at the site of RuBisCO, thereby reducing the oxygenation and photorespiration. RuBisCO can function well at elevated CO_2 levels in the bundle sheath cells of C_4 plants, thus increasing its photosynthetic efficiency when compared with C_3 plants (Ainsworth et al., 2002; Ainsworth et al., 2004; Brainbridge et al., 1995).

C_4 plants evolved first in grasses during the Oligocene era and later in dicots in arid regions due to abiotic stressors like heat, drought, and salinity. These plants thus dominate in grasslands and results in the biomass production in tropical and subtropical regions (Edwards and Smith, 2010; Ashton and Hatch, 1983). Depletion in CO_2 levels and high temperatures was found to be the predominant reason for the evolution of C_4 photosynthetic pathway. The carboxylation reaction of RuBisCO is limited by the concentration of its substrate CO_2 when it drops down below 70 Pa at its active site. Water potential is another reason for the evolution of C_4 plants. Moreover, in arid regions where loss of water is inevitable, the stomata close partially to conserve water, thereby reducing the CO_2 supply to photosynthesis and decreasing transpiration resulting in reduced latent heat loss and raised leaf temperature (Ashton et al., 1984).

9.4 ABIOTIC STRESS AND ITS EFFECT ON PHOTOSYNTHESIS

Water deficit is a state that underlies in predominantly all types of abiotic stressors. However, water is a limiting factor for photosynthesis. This water loss leads to reduced heat on the leaf surface. The factors that affect photosynthesis at elevated temperatures are optimum temperature, respiration rate, and membrane-bound electron transport process. Various other physiological parameters are affected due to high temperature and photosynthesis is inhibited much before all the other cell functions (Berry and Bjo¨rkman, 1980). It also affects photosynthesis by altering the excitation energy distribution by changing the structure of thylakoids (Weis and Berry, 1988).

9.4.1 *ABIOTIC STRESS AND C_3 AND C_4 PHOTOSYNTHESIS*

The rise in atmospheric temperature leading to an increased frequency in drought is a resultant of the increased CO_2 in the beginning of the 18th century (IPCC, 2001). Higher CO_2 is said to increase the water use efficiency in C_3 plants because it results in a reduction on transpiration rate and an increased CO_2 assimilation (Knapp and Medina, 1999; Long, 1999; von Caemmerer *et al*, 1994). Moreover, it has been suggested that C_3 plants are more benefitted by an increase in CO_2 under drought stress conditions, though at severe drought, C_4 plants are benefitted more due to the concentrating mechanism of CO_2 in its structure (Ward et al., 1999; Dengler and Taylor, 2000). The extent of the benefit gained by C_3 and C_4 plants under elevated CO_2 levels are still to be investigated elaborately. C_4 plants are native to tropical areas and warm temperate regions where the temperature and light intensity is higher. In this condition, C_4 plants have elevated photosynthetic and growth rates owing to increased water, carbon, and nitrogen use efficiencies (Sage and Monson, 1999; Piedade et al., 1991; Long, 1999). Photosynthesis is a thermosensitive process (Havaux, 1993). In general, the photosynthetic machinery is affected at elevated temperature as each plant has an optimum temperature, which is necessary for its normal functioning (Bhagat et al., 2014). The respiration rate is also affected. At low temperatures, the stomatal conductance, photochemical efficiency pf PSII, carbon metabolism, thylakoid electron transport rate, and the photosynthetic pigment complex systems are hampered (Suzuki et al., 2008; Ramalho et al., 2003; Silva et al., 2004). However, photosynthesis is revived after the process of acclimatization (Berry and Bjo¨rkman, 1980). When plants are subjected to waterlogging, photosynthesis may be affected due to reduced CO_2 transfer conductance from substomatal cavities to the site of carboxylation and reduced CO_2 fixation by RuBisCO (Massacci et al., 1996). Other effects are reduced respiration rates and chlorophyll content in leaves (Parolin, 2000). The photosynthesis in C_3 plants is less affected when subjected to water stress as the CO_2 assimilation depends on the leaf water content. When the relative water content is greater than 70%, the reduced CO_2 assimilation is a result of reduced intercellular CO_2 concentration, which is a resultant of the decreased stomatal conductance (Flexas et al., 2004). In addition, it has been found that under water stress, the photosynthesis may decline, while the respiration rate will either decrease or increase but will seldom be impaired (Flexas et al., 2004).

There is a decline in photosynthetic rates when plants are subjected to salinity due to stomatal closure and the extent of CO_2 fixation. Alternatively, the two main factors responsible for the loss in photosynthesis are the osmotic effect and the specific ion effect of the accumulated salt (Kalaji and Nalborczyk, 1991). The reduced effect of O_3 on photosynthesis is due to the stomatal closure and reduced rate of RuBisCO activity (Farage et al., 1991; Robinson et al., 1997). The strong oxidizing effect of ozone has a deleterious effect on photosynthesis (Reich, 1987).

C_4 plants as already mentioned in this chapter, have evolved from C_3 plants due to high CO_2 environment resulting in an increase in nitrogen and water use efficiencies compared with C_3 plants. C_4 plants show increased CO_2 assimilation for a given leaf nitrogen (Ghannoum et al., 2011). The leaf nitrogen content per unit area is less in C_4 when compared with C_3 plants. However, the nitrogen concentration per unit dry mass is similar for both types of plants. The allocation of nitrogen to RuBisCO by C_4 plants is less than that allocated to other soluble proteins and thylakoid components. In C_3 plants, 30% of leaf nitrogen is allocated to RuBisCO when compared with 21% in C_4 plants. Thus, the lower requirement of N_2 in C_4 plants is due to the CO_2 concentrating mechanism in bundle sheath. This mechanism is lacking in C_3 plants and hence RuBisCO functions at only 25% of its capacity (Sage and Monson, 1987) and loses 25% to photorespiration (Ludwig and Canvin, 1971).

Hence, to achieve similar photosynthetic rates as in C_4 plants, C_3 leaves should contribute more N_2 to RuBisCO. At elevated temperatures, the specificity of RuBisCO to CO_2 decreases and photosynthetic N_2 use efficiency is also higher (Long, 1999). The improved water use efficiency in C_4 plants is due to higher photosynthetic rate per unit area of leaf and reduced stomatal conductance (Ghannoum et al., 2011).

Photorespiration is eliminated in C_4 plants but the additional energy is required in the form of ATP for the completion of C_4 cycle. However, in C_3 plants when the leaf temperature exceeds 25°C, the amount of light energy allocated for photorespiration exceeds the additional energy required for CO_2 assimilation in C_4 plants. Thus, under light limiting conditions, when temperature is below 24°C, C_4 photosynthesis is less efficient than C_3 photosynthesis (Hatch and Burnell, 1990; Long, 1999). C_4 plants in cold environment tend to show cold adaptation requiring warm periods during the day for its existence. Thus, they are less competent than their C_3 counterparts in cold environment (Sage and Mckown, 2006; Sage and Pearcy, 2000).

The Intergovernmental Panel on Climate Change (2001, 2007) has projected an increase in global mean surface temperatures to a tune of 1.4°C–5°C by 2100, which is due to the elevated CO_2 levels caused by human intervention. This is further predicted to reduce the annual precipitation to 20% per year amounting to a reduction in soil moisture by 20% (Schiermeier, 2008). This will result in increased heat and drought stress (Ciais et al., 2005; Thomas et al., 2004). Moreover, higher CO_2 levels also increases the water use efficiency by decreasing stomatal conductance and transpiration (Ainsworth et al., 2002; Leakey et al., 2009; Wall et al., 2001).

Elevated CO_2 levels have been found to increase photosynthesis to a tune of 40–45% in C_3 plants and 20–25% in C_4 plants (Ghannoum et al., 2005). An increase in CO_2 levels results in increased photosynthesis and decreased photorespiration. Hence, C_3 plants are said to perform well at less optimal CO_2 levels (Bernacchi et al., 2003; Long et al., 2004). However, the response of C_4 plants to elevated CO_2 levels is not well understood barring the role of PEP carboxylase that renders the carbon fixation in the C_4 plants less effective even at elevated CO_2 levels (Reddy et al., 2010; Bernacchi et al., 2007). Thus, it can be concluded that though both C_3 and C_4 plants are far better under elevated CO_2 levels, the latter perform better in severe drought stress conditions, which is also accompanied by elevated temperatures and CO_2 levels.

In C_3 plants, respiratory rates are reduced at elevated CO_2 leading to increased biomass yield and not much is known about this parameter in C_4 plants (Reddy et al., 2010). When plants are grown at optimal conditions but at elevated CO_2 levels, photosynthetic rate has been found to increase by 50% when compared with plants grown at normal CO_2 levels (Kirschbaum, 2010). These growth responses are attributed to the sink capacity and nutrient status of the plant. The C_4 plants grown under free air carbon enrichment exhibited increased photosynthetic rates during drought or under atmospheric vapor pressure deficit (Cousins, 2002; Leakey et al., 2009; Wall et al., 2001; Conley et al., 2001). Hence, C_4 plants perform well at elevated CO_2 levels especially under drought conditions (Ward et al., 1999).

9.4.2 ENZYMES IN C_4 PHOTOSYNTHESIS

There are about 120 genes involved in chloroplast protein synthesis and photosynthesis (Shinozaki and Sugiura, 1986). The photosynthetic enzymes are located in the mesophyll cells in C_3 plants and in mesophyll and bundle sheath cells in C_4 plants. The venation in C_4 plants is unique as it displays the Kranz anatomy where a ring of bundle sheath cells surrounds each vein and an outer ring of mesophyll cells surrounds the bundle sheath. Phosphoenol pyruvate carboxylase (PEPC) in the C_4 pathway helps in fixing the CO_2 in the mesophyll cells and gets converted to oxaloacetic acid, which is a 4-carbon acid that is reduced to malic acid by NADP-MDH (malate dehydrogenase) or transaminated to aspartic acid by AspAT (Matsuoka et al., 2001). This is then transported to the bundle sheath cells and gets decarboxylated to release CO_2. This occurs in the vicinity of RuBisCO. This decarboxylation is carried out by three enzymes NADP-ME, NAD-ME, and PEP-CK. During this decarboxylation, a 3-carbon acid is formed that is transported to the mesophyll cells to regenerate PEP with the aid of pyruvate, orthophosphate dikinase (PPDK) (Table 9.1).

TABLE 9.1 Enzymes Involved in C_4 Photosynthesis

S. no	Photosynthetic enzyme	Origin
1	Ribulose-1,5-biphosphate carboxylase (RuBisCO)	C_3
2	Phosphoenol pyruvate carboxylase (PEPC)	C_4
3	Pyruvate, orthophosphate dikinase (PPDK)	C_4
4	NADP-Malate dehydrogenase (NADP-MDH)	C_4
5	NADP-Malic enzyme (NADP-ME)	C_4
6	NAD-Malic enzyme (NAD-ME)	C_4
7	Aspartate aminotransferase (AspAT)	C_4
8	Phosphoenolpyruvate carboxykinase (PEP-CK)	C_4
9	CA (Carbonic anhydrase)	C_4

At the molecular level, there are three features that differentiate C_4 and C_3 photosynthesis: (1) it is the differentiation into two cell and chloroplast types, (2) presence of additional set of genes, and (3) a mechanism to regulate these additional genes (Furbank and Taylor, 1995). It has been postulated that the cells closer to the veins tend to become the bundle sheath cells and the slightly one cell away from the veins become the mesophyll cells (Nelson and Langdale, 1992). The genes that encode for C_4 enzymes belong to small gene families and they are expressed at high levels in the bundle sheath and mesophyll cells (Furbank and Taylor, 1995).

9.4.3 EVOLUTION OF C_4 PHOTOSYNTHETIC GENES

The transfer of C_4 traits in C_3 plants to enhance its photosynthetic rate have been under study from the time C_4 pathway was discovered. However, in the initial stages, conventional breeding methods were adopted to obtain crosses between C_4 and C_3 plants (Miyao, 2003). This method was not successful as the C_3–C_4 hybrids were infertile (Brown and Bouton, 1993). The application of recombinant DNA technology to transform C_4 genes to C_3 plants started in the early 90s. This has led to the overexpression of C_4 photosynthetic genes in C_3 plants. C_4 photosynthetic genes evolved from C_3 genes, though their kinetic properties and expression levels vary (Ku et al., 1996). The photosynthetic genes in C_4 and C_3 plants are classified into C_4-specific genes that are located in C_4 plants only and C_4-like genes that are present in C_3 plants (Monson, 1999). Apart from these two groups of genes, there are other housekeeping genes present in both C_3 and C_4 plants.

The photosynthetic machinery of maize has been well studied. Maize is said to have three isoforms of PPDK, one chloroplastic isoform that is involved in C_4 pathway and two cytosolic isoforms Sheen, 1991). They are encoded by a single gene having a dual promoter system (Glackin and Grula, 1990). Moreover, this gene has two transcription initiation sites and the transcription is regulated by promoters at its 5′-flanking regions. These chloroplastic and cytosolic isoforms are C_4 and C_3 specific, respectively (Sheen, 1991). In rice there again exists three isoforms of PPDK. Two different genes are also identified among which one has a dual promoter system encoding one cytosolic and one choroplastic isoform (Imaizumi et al., 1997) and the other encodes for the cytosolic isoform (Moons et al., 1998).

Therefore, the two genes encoding the cytosolic isoform were derived from a single ancestral gene, a *PPDK2* gene that later evolved to become the *PPDK2* genes of both C_3 and C_4 plants. The other gene was the ancestral *PPDK1* gene with a single promoter. This later evolved and became a *PPDK1* gene with a dual promoter in the ancestral C_3 plants, but later after acquiring a sequence for a transit peptide, it evolved to become the *PPDK1* gene in C_4 plants with high expression levels with the aid of certain *cis*-acting elements. It has also been found that the other photosynthetic genes have also evolved in a similar manner (Miyao, 2003).

9.4.4 OVEREXPRESSION OF C4 GENES IN C_3 PLANTS

The carbon and nitrogen metabolism occurs in the mesophyll cells of C_3 plants. Hence, to increase the metabolic rates, C_4 genes need to be overexpressed in the mesophyll cells of C_3 plants. A very simple way of overexpression of C_4 genes in C_3 plants was anticipated in view of the evolutionary origin of the C_4-specific genes in C_3 plants leading to a high-level cell-specific expression of C_4 enzymes. Moreover, the transfer of enzymes specific to bindle sheath cells were possible with conventional methods involving strong promoters (Miyao, 2003). Ku et al. (1999) is credited in conducting the first trial of expressing an intact maize *PEPC* gene in transgenic rice plants. The intact maize *PEPC* gene (8.8 Kbp) was transferred into rice plants. It resulted in a 110-fold activity of *PEPC* gene than the nontransformants (Fukayama et al., 2001; Agarie et al., 2002). The PEPC protein accounted for 12% of the total leaf protein content.

Similarly, the introduction of an intact maize *PPDK1* gene (7.3 Kbp) into rice plants resulted in a 40-fold increase in activity of PPDK1 in transgenic rice plants when compared with the nontransformants. This amounted for 35% of the total leaf soluble protein content. Moreover, *PPDK* gene expression was found to be organ specific as in maize plants. The C_3-specific *PPDK1* gene was expressed in the grains of rice plants though the C_4-specific *PPDK1* gene was expressed in the leaves of transgenic rice plants. It was also emphasized that the reason for the high-level expression might be due to the additive effect of the introns and the terminator sequences along with the promotor sequences. The levels of transcript, protein, and activity of C_4-specific PEPC and PPDK genes were equal to the copy number of the introduced gene (Fukayama et al., 2001; Ku et al., 1999).

However, the high-level expression of C_4 photosynthetic genes was successful in phylogenetically closely related C_3 plants. The intact C_4-specfic maize *PEPC* gene was not expressed at a higher level in tobacco leaves due to incorrect transcription initiation (Hudspeth et al., 1992). Apart from incorrect transcription initiation resulting in termination of transcription, incorrect splicing might also occur when genes are transferred from monocots to dicots (Goodall and Filipowicz, 1991). The transfer of C_4-specific genes by conventional methods of overexpression using strong promoters is not effective (Matsuoka et al., 2001).

In addition, the C_4-specific genes that express in the bundle sheath cells cannot be expressed to a greater extent in the mesophyll cells of C_3 plants. Hence, methods that utilize strong promoter like the rice *Cab* promoter that is used to drive the expression of C_4-specific NADP-ME in rice plants has been successful. This method has resulted in 30–70-fold increase in NADP-ME activity in rice leaves than the nontransformants (Takeuchi et al., 2000; Tsuchida et al., 2001). Efforts were taken to make these genes to express at different target sites other than the mesophyll cells. A target peptide was used to target the expression in other sites along with a C_4-specific promoter of a C_4-specific photosynthetic gene like *PPDK1* or *PEP-CK* (Suzuki et al., 2000).

9.4.5 FACTORS AFFECTING THE EXPRESSION LEVEL OF TRANSGENES

There are three factors that affect the expression level of transgenes, namely, the positional effect, silencing effect, and the rearrangement effects. Screening has to be done to achieve a desired expression level in order to confirm the enzyme location in C_3 plants. A significant amount of rice transformants with maize C_4-specific genes showed enzyme activity to a similar extent when compared with the nontransformants (Ku et al., 1999; Fukayama et al., 2001). This was due to certain rearrangements in the form of partial deletions or linking (Fukayama et al., 2001). Moreover, certain *cis*-acting elements or transit peptides might be deleted from the gene thereby altering the location or the extent of expression.

The main purpose of overexpression of C_4-specific genes in C_3 plants is to enhance the photosynthetic efficiency of C_3 plants (Edwards and Andreo, 1992). Initially, overexpressing one C_4-specific gene in C_3 mesophyll cells was investigated but it was not that fruitful as none of the merits of C_4 photosynthesis was observed in C_3 plants (Häusler et al., 2002). Later, the transfer of more than one C_4-specific gene in the mesophyll cells of C_3 plants was found to be successful. However, there is still a controversy whether the C_4 pathway is really successful in C_3 plants (Edwards, 1999; Leegood, 2002; Häusler et al., 2002). On the contrary, overproduction of C_4 photosynthetic enzyme has led to an increase in the yield of tobacco and also enhanced the tolerance to aluminium for root elongation in transgenic rice plants (Miyao et al., 2001). This shows that the overproduction of C_4 enzymes in C_3 plants was able to contribute to the improvement of other characters in C_3 plants. Further, it has been suggested that the C_4-like pathway can actually function successfully in C_3 plants under severe abiotic stress conditions especially during drought in which case the CO_2 availability in C_3 plants will be limiting (Leegood, 2002). The research on the development of a transgenic C_3 plant with increased photosynthetic rate has been on for more than 25 years but not much viable conclusions have been made so far (Long et al., 2015; Ort et al., 2015; Sowjanya et al., 2019). The engineering of the Kranz anatomy and the integration of the C_4 cycle may be an option toward a positive development in the research on the conversion of a C_3 plant into a C_4 plant (Schuler et al., 2016). The overexpression of maize C_4-specific *PEPC* gene and *PPDK* gene in transgenic rice resulted in mixed characters like stunted growth, increased yield, increased CO_2 assimilation, and delay in leaf senescence (Kumari et al., 2017).

9.5 CONCLUSIONS

The increased CO_2 level in the atmosphere due to burning of fossil fuels, industrialization, and reduced cultivable land has led to climate change in recent years by way of global warming. Plants have survived all these climatic changes for millions of years by way of adaptation. This chapter highlighted the salient features of C_4 and C_3 plants, their merits and demerits, their mechanism of photosynthesis and their ways to cope up with elevated CO_2 levels in the atmosphere. Here, we also get to know about the efforts taken by other researchers on the conversion of C_3 plants to C_4 plants by the transfer of C_4-specific photosynthetic genes into C_3 mesophyll cells using genetic engineering. This approach has been quite challenging as till date, no satisfactory conclusions in this regard have been arrived yet. There have been some reports indicating that the transgenic C_3 plants with C_4-specific photosynthetic genes showed increased yield but the results regarding the enhanced photosynthetic rates in transgenic C_3 plants are still controversial.

KEYWORDS

- **carbon fixation**
- **C_3 plants**
- **C_4 plants**
- **photosynthesis**
- **abiotic stress**
- **carbon dioxide**
- **transgenic C_3 plants**

REFERENCES

Agarie, S.; Miura, A.; Sumikura, R.; Tsukamoto, S.; Nose, A.; Arima, S.; Matsuoka, M.; Miyao-Tokutomi, M. Overexpression of C4 PEPC Caused O2-Insensitive Photosynthesis in Transgenic Rice Plants. *Plant Sci.* **2002,** *162*, 257–265.

Ainsworth, E. A.; Davey, P. A.; Bernacchi, C. J. A Metaanalysis of Elevated [CO_2] Effects on Soybean (*Glycine max*) Physiology, Growth and Yield. *Global Change Biol.* **2002,** *8*, 695–709.

Ainsworth, E. A.; Rogers, A.; Nelson, R.; Long, S. P. Testing the 'Source-Sink' Hypothesis of Down-Regulation of Photosynthesis in Elevated CO_2 in the Field with Single Gene Substitutions in *Glycine max*. *Agric. Forest Meteorol.* **2004,** *122*, 85–94.

Anderson, L. E. Chloroplast and Cytoplasmic Enzymes. II. Pea Leaf Triose Phosphate Isomerases. *Biochim. Biophys. Acta* **1971,** *235*, 237–244.

Ashton, A. R.; Burnell, J. N.; Hatch, M. D. Regulation of C, Photosynthesis: Inactivation of Pyruvate, Pi Dikinase by ADP Dependent Phosphorylation and Activation by Phosphorolysis. *Arch. Biochem. Biophys.* **1984,** *230*, 492–503.

Ashton, A. R.; Hatch, M. D. Regulation of C4 Photosynthesis: Regulation of Pyruvate, Pi Dikinase by ADP-Dependent Phosphorylation and Dephosphorylation. *Biochem. Biophys. Res. Commun.* **1983,** *115*, 53–60.

Beadle, C. L.; Long, S. P. Photosynthesis—Is It Limiting to Biomass Production? *Biomass* **1985,** *8*, 119–168.

Beale, C. V.; Long, S. P. Can Perennial C-4 Grasses Attain High Efficiencies of Radiant Energy-Conversion in Cool Climates? *Plant, Cell Environ.* **1995,** *18*, 641–650.

Beale, C. V.; Morison, J. I. L.; Long, S. P. Water Use Efficiency of C-4 Perennial Grasses in a Temperate Climate. *Agric. Forest Meteorol.* **1999**, *96*, 103–115.

Bernacchi, C. J.; Kimball, B. A.; Quarles, D. R.; Long, S. P.; Ort, D. R. Decreases in Stomatal Conductance of Soybean under Open Air Elevation of CO_2 Are Closely Coupled with Decreases in Ecosystem Evapotranspiration. *Plant Physiol.* **2007,** *143*, 134–144.

Bernacchi, C. J.; Pimentel, C.; Long, S. P. In Vivo Temperature Response Functions of Parameters Required to Model RuBP-Limited Photosynthesis. *Plant Cell Environ.* **2003,** *26*, 1419–1430.

Berry, J.; Bjo¨rkman, O. Photosynthetic Response and Adaptation to Temperature in Higher Plants. *Annu. Rev. Plant Physiol.* **1980,** *31*, 491–543.

Bhagat, K. P.; Arun Kumar, R.; Ratnakumar, P.; Kumar, S.; Bal, S. K.; Agrawal, P. K. Photosynthesis and Associated Aspects under Abiotic Stresses Environment. In *Approaches to Plant Stress and their Management*; Gaur, R. K., Sharma, P., Eds.; 2014; pp 191–205.

Bonner, W.; Bonner, J. The Role of Carbon Dioxide in Acid Formation by Succulent Plants. *Am. J. Bot.* **1948,** *35* (2), 113.

Brainbridge, G.; Madgwick, P.; Parmar, S.; Mitchell, R.; Paul, M.; Pitts, J.; Keys, A. J.; Parry, M. A. J. Engineering Rubisco to Change Its Catalytic Properties. *J. Exp. Bot.* **1995,** *46*, 1269–1276.

Ciais, P. H.; Reichstein, M.; Viovy, N.; Granier, A.; Ogee, J.; Allard, V. Europe-Wide Reduction in Primary Productivity Caused by the Heat and Drought in 2003. *Nature* **2005,** *437*, 529–533.

Conley, M. M.; Kimball, B. A.; Brooks T. J. CO2 enrichment increases water-use efficiency in sorghum. *New Phytologist* **2001,** *151*, 407–412.

Cousins, A. B. Photosystem II energy use, non-photochemical quenching and the xanthophyll cycle in Sorghum bicolor grown under drought and free-air CO2 enrichment (FACE) conditions. *Plant Cell and Environment* **2002,** *25*, 1551–1559.

Dengler, N. G.; Taylor, W. C. Developmental Aspects of C4 Photosynthesis. In *Advances in Photosynthesis, Vol. 9, Photosynthesis: Physiology and Metabolism*; Leegood, R. C., Sharkey, T. D., von Caemmerer, S., Eds.; Kluwer: Dordrecht, 2000; pp 471–495, 624 pp.

Edwards, E. J.; Smith, S. A. Phylogenetic Analyses Reveal the Shady History of C4 Grasses. *Proc. Natl. Acad. Sci. USA* **2010,** *107*, 2532–2537.

Edwards, G. E.; Andreo, C. S. NADP-Malic Enzyme from Plants. *Photochemistry* **1992,** *31*, 1845–1857.

Edwards, G. Tuning Up Crop Photosynthesis. *Nat. Biotechnol.* **1999,** *17*, 22–23.

Ehleringer, J. R.; Monson, R. K. Evolutionary and Ecological Aspects of Photosynthetic Pathway Variation. *Annu. Rev. Ecol. Syst.* **1993,** *24*, 411–439.

Farage, P. K.; Long, S. P.; Lechner, E. G.; Baker, N. R. The Sequence of Change Within the Photosynthetic Apparatus of Wheat Following Short-Term Exposure to Ozone. *Plant Physiol.* **1991,** *95*, 529–535.

Flexas, J.; Bota, J.; Loreto, F.; Cornic, G.; Sharkey, T. D. Diffusive and Metabolic Limitations to Photosynthesis under Drought and Salinity in C3 Plants. *Plant Biol.* **2004,** *5*, 1–11.

Fukayama, H.; Tsuchida, H.; Agarie, S. Significant Accumulation of C4-Specific Pyruvate, Orthophosphate Dikinase in a C3 Plant, Rice. *Plant Physiol.* **2001,** *127*, 1136–1146.

Furbank, R. T.; Taylor, W. C. 1995. Regulation of Photosynthesis in C3 and C4 Plants: A Molecular Approach. *Plant Cell* **1995,** *7*,797–807.

Ghannoum, O.; Evans, J. R.; Chow, W. S.; Andrews, T. J.; Conroy, J. P.; von Caemmerer, S. Faster Rubisco Is the Key to Superior Nitrogen-Use Efficiency in NADP-Malic Enzyme Relative to NAD-Malic Enzyme C4 Grasses. *Plant Physiol.* **2005,** *137*, 638–650.

Ghannoum, O.; Evans, J. R.; von Caemmerer, S. Nitrogen and Water Use Efficiency of C4 Plants. In *C4 Photosynthesis and Related CO2 Concentrating Mechanisms*; Raghavendra, A. S., Sage, R. F., Eds.; Springer: Dordrecht, The Netherlands, 2011; pp 129–146.

Gibbs, M.; Black, C. C.; Kok, B. Factors Affecting CO_2 Fixation by Chloroplasts. *Biochimica et Biophysics Acta.* **1961,** *52* (3), 474–477.

Glackin, C. A.; Grula, J. W. 1990. Organ-speci®c Transcripts of Different Size and Abundance Derive from the Same Pyruvate, Orthophosphate Dikinase Gene in Maize. *Proc. Natl. Acad. Sci. USA* **1990,** *87*, 3004–3008.

Goodall, G. J.; Filipowicz, W. Different Effects of Intron Nucleotide Composition and Secondary Structure on Pre-mRNA Splicing in Monocot and Dicot Plants. *EMBO J.* **1991,** *10*, 2625–2644.

Hatch, M. D. C4 Photosynthesis: A Unique Blend of Modified Biochemistry, Anatomy and Ultrastructure. *Biochim. et Biophys. Acta* **1987,** *895*, 81–106.

Hatch, M. D.; Burnell, J. N. Carbonic Anhydrase Activity in Leaves and Its Role in the First Step of C_4 Photosynthesis. *Plant Physiol.* **1990,** 93, 825– 828.

Häusler, R. E.; Hirsch, H. J.; Kreuzaler, F.; PeterhaÈnsel, C. Overexpression of C4-Cycle Enzymes in Transgenic C3 Plants: A Biotechnological Approach to Improve C3-Photosynthesis. *J. Exp. Bot.* **2002,** *53*, 591–607.

Häusler, R. E.; Kleines, M.; Uhrig, H.; Hirsch, H. J.; Smets, H. Overexpression of Phosphoenolpyruvate Carboxylase from Corynebacterium Glutamicum Lowers the CO_2 Compensation Point (G*) and Enhances Dark and Light Respiration in Transgenic Potato. *J. Exp. Bot.* **1999,** *50*, 1231–1242.

Havaux, M. Characterization of Thermal Damage to the Photosynthetic Electron Transport System in Potato Leaves. *Plant Sci.* **1993,** *94*, 19–33.

Hudspeth, R. L.; Grula, J. W.; Dai, Z.; Edwards, G. E.; Ku, M. S. B. Expression of Maize Phosphoenolpyruvate Carboxylase in Transgenic Tobacco. *Plant Physiol.* **1992,** *98*, 458–464.

Imaizumi, N.; Ku, M. S. B.; Ishihara, K.; Samejima, M.; Kaneko, S.; Matsuoka, M. Characterization of the Gene for Pyruvate, Orthophosphate Dikinase from Rice, a C3 Plant, and a Comparison of Structure and Expression between C3 and C4 Genes for This Protein. *Plant Mol. Biol.* **1997,** *34*, 701–716.

IPCC. *Climate Change 2001: The Scientific Basis. Intergovernment Panel on Climate Change*; Cambridge University Press: Cambridge, 2001.

Kalaji, H.; Nalborczyk, E. Gas-Exchange of Barley Seedlings Growing under Salinity Stress. *Photosynthetica* **1991,** *25*, 197–202.
Kapralov, M. V.; Filatov, D. A. Widespread Positive Selection in the Photosynthetic Rubisco Enzyme. *BMC Evol. Biol.* **2007,** *7*, 73.
Kirschbaum, M. U. F. Does Enhanced Photosynthesis Enhance Growth? Lessons Learned from CO2 Enrichment Studies. *Plant Physiology* **2010,** *155*, 117–124.
Knapp, A. K.; Medina, E. Success of C4 Photosynthesis in the Field: Lessons from Communities Dominated by C4 Plants. In *C4 Plant Biology*; Sage, R. F., Monson, R. K., Eds.; Academic Press: London, 1999; pp 251–283.
Ku, M. S. B.; Kano-Murakami, Y.; Matsuoka, M. Evolution and Expression of C4 Photosynthesis Genes. *Plant Physiol.* **1996,** *111*, 949–957.
Kumari, R.; Kumar, P.; Sharma, V. K.; Kumar, H. Genetic Engineering to Establish a C_4-Metabolism in C3 Plants for Crop Improvement: A Review. *J. Cell Tissue Res.* **2017,** *17* (2), 6179–6184.
Leakey, A. D. B.; Ainsworth, E. A.; Bernacchi, C. J.; Rogers, A.; Long, S. P.; Ort, D. R. Elevated CO_2 Effects on Plant Carbon, Nitrogen and Water Relations: Six Important Lessons from FACE. *J. Exp. Bot.* **2009,** *60*, 2859–2876.
Leegood, R. C. C4 Photosynthesis: Principles of CO_2 Concentration and Prospects for Its Introduction into C3 Plants. *J. Exp. Bot.* **2002,** *53*, 581–590.
Long S. P. Environmental responses. In *The Biology of C4 Photosynthesis*, Sage, R. F., Monson, R. K., Eds.; Academic Press: San Diego, CA, 1999; pp 209–243..
Long, S. P.; Marshall-Colon, A.; Zhu, X. Meeting the Global Food Demand of the Future by Engineering Crop Photosynthesis and Yield Potential. *Cell* **2015,** *161*, 56–66.
Long, S. P.; Ainsworth, E. A.; Rogers, A.; Ort, D. R. (2004) Rising Atmospheric Carbon Dioxide: Plants Face Their Future. *Annu. Rev. Plant Biol.* **2004,** *55*, 591–628.
Ludwig, L. J.; Canvin, D. T. The Rate of Photorespiration during Photosynthesis and the Relationship of the Substrate of Light Respiration to the Products of Photosynthesis in Sunflower Leaves. *Plant Physiol.* **1971,** *48*, 712–719.
Massacci, A.; Battistelli, A.; Loreto, F. Effect of Drought Stress on Photosynthetic Characteristics, Growth and Sugar Accumulation of Field Grown Sweet Sorghum. *Aust. J. Plant Physiol.* **1996,** *23*, 331–340.
Matsuoka, M.; Robert Furbank, T.; Fukayama, H.; Miyao, M. Molecular Engineering of C4 Photosynthesis. *Annu. Rev. Plant Physiol. Plant Mol. Biol.* **2001,** *52*, 297–314.
Matsuoka, M.; Tada, Y.; Fujimura, T.; Kano-Murakami, Y. 1993. Tissue-Specific Light-Regulated Expression Directed by the Promoter of a C4 Gene, Maize Pyruvate, Orthophosphate Dikinase, in a C3 Plant, Rice. *Proc. Natl. Acad. Sci., USA* **1993,** *90*, 9586–9590.
Miyao, M. Molecular Evolution and Genetic Engineering of C4 Photosynthetic Enzymes. *J. Exp. Bot.* **2003,** *54*, 179–189.
Miyao, M.; Fukayama, H.; Tamai, T.; Matsuoka, M. High Level Expression of C4 Photosynthesis Enzymes in Transgenic Rice. In *Proceedings of 12th International Congress on Photosynthesis*, CSIRO Publishing: Canberra, Australia, 2001; pp S39–001.
Monson, R. K. The Origins of C_4 Genes and Evolutionary Pattern in the C_4 Metabolic Phenotype. In *C_4 Plant Biology*; Sage, R. F., Monson, R. K., Eds.; Academic Press: San Diego, CA, 1999; pp 377–410.

Moons, A.; Valcke, R.; van Montagu, M. Low-Oxygen Stress and Water Deficit Induce Cytosolic Pyruvate Orthophosphate Dikinase (PPDK) Expression in Roots of Rice, a C3 Plant. *Plant J.* **1998,** *15*, 89–98.

Normile, D. Round and Round: A Guide to the Carbon Cycle. *Science* **2009,** *325*, 1642–1643.

Ort, D. R.; Merchant, S. S.; Alric, J.; Barkan, A.; Blankenship, R. E.; Bock, R.; Croce, R.; Hanson, M. R.; Hibberd, J. M.; Long, S. P. Redesigning Photosynthesis to Sustainably Meet Global Food and Bioenergy Demand. *Proc. Natl. Acad. Sci. USA* **2015,** *112*, 8529–8536.

Parolin, P. Phenology and CO2-Assimilation of Trees in Central Amazonian Floodplains. *J. Trop. Ecol.* **2000,** *16*, 465–473.

Piedade, M. T. F.; Junk, W. J.; Long, S. P. The Productivity of the C4 Grass Echinochloa-Polystachya on the Amazon Floodplain. *Ecology* **1991,** *72*, 1456–1463.

Poorter, H.; Navas, M. L. Plant Growth and Competition at Elevated CO2: On Winners, Losers and Functional Groups. *New Phytol.* **2003,** *157*, 175–198.

Ramalho, J. C.; Quartin, V. L.; Leitao, E.; Campos, P. S.; Carelli, M. L. C.; Fahl, J. I.; Nunes, M. A. Cold Acclimation Ability and Photosynthesis among Species of the Tropical Coffea Genus. *Plant Biol.* **2003,** *5*, 631–641.

Reddy, A. R.; Rasineni, G. K.; Raghavendra, A. S. The Impact of Global Elevated CO_2 Concentration on Photosynthesis and Plant Productivity. *Curr. Sci.* **2010,** *99*, 46–57.

Reece, J.; Urry, L.; Cain, M.; Wasserman, S.; Minorsky, P.; Jackson, R. *Biology*, International ed.; Pearson Education: Upper Saddle River, NJ, 2011, pp 235–244.

Reich, P. B. Quantifying Plant Response to Ozone: A Unifying Theory. *Tree Physiol.* **1987,** *3*, 63–91.

Robinson, M. F.; Heath, J.; Mansfield, T. A. Disturbances in Stomatal Behaviour Caused by Air Pollutants. *J. Exp. Bot.* **1997,** *49*, 461–469.

Sage, R. F. Variation in the k(cat) of Rubisco in C (3) and C (4) Plants and Some Mplications for Photosynthetic Performance at High and Low Temperature. *J. Exp. Bot.* **2002,** *53*, 609–620.

Sage, R. F.; McKown, A. D. Is C4 Photosynthesis Less Phenotypically Plastic Than C3 Photosynthesis? *J. Exp. Bot.* **2006,** *57* (2), 303–317.

Sage, R. F.; Monson, R. K., Eds. *C4 Plant Biology*; Academic Press: San Diego, 1987; pp 377–410.

Sage, R. F.; Monson, R. K., Eds. *Plant Biology*; Academic Press: San Diego, 1999; 596 pp.

Sage, R. F.; Pearcy, R. W. The physiological ecology of C4 photosynthesis. In *Advances in Photosynthesis, Vol. 9, Photosynthesis: Physiology and Metabolism*; Leegood, R. C., Sharkey, T. D., von Caemmerer, S., Eds.; 2000; pp 497–532.

Schiermeier, Q. Water: A Long Dry Summer. *Nature* **2008,** *452*, 270–273.

Schuler Mara, L.; Mantegazza, O.; Andreas Weber, P. M. Engineering C4 Photosynthesis into C3 Chassis in the Synthetic Biology Age. *Plant J.* **2016,** *87* (1), 51–65.

Sheen, J. Molecular Mechanisms Underlying the Differential Expression of Maize Pyruvate, Orthophosphate Dikinase Genes. *Plant Cell* **1991,** *3*, 225–245.

Shinozaki, K.; Sugiura, M. Organization of Chloroplast Genomes. *Adv. Biophys.* **1986,** *21*, 57–78.

Silva, E. A.; Damatta, F. M.; Ducatti, C.; Regazzi, A. J.; Barros, R. S. Seasonal Changes in Vegetative Growth and Photosynthesis of Arabica Coffee Trees. *Field Crops Res.* **2004,** *89*, 349–357.

Sowjanya, B. A.; Narayana, B. D.; Shreyas S. Enhancement of Photosynthetic Efficiency of C3 Plants. *Int. J. Curr. Microbiol. App. Sci.* **2019,** *8* (3), 775–786.

Suzuki, K.; Nagasuga, K.; Okada, M. The Chilling Injury Induced by High Root Temperature in the Leaves of Rice Seedlings. *Plant Cell Physiol.* **2008,** *49*, 433–442.
Suzuki, S.; Murai, N.; Burnell, J. N.; Arai, M. Changes in Photosynthetic Carbon ¯ow in Transgenic Rice Plants That Express C4-Type Phosphoenolpyruvate Carboxykinase from Urochloa Panicoides. *Plant Physiol.* **2000,** *124*, 163–172.
Takeuchi, Y.; Akagi, H.; Kamasawa, N.; Osumi, M.; Honda, H. Aberrant Chloroplasts in Transgenic Rice Plants Expressing a High Level of Maize NADP-Dependent Malic Enzyme. *Planta* **2000,** *211*, 265–274.
Thomas, C. D.; Cameron, A.; Green, R. E.; Bakkenes, M.; Beaumont, L. J.; Collingham, Y. C. Extinction Risk from Climate Change. *Nature* **2004,** *427*, 145–158.
Tsuchida, H.; Tamai, T.; Fukayama, H. High Level Expression of C4-Specific NADP-Malic Enzyme in Leaves and Impairment of Photoautotrophic Growth of a C3 Plant, Rice. *Plant Cell Physiol.* **2001,** *42*, 138–145.
von Caemmerer, S.; Evans, J. R.; Hudson, G. S.; Andrews, T. J. The Kinetics of Ribulose-1,5-Bisphosphate Carboxylase/Oxygenase In Vivo Inferred from Measurements of Photosynthesis in Leaves of Transgenic Tobacco. *Planta* **1994,** *195*, 88–97.
Wall, G. W.; Brooks, T. J.; Adam, R. 2001 Elevated Atmospheric CO_2 Improved Sorghum Plant Water Status by Ameliorating the Adverse Effects of Drought. *New Phytol.* **2001,** *152*, 231–248.
Ward, J. K.; Tissue, D. T.; Thomas, R. B.; Strain, B. R. Comparative Responses of Model C3 and C4 Plants to Drought in Low and Elevated CO_2. *Glob. Change Biol.* **1999,** *5,* 857–867.
Weis, E.; Berry, J. A. Plants and High Temperature Stress. *Symp. Soc Expt. Biol.* **1988,** *42*, 329–346.

CHAPTER 10

Phytohormones and Antioxidants in the Improvement of Photosynthesis and Respiration under Environmental Stress

ALLAH NAWAZ[1*], NAQSHE ZUHRA[1], MUHAMMAD ASHAR AYUB[1*], AMARA FAROOQ[1], AYMAN E. L. SABAGH[2], MUHAMMAD IMRAN ASHRAF[1], ABBAS SHOUKAT[1], WAJID UMAR[3], and MUHAMMAD ZOHAIB[1]

[1]*Institute of Soil and Environmental Sciences, University of Agriculture, Faisalabad, Punjab, Pakistan*

[2]*Department of Agronomy, Faculty of Agriculture, University of Kafrelsheikh, Egypt*

[3]*Doctoral School of Environmental Science, the Hungarian University of Agricultural and Life Sciences, Gödöllő, Hungary*

[*]*Corresponding author. E-mail: an_uaf99@yahoo.com; muhammadasharayub@gmail.com*

ABSTRACT

Plants are exposed to various abiotic (salinity, temperature extreme, water deficit, nitrogen loss, and heavy metals) and biotic (microbial and weeds) stresses. Plant biochemical pathways are impaired upon environmental stresses, and it becomes hard for plants to tolerate stresses. Photosynthesis and respiration are the most important processes on Earth and need to be regulated in an altered way to tolerate stress conditions. Plants withstand

Photosynthesis and Respiratory Cycles during Environmental Stress Response in Plants. Aryadeep Roychoudhury (Ed.)

stress conditions by altering these mechanisms other than at normal conditions. Phytohormones and antioxidants display a significant role in the plant defense system. Phytohormones (auxins, abscisic acid, gibberellins, jasmonic acid, and salicylic acid) help plants to survive under various stresses by changing their endogenous levels, while the built-in antioxidant defense systems support the plants to tolerate environmental stresses by removing reactive oxygen species and reactive nitrogen species produced under such harsh conditions. An improved understanding of plant strategies in stress conditions is needed to make plants able to withstand changing environmental conditions to reduce massive damages to crops. This chapter will highlight the role of phytohormones and antioxidants in mitigating the ill effects of different abiotic and biotic stresses, especially with reference to photosynthesis and respiration.

10.1 INTRODUCTION

The conditions at which normal development, growth, and plant metabolism are stopped are recognized as stress (Rhodes and Nadolska-Orczyk, 2001). The rising population continuously affects the climate and generates negative environmental stress, which hamper plant production (Denby and Gehring, 2005). It is predicted that increasing global surface temperature from 3°C to 5°C due to natural events (drought and heatwaves) will become frequent. Salinization increases as sea-level rise, decreasing the agricultural land area, and changing the seasonal pattern along with decreasing the growing season (Easterling et al., 2000; IPCC, 2007, 2008). The changes in temperature and rainfall patterns affect the nutritional value of crops and rate of pathogen attack, ultimately reducing plant growth (Porter and Semenov, 2005). Stresses are of two forms, biotic that is due to bacteria, viruses, and fungi (Luck et al., 2011; Alvarez and Lamb, 1997), and abiotic, that is, temperature, salinity, drought, ultraviolet radiations, ozone, and herbicides (Mohamed et al., 2003; Von Wiren et al., 2006; O'Rourke et al., 2007). These stresses cause oxidative damages due to reactive oxygen species (ROS) production in thylakoids, which affect chlorophyll and both photosynthesis and respiration processes (Roach and Krieger-Liszkay, 2014). The respiration process uses the sugars and oxygen produced from photosynthesis to generate energy for plant growth. To a greater degree, respiration is the opposite of photosynthesis. Photosynthesis as well as respiration are essential metabolizing mechanisms to satisfy the cell energy requirement for maintenance and development by forming adenosine triphosphates (ATPs) and redox equivalents, namely, nicotinamide adenine dinucleotide hydrogen (NADH) and nicotinamide adenine dinucleotide phosphate hydrogen (Noguchi and Yoshida, 2008). Plant development is impacted by environmental conditions, contributing to many modifications in the physiology and morphology of plants. To cope with stressful conditions, plants can produce several phytohormones, for example, auxin, cytokinin (CK), gibberellins (GA), salicylic acid (SA), abscisic acid (ABA), jasmonic acid, and ethylene (ET) (Bari and Jones, 2009; Nakashima and Yamaguchi-Shinozaki, 2013). They can participate in different functions, for example, growth, reproduction, and metabolism in the receptive tissues (Santner et al., 2009; Davies, 2010) and in plant protection against many pathogens and pesticides as well (Bari and Jones, 2009). In chloroplast, mitochondria, and peroxisomes, various ROS are produced continuously during the respiration and photosynthesis of plants under natural circumstances. The enzymatic and nonenzymatic antioxidants can maintain plant homeostasis (Wu et al., 2017). The antioxidants that improve plant growth under stressed environment are superoxide dismutase (SOD), ascorbate peroxidase (APX), monodehydroascorbate reductase (MDAR), catalase (CAT), glutathione S-transferase (GST), glutathione reductase (GR), dehydroascorbate reductase (DHAR), etc. (Wu et al., 2017).

10.2 FORMS OF STRESSES

10.2.1 ABIOTIC STRESSES

Abiotic stresses like salinity, water stress (drought), cold, frost, and heat (extreme temperatures), waterlogging (excessive watering), and mineral toxicity negatively impact plant growth (Mohamed et al., 2003; Von Wiren et al., 2006; O'Rourke et al., 2007). For example, high salinity disrupts the sodium-potassium ion fraction in the cytoplasm (Jiang et al., 2013). Water stress is another significant environmental abiotic stress exacerbated by a lack of available water in the soil due to evapotranspiration at high temperatures. This condition lowers the turgor pressure and water potential that leads to the closure of stomata. It slows down the plant metabolic processes (photosynthesis and respiration). Excess water shortage lowers the photosynthetic rate, ultimately slowing plant growth, and eventually leading to death (Fahad et al., 2017). Water stress leads to negative effects like inhibiting cell enlargement, cell division, and seed germination. It causes oxidative damages due to ROS production in thylakoid. It also affects chlorophyll and carotenoids and disrupts the photosystem (Roach and Krieger-Liszkay, 2014). Cold and frost also adversely impact plant reproduction, lower the metabolic functions and reduce the yield. Cold temperature prompts pollen and ovule infertility, and flower abortion causes failure of fertilization, and disturbs seed filling, caused due to low seed set and eventually low grain yield (Thakur et al., 2010). Waterlogging is also a damaging environmental stress for plant growth. It causes disturbance in stomatal conductance (Folzer et al., 2006). It limits water uptake caused by internal water deficiency and lowers the oxygen level, reducing hydraulic conductivity by hampering the root permeability. It also decreases photosynthesis and transpiration rate due to stomatal closure (Ashraf and Arfan, 2005). Mineral stress is also important environmental stress defined as the excess availability of essential nutrients and nonessential nutrients at a toxic level, for example, Al, Mn, Na, Cl, and other heavy metals. It affects transpiration in roots, such as Ca and Si altered root architecture on getting hold of immobile nutrients, particularly P (Lynch and Clair, 2004).

10.2.2 BIOTIC STRESSES

Biotic stress is caused by plant parasites and pathogens, which lead to various plant diseases. Fungal parasites are classified into biotrophs and necrotrophs (which use toxins to destroy the host cells) (Abood and Losel, 2003). Bacteria may enter the plants and cause vascular wilts, leaf spots, and cankers. Nematodes remove the material of plant cells and prey on the plants in all pieces, but plant parasites cause diseases due to soil transmission and degradation of the root systems. They produce signs such as wilting or prominent nutrient deficiency. Viruses cause lesions and damage to the system, resulting in stunts, chlorosis, and malformations in various plant areas, while their hosts are scarcely killed (Schumann and D'Arcy, 2006). Pests like flies and mites disrupt plant growth by feeding plant parts or laying eggs. Piercing mosquitoes act as vectors of the virus spreads to crop plants. Overall, such biotic stress causes a deficiency of water and nutrients, mechanical wounding, damage to the cell wall, intracellular osmotic stress, and damage to both photosynthesis and respiration (Abood and Losel, 2003).

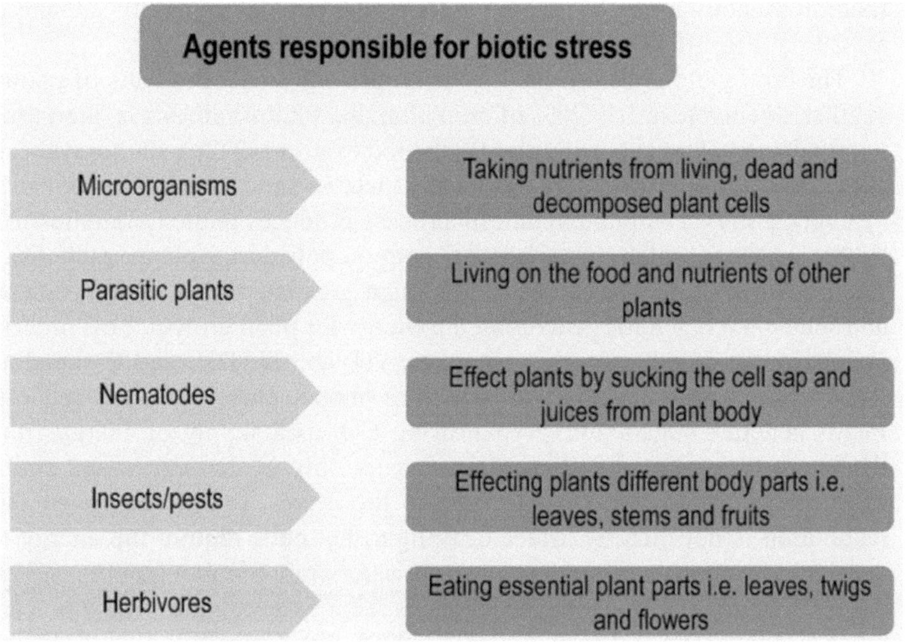

FIGURE 10.1 Different agents responsible for biotic stress and their mode of actions in plants.

10.3 PHYSIOLOGICAL PROCESSES: PHOTOSYNTHESIS AND RESPIRATION

Photosynthesis is an important physiological process of plants in which solar energy is used for biochemical reactions and convert carbon dioxide (CO_2) and water (H_2O) into carbohydrates. This process is very crucial to sustain life on earth (Nickelsen, 2015). Photosystems I and II harvest light to drive photosynthesis (Wientjes et al., 2017). The efficiency of photosynthetic machinery is affected by various environmental stresses like water stress, salinity, high or low temperatures, and excessive ROS (Powles, 1984; Oquist and Huner, 1991; Allakhverdiev et al., 2001). The ROS overproduction is due to high light intensity (Blokhina and Fagerstedt, 2010) that damage the photosystems I and II termed photoinhibition. It also directly affects photosystem II for the overproduction of hydrogen peroxide (H_2O_2), which ultimately affect the photosynthetic process (Stefanov et al., 2018). The ROS cause oxidative damage and inactivate some key enzymes that reduce CO_2

fixation capacity in the Calvin cycle (Scheller and Haldrup, 2005; Zhang et al., 2011) (Fig. 10.1).

The crop yield is determined by plant growth, while products of photosynthesis generate nearly 90% of crop biomass. Photosynthesis is, therefore, the fundamental mechanism to provide food (Yamori, 2020). Photosynthesis occurs in chlorophyll, and the respiratory mechanism occurs in mitochondria (Yamori, 2020). The photosynthesis process produces photoassimilates like starch and sugar, which are a source of energy supply during plant metabolism. The plant respiration mechanism uses sugar generated from photosynthesis and skeleton oxygen for generating the energy for plant growth. This process is carried out in mitochondria to create ATP and NADH, and C-skeleton. Respiration is thought to be the reverse of photosynthesis in many regards. Plants generate sugars and oxygen using CO_2 as a supply of energy from the atmosphere (O_2). Photosynthesis occurs only in the leaves and stalks of the plant, while respiration occurs in the leaves, stalks, and roots. The respiration is not directly affected by light, but cells require the energy to survive continuously. Therefore, the respiratory rate of mitochondria is 50% or less at night, as mitochondrial enzymes seem to be using less light and are quickly inactivated (Atkin et al., 1998, 2000). Plant growth is directly linked to respiration and photosynthesis (Noguchi and Yoshida, 2008).

10.4 CLASSIFICATION, BIOSYNTHESIS, AND ROLE OF PHYTOHORMONES AND ANTIOXIDANTS IN PLANTS

Auxins (IAA), ABA, CKs, GA, and ET are the five groups of phytohormones. Hormones are chemical compounds released in small quantities in certain tissues that regulate various functions in the receptive tissue related to the development, reproduction, and metabolism (Davies 2010; Santner et al., 2009). IAA encourage plant growth and development, for example, cell growth stimulation, elongation of the stem and root, apical dominance, differentiation of the vascular tissues, and lateral and adventitious root initiation (Davies, 2010). Indole-3-acetic acid (IAA) is the common form of auxin. In the biosynthesis of auxin, two routes have emerged, one via indole-3-pyruvate (IPyA) intermediates and the other via indole 3-acetamide intermediate (IAM) (Spaepen et al., 2007). Tryptophan monooxygenase transfers tryptophan to IAM and thus tryptophan is catalyzed by the enzyme, IAM hydrolase in the IAM pathway. Tryptophan is converted to IPyA by aromatic aminotransferase, and the IPyA is converted to IAAld via

IPyA decarboxylation reaction (*ipdC* gene-encoded IPDC). IAAld is then transformed to IAA (Spaepen et al., 2007). CK is an important phytohormone produced from adenine. Substitutions reformed it at the N6, such as ribosides, ribotides, and glycosides. Cell division is aided by CKs, which are also involved in a number of plant processes such as organ formation, senescence, leaf expansion, and root hair development (Davies, 2010). The CK biosynthesis in plants begins with the conversion of isoprenoid moiety mostly known as dimethylallyl pyrophosphate to adenosine phosphate, catalyzed by an enzyme called adenosine phosphate-isopentenyl transferase. The side-chain of the isoprenoid-resultant shifts as a result of hydroxylation. A phospho-ribohydrolase then hydrolyzes the CK compound to a free base (Frébort et al., 2011). GA is an extremely important phytohormone. The GA biosynthesis that begins from the GA12-aldehyde has been observed in *Gibberella fujikuroi*, which contains the commercial GA3 (Bömke and Tudzynski, 2009). It encourages developmental pathways such as germination and flowering (Davies, 2010). ABA causes stomatal closing and seed germination to be inhibited (Davies, 2010). In ABA biosynthesis, the carotenoid lycopene is converted to violaxanthin and then to xanthoxin by a dioxygenase. Aldehyde oxidase and dehydrogenase/reductase convert the latter compound to ABA (Nambara and Marion-Poll, 2005). ET is another dominant phytohormone with its key role in plant developmental and physiological processes, together with cell growth, senescence, and abscission. It is also known as the ripening hormone because it causes fruit to ripen (Davies 2010). ET is responsible for resistance against plant pathogens. Methionine is converted to S-adenosylmethionine (SAM), with the help of an enzyme SAM synthase and ACC synthase produces 1-aminocyclopropane-1-carboxylate (ACC) and 5'-deoxy-methylthioadenosine. Finally, ACC oxidase carries out the conversion of ACC into ET (Tsavkelova et al., 2006).

Higher plants are likely to exhibit a greater capacity to scavenge the harmful effects of activated oxygen than others. Antioxidants are the first line of protection against free radicals. They are crucial to keep the plant cells healthy. Plants use a variety of antioxidant enzymes, peptides, and metabolites to scavenge activated oxygen. The antioxidants mentioned below have evolved to protect plants from oxidative damages. Among these, SOD, APX, CAT, GR, DHAR, GST, MDAR, are involved (Wu et al., 2017). Among the three kinds of SODs, Fe-SOD, Mn-SOD, and Cu/Zn-SOD are found in plants (Zelko et al., 2002, Fink and Scandalios, 2002). In the cytoplasm, mitochondria, peroxisomes, and chloroplasts, antioxidants present are named as Fe-SOD, Mn-SOD, and Cu/Zn-SOD (Wang et al., 2016). *SOD* genes have been cloned from monocotyledons and dicotyledons, with maize being the first plant source (Cannon et al., 1987, Sunkar et al., 2006). CAT is a special antioxidant enzyme found in all aerobic species that catalyzes the conversion process of H_2O_2 into water as well as oxygen during various environmental stresses. It can be found in chloroplasts, cytosol, peroxisomes, and mitochondria, where H_2O_2 is generated (Sharma and Ahmad, 2014). The APX enzyme catalyzes the conversion of H_2O_2 to H_2O and O_2; it can be cytosolic (cAPX), stroma-chloroplastic (sAPX), mitochondrial (mitAPX), and thylakoid-chloroplastic (tAPX), depending on where it is found in the cell (Shigeoka et al. 2002). MDHAR is an important antioxidant that could directly regenerate the APX. MDHAR activity is frequently noted in the mitochondria, chloroplast, cytosol, and peroxisomes (Hossain et al., 1984; Dalton et al., 1993; Jiménez et al., 1997). Glutathione (GSH) is a nonenzymatic antioxidant and contributes to a redox control for lowering of ROS levels in plants (Foyer and Noctor, 2005). The majority of GSH is found in plastids and cytosol (Cohen et al., 2020). DHAR is considered an effective antioxidant enzyme in preserving the ascorbate redox state and the pool of DHARs (Ding et al., 2020). Herbicides can be detoxified by the plant GSTs. The GSH conjugation with herbicide molecules is catalyzed to GSH S conjugates, which are then transferred to the vacuole, thus preventing herbicide damages to the plants (Reade et al., 2004).

10.5 ROLE OF PHYTOHORMONES IN THE MODULATION OF PLANT PHOTOSYNTHESIS AND RESPIRATION UNDER ENVIRONMENTAL STRESSES

Among the phytohormones, auxin, ABA, GA, SA, CKs, jasmonic acid, and ET perform significant role in plant defense mechanisms against biotic and abiotic stresses (Bari and Jones, 2009; Nakashima and Yamaguchi-Shinozaki, 2013). Phytohormones respond to abiotic and biotic stress differently. Jasmonic acid, SA, and ET regulate plant defense responses to pathogens and pests (Bari and Jones, 2009). They help plants in stress conditions by signaling networks. The fundamental mechanism of the signaling network is a condition known as phytohormone crosstalk. Therefore, plants have progressed under the influence of a particular hormone in various stress reactions. However, hormonal crosstalk is also controlled within the plant population and plays roles at different stages (Diezel et al., 2009).

10.5.1 RESPONSE UNDER BIOTIC STRESS

ABA has a substantial role in plant defense against biotic stresses. ABA -regulated stomatal closure is well known; however, stomata are the main entry route of pathogens (Zeng et al., 2010). A decrease in stomatal density is related to increased defense against pathogens (Casson and Hetherington, 2014). Plants can adjust stomatal apertures in response to pathogens. In that way, regulation of the transpiration process also controls pathogen movement in plants. Furthermore, callose accumulation in the cell wall entails the synthesis of ABA at sites where pathogens enter (Ton et al., 2004).

Pathogen attack on plant tissues triggers ET production (Di Lucchio et al., 2018). Increased ET production induces defense-related genes. Several proteins are produced that correspond to ET production and make physical barriers to pathogens, for example, hydroxyproline-rich proteins strengthen cell walls upon treatment with ET. ET results in the creation of antimicrobial secondary metabolites named as phytoalexins. These are derived from the phenylpropanoid pathway that is induced by ET under pathogen attack (Thomma et al., 2001) (Table 10.1). Pathogenesis-related genes are known to exert antimicrobial activity against several fungi and bacteria (Tsuchisaka and Theologis, 2004). Positive roles of ET are widespread in resistance against necrotrophs (Glazebrook, 2005), and biotrophic and hemibiotrophic pathogens (Shibata et al., 2010).

Auxin is involved in expressing auxin-responsive genes, for example, GH_3 gene family that plays numerous functions in biotic stresses (Dowd et al., 2004; Wang et al., 2007). Like other phytohormones, CKs induce neoplastic growth in host plants and impart resistance against pathogens (Walters and McRoberts, 2006; Giron et al., 2013). CKs enhance resistance against the fungus and viruses (Siddique et al., 2015; Dowd et al., 2017). GA regulate plant immunity and enhance biotrophic pathogen resistance (Walters 2011; Denancé et al., 2013). It increases defense-related gene expression (Kamali et al., 2015; El-Beltagi et al., 2012). GA also enhance resistance mechanisms for plant defense against nematode infection (Blok and Powers, 2009).

Pathogen-induced SA synthesis is a well-known phenomenon. Protective mechanism of SA involves the expression of genes related to pathogenesis. Many of these genes encode proteins with antimicrobial activity, with increasing immunity to a wide range of pathogens. Because of the mechanism of programmed cell death, the hypersensitive reaction to SA results in necrotic lesions. This mechanism suppresses the spread of infection to undamaged tissues. Systemic acquired resistance is considered as a broad spectrum and long-lasting resistance process (Table 10.1). SA is a positive immunity regulator against biotrophic pathogens, which feed on living hosts and hemibiotrophic diseases and display a biological process early in the infections (Glazebrook, 2005). Enhanced resistance to pathogens was observed with SAs, such as for tobacco mosaic virus. SA accumulates in high levels at the site of infection. In this way, uninfected tissues are secured from getting infected by the minute accumulation of SA. Pathogenesis-related genes restrict the spread of the tobacco mosaic virus to plant tissues surrounding the entry point and where a necrotic lesion appears later. Consequently, a secondary infection results in much smaller lesions as compared to the primary infection. So, the process of necrotic lesion formation in photosynthetic tissues is a defense response in plants; photosynthesis in that specified necrotic region stops, and an overall reduction in net photosynthesis takes place (Anith et al., 2004).

The importance of jasmonic acid against pathogens is shown by a rise in its levels in response to pathogen infection. Immunity against necrotrophic bacteria is positively regulated by jasmonic acid (Howe and Jander, 2008; Burra et al., 2015). In plant tissues where bacteria have infected, jasmonic acid causes programmed cell death or leads to necrotic spots on leaves or other affected parts. In this way, jasmonic acid makes the infection localized to the specific parts, and neighboring tissues are secured from infection (Fonseca et al., 2009) (Table 10.1). This adaptability affects photosynthesis indirectly by localized cell death of leaves. Necrotic tissues also stop photosynthesis (Radhika et al., 2012). Less CO_2 fixation and suppressed photosynthesis results in less formation of glucose, and thus respiration process is affected. Respiration is also affected due to a reduction in the gaseous exchange through stomata of the affected cells. Volatile organic compounds are produced by jasmonic acid (Radhika et al., 2012). They might attract the natural predators of herbivores and act as defense molecules (Gulsen et al., 2010) (Table 10.1).

Interestingly, volatile organic compounds act as messengers in plants in anticipation of dangers (Okada et al., 2015). Indeed, several jasmonic acid-producing plants exhibit increased resistance to necrotrophic fungi. The jasmonic acid synthesis increases upon chewing and wounding triggered by pest attack. Jasmonic acid appears at injured sites as well as on entire leaves to suppress or prevent herbivores. It does so by activating the expression of defense-related genes. This evidence reveals that jasmonic acid regulates plant defense against a broad spectrum of insect herbivores. However, jasmonic acid-mediated defense responses are triggered by initial signaling events involving calcium flux and membrane potential upon infection or insect attack (Scholz et al., 2014).

10.5.2 RESPONSE UNDER ABIOTIC STRESS

In times of reduced water availability, especially during water stress, ABA builds up in the leaves and decreases the stomatal cavity pore. This synchronization is directly linked to the process of photosynthesis (Sharma, 2014). Even though it is known that ABA signaling for the closing of stomata is regulated by calcium (Ca) ions and channels, proper functioning depends on external stimuli, especially water availability (Mori et al., 2006). The photosynthesis process stops upon the closing of stomata. When talking about plant respiration under stress conditions, a special feature of mitochondria is the "altered electron pathway." Without the aid of cytochrome oxidase, this mechanism pulls electrons from the ubiquinone reservoir. In a single cycle of four-electron transfer, the mitochondrial alternative oxidase reduces molecular oxygen to water. Salicylic hydroxamic acid and n-propyl gallate block this alternative mechanism. Increased mitochondrial alternative oxidase dosage can influence active oxygen species formation (Umbach et al., 2005).

IAA cause cell elongation, vascular tissue development, and apical dominance. Auxin response factor 6A plays a role in photosynthesis, but under stress conditions, auxin biosynthesis is altered (Bari et al., 2009). CKs are considered master regulators during growth, cell division, chloroplast biogenesis, apical dominance, and photomorphogenetic development. GA regulate osmotic stress, interact with other hormones, participate in sugar signaling, causes enhanced efficiency of nutrients, increases sink strength, and regulates water use efficiency (Leadbeater and Staub, 2014).

ET could play an important role during abiotic stresses, especially under waterlogging and drought stress resulting in reduced photosynthesis (Rajala and Sainio, 2001). In addition, ET inhibits root growth, reduces shoot/leaf expansion, and decreases grain yield by grain abortion (Pierik et al., 2007; Wilkinson and Davies, 2010). It regulates gene expression accompanying leaf senescence (Table 10.1). Fermentation pathways, aerenchyma formation, and glycolysis are also affected by ET (Pistelli et al., 2012). Another ET-related modulation is leaf abscission (Table 10.1). In *Citrus*, water stress conditions result in the induction of leaf abscission associated with ACC expression associated with an increase in ET. Hypoxic conditions in the root zone results in ET accumulation in leaves, increasing leaf abscission (Gil et al., 2009). Numerous abiotic factors modify enzymes implicated in the biosynthesis of SA (Li et al., 2010; Horváth et al., 2007). In *Arabidopsis*, SA is related to induce heat stress tolerance. Certain heat shock protein-encoded genes are involved in thermotolerance induction by SA (Clarke et al., 2009). SA biosynthesis-related enzymes (isochorismate synthase and phenylalanine ammonia-lyase) are the key regulators of SA functions (Ogawa et al., 2005). SA plays an indirect role in the tolerance mechanism of environmental stress in plants by generation of compatible solutes and antioxidants. Thus, photosynthetic machinery and cellular respiration processes are also affected due to SA-regulated antioxidant production.

Abiotic environmental stresses result in poor plant growth. Plants produce endogenous jasmonic acid upon stress conditions (Zhang et al., 2017). Jasmonic acid helps plants cope with abiotic stresses, including salt, drought, toxic metals, cold stress, and micronutrient toxicity. Environmental stresses decrease or inhibit photosynthesis and respiration directly or indirectly. The involvement of jasmonic acid in antioxidant systems modulates photosynthesis and respiration and protects these two processes to normalize plant biochemical reactions during stress. Jasmonic acid can alleviate stress by the antioxidative system activated in response to unstable ROS. Stress tolerance regulated by jasmonic acid activates enzymatic and nonenzymatic antioxidative systems (Wu et al., 2012; Faghih et al., 2017). Oxidative stress is diminished indirectly by jasmonic acid through the expression of genes encoding antioxidants and secondary metabolites (Bali et al., 2019). Jasmonic acid could alleviate stresses by creating heat shock proteins as well as phenolic compounds. Jasmonic acid accumulates in response to mechanical injury or wounding, which is reflected as a mutual physiological feedback mechanism in all plants (Isah, 2019). It also protects chlorophyll and mitochondria, and conclusively, it regulates photosynthesis and respiration processes during environmental stresses.

TABLE 10.1 Role of Phytohormones in the Modulation of Photosynthesis and Respiration in Plants under Environmental Stresses.

Phytohormones	Stress	
	Under abiotic stress	**Under biotic stress**
Ethylene	Increased leaf abscission/ acts against heavy metal accumulation, excess salt accumulation in leaves, or reduction in water loss/ decreased stomatal conductance (Gil et al., 2009)	Physical barriers to pathogens/ antimicrobial metabolite synthesis/ Pathogenesis related protein synthesis (Yu et al., 2017)
Abscisic acid (ABA)	Stomatal closure to conserve water (Depuydt and Hardtke, 2011)	Stomatal closure to stop pathogen entry/ virus/ fungal hyphae entrapment hindrance (Lind et al., 2015)
Salicylic Acid (SA)	Production of osmotic solutes and antioxidants (Flexas et al., 2005)	Systemic acquired resistance, programmed cell death and hypersensitive response (Glazebrook, 2005)
Jasmonic Acid	Activation of some defense compounds such as heat shock proteins, phenolic compounds, synthesis of antioxidants, expression of defense-related genes (Zhang et al., 2012)	Formation of a physical barrier (such as trichome), generation of volatile organic compounds which attract predators or parasitic enemies of pest; production of toxic secondary chemicals against pathogens, induced systemic resistance, activation of defense-related genes (Okada et al., 2015; Campos et al., 2014)
Auxins	Cell elongation, vascular tissue development, apical dominance. Auxin response factor 6A plays a role in photosynthesis (Bari et al., 2009)	Differential expression of auxin-responsive genes (GH3 gene family) (Dowd et al., 2004; Wang et al., 2007)
Cytokinins	Master regulator during growth and development, cell division, apical dominance, chloroplast biogenesis and photomorphogenesis development	Induce neoplastic growth in host plant, enhanced resistance against fungus and viruses (Walters and McRoberts, 2006; Dowd et al., 2017; Siddique et al., 2015)

TABLE 10.1 *(Continued)*

Phytohormones	Stress	
	Under abiotic stress	**Under biotic stress**
Gibberellins	Regulate plant growth and development, osmotic stress, interact with other hormones for performing functions, sugar signaling, enhanced efficiency of nutrients, increasing sink strength, role in water use efficiency (Leadbeater and Staub, 2014)	Regulates plant immunity and enhances resistance to biotrophs. Increased expression of the defense-related gene. Enhances PAL, POX and PPO activity for plant defense after nematode infection (Blok and Powers 2009; Xing et al., 2008). It increases resistance against cyst nematode (Beneventi et al., 2013)

10.6 ROLE OF ANTIOXIDANTS IN THE MODULATION OF PLANT PHOTOSYNTHESIS AND RESPIRATION UNDER ENVIRONMENTAL STRESSES

To sustain growth and metabolism, plants also produce numerous phytochemicals and enzymes as antioxidant defenses. Antioxidants are generated in the leaves and by quenching free radicals, protect the plant from damages (Zomeno et al., 2013) (Table 10.2). Plants with high antioxidant potential are more resistant to stress.

10.6.1 RESPONSE UNDER BIOTIC STRESS

Biotic stress occurs since different living organisms harm plants. Biotic agents causing plant stresses include microbial pathogens (bacteria, viruses, and fungi), nematodes, insects, herbivores, and plant pests. These agents result in mechanical injury, nutrient and water loss, intracellular osmotic stress, and cell wall damage (Abood and Losel, 2003). Plant defense against pathogen invasion and pest attack requires the formation of antioxidants in response to high ROS at the site of the attempted attack (Apel and Hirt, 2004). Antioxidants act as antimicrobial agents. Plant cells can regulate chloroplast-targeted ROS enzymes and ROS development in response to biotic stress (Apel and Hirt, 2004) (Table 10.2). ROS is therefore used as a signaling molecule to reduce pathogenic invasion implying a delicate spatial balance between ROS production and ROS scavenging pathways.

10.6.2 RESPONSE UNDER ABIOTIC STRESS

The development and growth of plants are affected by various environmental stresses such as temperature, drought, salt, and high light. The biochemical, physiological, structural, and developmental progress of plants are adversely affected when plants endure the stress conditions and when the optimal range of environmental factors is exceeded (Tuteja et al., 2009). ROS are formed when high potential electrons are transported to molecular oxygen (O_2) during stress conditions. The most toxic molecules of ROS are superoxide ions (O_2^-), singlet oxygen ($^1O_2$), and hydrogen peroxide (Ahmad et al., 2010).

The ROS is antagonized by the enzymatic antioxidant system. Antioxidant system consists of numerous enzymes such as CAT, glutathione peroxidase (GPX), APX, SOD, and GST (Gill et al., 2011). In cell organelles like chloroplast, mitochondria, and peroxisomes, various ROS are produced continuously during respiration and photosynthesis of plants under natural circumstances. So, using two unlike detoxification methods namely enzymatic and nonenzymatic antioxidants, plant homeostasis is maintained (Wu et al., 2017). The crop productivity is affected under drought conditions by the disturbances in the photosynthetic pigments and disrupting the balance of the ROS generation and antioxidants (Gong et al., 2005). Due to the increase in SOD and CAT activities, root water uptake ability and drought stress tolerance are increased due to silicon supplementation in tomato plants (Shi et al., 2016). To study the antioxidative reaction in the plants, the arbuscular mycorrhizal (AM) fungi are also used. The cold, salt, and drought stresses were better endured due to the regulation in hydraulic characteristics by AM symbiosis with *Phaseolus vulgaris* (Pedranzani et al., 2016).

TABLE 10.2 Role of Antioxidants in the Modulation of Photosynthesis and Respiration in Plants under Environmental Stresses.

Antioxidants	Stress	
	Under abiotic stress	**Under biotic stress**
Superoxide dismutase (SOD)	Improved tolerance to salt, heavy metal and water stress; development of chloroplast antioxidant system; Low TBARS, lowering of ion leakage and chlorophyll degradation (Badawi et al., 2004)	Combat over-produced ROS during metabolic processes (Berwal and Ram, 2018); Help in the defense responses of the plant (herbicide-induced oxidative stress) (Poor et al., 2018)

TABLE 10.2 *(Continued)*

Antioxidants	Stress	
	Under abiotic stress	**Under biotic stress**
Catalase (CAT)	Low temperature stress resistance because of successful detoxification of H_2O_2; Cd stress resilience, better seedling development, and longer roots; Overcoming of salt stress by scavenging hydrogen peroxide (Matsumura et al., 2002; Gichner et al., 2004; Al-Taweel et al., 2007)	Counter the phagocyte oxidative burst (Ng et al., 2019) Maintain reactive oxygen homeostasis under pathogenic (fungal) attack (Magbanua et al., 2007)
Ascorbate peroxidase (APX)	Enhanced tolerance to UV-B, heat, drought and chilling stresses via increased APX activity (Wang et al., 2018). Water deficit tolerance with higher photosynthesis (Yan et al., 2003), Protection of different stresses from NaCl, MV, H_2O_2 (Kim et al., 2008)	Maintain cellular homeostasis under stress, Control bacterial blight in rice (Jiang et al., 2016; Panday et al., 2017)
Glutathione reductase (GR)	Chilling stress resistance and photoprotection (Kornyeyev et al., 2003)	GR is mainly involved in the reduction of glutathione for defense in plants, Maintaining hormonal homeostasis in plants, detoxification of herbicides, and hydrogen peroxide. Detoxify those compounds which are involved in damaging DNA, RNA or proteins (Afzal et al., 2014)
Monodehydroascorbate reductase (MDHAR)	Ozone, salt and PEG stress tolerance due to higher MDHAR activity and a higher level of reduced ascorbate (Eltayeb et al., 2007)	Protects plants against viral diseases such as sugarcane mosaic virus (SCMV) by scavenging ROS (Akbar et al., 2020)
Dehydroascorbate reductase (DHAR)	Tolerance to low-temperature, H_2O_2, and NaCl stress (Kwon et al., 2005), Drought and salt tolerance with higher DHAR activity and reduced ascorbate content (Eltayeb et al., 2007).	Balance the ascorbate-glutathione relative content under biotic stress and enhance plant resistance (Kwon et al., 2003; Lin et al., 2016)

TABLE 10.2 *(Continued)*

Antioxidants	Stress	
	Under abiotic stress	**Under biotic stress**
Glutathione S-transferase (GST)	Protect against Al toxicity and oxidative stress (Ezaki et al., 2001), Increased thermal or salt-stress tolerance due to glutathione and ascorbate content (Jogeswar et al., 2006)	Induce a systemic resistance response to subsequent pathogen infections (bacteria and fungi), Detoxify toxic lipid hydroperoxides that accumulate during pathogen attack and participate in intracellular hormone (auxins) transport and ligand functions (Gullner et al., 2018), (Duldler et al., 1991)

10.7 CONCLUSION

Abiotic stresses are mainly responsible for higher ROS production, membrane instability, and oxidative damages to plants. However, it is possible to increase stress tolerance by genetic manipulation of antioxidants, by understanding ROS signaling, and regulatory pathways of stress. Abiotic stresses reduce the photosynthetic activity and inhibit the PSII by damaging the repair system of PSII. Phytohormones like auxin, GA, CK, ABA, and ET play significant role in regulating photosynthetic and respiration mechanism under biotic and abiotic stress conditions. Phytohormones activate the defense mechanisms by changing the morphology, physiology, and biochemical makeup of plants. Similarly, antioxidant defense system of plants is also involved in the protection of plant photosynthetic machinery. Different plant antioxidants like SOD, CAT, APX, and others involved in the scavenging of ROS also play significant role in the repairing of PSII.

KEYWORDS

- **photosynthesis**
- **phytohormones**
- **respiration**
- **antioxidants**
- **reactive oxygen species**

REFERENCES

Abood, J. K.; Losel, M. Changes in Carbohydrate Composition of Cucumber Leaves during the Development of Powdery Mildew Infection. *Plant Pathol.* **2003,** *52*, 256–265.

Afzal, F.; Khurshid, R.; Ashraf, M.; Kazi, A. G. Oxidative Damage to Plants Antioxidant Networks and signaling. In *Reactive Oxygen Species and Antioxidants in Response to Pathogens and Wounding*; 2014; pp 397–424.

Allakhverdiev, S. I.; Kinoshita, M.; Inaba, M.; Suzuki, I.; Murata, N. Unsaturated Fatty Acids in Membrane Lipids Protect the Photosynthetic Machinery against Salt-Induced Damage Insynechococcus. *Plant Physiol.* **2001,** *125* (4), 1842–1853.

Alvarez, M. E.; Lamb, C. Oxidative Burst Mediated Defense Responses in Plant Disease Resistance See Ref. **1997,** *185a*, 815–839.

Anith, K. N.; Momol, M. T.; Kloepper, J. W.; Marois, J. J.; Olson, S. M.; Jones, J. B. Efficacy of Plant Growth-Promoting Rhizobacteria, Acibenzolar-S-methyl, and Soil Amendment for Integrated Management of Bacterial Wilt on Tomato. *Plant Disease* Jun **2004,** *88*, 669–673.

Ashraf, M.; Arfan, M. Gas Exchange Characteristics and Water Relations in Two Cultivars of Hibiscus Esculentus under Waterlogging. *Biol. Plant* **2005,** *49*, 459–462.

Badawi, G. H.; Yamauchi, Y.; Shimada, E.; Sasaki, R.; Kawano, N.; Tanaka, K.; Tanaka, K. Enhanced Tolerance to Salt Stress and Water Deficit by Overexpressing Superoxide Dismutase in Tobacco (*Nicotiana tabacum*) Chloroplasts. *Plant Sci.* **2004,** *166*, 919–928.

Bali, S.; Jamwal, V. L.; Kaur, P.; Kohli, S. K.; Ohri, P.; Gandhi, S. G.; Bhardwaj, R.; Al-Huqail, A. A.; Siddiqui, M. H.; Ahmad, P. Role of P-type ATPase Metal Transporters and Plant Immunity Induced by Jasmonic Acid against Lead (Pb) Toxicity in Tomato. *Ecotoxicol. Environ. Saf.* Jun 15, **2019,** *174*, 283–294.

Bari, R.; Jones, J. D. Role of Plant Hormones in Plant Defence Responses. *Plant Mol. Biol.* Mar 1, **2009,** *69* (4), 473–488.

Bates, B. C.; Kundzewicz, Z. W.; Wu, S.; Palutikof, J. P. Climate Change and Water Technical Paper of the Intergovernmental Panel on Climate Change (Geneva: IPCC Secretariat). *Climate Change* **2008,** *95*, 96.

Berwal, M. K.; Ram, C. Superoxide Dismutase: A Stable Biochemical Marker for Abiotic Stress Tolerance in Higher Plants, 2018. DOI: http://dx.doi.org/10.5772/intechopen.82079.

Blok, V. C.; Powers, T. O. Biochemical and Molecular Identification. *Root-Knot Nematodes.* Oct 13, **2009,** 98–118.

Blokhina, O.; Fagerstedt, K. V. Oxidative Metabolism, ROS and NO under Oxygen Deprivation. *Plant Physiol. Biochem.* **2010,** *48* (5), 359–373.

Burra, D. D.; Mühlenbock, P.; Andreasson, E. Salicylic and Jasmonic Acid Pathways Are Necessary for Defence against Dickeyasolani as Revealed by a Novel Method for Blackleg Disease Screening of In Vitro Grown Potato. *Plant Biol.*. Sep **2015,** *17* (5), 1030–1038.

Campos, M. L.; Kang, J. H.; Howe, G. A. Jasmonate-Triggered Plant Immunity. *J. Chem. Ecol.* Jul **2014,** *40* (7), 657–675.

Cannon, R. E.; White, J. A.; Scandalios, J. G. Cloning of cDNA for Maize Superoxide Dismutase 2 (SOD2). *Proc. Natl. Acad. Sci.* **1987,** *84* (1), 179–183.

Casson, S. A.; Hetherington, A. M. Phytochrome B Is Required for Light-Mediated Systemic Control of Stomatal Development. *Curr. Biol.* **2014,** *24*, 1216–1221.

Clarke, S. M.; Cristescu, S. M.; Miersch, O.; Harren, F. J.; Wasternack, C.; Mur, L. A. Jasmonates Act with Salicylic Acid to Confer Basal Thermotolerance in *Arabidopsis thaliana*. *New Phytol.* Apr **2009,** *182* (1), 175–187.

Cohen, A.; Hacham, Y.; Welfe, Y.; Khatib, S.; Avice, J. C.; Amir, R. Evidence of a Significant Role of Glutathione Reductase in the Sulfur Assimilation Pathway. *Plant J.* **2020,** *102* (2), 246–261.

Dalton, D. A.; Baird, L. M.; Langeberg, L.; Taugher, C. Y.; Anyan, W. R.; Vance, C. P.; Sarath G. Subcellular Localization of Oxygen Defense Enzymes in Soybean (*Glycine max* [L.] Merr.) Root Nodules. *Plant Physiol.* **1993,** *102*, 481–489.

Davies, P. J. Plant Hormones: Biosynthesis, Signal Transduction, Action!; Springer: Dordrecht, 2010.

Denancé, N.; Sánchez-Vallet, A.; Goffner, D.; Molina, A. Disease Resistance or Growth: The Role of Plant Hormones in Balancing Immune Responses and Fitness Costs. *Front. Plant Sci.* May 24, **2013,** *4*, 155.

Depuydt, S.; Hardtke, C. S. Hormone Signaling Crosstalk in Plant Growth Regulation. *Curr. Biol.* May 10, **2011,** *21* (9), R365–R373.

Di Lucchio, L. M.; Fensholt, R.; Markussen, B.; Ræbild, A. Leaf Phenology of Thirteen African Origins of Baobab (Adansonia digitata (L.)) as Influenced by Daylength and Water Availability. *Ecol. Evol.* Nov **2018,** *8* (22), 11261–11272.

Diezel, C.; von Dahl, C. C.; Gaquerel, E.; Baldwin, I. T. Different Lepidopteran Elicitors Account for Cross-Talk in Herbivory-Induced Phytohormone Signaling. *Plant Physiol.* Jul 1, **2009,** *150* (3), 1576–1586.

Ding, H.; Wang, B.; Han, Y.; Li, S. The Pivotal Function of Dehydroascorbate Reductase in Glutathione Homeostasis in Plants. *J. Exp. Bot.* **2020,** *71* (12), 3405–3416.

Dowd, C.; Wilson, I. W.; McFadden, H. Gene Expression Profile Changes in Cotton Root and Hypocotyl Tissues in Response to Infection with Fusarium Oxysporum f. sp. Vasinfectum. *Mol. Plant-Microbe Interact.* Jun **2004,** *17* (6), 654–667.

Dowd, C. D.; Chronis, D.; Radakovic, Z. S.; Siddique, S.; Schmülling, T.; Werner, T.; Kakimoto, T.; Grundler, F. M.; Mitchum, M. G. Divergent Expression of Cytokinin Biosynthesis, Signaling and Catabolism Genes Underlying Differences in Feeding Sites Induced by Cyst and Root-Knot Nematodes. *Plant J.* Oct **2017,** *92* (2), 211–228.

Duldler, R.; Hertig, C.; Rebmann, G.; Bull, J.; Mauch, F. A Pathogen-Induced Wheat Gene Encodes a Protein Homologous to Glutathione-S-Transferases. *Mol. Plant-Microbe. Int.* **1991,** *4*, 14–18.

Easterling, D. R.; Meehl, G. A.; Parmesan, C.; Changnon, S. A.; Karl, T. R.; Mearns, L. O. Climate Extremes: Observations, Modeling, and Impacts. *Science* **2000,** *289*, 2068–2074.

El-Beltagi H. S.; Farahat A. A.; Alsayed A. A.; Mahfoud, N. M. Response of Antioxidant Substances and Enzyme Activities as a Defense Mechanism against Root-Knot Nematode Infection. *Notulae Botanicae Horti Agro botanici Cluj-Napoca* May 14, **2012,** *40* (1), 132–142.

Eltayeb, A. E.; Kawano, N.; Badawi, G. H.; Kaminaka, H.; Sanekata, T.; Shibahara, T.; lnanaga, S.; Tanaka, K. Overexpression of Monodehydroascorbate Reductase in Transgenic Tobacco Confers Enhanced Tolerance to Ozone, Salt and Polyethylene Glycol Stresses. *Planta* **2007,** *225*, 1255–1264.

Ezaki, B.; Katsuhara, M.; Kawamura, M.; Matsumoto, H. Different Mechanisms of Four Aluminum (Al)-Resistant Transgenes for Al Toxicity in Arabidopsis, 2001. DOI: https://doi.org/10.1104/pp.010399.

Faghih, S.; Ghobadi, C.; Zarei, A. Response of Strawberry Plant cv. 'Camarosa' to Salicylic Acid and Methyl Jasmonate Application under Salt Stress Condition. *J. Plant Growth Regulation.* Sep **2017,** *36* (3), 651–659.

Fahad, S.; Bajwa, A. A.; Nazir, U.; Anjum, S. A.; Farooq, A.; Zohaib, A.; Sadia, S.; Nasim, W.; Adkins, S.; Saud, S.; Ihsan, M. Z.; Alharby, H.; Wu, C.; Wang, D. P.; Haung, J. L. Crop Production under Drought and Heat Stress: Plant Responses and Management Options. *Front. Plant Sci.* **2017,** *8*, 1147.

Fink, R. C.; Scandalios, J. G. Molecular Evolution and Structure–Function Relationships of the Superoxide Dismutase Gene Families in Angiosperms and Their Relationship to Other Eukaryotic and Prokaryotic Superoxide Dismutases. *Arch. Biochem. Biophys.* **2002,** *399* (1), 19–36.

Flexas, J.; Galmes, J.; Ribas-Carbo, M.; Medrano, H. The Effects of Water Stress on Plant Respiration. In *Plant Respiration*; Springer: Dordrecht, 2005; pp 85–94.

Folzer, H.; Dat, J.; Capelli, N.; Rieffel, D.; Badot, P. M. Response to Flooding of Sessile Oak: An Integrative Study. *Tree Physiol.* **2006,** *26*, 759–766.

Fonseca, S.; Chini, A.; Hamberg, M.; Adie, B.; Porzel, A.; Kramell, R.; Miersch, O.; Wasternack, C.; Solano, R. (+)-7-iso-Jasmonoyl-L-Isoleucine Is the Endogenous Bioactive Jasmonate. *Nat. Chem. Biol.* May **2009,** *5* (5), 344–350.

Foyer, C. H.; Noctor, G. Redox Homeostasis and Antioxidant Signaling: A Metabolic Interface between Stress Perception and Physiological Responses. *Plant Cell.* **2005,** *17*, 1866–1875.

Foyer, C. H.; Noctor, G. Ascorbate and Glutathione: The Heart of the Redox Hub. *Plant Physiol.* **2011,** *155* (1), 2–18.

Foyer, C. H.; Shigeoka, S. Understanding Oxidative Stress and Antioxidant Functions to Enhance Photosynthesis. *Plant Physiol..* **2011,** *155* (1) 93–100.

Frébort, I.; Kowalska, M.; Hluska, T. et al. Evolution of Cytokinin Biosynthesis and Degradation. *J. Exp. Bot.* **2011,** *62*, 2431–2452.

Gichner, T.; Patkova, Z.; Szakova, J.; Demnerov, K. Cadmium Induces DNA Damage in Tobacco Roots, But No DNA Damage, Somatic Mutations or Homologous Recombination in Tobacco Leaves. *Mutation Res/Genet. Toxiol. Environ. Mutagenesis* **2004,** *559* (1–2), 49–57.

Gil, P. M.; Gurovich, L.; Schaffer, B.; García, N.; Iturriaga, R. Electrical Signaling, Stomatal Conductance, ABA and Ethylene Content in Avocado Trees in Response to Root Hypoxia. *Plant Signal. Behav.* Feb 1, **2009,** *4* (2), 100–108.

Gill, S. S.; Khan, N. A.; Anjum, N. A.; Tuteja, N. Amelioration of Cadmium Stress in Crop Plants by Nutrients Management: Morphological, Physiological and Biochemical Aspects. *Plant. Stress* **2011,** *5*, 1–23.

Giron, D.; Frago, E.; Glevarec, G.; Pieterse, C. M.; Dicke, M. Cytokinins as Key Regulators in Plant–Microbe–Insect Interactions: Connecting Plant Growth and Defence. *Funct. Ecol.* Jun **2013,** *27* (3), 599–609.

Glazebrook, J. Contrasting Mechanisms of Defense against Biotrophic and Necrotrophic Pathogens. *Annu. Rev. Phytopathol.* Jul 28, **2005,** 43, 205–227.

Gong, H.; Zhu, X.; Chen, K.; Wang, S.; Zhang, C. Silicon Alleviates Oxidative Damage of Wheat Plants in Pots under Drought. *Plant Sci.* **2005,** *169*, 313–321.

Li, Z.; Wakao, S.; Fischer, B. B.; Niyogi, K. K. Sensing and Responding to Excess Light. *Annu. Rev. Plant Biol.* **2009,** *60*, 239–260.

Gullner, G.; Komives, T.; Király, L.; Schröder, P. Glutathione s-Transferase Enzymes in Plant-Pathogen Interactions. *Front. Plant Sci.* **2018.** https://doi.org/10.3389/fpls.2018.01836.

Gulsen, O.; Eickhoff, T.; Heng-Moss, T.; Shearman, R.; Baxendale, F.; Sarath, G. et al. Characterization of Peroxidase Changes in Resistant and Susceptible Warm-Season Turf Grasses Challenged by Blissusocciduus. *Arthropod-Plant Interact.* **2010,** *4*, 45–55.

Horváth, E.; Szalai, G.; Janda, T. Induction of Abiotic Stress Tolerance by Salicylic Acid Signaling. *J. Plant Growth Regul.* **2007,** *26* (3), 290–300. DOI: 10.1007/s00344-007-9017-410.1007/s00344-007-9017-4.

Hossain, M. A.; Nakano, Y.; Asada, K. Monodehydroascorbate Reductase in Spinach Chloroplasts and Its Participation in Regeneration of Ascorbate for Scavenging Hydrogen Peroxide. *Plant Cell Physiol.* **1984,** *25*, 385–395.

Howe, G. A.; Jander, G. Plant Immunity to Insect Herbivores. *Annu. Rev. Plant Biol.* Jun 2, **2008,** *59*, 41–66.

Isah, T. Stress and Defense Responses in Plant Secondary Metabolites Production. *Biol. Res.* **2019,** 1–25.

Jiang, C.; Belfield, E. J.; Cao, Y.; Smith, J. A. C.; Harberd, N. P. An Arabidopsis Soil-Salinity–Tolerance Mutation Confers Ethylene-Mediated Enhancement of SODium/Potassium Homeostasis. *Plant Cell* **2013,** *25* (9), 3535–3552.

Jiménez, A.; Hernández, J. A.; del Río, L. A.; Sevilla, F. Evidence for the Presence of the Ascorbate-Glutathione Cycle in Mitochondria and Peroxisomes of Pea Leaves. *Plant Physiol.* **1997,** *114*, 275–284.

Jogeswar, G.; Pallela, R. Jakka, N. m.; Reddy, P. S.; Rao, J. V.; Sreenivasulu, N.; Kishor, P. B. K. Antioxidant Response in Different Sorghum Species under Short-Term Salinity Stress. *Acta. Physiol. Plant.* **2006,** *28*, 465–475.

Kamali, N.; Pourjam, E.; Sahebani, N. Elicitation of Defense Responses in Tomato against Meloidogyne Javanica and Fusarium Oxysporum f. sp. Lycopersici Wilt Complex. *J. Crop Protection* Mar 10, **2015,** *4* (1), 29–38.

Kim, Y. H.; Kim, C. Y.; Song, W. K.; Park, D. S.; Kwon, S. Y.; Lee, H. S.; Bang, J. W.; Kwak, S. S. Overexpression of Sweet Potato SWPA4 Peroxidase Results in Increased Hydrogen Peroxide Production and Enhances Stress Tolerance in Tobacco. *Planta* **2008,** *227*, 867e881.

Kornyeyev, D.; Logan, B. A.; Payton, P. R.; Allen, R. D. Elevates Chloroplastic Glutathione Reductase Activities Decrease Chilling Induced Photoinhibition by Increasing Rates of Photochemistry, But Not Thermal Energy Dissipation, in Transgenic Cotton. *Funct. Plant. Bio.* **2003,** *30* (1). DOI: 10.1071/FP02144

Kwon, S.; Choi, Y.; Ahn, H.; Lee, H.; Park, Y. L.; Kwak, S. Enhanced Stress-Tolerance of Transgenic Tobacco Plants Expressing a Human Dehydroascorbate Reductase Gene. *J. Plant Physiol.* **2003,** *160*, 347–353.

Lars, D.; Pfannschmidt, T. Photosynthetic Acclimation to Light Gradients in Plant Stands Comes Out of Shade: H_2O_2- Induced Changes in Primary Metabolic Pathways. *Plant Signal. Behav.* **2008,** *3* (12), 1116–1118.

Leadbeater, A.; Staub, T. Exploitation of Induced Resistance: A Commercial Perspective. In *Induced Resistance for Plant Defense: A Sustainable Approach to Crop Protection*; Walters, D. R., et al., Eds.; Blackwell: Oxford, 2014; pp 300–315.

Li, S.; Zhou, X.; Chen, L.; Huang, W.; Yu, D. Functional Characterization of *Arabidopsis thaliana* WRKY39 in Heat Stress. *Mol. Cells* May 1, **2010,** *29* (5), 475–483.

Lin, S.; Chiou, C.; Chu, Y.; Hsiao, Y.; Tseng, Y.; Chen, Y.; Chen, H.; Changa, H.; Lee, T. Enhanced Ascorbate Regeneration via Dehydroascorbate Reductase Confers Tolerance to Photo-Oxidative Stress in *Chlamydomonas reinhardtii. Plant Cell Physiol.* **2016,** *57*, 2104–2121.

Lind, C.; Dreyer, I.; López-Sanjurjo, E. J.; von Meyer, K.; Ishizaki, K.; Kohchi, T.; Lang, D.; Zhao, Y.; Kreuzer, I.; Al-Rasheid, K. A.; Ronne, H. Stomatal Guard Cells Co-opted an

Ancient ABA-Dependent Desiccation Survival System to Regulate Stomatal Closure. *Curr. Biol.* Mar 30, **2015,** *25* (7), 928–935.

Luck, J.; Spackman, M.; Freeman, A.; Trebicki, P.; Griffiths, W.; Finlay, K.; Chakraborty, S. Climate Change and Diseases of Food Crops. *Plant Pathol.* **2011,** *60*, 113–121.

Lynch, J. P.; Clair, S. B. S. Mineral Stress: The Missing Link in Understanding How Global Climate Change Will Affect Plants in Real World Soils. *Field Crops Res.* **2004,** *90* (1), 101–115.

Mahajan, S.; Tuteja, N. Cold, Salinity and Drought Stress: An Overview. *Arch. Biochem. Biophys.* **2005,** *444* (2), 139–158.

Matsumura, T.; Tabayashi, N.; Kamagata, Y.; Souma, C.; Saruyama, H. Wheat Catalase Expressed in Transgenic Rice Can Improve Tolerance against Low Temperature Stress. *Physiol. Plant.* **2002,** *116*, 317e327.

Mohamed, A.; Amal, F. K.; El-Baz; Khalifa, R. H. M. Genotypic Differences of Two Wheat Cultivars for Enzymes Activity, Amino Acids and Protein Profile under Fe- Deficiency. *J. Biol. Sci.* **2003,** *3* (10), 864–874.

Mori, I. C.; Murata, Y.; Yang, Y.; Munemasa, S.; Wang, Y. F.; Andreoli, S.; Tiriac, H.; Alonso, J. M.; Harper, J. F.; Ecker, J. R.; Kwak, J. M. CDPKs CPK6 and CPK3 Function in ABA Regulation of Guard Cell S-Type Anion-and Ca 2+-Permeable Channels and Stomatal Closure. *PLoS Biol.* Oct 10, **2006,** *4* (10):e327.

Nakashima, K.; Yamaguchi-Shinozaki, K. ABA Signaling in Stress-Response and Seed Development. *Plant Cell Rep.* Jul **2013,** *32* (7), 959–970.

Ng, V. H.; Cox, J. S.; Sousa, A. O.; MacMicking, J. D.; McKinney, J. D. Role of KatG Catalase-Peroxidase in Mycobacterial Pathogenesis: Countering the Phagocyte Oxidative Burst. *Mol. Microbiol.* 2019. https://doi.org/10.1155/2019/9613090.

Nickelsen, K. *Explaining Photosynthesis: Modelling Biochemical Mechanisms 1840–1960* (Series: History, Philosophy and Theory of the Life Sciences, Vol. 8) Springer: New York et al., 2015.

Ogawa, D.; Nakajima, N.; Sano, T.; Tamaoki, M.; Aono, M.; Kubo, A.; Kanna, M.; Ioki, M.; Kamada, H.; Saji, H. Salicylic Acid Accumulation under O3 Exposure Is Regulated by Ethylene in Tobacco Plants. *Plant Cell Physiol.* Jul 1, **2005,** *46* (7), 1062–1072.

Okada, K.; Abe, H.; Arimura, G. I. Jasmonates Induce Both Defense Responses and Communication in Monocotyledonous and Dicotyledonous Plants. *Plant Cell Physiol.* Jan 1, **2015,** *56* (1), 16–27.

Oquist, G.; Huner, N. P. A. Effects of Cold Acclimation on the Susceptibility of Photosynthesis to Photoinhibition in Scots Pine and in Winter and Spring Cereals: A Fluorescence Analysis. *Funct. Ecol.* **1991,** 91–100.

Ò-Rourke, J. A.; Dirk, V.; Delkin, O.; Graham, A.; Cianzio, R. Micro Array Analysis of Iron Deficiency Chlorosis in Nearisogenic Soybean Lines. *BMC Genom.* **2007,** *8*, 476–483.

Parvaiz, A.; Jaleel, C. A.; Salem, M. A.; Nabi, G.; Sharma, S. Role of Enzymetic and Nonenzymetic Antioxidants in Plants DURING Abiotic Stress. *Crit. Rev. Biotechnol.* **2010,** *30* (3), 161–175.

Pedranzani, H.; Rodríguez-Rivera, M.; Gutiérrez, M.; Porcel, R.; Hause, B.; Ruiz-Lozano, J. M. Arbuscular Mycorrhizal Symbiosis Regulates Physiology and Performance of Digitariaeriantha Plants Subjected to Abiotic Stresses by Modulating Antioxidant and Jasmonate Levels. *Mycorrhiza* **2016,** *26*, 141–152.

Pierik, R.; Sasidharan, R.; Voesenek, L. A. Growth Control by Ethylene: Adjusting Phenotypes to the Environment. *J. Plant Growth Regul.* Jun **2007,** *26* (2), 188–200.

Pistelli, L.; Iacona, C.; Miano, D.; Cirilli, M.; Colao, M. C.; Mensuali-SODi, A.; Muleo, R. Novel Prunus Rootstock Somaclonal Variants with Divergent Ability to Tolerate Waterlogging. *Tree Physiol.* Mar 1, **2012,** *32* (3), 355–368.

Poor, P.; Ördog, A.; Czekus, Z.; Borbely, P.; Takacs, Z.; Kovacs, J.; Tari, I. Regulation of the Key Antioxidant Enzymes by Developmental Processes and Environmental Stresses in the Dark. *Biol. Plant.* **2018,** *62* (2), 201–210.

Porter, J. R.; Semenov, M. A. Crop Responses to Climatic Variation. *Phil. Trans. R. Soc. B* **2005,** *360*, 2021–2035.

Powles, S. B. Photoinhibition of Photosynthesis Induced by Visible Light. *Annu. Rev. Plant Physiol* **1984,** *35* (1), 15–44.

Qiu-Fang, Z.; Yuan-Yuan, L.; Cai-Hong, P.; Cong-Ming, L.; Bao-Shan, W. NaCl Enhances Thylakoid-Bound SOD Activity in the Leaves of C3 Halophyte *Suaeda salsa* L. *Plant Sci.* **2005,** *168*, 423–430.

Racchi, M. L. Antioxidant Defenses in Plants with Attention to Prunus and Citrus spp. *Antioxidants* **2013,** *2*, 340–369.

Radhika, V.; Kost, C.; Bonaventure, G.; David, A.; Boland, W. Volatile Emission in Bracken Fern Is Induced by Jasmonates But Not by *Spodoptera littoralis* or Strongylogastermultifasciata Herbivory. *PLoS One*. Nov 20, **2020,** *7* (11), e48050.

Rajala, A.; Peltonen-Sainio, P. Plant Growth Regulator Effects on Spring Cereal Root and Shoot Growth. *Agron. J.* Jul **2001,** *93* (4), 936–943.

Reade, J. P. H.; Milner, L. J.; Cobb, A. H. A Role for Glutathione S-Transferases in Resistance to Herbicides in Grasses. *Weed Sci.* **2004,** *52*, 468–474.

Rhodes, D.; Nadolska-Orczyk, A. Plant Stress Physiology. *e LS*, 2001.

Santner, A.; Calderon-Villalobos, L. I. A.; Estelle, M. Plant Hormones Are Versatile Chemical Regulators of Plant Growth. *Nat. Chem. Biol.* **2009,** *5*, 301–307.

Scheller, H. V.; Haldrup, A. Photoinhibition of Photosystem I. *Planta* **2005,** *221* (1), 5–8.

Scholz, S. S.; Vadassery, J.; Heyer, M.; Reichelt, M.; Bender, K. W.; Snedden, W. A.; Boland W.; Mithöfer, A. Mutation of the Arabidopsis Calmodulin-Like Protein CML37 Deregulates the Jasmonate Pathway and Enhances Susceptibility to Herbivory. *Mol. Plant* Dec 1, **2014,** *7* (12), 1712–1726.

Schumann, G. L.; D'Arcy, C. J. *Essential Plant Pathology*; American Phytopathological Society (APS Press), 2006.

Sharma, R. Study of Glycosyltransferases and Other Stress-Related Proteins Encoded in Chickpea Genome and Analysis of Their Structures and Functions Using Molecular Modeling and Docking, 2014.

Sharma, I.; Ahmad, P. Catalase: A Versatile Antioxidant in Plants. In *Oxidative Damage to Plants*; Academic Press, 2014; pp 131–148.

Sharma, I.; Navdeep, K.; Pratap, K. P. Brassinosteroids: A Promising Option in Deciphering Remedial Strategies for Abiotic Stress Tolerance in Rice. *Front. Plant Sci.* Dec 20, 2017. https://doi.org/10.3389/fpls.2017.02151.

Shi, Y.; Zhang, Y.; Han, W.; Feng, R.; Hu, Y.; Guo, J.; Gong, H. Silicon Enhances Water Stress Tolerance by Improving Root Hydraulic Conductance in *Solanum lycopersicum L. Front. Plant Sci.* **2016,** *7*, 196.

Shibata, Y.; Kawakita, K.; Takemoto, D. Age-Related Resistance of Nicotiana Benthamiana against Hemibiotrophic Pathogen Phytophthora Infestans Requires Both Ethylene-and Salicylic Acid–Mediated Signaling Pathways. *Mol. Plant-Microb. Inter.* Sep **2010,** *23* (9), 1130–1142.

Shigeoka, S.; Ishikawa, T.; Tamoi, M.; Miyagawa, Y.; Takeda, T.; Yabuta, Y.; Yoshimura, K. Regulation and Function of Ascorbate Peroxidase Isoenzymes. *J. Exp. Bot.* **2002,** *53* (372), 1305–1319.

Siddique, S.; Radakovic, Z. S.; Carola, M.; Chronis, D.; Novák, O.; Ramireddy, E.; Holbein, J.; Matera, C.; Hütten, M.; Gutbrod, P.; Anjam, M. S. A Parasitic Nematode Releases Cytokinin That Controls Cell Division and Orchestrates Feeding Site Formation in Host Plants. *Proc. Natl. Acad. Sci.* Oct 13, **2015,** *112* (41), 12669–12674.

Soares, C.; Marcia, E. A.; Carvalhob, R. A.; Azevedob, F. F. Plants Facing Oxidative Challenges-A Little Help from the Antioxidant Networks. *Environ. Exp. Bot.* **2019,** *161*, 4–25.

Solomon, S.; Qin, D.; Manning, M.; Chen, Z.; Marquis, M.; Averyt, K. B.; Miller, H. L. Contribution of Working Group I to the Fourth Assessment Report of the Intergovernmental Panel on Climate Change, 2007.

Stefanov, M.; Yotsova, E.; Markovska, Y.; Apostolova, E. L. Effect of High Light Intensity on the Photosynthetic Apparatus of Two Hybrid Lines of Paulownia Grown on Soils with Different Salinity. *Photosynthetica* **2018,** *56* (3), 832–840.

Sunkar, R.; Kapoor, A.; Zhu, J. K. Posttranscriptional Induction of Two Cu/Zn Superoxide Dismutase Genes in Arabidopsis Is Mediated by Downregulation of miR398 and Important for Oxidative Stress Tolerance. *Plant Cell* **2006,** *18* (8), 2051–2065.

Thakur, P.; Kumar, S.; Malik, J. A.; Berger, J. D.; Nayyar, H. Cold Stress Effects on Reproductive Development in Grain Crops: An Overview. *Environ. Exp. Bot.* **2010,** *67* (3), 429–443.

Thomma, B. P.; Penninckx, I. A.; Cammue, B. P.; Broekaert, W. F. The Complexity of Disease Signaling in Arabidopsis. *Curr. Opin. Immunol.* Feb 1, **2001,** *13* (1), 63–68.

Ton, J.; Mauch-Mani, B. β-amino-Butyric Acid-Induced Resistance against Necrotrophic Pathogens Is Based on ABA-Dependent Priming for Callose. *Plant J.* Apr **2004,** *38* (1), 119–130.

Tsavkelova, E. A.; Klimova, S. Y.; Cherdyntseva, T. A. et al. Hormones and Hormone-Like Substances of Microorganisms: A Review. *Appl. Biochem. Microbiol.* **2006,** *42*, 229–235.

Tsuchisaka, A.; Theologis, A. Unique and Overlapping Expression Patterns among the Arabidopsis 1-Amino-Cyclopropane-1-Carboxylate Synthase Gene Family Members. *Plant Physiol.* Oct 1, **2004,** *136* (2), 2982–3000.

Tuteja, N.; Ahmad, P.; Panda, B. B.; Tuteja, R. Genotoxic Stress in Plants: Shedding Light on DNA Damage, Repair and DNA Repair Helicases. *Mutat. Res./Rev. Mutat. Res.* **2009,** *681*, 134–149.

Umbach, A. L.; Fiorani, F.; Siedow, J. N. Characterization of Transformed Arabidopsis with Altered Alternative Oxidase Levels and Analysis of Effects on Reactive Oxygen Species in Tissue. *Plant Physiol.* Dec 1, **2005,** *139* (4), 1806–1820.

Von Wiren, N.; Marschner, H.; Romheld, V. Uptake Kinetics of Iron–Phytosideorophores in Two Maize Genotypes Differing in Iron Efficiency. *Physiol Plant.* **2006,** *93*, 611–616.

Walters, D. R. *Sounding the Alarm: Signaling and Communication in Plant Defense. Plant Defense: Warding Off Attack By Pathogens, Herbivores, and Parasitic Plants*; Oxford: Blackwell Publishing, 2011; pp 77–124.

Walters, D. R.; McRoberts, N. Plants and Biotrophs: A Pivotal Role for Cytokinins? *Trends Plant Sci.* Dec 1, **2006,** *11* (12), 581–586.

Wang, D.; Pajerowska-Mukhtar, K.; Culler, A. H.; Dong, X. Salicylic Acid Inhibits Pathogen Growth in Plants through Repression of the Auxin Signaling Pathway. *Curr. Biol.* Oct 23, **2007,** *17* (20), 1784–1790.

Wang, Q.-L.; Chen, J.-H.; He, N.-Y.; Guo, F.-Q. Metabolic Reprogramming in Chloroplasts under Heat Stress in Plants. *Int. J. Mol. Sci.* **2018,** *19* (3), 849. https://doi.org/10.3390/ijms19030849.

Wang, W.; Xia, M. X.; Chen, J.; Yuan, R.; Deng, F. N.; Shen, F. F. Gene Expression Characteristics and Regulation Mechanisms of Superoxide Dismutase and Its Physiological Roles in Plants under Stress. *Biochemistry (Moscow)* **2016,** *81* (5), 465–480.

Wientjes, E.; Philippi, J.; Borst, J. W.; van Amerongen, H. Imaging the Photosystem I/Photosystem II Chlorophyll Ratio Inside the Leaf. *Biochim. Biophys. Acta (BBA)-Bioenerg.* **2017,** *1858* (3), 259–265.

Wilkinson, S.; Davies, W. J. Drought, Ozone, ABA and Ethylene: New Insights from Cell to Plant to Community. *Plant Cell Environ.* Apr **2010,** *33* (4), 510–525.

Wu, Y.; Zhang, D.; Chu, J. Y.; Boyle, P.; Wang, Y.; Brindle, I. D.; De Luca, V.; Després, C. The Arabidopsis NPR1 Protein Is a Receptor for the Plant Defense Hormone Salicylic Acid. *Cell Rep.* Jun 28, **2012,** *1* (6), 639–647. DOI: 10.1016/j.celrep.2012.05.008.

Wu, Z.; Liu, S.; Zhao, J.; Wang, F.; Du, Y.; Zou, S.; Li, H.; Wen, D.; Huang, Y. Comparative Responses to Silicon and Selenium in Relation to Antioxidant Enzyme System and the Glutathione-Ascorbate Cycle in Flowering Chinese Cabbage (*Brassica campestris L.* ssp. *chinensis var. utilis*) under Cadmium Stress. *Environ. Exp. Bot.* **2017,** *133*, 1–11.

Xing, Y.; Jia, W.; Zhang, J. AtMKK1 Mediates ABA-Induced CAT1 Expression and H2O2 Production via AtMPK6-Coupled Signaling in Arabidopsis. *Plant J.* May **2008,** *54* (3), 440–451.

Yamori, W. Photosynthesis and Respiration. In *Plant Factory*. Academic Press, 2020; pp 197–206.

Yan, J.; Wang, J.; Tissue, D.; Holaday, A. S.; Allen, R.; Zhang, H. Photosynthesis and Seed Production under Water-Deficit Conditions in Transgenic Tobacco Plants That Overexpress an Arabidopsis Ascorbate Peroxidase Gene. *Crop. Sci.* **2003,** *43*, 1477e1483.

Yu, Y.; Yang, D.; Zhou, S.; Gu, J.; Wang, F.; Dong, J.; Huang, R. The Ethylene Response Factor OsERF109 Negatively Affects Ethylene Biosynthesis and Drought Tolerance in Rice. *Protoplasma* Jan 1, **2017,** *254* (1), 401–408.

Zelko, I. N.; Mariani, T. J.; Folz, R. J. Superoxide Dismutase Multigene Family: A Comparison of the CuZn-SOD (SOD1), Mn-SOD (SOD2), and EC-SOD (SOD3) Gene Structures, Evolution, and Expression. *Free Radical Biol. Med.* **2002,** *33* (3), 337–349.

Zeng, W.; Melotto, M.; He, S. Y. Plant Stomata: A Checkpoint of Host Immunity and Pathogen Virulence. *Curr. Opin. Biotechnol.* Oct 1, **2010,** *21* (5), 599–603.

Zhang, H.; Zhang, Q.; Zhai, H.; Li, Y.; Wang, X.; Liu, Q.; He, S. Transcript Profile Analysis Reveals Important Roles of Jasmonic Acid Signalling Pathway in the Response of Sweet Potato to Salt Stress. *Sci. Rep.* Jan 13, **2017,** *7* (1), 1–2.

Zhang, X.; Sheng, J.; Li, F.; Meng, D.; Shen, L. Methyl Jasmonate Alters Arginine Catabolism and Improves Postharvest Chilling Tolerance in Cherry Tomato Fruit. *Postharvest Biol. Technol.* Feb 1, **2012,** *64* (1), 160–167.

Zhang, Z.; Jia, Y.; Gao, H.; Zhang, L.; Li, H.; Meng, Q. Characterization of PSI Recovery after Chilling-Induced Photoinhibition in Cucumber (*Cucumis sativus* L.) Leaves. *Planta* **2011,** *234* (5), 883–889.

CHAPTER 11

Regulation of Mitochondrial Respiration during Salt Stress

NAVEED UL MUSHTAQ[1], SEERAT SALEEM[1], AADIL RASOOL[1], WASIFA HAFIZ SHAH[1], INAYATULLAH TAHIR[2], and REIAZ UL REHMAN[1*]

[1]Department of Bioresources, School of Biological Sciences, University of Kashmir, Srinagar 190006, India

[2]Department of Botany, School of Biological Sciences, University of Kashmir, Srinagar 190006, India

[*]Corresponding author. E-mail: rreaizbiores@gmail.com

ABSTRACT

Eukaryotic cells are differentiated from prokaryotes by membrane-bound organelles which perform specific cellular functions. In eukaryotes, mitochondria perform respiration that is essential for adenosine triphosphate (ATP) generation, required for growth, development, and cellular maintenance. Respiration in the presence of oxygen (aerobic respiration) results in the generation of energy-rich molecules, driven by break down of respiratory substrates. Besides, the mitochondrial processes in plants are also required in carbon and nitrogen metabolism. They have a role in photosynthesis and regulation of abiotic stress. Understanding the machinery, organization, and parameter of respiratory routes is essential to know about plant growth. Reactive oxygen and nitrogen species are formed under stress conditions which can lead to oxidative damage affecting quantity and quality of plant products. The cellular organelles work in tandem with the help of network

Photosynthesis and Respiratory Cycles during Environmental Stress Response in Plants. Aryadeep Roychoudhury (Ed.)

of scavenging systems that help in quenching the reactive species. The effect of stress on mitochondria can impair the ATP supply of the cell and lead to oxidative injury leading to the production of reactive oxygen species (ROS) which when exceeding the capacity of cellular antioxidants leads to damage. Understanding the control and regulation of the respiratory processes is vital to explain plant growth and its variability in different environmental conditions. In this chapter, we aim to explain the regulation of mitochondrial respiration during environmental stress.

11.1 INTRODUCTION

Plants are exposed to abiotic as well as biotic stresses due to their static nature and the various abiotic stresses include high doses of heavy metal, salt stress, drought, and extreme temperature (Li, 2013). Due to these stresses, the growth and development is easily hampered because of the enhanced synthesis of reactive oxygen species (ROS) and singlet oxygen. The main ROS producing sites in plants are the peroxisomes, chloroplasts, and mitochondria (Choudhury et al., 2017). The excessive production of these reactive species leads to the imbalance in respiration and photosynthesis, for instance, the assembly of singlet species of oxygen in chloroplasts leads to programmed cell death (Kleine and Leister, 2016). In plants, there is internal mechanism to tolerate stress and threshold levels of ROS. The upregulation of various scavenging molecules and enzymes like proline, catalase, peroxidase, and superoxide dismutase (SOD) occurs to prevent damage. However, if the ROS levels exceed the threshold, it results in damaging the cellular organelles and their products (Huang et al., 2020). The regulation of different biochemical agents under salt stress is listed in Table 11.1.The communication of different cellular organelles for the mitigation of abiotic stress is shown in Figure 11.1.

Respiration and photosynthesis are two vital processes on earth to sustain both plant and animal life. In the process of respiration, sugars are consumed which plants produce during photosynthesis to produce energy currency for their growth and development. Thus, respiration can be treated as an opposite process when compared to photosynthesis. Photosynthesis cannot take place in animals; however, in all plant and animal cells, respiration has an important cellular scope-like production of ATP, etc. (Van Dongen et al., 2011). So, in this catabolic process, the substrates like sugars are converted into energy, H_2O, and CO_2 in the presence of oxygen.

11.2 RESPIRATION IN PLANTS

The process of respiration includes group of subprocesses like glycolysis, tricarboxylic acid (TCA) cycle, and electron transport chain (ETC) system (Van Dongen et al., 2011). Cytosol and plastid are the sites for glycolysis, mitochondrial matrix for TCA cycle and inner mitochondrial membrane for oxidative phosphorylation.

11.2.1 GLYCOLYSIS

Glycolysis is a catabolic (breakdown) pathway which occurs in the presence of oxygen and fulfills basic roles of generating ATP by oxidizing hexoses. The sequential change of hexoses to pyruvate occurs separately as put forward by Gustav Embden, Otto Meyerhof, and J. Parnas, in short called as glycolysis or Embden–Meyerhof–Parnas pathway (Akram, 2013) during which two molecules of pyruvic acid are formed from glucose. In photosynthesis, sucrose is produced, which is converted to glucose and fructose with the enzyme invertase. The monosaccharide molecules that enter the glycolytic pathway are first phosphorylated to produce glucose-6-phosphate with the help of the enzyme, hexokinase (Akram, 2013). Glycolysis is the 10-step reaction which is controlled by several enzymes till pyruvate is produced. The ATP is utilized at two steps, first when glucose is converted into glucose 6-phosphate and second in the conversion of fructose 6-phosphate to fructose 1,6-bisphosphate. Taking into account the ATP consumed in glycolysis, the net energy yield is 2 ATP, 2 NADH, and 2 pyruvates per molecule of glucose (Lenzen, 2014). The end product, that is, pyruvic acid produced in glycolysis can be utilized by fermentation, alcoholic fermentation, and aerobic respiration. Fermentation occurs without oxygen consumption, for example, in prokaryotes and unicellular eukaryotes. In aerobic respiration, organisms require oxygen supply for complete oxidation to CO_2 and H_2O. Aerobic respiration is known by various names like Krebs cycle, citric acid cycle, or TCA.

11.2.2 KREBS CYCLE

Acetyl coenzyme-A formed by glycolysis acts as fuel for the Krebs cycle. After the formation of pyruvates during glycolysis, it is converted into acetyl coenzyme-A (acetyl-CoA) with the help of an enzyme pyruvate dehydrogenase. The subsequent steps are carried out by the Krebs cycle, named after its discoverer, Hans Adolf Krebs (Preedy, 2016). Acetyl-CoA reacts in eight step cycle called the Krebs cycle that fully oxidizes it to two molecules of CO_2 and H_2O (Wiskich, 1980). The NADH and $FADH_2$ produced are used by the oxidative phosphorylation pathway. This cycle does not produce considerable ATP directly. As an alternative, NADH and $FADH_2$ is produced which enters respiratory cycle where ATP is produced. In the presence of the enzyme citrate synthase, the acetyl group condenses with oxaloacetic acid and H_2O to yield citric acid and CoA. With successive isomerization, citrate is then isomerized to isocitrate. It is chased by two consecutive steps of decarboxylation, resulting in the generation of α-ketoglutaric acid, then succinyl-CoA, malic acid, and at last oxaloacetic acid (Kumar and Dubey, 2019).

11.2.3 OXIDATIVE PHOSPHORYLATION AND ELECTRON TRANSPORT SYSTEM

In the electron transport system present in inner mitochondrial membrane, the electrons are transferred from one carrier to another carrier. The electrons formed in the Krebs cycle are oxidized by NADH dehydrogenase (called as complex I), and then transferred to ubiquinone in inner membrane (Lencina et al., 2018). With the help of complex II, ubiquinone in addition obtains reducing equivalents that are formed when succinate is oxidized in the Krebs cycle (Horsefield et al., 2004). The oxidation of ubiquinol (reduced) occurs due to electron transfer to cytochrome c via cytochrome bc1 complex (complex III) (Sun et al., 2018). Cytochrome c is present on outer surface of the inner membrane and performs electron transfer involving complex III and IV (cytochrome c oxidase) (Nath and Villadsen, 2015). In electron chain, transfer of electrons takes place from complex I to IV and finally to complex V (ATP synthase), resulting in the production of ATP (Beyer and Klingenberg, 1985). Upon oxidation, single molecule of NADH results in the production of three ATP molecules, whereas one $FADH_2$ molecule produces two ATP molecules (Gautheron, 1984). Oxygen drives this process by acting as the final hydrogen receiver. Further, in respiration, the energy of oxidation and reduction is consumed and that is why this process is known as oxidative phosphorylation (Schmidt-Rohr, 2020).

11.3 MITOCHONDRIA AND REGULATION OF ABIOTIC STRESS

Abiotic stress regularly diminishes plant enlargement, maturity, and efficiency because it generates ROS; thus upsetting balance, damaging photosynthetic apparatus, respiration chain, shrinking membrane permeability, reducing stability of membrane proteins, enzymes, etc. (Carvalho, 2008). Both anabolic and catabolic processes in organisms are affected by environmental conditions. In stressful conditions, the mitochondrial respiration and respiratory complexes I, II, III become linked and produce ROS (Moller, 2001). Mitochondrion is cardinal for reducing the effect of photosynthetic reducing power, lowering oxidative injuries of thylakoids and other cellular machinery (Dinakar et al., 2010).

TCA cycle is inhibited by the products generated under stress conditions, causing change in metabolism of nitrogen and carbon (Suzuki et al., 2012). Oxidized metabolites and proteins have been considered as indicators of stress; however, such agents are found to function as signaling molecules at lower concentrations under oxidative stress conditions (Mueller and Berger, 2009). In both chloroplasts and mitochondria, changes occur under abiotic stress (Takahashi and Murata, 2008); thus, elevated metabolic organization is necessary to sustain energy flow to avoid overproduction of ROS and oxidative injuries. ROS network and redox regulatory enzymes like glutathione reductase and glutathione peroxidase are unregulated under stress conditions (Mittler et al., 2004). Recent proteomic studies revealed that phosphorylation of mitochondrial proteins are having a major role in mitochondrial redox regulation (Ito et al., 2009). In mitochondria, there is an uneven distribution of genome suggesting that it is meant to prevent plant mtDNA from oxidative damage (Logan, 2010; Mandavilli et al., 2002). Retrograde signals from the mitochondrion to the nucleus act in harmonizing mitochondrial and cytosolic protein formation (Suzuki et al., 2012).

11.4 MITOCHONDRIAL ACTIVITIES UNDER SALT STRESS

The utilization of oxygen and oxidative phosphorylation in mitochondria is affected by NaCl stress, for example, in pea and barley plants (Brown et al., 1987; Jolivet et al., 1990). Leaves of pea contain three types of SODs, namely, Mn-SOD and Cu, Zn-SOD in mitochondria (Sandalio et al., 1987; Luis et al., 1985). Hernandez et al. (1993) studied *Pisum sativum* L. during NaCl toxicity and revealed harmful effects on mitochondria. In *Triticum aestivum*

L. seedling, respiration was reduced with respect to seedling growth under high salt levels. On the other hand, correlation among seedling development characters and respiration illustrated that supplementary respiration results in more vigorous seedling under salt stress. Similar study indicated the variation in respiration among cultivars of *Triticum aestivum* L. seedling which maintains respiration at elevated levels than expected for C3 plants (Moud and Maghsoudi, 2008). Salt treated *Triticum aestivum* L. leaves showed elevated respiration pace and broad modifications in metabolite (Che Othman et al., 2020). Salt stress also influences enzymes of TCA cycle like pyruvate dehydrogenase, and two oxoglutarate dehydrogenase. Besides, the Gamma aminobutyric acid (GABA) shunt is also active under salt stress. The TCA cycle is physiochemically inhibited by salt. This may be to overcome the increased GABA shunt activity, as it promotes the alternative carbon supply for mitochondria in order to avoid salt and increase respiration in wheat leaves. This shunt is cardinal in stress conditions and helps in regulating pH, ROS, and bypassing some steps of TCA (Che Othman et al., 2017; Carillo, 2018).

11.5 MITOCHONDRIAL RESPIRATION AND TOLERANCE TO SALINITY STRESS

The accumulation of biomass in plants is affected by salinity and the limit of injury is directly proportional to balance among respiration and photosynthesis (Earl and Tollenaar, 1998). However, responses to respiratory rate under salinity in plant cells are complex, with 37% of studies suggesting an increase, 34% suggesting reduction, and 29% suggesting no regular variation (Jacoby et al., 2011). In short, respiration rate can be advantageous or disadvantageous to plant, depending on plant part and level of tolerance. The benefits include vigorous production of ATP for growth, defense, sodium exclusion, and osmotic adjustments (Yeo, 1983). On the other hand, the disadvantages include less synthesis of new parts and tissues, less growth, etc. (Poorter et al., 1990). Evidences suggest that respiratory homeostasis in shoot is related to salt tolerance (Kasai et al., 1998). In roots of rice plant, high rate of respiration increases tolerance to salt stress by the progress of ion exclusion (Malagoli et al., 2008). In various halophytes, under respiration in roots during salt stress is decreased and this is due to the fact they deploy carbon resources to shoot so that they can maintain tissue tolerance (Hwang et al.; Morris, 1994). Further studies on salt stress and carbon balance suggest that there is slight slowing of respiration and photosynthesis (Richardson and McCree, 1985; Schwarz and Gale, 1981). Similarly, lots of other mitochondrial actions are activated under saline conditions, such as variations in proteins and protein turnover is shown by transcriptomic and proteomic studies (Jacoby et al., 2011).

11.6 ELECTRON TRANSPORT CHAIN AND SALT STRESS

Various complex processes are involved in ETC for the production of ATP and in plants it is more sensitive than animals. Some salts like $MgCl_2$ and KCl stimulate ETC in vitro (Krab et al., 2000), but at higher concentrations, ETC is inhibited. The inhibition can be due to denaturation of proteins or disassemblage of cardinal complex systems (Flowers, 1974). Thus, salinity stress alters the pace at which ETC receives or provides the electrons and this effect differs between different ETCs. In seedlings of durum wheat and barley, respiration rate of mitochondria under salt was found to be lower when compared with control (Flagella et al., 2006; Jolivet et al., 1990). This study also suggested that in durum wheat, the ionic component of salinity increased toxic effect on ETC than osmotic component (Flagella et al., 2006). In another study, it was revealed that salt stress reduces the activity of succinate dehydrogenase, NADH dehydrogenase and heat shock protein (Hamilton and Heckathorn, 2001).

11.7 EFFECT OF SALT STRESS ON GLYCOLYSIS

Like all physiological processes, glycolysis pathway is also subjected to variations under stress (Anderson and Carol, 2004). It was revealed that reduction in the activities of rate-limiting steps would reduce the glycolysis process (Jang et al., 1997). In *Bruguiera sexangula*, it was revealed that the two enzymes, namely, phosphofructokinase and pyruvate kinase involved in glycolytic pathway increased with long term salt stress (Suzuki et al., 2005). This may be due to their contribution in increased glucose catabolism for energy supply. Conversely, less salt stress appears to activate the glycolytic step that converts fructose-6-phosphate to fructose-1,6-bisphosphate. In another study, it was revealed that pyruvate kinase, phosphofructokinase, and phosphoenolpyruvate kinase were appreciably declined with salt stress applications for 7 days, signifying metabolic chaos arising during stress (Zhong et al., 2016). Pyruvate links two pathways, that is, glycolysis and

TCA cycle and in barely, pyruvate was found to have major role in increasing salt tolerance (Wu et al., 2013; Jardine et al., 2010). The exogenous usage of putrescine raised the pyruvate quantity under salt stress, providing supplementary raw substances to TCA cycle which in turn protected plants under stress (Zhong et al., 2016; Fernie et al., 2004).

11.8 MITOCHONDRIAL STRESS AND EFFECT ON PROLINE

Salt stress results in considerable adjustment in the intensities of stress-mitigating osmolytes like antioxidant enzymes, betaine, proline, and glycine. In response to salt stress, proline content is highly increased (Mushtaq et al., 2021; Shah et al., 2020). In vitro studies showed that native configurations of proteins are maintained in the presence of higher proline levels (Arakawa and Timasheff, 1985). Proline is an osmoprotectant which helps in scavenging of ROS (Chen and Dickman, 2005). Salt stressed durum shoots showed reduced activity of proline dehydrogenase which means that downregulation of this enzyme is essential for increased levels of proline (Flagella et al., 2006). Some studies also suggested that proline acts as an alternative carbon source under stress (Verslues and Sharp, 1999). Furthermore, the synthesis of proline from glutamate consumes nicotinamide adenine dinucleotide phosphate (reduced) (NADPH). On the other hand, its breakdown discharges $FADH_2$ and NADH in the mitochondrion; as a result, this metabolic process helps in balancing the subcellular site of electron carriers (Atkin and Macherel, 2009).

TABLE 11.1 Regulation of Different Biochemical Agents Under Salt Stress.

Plant species	Enzyme/protein/ osmolyte	Regulation	Result	References
Seteria italica L.	Proline, glycine betaine and antioxidant enzymes	Upregulated	Scavenging of ROS, maintaining configurations of proteins	Shah et al. (2020)
Panicum miliaceum L.	Antioxidant enzymes	Upregulated	Scavenging ROS	Rasool et al. (2020)
Triticum durum Desf.	Proline dehydrogenase	Downregulated	Essential for increased levels of proline	Flagella et al. (2006)

TABLE 11.1 *(Continued)*

Plant species	Enzyme/protein/ osmolyte	Regulation	Result	References
Bruguiera sexangula (Lour.) Poir.	Phosphofructokinase and pyruvate kinase	Upregulated	Increased glucose catabolism for energy supply	Suzuki et al.,. (2005)
Cucumis sativus L. cv.	Phosphoenolpyruvate pyruvate kinase	Downregulated	Metabolic chaos	Zhong et al. (2016)
Zea mays L.	Heat shock protein	–	Protects Complex I	Hamilton and Heckathorn (2001)
Zea mays L.	Proline and Betaine	Upregulated	Protects Complex II	Hamilton and Heckathorn (2001)

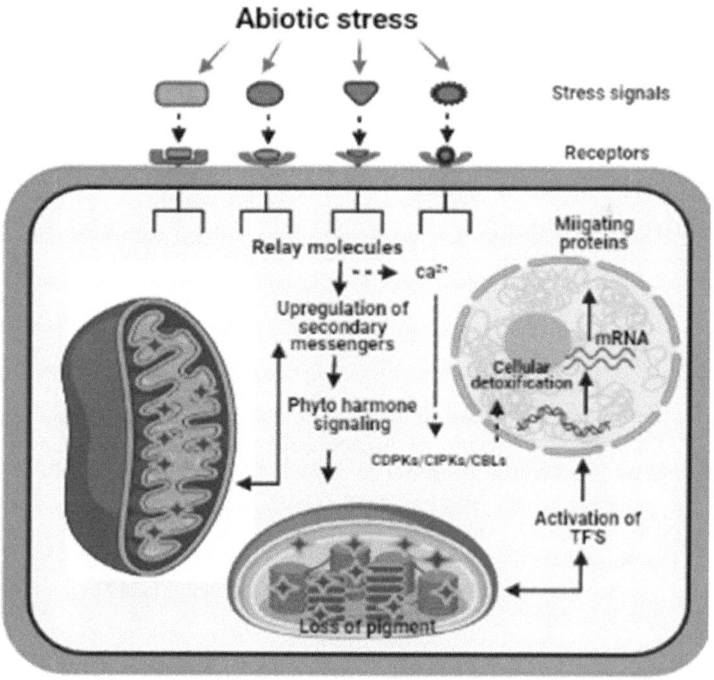

FIGURE 11.1 Communication of different cellular organelles for mitigation of abiotic stress. Red star indicates accumulation of ROS. CDPK (calcium-dependent protein kinase), CIPKs (CBL-interacting protein kinases), and CBLs (Calcineurin B-like proteins).

11.9 CONCLUSION

Mitochondria and mitochondrial respiration plays important role in management of abiotic stresses by ion exclusion, ionic and osmotic balance, production of ATP, and ROS scavenging. Through various studies, it is found that there is a positive correlation between salinity stress and mitochondrial respiration. The molecular mechanism, site of activation, and deactivation of various enzymes is still poorly defined. However, imaging technology, molecular mechanism studies and marker-based studies can reveal promising results. Studies focused on the role of proline as an alternate carbon source is promising and for plant physiologists. Further studies for measurements of respiration rate, and carbon balance are also vital for perception of abiotic stress response and would aid to clarify complex mechanisms for future screening for tolerance.

KEYWORDS

- **mitochondrial stress**
- **respiration**
- **environmental stress**
- **oxidative phosphorylation**
- **cellular antioxidants**

REFERENCES

Akram, M. Mini-Review on Glycolysis and Cancer. *J. Cancer Educ.* **2013,** *28* (3), 454–457.

Anderson, L. E.; Carol, A. A. Seven Enzymes of Carbon Metabolism, Including Three Calvin Cycle Isozymes, Are Present in the Secondary Cell Wall Thickenings of the Developing Xylem Tracheary Elements in Pea Leaves. *Int. J. Plant Sci.* **2004,** *165* (2), 243–256.

Arakawa, T.; Timasheff, S. N. The Stabilization of Proteins by Osmolytes. *Biophys. J.* **1985,** *47* (3), 411–414.

Atkin, O. K.; Macherel, D. The Crucial Role of Plant Mitochondria in Orchestrating Drought Tolerance. *Ann. Bot.* **2009,** *103* (4), 581–597.

Beyer, K.; Klingenberg, M. ADP/ATP Carrier Protein from Beef Heart Mitochondria Has High Amounts of Tightly Bound Cardiolipin, as Revealed by Phosphorus-31 Nuclear Magnetic Resonance. *Biochemistry* **1985,** *24* (15), 3821–3826.

Brown, S.; Day, D. A.; Critchley, C. Salt Tolerance -Does Leaf Respiration Have a Contribution to Make? In *Plant Mitochondria: Structural, Functional and Physiological*

Aspects; Moore, A. L.; Bechey, R. B., Eds.; Plenum Press: New York, 1987; pp 393–396. ISBN 0-306- 42572-6

Carillo, P. GABA Shunt in Durum Wheat. *Front. Plant Sci.* **2018,** *9*, 100.

Carvalho, M. D. Drought Stress and Reactive Oxygen Species. *Plant Signal. Behav.* **2008,** *3* (3), 156–165.

Chen, C.; Dickman, M. B. Proline Suppresses Apoptosis in the Fungal Pathogen *Colletotrichum trifolii*. *Proc. Natl. Acad. Sci.* **2005,** *102* (9), 3459–3464.

Che-Othman, M. H.; Jacoby, R. P.; Millar, A. H.; Taylor, N. L. Wheat Mitochondrial Respiration Shifts from the Tricarboxylic Acid Cycle to the GABA Shunt under Salt Stress. *New Phytol.* **2020,** *225* (3), 1166–1180.

Che-Othman, M. H.; Millar, A. H.; Taylor, N. L. Connecting Salt Stress Signalling Pathways with Salinity-Induced Changes in Mitochondrial Metabolic Processes in C 3 Plants. *Plant Cell Environ.* **2017,** *40* (12), 2875–2905.

Choudhury, F. K.; Rivero, R. M.; Blumwald, E.; Mittler, R. Reactive Oxygen Species, Abiotic Stress and Stress Combination. *Plant J.* **2017,** *90* (5), 856–867.

Dinakar, C.; Abhaypratap, V.; Yearla, S. R.; Raghavendra, A. S.; Padmasree, K. Importance of ROS and Antioxidant System during the Beneficial Interactions of Mitochondrial Metabolism with Photosynthetic Carbon Assimilation. *Planta* **2010,** *231* (2), 461–474.

Earl, H. J.; Tollenaar, M. Differences among Commercial Maize (*Zea mays* L.) Hybrids in Respiration Rates of Mature Leaves. *Field Crops Res.* **1998,** *59* (1), 9–19.

Fernie, A. R.; Carrari, F.; Sweetlove, L. J. Respiratory Metabolism: Glycolysis, the TCA Cycle and Mitochondrial Electron Transport. *Curr. Opin. Plant Biol.* **2004,** *7* (3), 254–261.

Flagella, Z.; Trono, D.; Pompa, M.; Di Fonzo, N.; Pastore, D. Seawater Stress Applied at Germination Affects Mitochondrial Function in Durum Wheat (*Triticum durum*) Early Seedlings. *Funct. Plant Biol.* **2006,** *33* (4), 357–366.

Flowers, T. J. Salt Tolerance in *Suaeda maritima* (L.) Dum: A Comparison of Mitochondria Isolated from Green Tissues of *suaeda* and *pisum*. *J. Exp. Bot.* **1974,** *25* (1), 101–110.

Gautheron, D. C. Mitochondrial Oxidative Phosphorylation and Respiratory Chain: Review. In *Organic Acidurias;* Addison, G. M., Chalmers, R. A., Divry, P., Harkness, R. A., Pollitt, R. J., Eds.; Springer: Dordrecht, 1984. https://doi.org/10.1007/978-94-009-5612-4_13.

Hamilton, E. W.; Heckathorn, S. A. Mitochondrial Adaptations to NaCl. Complex I Is Protected by Anti-Oxidants and Small Heat Shock Proteins, Whereas Complex II Is Protected by Proline and Betaine. *Plant Physiol.* **2001,** *126* (3), 1266–1274.

Hernandez, J. A.; Corpas, F. J.; Gomez, M.; del Rio, L. A.; Sevilla, F. Salt-Induced Oxidative Stress Mediated by Activated Oxygen Species in Pea Leaf Mitochondria. *Physiol. Plant.* **1993,** *89* (1), 103–110.

Horsefield, R.; Iwata, S.; Byrne, B.; Complex II from a Structural Perspective. *Curr Protein Pept Sci.* Apr **2004,** *5* (2), 107–118.

Huang, D.; Huo, J.; Liao, W. Hydrogen Sulfide: Roles in Plant Abiotic Stress Response and Crosstalk with Other Signals. *Plant Sci.* **2020,** 110733.

Hwang, Y. H.; Morris, J. T. Whole-Plant Gas Exchange Responses of *Spartina alterniflora* (Poaceae) to a Range of Constant and Transient Salinities. *Am. J. Bot.* **1994,** *81* (6), 659–665.

Ito, J.; Taylor, N. L.; Castleden, I.; Weckwerth, W.; Millar, A. H.; Heazlewood, J. L. A Survey of the *Arabidopsis thaliana* Mitochondrial Phosphoproteome. *Proteomics* **2009,** *9*, 4229–4240.

Jacoby, R. P.; Taylor, N. L.; Millar, A. H. The Role of Mitochondrial Respiration in Salinity Tolerance. *Trends Plant Sci.* **2011,** *16* (11), 614–623.

Jang, J. C.; León, P.; Zhou, L.; Sheen, J. Hexokinase as a Sugar Sensor in Higher Plants. *Plant Cell* **1997,** *9* (1), 5–19.

Jardine, K. J.; Sommer, E. D.; Saleska, S. R.; Huxman, T. E.; Harley, P. C.; Abrell, L. Gas Phase Measurements of Pyruvic Acid and Its Volatile Metabolites. *Environ. Sci. Technol.* **2010,** *44* (7), 2454–2460.

Jolivet, Y.; Pireaux, J. C.; Dizengremel, P. Changes in Properties of Barley Leaf Mitochondria Isolated from NaCl-Treated Plants. *Plant Physiol.* **1990,** *94* (2), 641–646.

.;.;Kasai, K.; Fukayama, H.; Uchida, N.; Mori, N.; Yasuda, T.; Oji, Y.; Nakamura, C. Salinity Tolerance in *Triticum aestivum*-Lophopyrum Elongatum Amphiploid and 5E Disomic Addition Line Evaluated by NaCl Effects on Photosynthesis and Respiration. *Cereal Res. Commun.* **1998,** *26* (3), 281–287.

Kleine, T.; Leister, D. Retrograde Signaling: Organelles Go Networking. *Biochim. Biophys. Acta* **2016,** *1857* (8), 1313–1325.

Krab, K.; Wagner, M. J.; Wagner, A. M.; Moller, I. M. Identification of the Site Where the Electron Transfer Chain of Plant Mitochondria Is Stimulated by Electrostatic Charge Screening. *Eur. J. Biochem.* **2000,** *267* (3), 869–876.

Kumar, P.; Dubey, K. K. Citric Acid Cycle Regulation: Back Bone for Secondary Metabolite Production. In *New and Future Developments in Microbial Biotechnology and Bioengineering*; Gupta, V. K., Pandey, A., Eds.; 2019; pp 165–181. ISBN 978044463**5044**. https://doi.org/10.1016/B978-0-444-63504-4.00013-X.

Lencina, A. M.; Franza, T.; Sullivan, M. J.; Ulett, G. C.; Ipe, D. S.; Gaudu, P.; Schurig-Briccio, L. A. Type 2 NADH Dehydrogenase Is the Only Point of Entry for Electrons into the *Streptococcus agalactiae* Respiratory Chain and Is a Potential Drug Target. *MBio* **2018,** *9* (4).; 1–15.

Lenzen, S. A Fresh View of Glycolysis and Glucokinase Regulation: History and Current Status. *J. Biol. Chem.* **2014,** *289* (18), 12189–12194.

Li, Z. G. Hydrogen Sulfide: A Multifunctional Gaseous Molecule in Plants. *Russian J. Plant Physiol.* **2013,** *60* (6), 733–740.

Logan, D. C. Mitochondrial Fusion, Division and Positioning in Plants. *Biochem. Soc. Trans.* **2010,** *38* (3), 789–795.

Luis, A.; Sandalio, L. M.; Yáñez, J.; Gómez, M. Induction of a Manganese-Containing Superoxide Dismutase in Leaves of *Pisum sativum* L. by High Nutrient Levels of Zinc and Manganese. *J. Inorg. Biochem.* **1985,** *24* (1), 25–34.

Malagoli, P.; Britto, D. T.; Schulze, L. M.; Kronzucker, H. J. (2008). Futile Na^+ Cycling at the Root Plasma Membrane in Rice (*Oryza sativa* L.): Kinetics, Energetics, and Relationship to Salinity Tolerance. *J. Exp. Bot.* **2008,** *59* (15), 4109–4117.

Mandavilli, B. S.; Santos, J. H.; Van Houten, B. Mitochondrial DNA Repair and Aging. *Mutat. Res./Fundamental Mol. Mech. Mutagen.* **2002,** *509* (1–2), 127–151.

Mittler, R.; Vanderauwera, S.; Gollery, M.; Van Breusegem, F. Reactive Oxygen Gene Network of Plants. *Trends Plant Sci.* **2004,** *9* (10), 490–498.

Moller, I. M. Plant Mitochondria and Oxidative Stress: Electron Transport, NADPH Turnover, and Metabolism of Reactive Oxygen Species. *Annu. Rev. Plant Biol.* **2001,** *52* (1), 561–591.

Moud, A. M.; Maghsoudi, K. Salt Stress Effects on Respiration and Growth of Germinated Seeds of Different Wheat (*Triticum aestivum* L.) Cultivars. *World J. Agric. Sci.* **2008,** *4* (3), 351–358.

Mueller, M. J.; Berger, S. Reactive Electrophilic Oxylipins: Pattern Recognition and Signalling. *Phytochemistry* **2009,** *70* (13–14), 1511–1521.

Mushtaq, N. U.; Saleem, S.; Rasool, A.; Shah, W. H.; Hakeem, K. R.; Rehman, R. U. Salt Stress Threshold in Millets: Perspective on Cultivation on Marginal Lands for Biomass. *Phyton* **2021,** *90* (1), 51.

Nath, S.; Villadsen, J. Oxidative Phosphorylation Revisited. *Biotechnol. Bioeng.* **2015,** *112* (3), 429–437.

Poorter, H.; Remkes, C.; Lambers, H. Carbon and Nitrogen Economy of 24 Wild Species Differing in Relative Growth Rate. *Plant Physiol.* **1990,** *94* (2), 621–627.

Preedy, V. R. (Ed.). *Neuropathology of Drug Addictions and Substance Misuse Volume 1: Foundations of Understanding, Tobacco, Alcohol, Cannabinoids and Opioids*; Academic Press, 2016.

Rasool, A.; Shah, W. H.; Tahir, I.; Alharby, H. F.; Hakeem, K. R. et al. Exogenous Application of Selenium (Se) Mitigates NaCl Stress in Proso and Foxtail Millets by Improving Their Growth, Physiology and Biochemical Parameters. *Acta Physiol. Plant.* **2020,** *42* (7), 1–13. DOI 10.1007/s11738-020-03109-w.

Richardson, S. G.; McCree, K. J. Carbon Balance and Water Relations of *sorghum* Exposed to Salt and Water Stress. *Plant Physiol.* **1985,** *79* (4), 1015–1020.

Sandalio, L. M.; Palma, J. M.; Del Rio, L. A. Localization of Manganese Superoxide Dismutase in Peroxisomes Isolated from *Pisum sativum* L. *Plant Sci.* **1987,** *51* (1), 1–8.

Schmidt-Rohr, K. Oxygen is the High-Energy Molecule Powering Complex Multicellular Life: Fundamental Corrections to Traditional Bioenergetics. *ACS Omega* **2020,** *5* (5), 2221–2233.

Schwarz, M.; Gale, J. Maintenance Respiration and Carbon Balance of Plants at Low Levels of Sodium Chloride Salinity. *J. Exp. Bot.* **1981,** 933–941.

Shah, W. H.; Rasool, A.; Tahir, I.; Rehman, R. U. Exogenously Applied Selenium (Se) Mitigates the Impact of Salt Stress in *Setaria italica* L. and *Panicum miliaceum* L. *Nucleus* **2020,** *63* (3), 327–339.

Sun, C.; Benlekbir, S.; Venkatakrishnan, P.; Wang, Y.; Hong, S.; Hosler, J.; & Gennis, R. B. Structure of the Alternative Complex III in a Supercomplex with Cytochrome Oxidase. *Nature* **2018,** *557* (7703), 123–126.

Suzuki, M.; Hashioka, A.; Mimura, T.; Ashihara, H. Salt Stress and Glycolytic Regulation in Suspension-Cultured Cells of the Mangrove Tree, *Bruguiera sexangula*. *Physiol. Plant.* **2005,** *123* (3), 246–253.

Suzuki, N.; Koussevitzky, S. H. A. I.; Mittler, R. O. N.; Miller, G. A. D. ROS and Redox Signalling in the Response of Plants to Abiotic Stress. *Plant Cell Environ.* **2012,** *35* (2), 259–270.

Takahashi, S.; Murata, N. How Do Environmental Stresses Accelerate Photoinhibition? *Trends Plant Sci.* **2008,** *13* (4), 178–182.

Van Dongen, J. T.; Gupta, K. J.; Ramírez-Aguilar, S. J.; Araújo, W. L.; Nunes-Nesi, A.; Fernie, A. R. Regulation of Respiration in Plants: A Role for Alternative Metabolic Pathways. *J. Plant Physiol.* **2011,** *168* (12), 1434–1443.

Verslues, P. E.; Sharp, R. E. Proline Accumulation in Maize (*Zea mays* L.) Primary Roots at Low Water Potentials. II. Metabolic Source of Increased Proline Deposition in the Elongation Zone. *Plant Physiol.* **1999,** *119* (4), 1349–1360.

Wiskich, J. T. Control of the Krebs Cycle. *Metabolism and Respiration*; Davies, D. D.; Academic Press, 1980; pp 243–278. ISBN 9780126754025. https://doi.org/10.1016/B978-0-12-675402-5.50012-X.

Wu, D.; Cai, S.; Chen, M.; Ye, L.; Chen, Z.; Zhang, H.; Zhang, G. Tissue Metabolic Responses to Salt Stress in Wild and Cultivated Barley. *PLoS One* **2013,** *8* (1), e55431.

Yeo, A. R. Salinity Resistance: Physiologies and Prices. *Physiol. Plant.* **1983,** *58* (2), 214–222.

Zhong, M.; Yuan, Y.; Shu, S.; Sun, J.; Guo, S.; Yuan, R.; Tang, Y. Effects of Exogenous Putrescine on Glycolysis and Krebs Cycle Metabolism in Cucumber Leaves Subjected to Salt Stress. *Plant Growth Regul.* **2016,** *79* (3), 319–330.

CHAPTER 12

Consequences of Oxidative Stress on Glycolysis and Krebs Cycle

KOLLURU VISWANATHA CHAITANYA[1] and CHALLA SUREKHA[2*]

[1]*Department of Microbiology and Food Science Technology, GITAM School of Science, GITAM Deemed to be University, Visakhapatnam-530045, India*

[2]*Department of Biochemistry and Bioinformatics, GITAM School of Science, GITAM Deemed to be University, Visakhapatnam-530045, India*

[*]*Corresponding author. E-mail: schalla@gitam.edu*

ABSTRACT

Respiration is a ubiquitous process involved in the oxidation of carbohydrates through glycolysis and Krebs's cycle, and is responsible for the emergence of life on this planet. Respiration is important for the acclimation of plants against different types of abiotic stress such as drought, salinity, and extreme temperature, which brings changes in the energy levels as well as energy requirements of plants, eventually inducing the changes in the respiration process. One of the adverse consequences of these changes is the production of lethal toxic reactive oxygen species (ROS), leading to oxidative stress in plants. Mitochondria are known to fulfill a number of functions in plants, serve as the main sources of ROS production with complex I and III of the respiratory chain as the main centers. Antioxidative defense system acts against the toxicity of the ROS and its intermediates either by direct quenching or by indirect prevention of free radical formation, protecting the plant cells and its vital organelle

Photosynthesis and Respiratory Cycles during Environmental Stress Response in Plants. Aryadeep Roychoudhury (Ed.)

from the oxidative damages. The range of oxidative damages incurred to the glycolysis and Krebs's cycle pathways and the antioxidant properties of the enzymes, substrates, and cofactors of these pathways protecting these metabolisms are discussed in this chapter.

12.1 INTRODUCTION

12.1.1 PLANT RESPIRATION AND MITOCHONDRIA

Plants are autotrophic organisms capable of using sunlight for splitting of water molecules and reduction of carbon dioxide and finally storing food reserve as insoluble polysaccharides or for their utilization in the synthesis of other compounds. Plants produce glucose as their main substrate for the respiration involving the oxidation of carbohydrates through two vital biochemical pathways, glycolysis, and tricarboxylic acid cycle. End products of these two pathways are carbon dioxide (CO_2) and reduced NAD(P) H_2 and $FADH_2$, used in the oxidative phosphorylation by transferring their electrons to the terminal oxidase, with oxygen as the final electron acceptor, generating high energy phosphate bonds in the form of ATP (Toro and Pinto, 2015). ATP is consumed most efficiently for obtaining energy, utilized in the synthesis of biomolecules, maintenance of cellular structures, uptake of ions, intra and intercellular transport of photosynthesis products, regulation of protein turnover, and assimilation of macro and micronutrients.

Respiration plays a pivotal role in the acclimation of plants to environmental stress especially in conferring tolerance to the abiotic stress factors including drought, flood, high and low temperature, and salinity. These stress factors have a huge impact on the energy requirements of plants, inducing changes in respiratory metabolism, electron transport, and redox gradient formation. Krebs cycle is the process liable for the development of life on Earth. Two types of Krebs cycle have been identified; (1) reduction Krebs cycle, which is of ancient type traced in acetogenic and methanogenic microbes, generates Krebs cycle intermediates through acetogenesis and methanogenesis methods. CO_2 fixation produces acetate, which is enzymatically transformed to oxaloacetic acid and pyruvic acid, vital elements required to generate building blocks of life; and (2) Oxidative Krebs cycle, evolutionarily modern process found to occur among aerobic organisms, and fixes CO_2 for ATP production (Ryan, 2015). Transformation of the Krebs cycle fixes CO_2 into an oxidative pathway from a reductive pathway, and removes electrons from carbon-emitting CO_2. This must have occurred after the Great Oxygenation incident, 2.2 billion years ago (Crowe et al., 2013). By the incorporation of oxygen into metabolism, modern organisms were able to preserve/conserve a lot of energy from oxidation of food. Machinery needed for the oxidative metabolism has been incorporated in different compartments of mitochondria, which has substantially conserved the energy (Siebels and Dröse, 2013). Plant mitochondria are involved in the metabolic processes that provide relief to the plants against abiotic stresses. Interaction of mitochondria with chloroplasts and peroxisomes will reduce the excess equivalents produced during photosynthesis (Kro¨mer, 1995). Mitochondria are also involved in the accumulation of proline during osmotic stress in plants as its oxidative enzyme, proline dehydrogenase is synthesized in the inner mitochondrial membrane (Kiyosue et al., 1996). Plant mitochondria are capable of sensing the cellular stress, much in advance, and regulate programmed cell death (Rhoads et al., 2006). Mitochondria may also play an important role in the interorganelle cross talk during environmental stress conditions and induces altered gene expression through mitochondrial retrograde regulation (Donato et al., 2007).

12.1.2 OXIDATIVE STRESS IN PLANTS

Plants, being sessile, have to cope with various abiotic stresses such as drought, salinity, extreme temperatures, etc. that may act on the plant systems internally or externally. The changes include retardation (Navabpour et al., 2003), loss of water potential, osmotic imbalance, ionic imbalance, and oxidative stress, leading to the generation of Reactive Oxygen Species (ROS) (Kumar et al., 2009a, 2009b; Khare et al., 2015). ROS such as hydrogen peroxide (H_2O_2), superoxide radical (O_2•-), and hydroxyl radicals (OH•), oxidize lipid membranes, and damage essential biomolecules such as nucleic acid (i.e., DNA and RNA), pigments, fatty acids, and proteins. This damage subsequently results in cell damage (Kumar et al., 2009a, 2009b; Khare et al., 2015). The chloroplasts being the prime source for the ROS production, significant amounts of superoxide radicals (O_2•-) and singlet oxygen ($^1O_2$) are formed during stress conditions (Fig. 12.1) (Zimmermann and Zentgraf, 2005).

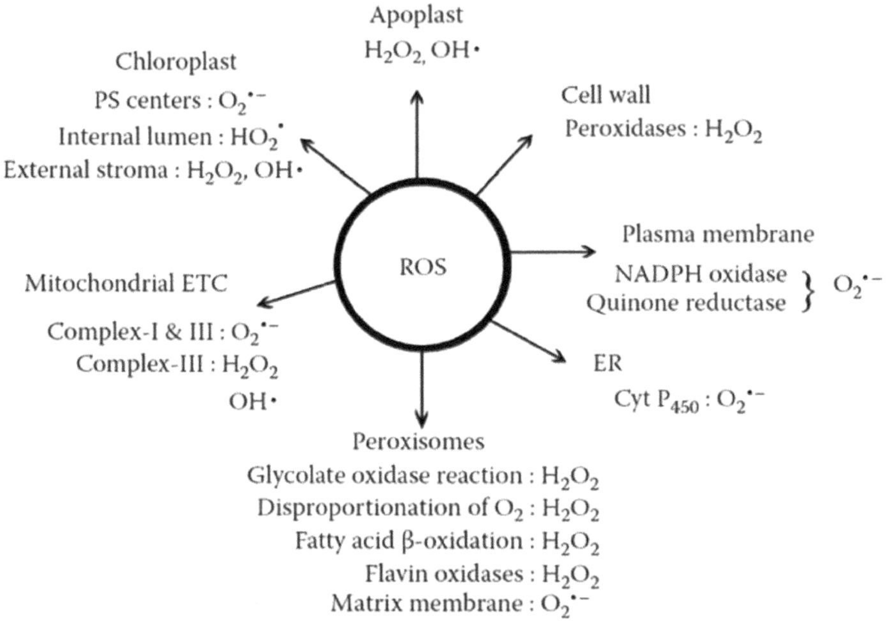

FIGURE 12.1 ROS and antioxidative defense system (Reddy and Surekha 2015; Book ISBN: 9780429161339).

ROS is a product of chloroplast and mitochondrial metabolism during external impact and proper protective mechanisms are to be included. Otherwise, these ROS can cause oxidative damage to other cellular components (Apel and Hirt 2004). Maintaining equipoise between ROS and antioxidative defense system determines the ultimate fate of a plant (Arora et al., 2002). Plants have developed various strategies like antioxidant machinery and activation of these cascades of complex signaling pathways to protect themselves (Jajic et al., 2015). Antioxidant machinery includes detoxification of ROS by superoxide dismutase, catalase, peroxidase, glutathione reductase, and ascorbate peroxidase (enzymatic antioxidants) and ascorbate, glutathione, flavonoids, and phenolic compounds (nonenzymatic antioxidants) (Khare et al., 2015).

12.2 ROS PRODUCTION IN MITOCHONDRIA

Mitochondrial electron chain transport complexes I and III are the most probable locations in mitochondria for ROS production (Reimer et al.,

2015). Electrons that are eliminated from the nutrients are transferred through respiratory complexes that reduce oxygen prematurely producing O_2^-. Generation of O_2^- radicals in mitochondria is the consequence of carbon metabolism that occurs in mitochondria. Flavins also contribute to the production of O_2^- known as flavin radicals at variable rates (Massy, 1994). Rate of ROS production in mitochondria is also accelerated by the protein micro-environment surrounding it. Blockage of electron transfer in the mitochondrial respiratory chain complex results in the reduction of respiratory complexes. It then hyperpolarizes the mitochondrial inner membrane subsequently increasing the ROS production (Yun and Finkel, 2014). Recent studies have shown that the Krebs cycle enzymes, 2-oxoglutarate dehydrogenase, succinate dehydrogenase, and pyruvate dehydrogenase also produce significant amounts of ROS (Fisher-Wellman et al., 2015). 2-oxoglutarate dehydrogenase and pyruvate dehydrogenase produce high concentrations of ROS while metabolizing the nutrients, which generate high amounts of NADH (Quinlan et al., 2014). Succinate dehydrogenase also generates ROS during succinate metabolism and reverses electron transfer from the Q pool (Quinlan et al., 2013). Metabolites such as proline, and glycerol-3-phosphate bypass mitochondrial respiratory chain complex I and supply electrons directly into the Q pool. Depending on the efficiency, these electrons are either transferred into the mitochondrial respiratory complex III or moved to succinate dehydrogenase generating ROS (Stein and Imai, 2012). Inevitable production of ROS by the respiratory complex I and III as major sites is a vital feature of the mitochondria under high matrix $NADH^+/NAD$ conditions. Increased membrane potentials will considerably reduce electron transport chain (ETC) components, increasing the possibility of single-electron leak to oxygen forming O_2^-. This O_2^- radical in turn will act as the substrate for the production of secondary ROS such as H_2O_2 and OH·.

12.3 CONSEQUENCES OF ROS PRODUCTION IN MITOCHONDRIA

Increased ROS levels in mitochondria lead to organellar damage. Peroxidation of polyunsaturated fatty acids in the mitochondrial membrane by the hydroxyl radicals leads to the formation of alkenals and hydroxyalkenals such as malondialdehyde and cytotoxic lipid aldehydes (Winger et al., 2005). Products of lipid peroxidation react with proteins and nucleic acids, and induce cellular damages. ROS also damage some important plant mitochondrial enzymes that are involved in ETC, membrane transport, protein folding, and assembly (Miller et al., 2001).

12.4 ROS DEFENSE BY PLANT MITOCHONDRIA

Mitochondria are receiving enormous interest due to their ability of defending against excess ROS. Proteomics studies indicate that mitochondria represent a major source of ROS production in plant cell with consequent damage. ROS production is particularly high under abiotic stress, especially during drought and salinity, suggesting the role of mitochondria in amelioration of ROS during environmental stress. Three strategies have been identified in mitochondria for defense against ROS. (1) avoiding the ROS production by retaining the ETC sufficiently oxidized; (2) detoxification of ROS; and (3) repairing the damages rendered by ROS. Among these three strategies, avoidance of ROS production by plant mitochondria has been studied by understanding the bioenergetics of mitochondria isolated from drought stress-resistant durum wheat, possessing three energy dissipating systems for efficient control of ROS production, namely, ATP-sensitive plant mitochondrial potassium channel, plant uncoupling protein, and alternative oxidase (AOX) (Pastore et al., 2003). Role of cold shock protein, CSP310 that acts as nonphosphorylating circumvent of the mitochondrial respiratory chain, has been elucidated in cereals by Kolesnichenko et al. (2005).

For the detoxification of ROS and maintaining the cellular redox homeostasis for controlling the oxidative damage, plant mitochondria possess an efficient antioxidative defense system comprising of enzymes such as manganese superoxide dismutase (MnSOD), enzymes involved in ascorbate glutathione cycle and enzymes of thioredoxins, peroxiredoxins, and sulfiredoxins (Trx/Prx/Srx). These enzymes along with antioxidant molecules such as ascorbate and glutathione regulate the level of ROS and modulate the redox signaling in plant mitochondria (Marti et al., 2012).

In plants, MnSOD is a tetrameric isoenzyme, whose expression is regulated by processes such as leaf senescence. This enzyme plays a significant role in protection of mitochondria against oxidative stress by preventing the formation of toxic H_2O_2 and OH· radicals. H_2O_2 formed in mitochondria is decomposed by activities of heme-containing ascorbate peroxidase and glutathione peroxidase, depending on the availability of the antioxidants, ascorbate, and glutathione. Oxidized ascorbate can be reduced by the FAD containing monodehydroascorbate reductase using NAD(P)H and dehydroascorbate reductase using GSH as electron donor. Glutathione oxidized during this process is reduced by the flavoprotein, glutathione reductase in a NADPH-dependent manner (Fig. 12.2). Antioxidant system in plant mitochondria is kept active by the adequate supply of NAD(P)H, maintained by the transhydrogenases, located on the mitochondrial membrane and isocitrate dehydrogenase and malate dehydrogenase, located in the mitochondrial matrix (Juan et al., 2013). Ascorbate peroxidase activity was found to be located outside of inner mitochondrial membrane, monodehydroascorbate reductase was found on the mitochondrial membrane, and DHAR and GR were found to be located in the mitochondrial matrix along with ascorbate and GSH (Shaobai et al., 2016). Location of APX and MDAR on membrane suggests a dual complementary function of these enzymes. They can reoxidize the endogenous NADH for meeting the constant supply of NAD^+ and protection of the organelle against the toxic effect of H_2O_2 (Chew et al., 2003).

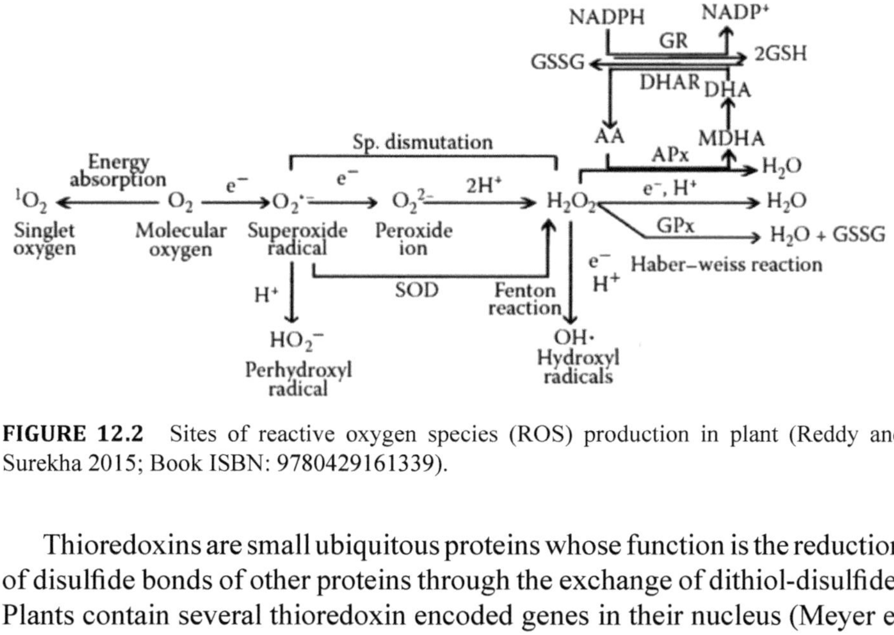

FIGURE 12.2 Sites of reactive oxygen species (ROS) production in plant (Reddy and Surekha 2015; Book ISBN: 9780429161339).

Thioredoxins are small ubiquitous proteins whose function is the reduction of disulfide bonds of other proteins through the exchange of dithiol-disulfide. Plants contain several thioredoxin encoded genes in their nucleus (Meyer et al., 2012). Plant mitochondrial thioredoxins were identified, and their function has been attributed to mitochondrial redox regulation and antioxidative defense system (Yoshida et al., 2013). Mitochondria thioredoxins perform antioxidative function by activating two enzymes, Prx11F and AOX (Marti et al., 2009). Peroxiredoxins are thiol-based peroxidases along with sulfiredoxins involved in the reduction of H_2O_2 and alkyl hydroperoxides to water and corresponding alcohol (Wood et al., 2003). Transgenic tobacco plants over- and under-expressing AOX enzyme has been linked with their role in regulating the level of ROS production in plant mitochondria (Maxwell et al., 1999). AOX enzyme has a unique energy dissipating property, which reduces the ROS regeneration in isolated plant mitochondria, cultured cells, and whole plant (Umbach et al., 2005).

12.5 CONCLUSION

In the past two decades, the role of antioxidative defense system in protecting the plant mitochondria, and avoiding of ROS production has been investigated by many research groups by applying classical biochemical analysis on isolated mitochondria, and utilizing sophisticated approaches such as gene expression analysis, proteomics analysis, mitochondrial genome analysis, isotope fractionation and in vivo microscopy on cultured cells and whole plants of Arabidopsis, tobacco, pea, potato, tomato, and cereals. Till date, most of the details remain unknown; particularly, the bioenergetics of mitochondria need to be studied in the plants adapted to abiotic stress. Such studies might provide first-hand information regarding the role of mitochondrial defense systems in relieving the plants against toxic ROS.

ACKNOWLEDGMENT

The authors are grateful to GITAM (Deemed to be University) for providing the necessary facilities to carry out review work and for extending constant support in writing this chapter.

KEYWORDS

- **oxidative stress**
- **reactive oxygen species**
- **respiration**
- **Kreb's cycle**
- **detoxification**

REFERENCES

Ajay Arora, R.; Sairam, K.; Srivastava, G. C. Oxidative Stress and Antioxidative System in Plants. *Curr. Sci.* **2002,** *82* (10), 1227–1238.

Apel, K.; Hirt, H. Reactive Oxygen Species: Metabolism, Oxidative Stress, and Signal Transduction. *Annu. Rev. Plant Biol.* **2004,** *55*, 373–399.
Chew, O.; Whelan, J.; Millar, A. H. Molecular Definition of the Ascorbate-Glutathione Cycle in Arabidopsis Mitochondria Reveals Dual Targeting of Antioxidant Defenses in Plants. *J. Biol. Chem.* **2003,** *278*, 46869–46877. DOI: 10.1074/jbc.M307525200.
Crowe S. A.; Døssing, L. N.; Beukes, N. J.; Bau, M.; Kruger, S. J.; Frei, R.; Canfield, D. E. Atmospheric Oxygenation Three Billion Years Ago. *Nature*. Sep 26, **2013,** *501* (7468), 535–538.
Fisher-Wellman, K. H.; Lin, C. T.; Ryan, T. E.; Reese, L. R.; Gilliam, L. A. et al. Pyruvate Dehydrogenase Complex and Nicotinamide Nucleotide Transhydrogenase Constitute an Energy Consuming Redox Circuit. *Biochem. J.* **2015,** *467*, 271–280.
Huang, S.; Van Aken, O.; Schwarzländer, M.; Belt, K.; Millar, A. H. The Roles of Mitochondrial Reactive Oxygen Species in Cellular Signaling and Stress Response in Plants. *Plant Physiol.* July **2016,** *171*, 1551–1559.
Jajic, I.; Sarna, T.; Strzalka, K. Senescence, Stress, and Reactive Oxygen Species. *Plants* **2015,** *4*, 393–411. https://doi. org/10.3390/plants4030393
Khare, T.; Kumar, V.; Kishor, P. B. K. Na+ and Cl- Ions Show Additive Effects under NaCl Stress on Induction of Oxidative Stress and the Responsive Antioxidative Defense in Rice. *Protoplasma* **2015,** *252*, 1149–1165. https://doi. org/10.1007/s00709-014-0749-2.
Kolesnichenko, V.; Grabelnych, O. I.; Pobezhimova, T. P.; Voinikov, V. K. Non-Phosphorylating Bypass of the Plant Mitochondrial Respiratory Chain by Stress Protein CSP 310. *Planta* **2005,** *221*, 113–122.
Kumar, N.; Pal, M.; Srivastava, G. C. Proline Metabolism in Senescing Rose Petals (*Rosa hybrida* L.—First Red). *J. Hortic. Sci. Biotechnol.* **2009a,** *84*, 536–540. https://doi.org/10. 1080/14620316.2009.11512562.
Kumar, V.; Shriram, V.; Nikam, T. D.; Jawali, N.; Shitole, M. G. Antioxidant Enzyme Activities and Protein Profiling under Salt Stress in Indica Rice Genotypes Differing in Salt Tolerance. *Arch. Agron. Soil Sci.* **2009b,** *55*, 379–394. https://doi.org/10.1080/03650340802595543
Lázaro, J. J.; Jiménez, A.; Camejo, D.; Iglesias-Baena, I.; del Carmen Martí, M.; Lázaro-Payo, A.; Barranco-Medina S.; Sevilla, F. Dissecting the Integrative Antioxidant and Redox Systems in Plant Mitochondria. Effect of Stress and *S*-Nitrosylation. *Front. Plant Sci.* **2013,** *4* (460). https://doi.org/10.3389/fpls.2013.00460
Mailloux, R. J. Still at the Center of It All; Novel Functions of the Oxidative Krebs Cycle. *Bioenergetics* **2015,** *4*, 122. DOI:10.4172/2167-7662.1000122.
Martí, M. C.; Florez-Sarasa, I.; Camejo, D.; Pallol, B.; Ortiz, A.; Ribas-Carbó, M. et al. Response of Mitochondrial Antioxidant System and Respiratory Pathways to Reactive Nitrogen Species in Pea Leaves. *Physiol. Plant.* **2012,** *147*, 194–206. DOI: 10.1111/j.1399-3054.2012.01654.
Massey, V. Activation of Molecular Oxygen by Flavins and Flavoproteins. *J. Biol. Chem.* **1994,** *269*, 22459–22462.
Maxwell, D. P.; Wang, Y.; McIntosh, L. The Alternative Oxidase Lowers Mitochondrial Reactive Oxygen Production in Plant Cells. *Proc. Natl. Acad. Sci. USA* **1999,** *96*, 8271–8276. DOI: 10.1073/pnas.96.14.8271.
Navabpour, S.; Morris, K.; Allen, R.; Harrison, E.; A-H-Mackerness, S.; Buchanan-Wollaston, V. Expression of Senescence-Enhanced Genes in Response to Oxidative Stress. *J. Exp. Bot.* **2003,** *54*, 2285–2292. https://doi.org/10.1093/ jxb/erg267.

Pastore, D.; Di Pede, S.; Passarella, S. Isolated Durum Wheat Mitochondria and Potato Cell Mitochondria Oxidize Externally Added NADH Mostly via the Malate/Oxaloacetate Shuttle with a Rate That Depends on the Carrier-Mediated Transport. *Plant Physiol.* **2003,** *133*, 2029–2039.

Quinlan, C.; Goncalves, R.; Hey-Mogensen, M.; Yadava, N.; Bunik, V. et al. The 2-Oxoacid Dehydrogenase Complexes in Mitochondria Can Produce Superoxide/Hydrogen Peroxide at Much Higher Rates Than Complex I. *J. Biol. Chem.* **2014,** *289*, 8312–8325.

Quinlan, C.; Perevoshchikova, I.; Hey-Mogensen, M.; Orr, A.; Brand, M. Sites of Reactive Oxygen Species Generation by Mitochondria Oxidizing Different Substrates. *Redox Biol.* **2013,** *1*, 304–312.

Riemer, J.; Schwarzländer, M.; Conrad, M.; Herrmann, J. M. Thiol Switches in Mitochondria: Operation and Physiological Relevance. *Biol. Chem.* **2015,** *396*, 465–482.

Siebels, I.; Dröse, S. Q-Site Inhibitor Induced ROS Production of Mitochondrial Complex II Is Attenuated by TCA Cycle Dicarboxylates. *Biochim. Biophys. Acta* **2013,** *1827*, 1156–1164.

Stein, L. R.; Imai, S. The Dynamic Regulation of NAD Metabolism in Mitochondria. *Trends Endocrinol. Metab.* **2012,** *23*, 420–428.

Umbach, A. L.; Fiorani, F.; Siedow, J. N. Characterization of Transformed Arabidopsis with Altered Alternative Oxidase Levels and Analysis of Effects on Reactive Oxygen Species in Tissue. *Plant Physiol.* **2005,** *139*, 1806–1820.

Winger, A. M.; Harvey Millar, A.; Day, D. A. Sensitivity of Plant Mitochondrial Terminal Oxidases to the Lipid Peroxidation Product 4-Hydroxy-2-Nonenal (HNE). *Biochem. J.* **2005,** *387*, 865–870.

Wood, Z. A.; Poole, L. B.; Karplus, P. A. Peroxiredoxin Evolution and the Regulation of Hydrogen Peroxide Signalling. *Science* **2003,** *300*, 650–653. DOI: 10.1126/science.1080405.

Yoshida, K.; Noguchi, K.; Motohashi, K.; Hisabori, T. Systematic Exploration of Thioredoxin Target Proteins in Plant Mitochondria. *Plant Cell Physiol.* **2013,** *54*, 875–892. DOI: 10.1093/pcp/pct037

Yun, J.; Finkel, T. Mitohormesis. *Cell Metab.* **2014,** *19* (5), 757–766.

Zimmermann, P.; Zentgraf, U. The Correlation between Oxidative Stress and LEAF. *Cell. Mol. Biol. Lett.* **2005,** *10*, 515–534.

CHAPTER 13

Regulation of Glycolysis and Krebs Cycle during Biotic and Abiotic Stresses

SOMALI DHAL AND HARSHATA PAL*

Amity Institute of Biotechnology, Amity University Kolkata, Major Arterial Road (South-East); Action Area II, Newtown, Kolkata 700135, West Bengal, India

**Corresponding author. E-mail: hpal@kol.amity.edu*

ABSTRACT

Plants are exposed to various environmental stresses due to their sessile nature. Stress at the molecular level results in reactive oxygen species (ROS) or active oxygen species (AOS) production, which are known to alter the physiological and biological processes in the plants. This chapter outlines the regulation of the important metabolic processes, that is, glycolysis and Krebs cycle that contributes to plant development and acclimation to environmental stresses. The changes made in the metabolic processes are an important survival strategy for the plants to overcome stress conditions. This can affect the activity of enzymes and might bring intricate changes in metabolite pools during carbon metabolism. Abiotic stress factors leading to oxidative stress conditions can inhibit or activate the enzymes that are involved in glycolysis and Krebs cycle. It results in the accumulation of specific compounds like raffinose, sucrose, proline, and γ-aminobutyric acid that are capable of defending cell membranes, which is a requirement for survival under stress conditions. An important enzyme of glycolysis, hexokinase, has been known to play a dual role by regulating the entry of glucose molecules into glycolysis and also the regulation of expression of sugar-responsive genes.

Photosynthesis and Respiratory Cycles during Environmental Stress Response in Plants. Aryadeep Roychoudhury (Ed.)

During biotic stress, the oxidative bursts limit the spread of the pathogen. Following infection, there are localized accumulations of carbohydrate and source-sink imbalance leading to the altered levels of metabolic enzymes in the plant. The secondary messenger cAMP furthermore plays an imperative role in eliciting plant responses to various stresses that can affect the abundance levels of enzymes involved in the glycolysis and Krebs cycle. The chapter finally highlights some of the important happenings during the interaction between the abiotic and biotic stresses that have both positive and negative effects on the plants.

13.1 INTRODUCTION

Plants, being sessile organisms, have to face various kinds of environmental stresses and thrive to survive the harsh conditions of nature. The stress factors affect plants and crops and have a devastating effect on their growth and yield through biochemical, morphological, physiological, and molecular variations. The fluctuating environment directed the evolution of highly flexible metabolism and development in the higher plants to cope with the changes (Fraire-Velázquez et al., 2011; Shao et al., 2008). Environmental stress can be classified into three categories: abiotic stress, biotic stress, and adverse climatic changes (Fig. 13.1). The foremost abiotic stresses that adversely affect plant growth and productivity are drought, cold, heat, and saline conditions. The principal cause of crop yield loss is abiotic stress, which decreases the usual yields of major crops by more than 50% leading to economic losses (Suzuki et al., 2014). Abiotic stress includes radiation, water availability, temperature, nutrients, and chemicals. They are also known to weaken plant defense systems and make the plants susceptible to biotic stress like animals, pathogen infection, and weeds. Besides abiotic and biotic stresses, plants also get affected by adverse and prolonged climatic changes. The stress factors can act in combination and hence the plants undergo multiple changes to neutralize the conditions, for example, a higher temperature can facilitate pathogen spread into the plant tissues (Bale et al., 2002; Luck et al., 2011). It has been found that during the stress conditions, the plant generates reactive oxygen species (ROS), reactive nitrogen species (RNS), and reactive sulfur species (RSS). ROS is defined as oxygen-containing reactive species and free radicals derived from molecular oxygen. It includes superoxide ($O_2^{\bullet-}$), singlet oxygen ($^1O_2$), hydrogen peroxide (H_2O_2), hydroxyl radical ($OH^{\bullet}$), hypochlorous acid (HOCl), and ozone (O_3) like molecules (Saminathan et al., 2016). Thus, plants have evolved stress tolerance mechanisms that lead to the induction and expression of the genes involved in the various respiratory and metabolic pathways. Such mechanisms include genes involved in encoding enzymes for scavenging ROS, genes involved in osmolyte synthesis, genes for molecular chaperons or Heat Shock Proteins (HSPs), genes encoding the late embryogenesis protein (LEA), genes encoding heterologous enzymes with different temperature optima, genes encoding transcription factors (TFs), and proteins involved in ion homeostasis (Fraire-Velázquez et al., 2011). ROS has intricate downstream effects on both primary and secondary metabolisms (Forrester et al., 2018; Huang et al., 2019). In addition, the processing of superoxide plasma membranes is one of the key ways the higher plants relay information about changing environments (Shao et al., 2008). The RNS is the result of the interaction of nitric oxide (NO) with O_2 and $O^{2\bullet-}$ (Molassiotis and Fotopoulos, 2011) and the RSS originated from the one-electron oxidation of hydrogen sulphide (H_2S) (Olson and Straub, 2015). The close interaction of RNS with ROS and RSS is an essential part of the inherent line of defense against the stress conditions (Ijaz et al., 2019; Savvides et al., 2016). Hence, RONSS (reactive oxygen, nitrogen, and sulfur species) can act as chemicals for priming schemes to develop cross-tolerance against abiotic stresses like salt, heat, drought, and chilling stress (Terrile et al., 2020). The three major metabolic pathways are glycolysis (in cytosol and plastids), the Krebs cycle /mitochondrial tricarboxylic acid (mTCA), and the mitochondrial electron transport chain (mETC). The changes made in the metabolic processes are an important survival strategy for the plants to overcome the environmental fluctuations. This can affect the activity of enzymes of the Krebs cycle as well as glycolysis and might induce complex variations in metabolite pools in carbon metabolism (Dumont and Rivoal, 2019). This chapter will discuss the varied impacts of stress on the plant glycolytic pathway and the mTCA cycle. Glycolysis or Embden-Meyerhof pathway is the formation of pyruvate from glucose or hexose molecules. The NADH (reduced form of NAD^+) generated in glycolysis is used to fuel mitochondrial ATP synthesis through oxidative phosphorylation (Armengaud et al., 2009) .Furthermore, the pyruvate molecules are oxidized and decarboxylated to acetyl-CoA which may enter in the Krebs cycle to reconstitute NAD^+ (nicotinamide adenine dinucleotide) (Fig. 13.2). TCA cycle is the ultimate pathway of generation of primary metabolites and oxidative breakdown of amino acids, carbohydrates, fatty acids. It constitutes a series of reactions that oxidizes the acetyl group of acetyl-CoA to produce CO_2 in such a way that conserves the free energy for consumption during ATP generation (Nolfi-Donegan et al., 2020). Increased carbon stream from glycolysis contributes to increased production of NADH, $FADH_2$, and eventually ATP. Oxidation of a sucrose molecule through glycolysis produces four ATP molecules, four pyruvate molecules, and four NADH molecules, all of which are included in the TCA cycle (Munns et al., 2020). The cytosolic glycolysis and mTCA cycle are the adaptive pathways whose regulation and enzyme activity change along with varied environmental stresses. The regulatory activity of glycolysis depends upon the three important enzymes, that is, phosphofructokinase (PFK), hexokinase (HK), and pyruvate kinase (PK). Likewise, Krebs cycle is typically regulated by the enzymatic action of citrate synthase (CS), isocitrate dehydrogenase (IDH), and α-ketoglutarate dehydrogenase (KDH). The biotic and abiotic agents that induce the formation of ROS can be fatal for the plant cells if present in excess amounts. Normally, ROS are produced as a by-product of mETC and detoxified sequentially and are useful as signaling agents and for the production of various phytohormones (Gautam et al., 2017). The link between the primary metabolism and the environmental stress factors has been elucidated mostly through transciptomic and metabolomic analysis (Mäkinen and De, 2019). Thus, knowledge regarding the biochemical and metabolic pathways by which plants withstand environmental stress is important for genetic engineering to enhance plant yield under stress. Through studying plants under stress, we will learn about the plasticity of metabolic pathways and the shortcomings of their working mechanisms.

13.2 ENVIRONMENTAL STRESS AND GLYCOLYTIC ENZYME GAPDH

Glyceraldehyde-3-phosphate dehydrogenase (GAPDH) is a housekeeping protein with conserved sequence and a significant enzyme of the glycolytic pathway where it reversibly converts glyceraldehyde-3-phosphate (GAP) combined with its cofactor NAD^+ and phosphate into 1,3-bisphosphoglycerate (1,3-BPG) and NADH (Yang and Zhai, 2017). Plants usually have three different types of *GAPDH* genes termed *GapA*, *GapB*, and *GapC*. The genes *GapA* and *GapB* encode for A and B subunits of chloroplastic GAPDH, that is, GAP A and GAP B, whereas *GapC* encodes C subunit of cytosolic GAPDH, that is, GAP C. All these proteins are encoded in the nucleus of plant cells (Schneider et al., 2018). GAPDH have a crucial function in plant defense mechanisms during abiotic and biotic stress conditions. In *Arabidopsis*, four different isoforms of GAPDH are found (Table 13.1) (Zaffagnini et al., 2013).

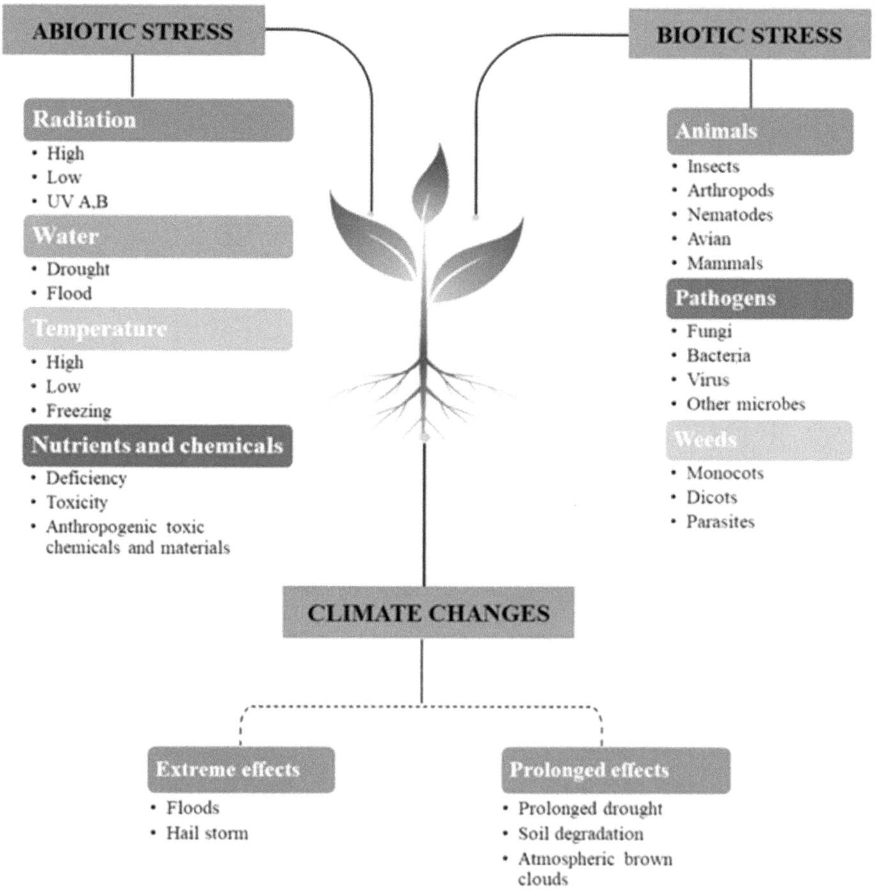

FIGURE 13.1 Different types of stresses faced by the plants due to their sessile nature.
Source: Adapted from Maheswari et al. (2012).

Overall reaction of Glycolysis:

$$Glucose(6C) + 2ADP + 2NAD^{+} + 2P_i \rightarrow 2Pyruvate(3C) + 2ATP + 2NADH + 2H^{+} + 2H_2O$$

Overall reaction of Krebs cycle:

$$Acetyl - CoA(2C) + 3NAD^{+} + FAD + GDP + 2H_2O + P_i \rightarrow CoA + 3NADH + FADH_2 + GTP + 2CO_2 + 3H^{+}$$

FIGURE 13.2 Net reaction of glycolysis and Krebs/TCA cycle.

TABLE 13.1 List of GAPDH Isoforms That Are Found in *Arabidopsis*.

Subfamily	Isoforms of GAPDHs	Phosphorylation	Proteins	Functions during stress	References
I	Chloroplastic and photosynthetic	Phosphorylating	GAPA1	Negatively regulates autophagy	Zaffagnini et al. (2013)
			GAPA2, GAPB	Shields the formation of free radicals	Zaffagnini et al. (2013)
II	Cytosolic and glycolytic	Phosphorylating	GAPC1	Negatively regulates autophagy, functions in Calvin cycle, and responds to bacterial flagellin	Zaffagnini et al. (2013)
			GAPC2	Functions in Calvin cycle	
III	Plastidic and glycolytic	Phosphorylating	GAPCp1, GAPCp2	An important role in drought and H_2O_2 stress	Li et al. (2019)
IV	NADP-dependent cytosolic	Non-Phosphorylating	NP-GAPDH	Involved in abiotic stress responses	Hýsková et al. (2017)

GAPA1, GAPA2, and GAPB utilize NADPH to produce nicotinamide adenine dinucleotide phosphate ($NADP^+$) that shields the formation of free radicals from the electron transport chain (ETC) through the dissipation of proton gradient at the thylakoid membrane of chloroplasts (Baalmann et al., 1995). This results in the maintenance of NAD^+/NADH or $NADP^+$/NADPH ratio in the cells. Oxidative stress can induce posttranslational modifications in GAPDH contributing to new GAPDH protein associations and regulation of primary metabolism in plants. "GAPDHs as pro-survival molecules in plants, negatively regulate cell death in response to pathogen challenge" (Henry et al., 2015). During the response to *Pseudomonas syringae* infection, the activity of GAPDH increases and it modulates the accumulation of ROS and cell demise. The ultimate product of glycolysis, pyruvate, is a scavenger of ROS, and hence increase in the glycolytic activities contributes to cell survival in wild plants during stress (Colell et al., 2007; Krawczyk et al., 2010). Additionally, the plant defense mechanism is an energy-consuming process, and the glycolytic pathway produces ATP. As a result, an increase in GAPDH activity during plant–pathogen interaction provides additional energy for overall cell reconfiguration for protection (Henry et al., 2015). The significance of GAPDH and its isoforms is seen in the potato plant, where GapC is regulated by biotic stress and elicitor action. A virulent strain of *Phytophthora infestans* elevates the expression of the GAPDH encoding gene in leaves. Even during abiotic stress (heat stress), many mRNAs encoding glycolytic enzymes get accumulated in the cells. Thus, the increase in the transcription rate becomes helpful when ROS might inactivate the activity of GAPDH, and there will be a sufficient supply of GAPDH enzymes to maintain the glycolysis level and respiration rate (Laxalt et al., 1996).

13.3 BIOTIC STRESS AND PRIMARY METABOLISM

"Plants are locked in an evolutionary arms race with their attackers, and faced with this onslaught have evolved myriad defenses. Once an invasion is recognised, plant metabolism must possibly compete in the defensive requirements against dynamic maintenance, growth, and reproduction requirements" (Bilgin et al., 2010). Plants scuffle with various biotic stresses triggered by living organisms like fungi, viruses, bacteria, nematodes, harmful insects, and parasites (Lata et al., 2018). They can cause numerous diseases, toxicities, and damage to plants and ultimately affect the yield and productivity. Thus, the plants have developed various mechanisms of regulation to combat these threats and get adapted to survive such conditions. Along with important roles in plant growth, the primary metabolic pathways are induced by several pathogen infections and can provide the energy and carbon skeleton required during the defense mechanism. The plant mitochondria also have a significant role during biotic stress responses. The signal molecules or pathways initiated by plant–pathogen interactions could directly or indirectly target mitochondrial enzymes and processes. The key effect of this targeting is the early deterioration of mitochondrial homeostasis leading to an increased generation of ROS in mitochondria (mROS) (Amirsadeghi et al., 2007). Biotic stresses contribute to the synthesis of superoxide ion ($O_2^{\bullet -}$) on plasma membranes (Nafees et al., 2019) and H_2O_2 during disease resistance in the plants (Huang et al., 2019); for example, insect vectors for bacteria and viruses can cause significant physiological damages in plants through allelopathy. The generation of ROS in plants has detrimental effects on metabolism, growth, and development and hence considered phytotoxic (Choudhury and Panda, 2004). A phenolic compound called salicylic acid (SA) is produced by plants as a signaling molecule during biotic stress resistance (Pál et al., 2020); for example, the endogenous levels of SA in fruits show an elevation during *Pseudomonas syringae* (Gram-negative bacteria) infection in the phloem sap of muskmelon, squash, and watermelon. It induces the development of metabolites for the protection of plants, among other phenolic and antioxidant compounds. Researchers have also testified that the exogenous administration of SA can enhance or induce the processes like glycolysis and Krebs cycle (Malamy and Klessig, 1992). Let us look upon the various types of pathogenic infections that induce stress and hence affect the primary metabolic processes.

13.3.1 VIRAL INFECTIONS

Plant viruses mostly have single-stranded RNA genome [e.g., Tobacco mosaic virus (TMV), tomato spotted wilt virus, Potato virus Y (PVY), etc.] (Roossinck, 2015), except a few that have double-strand DNA genome [e.g., Cauliflower mosaic virus (CaMV)] (Bak and Emerson, 2020). It is difficult to portray an unambiguous generalized image of the metabolic responses caused by plant viruses because of the uneven existence of viral infections in plants, various virus–host interactions, and the reversible and complex nature of metabolic pathways. However, recent theories and experiments have elucidated the influence of viruses on energy-associated processes like glycolysis, Krebs cycle, ETC, etc. During the infection period, plants show increased levels of carbohydrates and amino acids that fuel the Krebs cycle to keep up the high respiration rates (two- or three-fold) (Laxalt et al., 1996). The organic acids that are intermediates in the Krebs cycle display affirmative responses to viruses in different hosts (Llave, 2016). Infection with virulent/avirulent strains or pathogen-derived elicitors upregulated certain transcripts that are involved in energy productions such as the glycolysis and Krebs cycle (Rojas et al., 2014). It supports the theory that primary metabolism has been reconfigured or regulated to meet the energy requirements for plant safety and, possibly, for viral replication during infections. The expression of heat shock factors (HSF) during biotic stress also plays a role in defense mechanisms (von Koskull-Döring et al., 2007). The disease tolerance system of *Arabidopsis* plants is negotiated when viral infection (Turnip mosaic virus) is paired with abiotic stress (heat and drought), which is simultaneous with an increase in cytoplasmic protein response. This contributes to a favorable downregulation of transcripts and metabolites associated with glycolysis and the Krebs cycle. During these combined stress conditions, sucrose and hexose level increases and get accumulated in virus-infected plants (Prasch and Sonnewald, 2013). "*Systemic virus infection in plants relies on complex interactions between the invading virus and the host plant on the molecular level, which disrupts the normal biological processes of the host*" (Babu et al., 2008). Upregulation of genes related to the processes such as glycolysis, respiration, and starch accretion can be linked with the chlorosis formation on the leaf tissues of cucumber infected with the Cucumber mosaic virus (CMV) (Babu et al., 2008). Interaction of CMV and the cotyledons upsurges the enzymatic activity of ATP-dependent PFK, NAD-dependent malic enzyme, PEP carboxylase, NAD-dependent isocitrate dehydrogenase, and fumarate hydratase and a decline in the activity of enzymes such as GADPH and RuBisCO (Tecsi et al., 1996). A member of Potyvirus named potato virus Y infects the tobacco plants which results in dramatic changes and nonuniform perturbation in carbon metabolism and carbohydrate content of leaves. The reduced sugar export leads to the accumulation of soluble sugar such as sucrose, glucose, and fructose in the leaf tissues. Conversely, the steady-state levels of glycolytic and Krebs cycle intermediates (3-phosphoglycerate, phosphoenolpyruvate, citrate, malate, etc.), the hexose phosphates, and starch remain the same with no observable changes (Herbers et al., 2000). It is believed that the plant response during virus infection can be uneven concerning to various virus–plant interactions. The infection of maize plants with *rice black-streaked dwarf virus* supports the theory. During the infection period, GAPDH and fructose-1,6-bisphosphate aldolase show notable variations between virus-infected and normal plants (Li et al., 2011). The amalgamation of drought stress and viral infestation downregulates the genes involved in glycolysis, Krebs cycle, and other energy-producing pathways (Pandey et al., 2015).

13.3.2 FUNGAL INFECTIONS

Fungi are microscopic organisms whose association with the plant hosts can be symbiotic, mutualistic, parasitic, or endophytic depending on the survival strategies. The plant–fungi interactions are beneficial when it promotes plant growth and development, but it also can be harmful leading to survival stress and diseases in plants (Rai and Agarkar, 2016; Zeilinger et al., 2016). The fungal diseases include anthracnose, sclerotinia rots, botrytis rots, etc. and the common disease symptoms are leaf and stem rust, powdery mildew, etc. (Grau et al., 2004; Singh et al., 2019). The survival strategy of plants to resist the fungal infection includes the stimulation of physical and biochemical pathways such as the generation of ROS and changes in the primary metabolic pathways (Doughari, 2015). Plant respiration genes, such as glycolytic enzyme encoding genes, TCA cycle, ETC, are all upregulated in incompatible and compatible poplar-rust (*Melampsora* rust fungi) interactions. Among the regulatory enzymes of glycolysis, PFK is upregulated during the multi-pathogen resistance gene (Lr-34) expression in response to *Puccinia triticina* (rust fungus) infestation in wheat (Bolton et al., 2008). A similar kind of upregulation is also seen during infection caused by *Rhizoctonia solani* (sheath blight fungus) in genetically modified rice (Danson et al., 2000). The enhancement in respiration and mitochondrial activity can also be seen in the case of fungal elicitors (cryptogein from *Phytophthora cryptogea*) (Ashtamker et al., 2007) or fungal toxins (victorin from *Cochliobolus victoriae*) (Yao et al., 2002). The alternative respiratory pathways play a major role in biotic stress responses and it presumably enables the high flux of glycolysis and Krebs cycle when oxidative respiration is saturated (Doughari, 2015; Vanlerberghe, 2013). Major changes occur in the Krebs cycle of the Sorghum plant when it is infected with a necrotrophic fungus, *Bipolaris sorghicola*, that causes target leaf spot. The pathogen infiltration elicits the glyoxylate shunt, thus altering the function of TCA cycle from energy generation to the production of cell components. Genes of malate synthase and isocitrate lyase were shown to be strongly differentially expressed. These enzymes are active in the TCA cycle shunt pathway, the glyoxylate cycle. Succinate is used immediately in the TCA cycle, while glyoxylate comes into play in the TCA cycle preceeding malate synthase that converts it to malate (Fig. 13.3). The TCA cycle shunt pathway allows for improved carbon compound generation by bypassing the CO_2-generating stages of the TCA cycle and leads to the production of cell constituents including cell-wall polysaccharides, amino acids, and nucleotides (Mizuno et al., 2012).

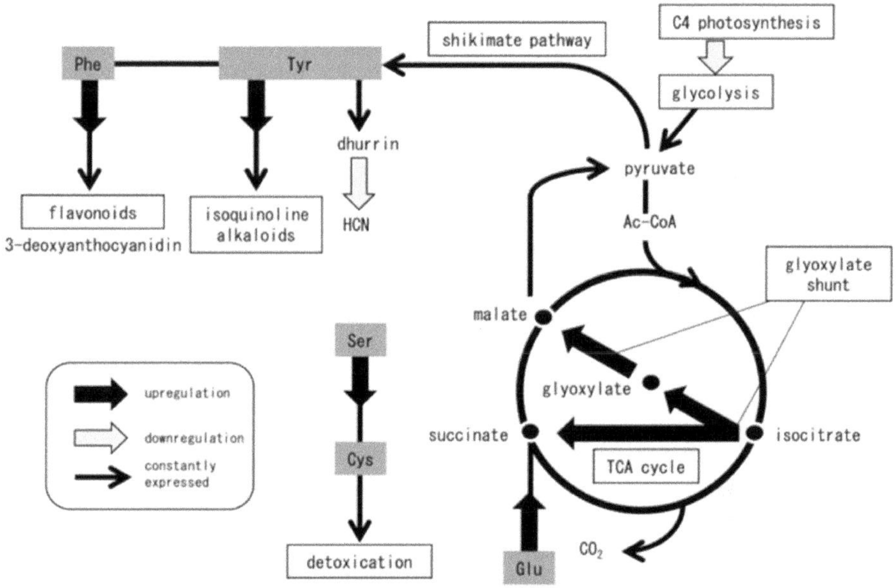

FIGURE 13.3 The figure represents the upregulation or downregulation of genes as well as the role of the glyoxylate shunt and amino acid metabolism during defense response. The TCA cycle's glyoxylate shunt bypasses the stages that produce CO_2. Afterward, tyrosine (Tyr) and phenylalanine (Phe) are produced through the shikimate pathway. Cys serves as a precursor for various sulfur-dependent detoxication.

Abbreviations: Cysteine (Cys), Glutamate (Glu), Hydrocyanic acid (HCN), Phenylalanine (Phe), Serine (Ser), Tyrosine (Tyr).

Source: Reprinted from Mizuno et al., 2012. Open access. https://creativecommons.org/licenses/by/2.0/

The Clubroot disease in the family Brassicaceae occurs due to a fungus-like protist called *Plasmodiophora brassicae*. The disease induces the formation of gall in the roots. Initially, there is a feeble upregulation of glycolysis and the Krebs cycle. In the later stages, when the galls are visible, the sugar and starch metabolism is upregulated steadily (Ludwig-Müller et al., 2009). Sheath blight disease of rice is caused by *Rhizoctonia solani Kuhn* which is a pathogenic fungus. The genetic resistance of the plants against *R. Solani* is quite remarkable. During the infection period, the activity of all glycolytic enzymes increases leading to an increase in the metabolite content in the leaves of rice plants. The metabolite content includes fructose-6-phosphate (F-6-P), fructose-1,6-bisphosphate (F-1,6-P2), dihydroxyacetone phosphate (DHAP), glyceraldehyde-3-phosphate (GAP), 3-phosphoglycerate (3-PG), phosphoenolpyruvate (PEP), and ultimately pyruvate and the enzymes include phosphoglucomutase (PGM), GAPDH, enolase, and PK. These changes suggest an uneven circulation of carbon amidst the upper and lower halves of glycolysis and the reactions carried out by these enzymes are nonequilibrium reactions in vivo (Mutuku and Nose, 2012). *Ustilago maydis* is a fungal pathogen that induces bulky tumors in maize plants and surprisingly downregulates the primary metabolism in general. The biotrophic relationship between the fungus and the host increases sucrose degradation in the infected leaves and ultimately transcriptional stimulation of glycolysis and Krebs cycle (Doehlemann et al., 2008). Postharvest pathogens such as *Penicillium digitatum* also cause changes in the primary metabolic pathways that involve the activation of ROS and pathogen-related (PR) genes (Cheng et al., 2020).

13.3.3 NEMATODE INFECTIONS

Plant-parasitic nematodes (PPNs) are the worms that feed and depend on the plant components and are common in agricultural soils. Among the most regulated pathways during nematode infestation are galactose and sucrose metabolism, CoA biosynthesis, amino acid metabolism, and phenylalanine metabolism. Infestation of tomato plant by parasitic roundworm, *Meloidogyne incognita*, also called root-knot nematode leads to the upregulation of soluble sugars like glucose, fructose, sucrose, and ribose in stems. On the other hand, melibiose gets consumed by an enzyme glycosyl hydrolase to yield glucose. The higher accumulation of the sugars leads to upregulation of glycolysis and thus Krebs cycle to maintain the sugar levels in the plant tissues. The primary metabolic pathways lead to the production of essential metabolites that eventually help in defense responses (Eloh et al., 2016). The nutritional necessities of nematode during infestation include structural and metabolic adaptations. In this situation, to allow effective plant cell metabolism, it is important to control the osmotic pressure of plant cells. The nematode, *M. incognita*, also infests the roots of *Medicago truncatula* plants and induces gall formations. The galls contain both nematode and plant tissues. As a result of the metabolic research, neither partner's metabolism can be distinguished. The metabolites of the Krebs cycle, citrate, as well as malate get accumulated in the vacuoles of cells that probably regulate the osmotic pressure (Baldacci-Cresp et al., 2015). *Heterodera schachtii*, a plant parasitic-nematode, induces the formation of syncytial feeding structures in *Arabidopsis* roots and creates robust sink tissues that are metabolically much active. A steady increase in glycolytic metabolites can be found during nematode development. This can be attributed to the substantial upregulation of the utmost of the glycolysis-related genes, for instance, At1g12900. Gene At1g12900 codes for an enzyme GAPDH, which is strongly upregulated and have oxidoreductase activity. The levels of most organic acids also get elevated in the succeeding Krebs cycle due to the upregulation of the genes like At5g14590 that encodes isocitrate dehydrogenase. This suggests a compensatory retort to the requirements for a high metabolic flux through this pathway (Hofmann et al., 2010). Another common plant nematode, *Hirschmanniella oryzae*, infests the flooded rice crops by penetrating the root cells, creating cavities resulting in root necrosis and cell death. Unlike others, *H. oryzae* invasion curbed the glycolysis and Krebs cycle as well as ATP synthesis (Kyndt et al., 2014).

13.3.4 BACTERIAL INFECTIONS

Bacterial infections in plants lead to the formation of galls, leaf spots, scabs, soft rots, spikes, tumors, and wilts. Unlike viruses within host cells, walled bacteria expand in spaces between cells and do not attack them (Leonard et al., 2017). The interaction between plant and bacteria leads to the production of soluble organic molecules, their utilization by the pathogen, and elevated metabolic activities. The diseased plant shows reduced photosynthesis, increased host catabolic activities, and decreased carbohydrate content (Nayudu, 1969). Reactive species such as $O_2^{\bullet-}$, H_2O_2, NO, and peroxynitrite are the signaling molecules to induce defense responses during the interactions. Bacterial pathogens can significantly compromise the mitochondrial function during the stress response of the plants. During environmental stress, aconitase, a Krebs cycle enzyme, gets impaired by the increased activity of monofluoroacetate which is an inhibitor of metabolic pathways

(Vanlerberghe, 2013). Studies on *Arabidopsis* plants have concluded that biotic stress can alter the levels of TCA/Krebs cycle intermediates, thus fulfilling the enhanced demand for carbon skeleton and it is understood that transcriptional variations caused by TCAs are unique to each metabolite (Finkemeier et al., 2013). The metabolites will therefore act as signaling molecules and clearly show that the TCA components have a favorable relationship to plant defenses. Induction of priming by citrate and fumarate can be detected for the period of infestation of Arabidopsis with *Pseudomonas syringae pv. Tomato* (Pst) DC3000 strain. Both citrate and fumarate are major constituents of the Krebs cycle. During the early infection, citrate induces SA and camalexin accretion trailed by an enhancement of the jasmonic acid (JA) pathway. Thus, changes in TCA richness can improve the *Arabidopsis* defense mechanism in stress situations (Balmer et al., 2018). When cucumber plants are exposed to *Pseudomonas syringae pv lachrymans* infection, several changes take place in the Krebs cycle. In the shoot and leaves, transient downregulation of phosphoenolpyruvate carboxylase (PEPC) leads to an increase in the metabolite content of the Krebs cycle (in particular citrate). In the roots, the NADP-malic enzyme has increased activity that leads to an increase in citrate and isocitrate dehydrogenase; however, the concentration of malic acid gets decreased (Chojak-Koźniewska et al., 2018). The application of combined stress (Psl and salt stress) in the cucumber plant leads to an upregulation of malate which employs the rhizobacterium, *Bacillus subtilis* for extending plant safety from disease (Rudrappa et al., 2008). A pathovar of *P. syringae* called *P. syringae pv. phaseolicola* (Pph) causes halo blight systemic disease in *Phaseolus vulgaris* (common bean) plants and triggers changes in the primary as well as secondary metabolism. Virulent pathogen *Phytophthora infestans* and pathogen-derived elicitors (Flg22, HrpZ) can bring about the expressions of transcription factors and metabolic enzymes in *Arabidopsis* (Less et al., 2011). Genome-wide studies in *Arabidopsis* and avirulent bacterial strain of *P. syringae pv.* reveal that hexokinase 1 can induce *PR* genes such as *PR-1* and *PR-5*. Conversely, the negative regulatory role of hexokinase1 is seen in *Nicotiana benthamiana* where HXK1 is downregulated by virus-induced gene silencing leading to the accumulation of H_2O_2 in the plant followed by downregulation of pyruvate decarboxylase. Thus, it can be comprehended that HXK1 has a dual role in defense responses (Fig. 13.4) (Rojas et al., 2014).

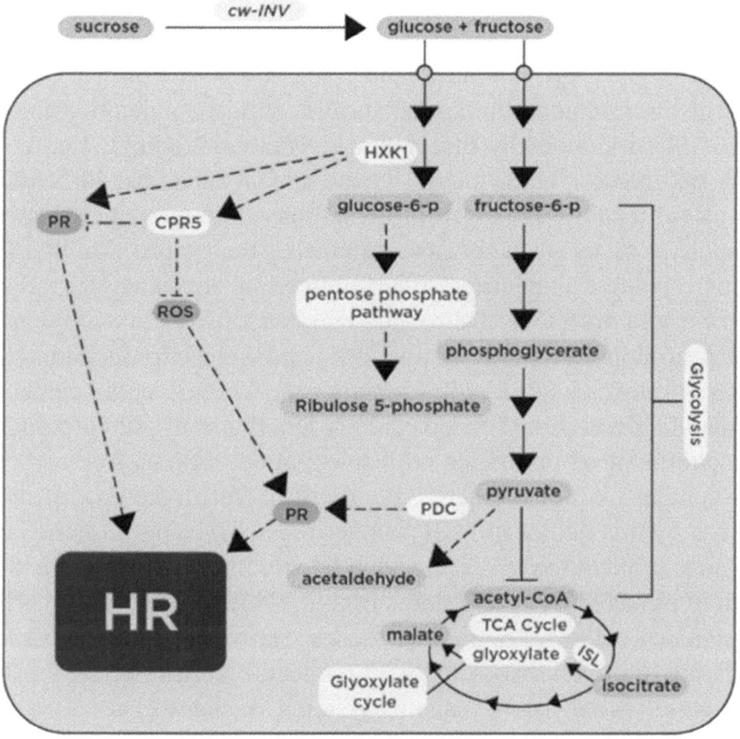

FIGURE 13.4 The figure depicts that induction of carbohydrate metabolic pathways upon plant exposure to avirulent pathogens and pathogen-derived elicitors subsidizes to positive as well as negative regulation of defense responses and resistances. The metabolic pathways such as glycolysis, pentose phosphate pathway, and TCA/Krebs cycle are shown in yellow, indicating upregulation during the presence of avirulent pathogens or pathogen-derived elicitors. Proteins including cell wall invertase (CwINV), hexokinase (HXK), constitutive expresser of *PR* genes 5 (CPR5), and pyruvate decarboxylase (PDC) are highlighted as components proposed to play a positive (arrows) or negative (blunt-ended lines) role in regulating known defense responses (shown in pink) such as the generation of ROS and the activation of pathogenesis-related (PR) proteins leading to the hypersensitive response (HR). ISL, isocitrate lyase. Dashed lines represent hypothetical regulations.

Source: Reprinted from Rojas et al., 2014. Open access. https://creativecommons.org/licenses/by/3.0/

13.3.5 INSECT HERBIVORY

Insect herbivory is characterized as the damage caused by a group of plant-feeding insects, which includes noneruptive and eruptive species with populations at "normal" low densities (Kozlov, 2015; Kozlov and Zvereva,

2017). Insect herbivory alters plant primary metabolism by remobilizing carbon and nitrogen supplies, increasing or decreasing photosynthetic ability, and changing plant growth rate (Zhou et al., 2015). The movement of current and freshly assimilated primary metabolites amidst condemned and systemic tissues is often induced by insect attack. Treatment/administration with the volatile defensive signal, methyl jasmonate, mechanical obliteration, and actual insect feeding overall upsurge carbohydrate movement toward treatment spots in many plant species, supporting the proposition that import of systemic resources encourages plant protection, for example, radioactively labeled carbohydrates were introduced into sink leaves in methyl jasmonate-treated Arabidopsis in hours of treatment and were introduced into defense-related complexes such as (*E*)-3-phenylprop-2-enoic acid (cinnamic acid) and phenolic glycosides (Ferrieri et al., 2012). Plants usually provide nitrogen as premanufactured resistance molecules, but carbon as free carbohydrates for the synthesis of local defence metabolites. Herbivore production may be hampered by this reallocation of primary metabolites because the nutritional importance of plant tissue is reduced. Carbohydrate reallocation is thought to be mainly powered by tissue sink strength, which is profoundly affected by the actions of cell wall-bound invertases. Herbivore-induced carbohydrate degradation has been associated with reduced invertase activity in the same tissues, which is consistent with this system (Machado et al., 2013; Robert et al., 2014). Root herbivory has a distinct impact on development, photosynthesis, and primary metabolism trends. Root herbivory can minimize nutrient and water uptake by reducing root biomass or disrupting water and nutrient hydraulics, exhaust resources stored belowground by the plant, enforce water deficits that reduce photosynthesis rates, and allow photoassimilates to be redirected belowground for root regrowth and repair (Fig. 13.5). This may result in the emergence of different stress conditions in terms of water, nutrients, and other necessities. Photoassimilates are frequently translocated to the roots for storage, particularly following incidents of shoot herbivory. However, limited evidence is available regarding the movement of primary metabolites in the opposite direction in reaction to root herbivory. In this context, Robert et al. (2014) demonstrated that root herbivores-infested maize plants assigned carbon to the stems as a precursor to root regrowth. Following root attack, nitrogen was diverted away from roots and into the shoots of knapweed and the stems of milkweed. Root herbivores, on the other hand, can manipulate host primary metabolites, such as carbon and phosphorus, belowground to develop host plant quality (Johnson et al., 2016).

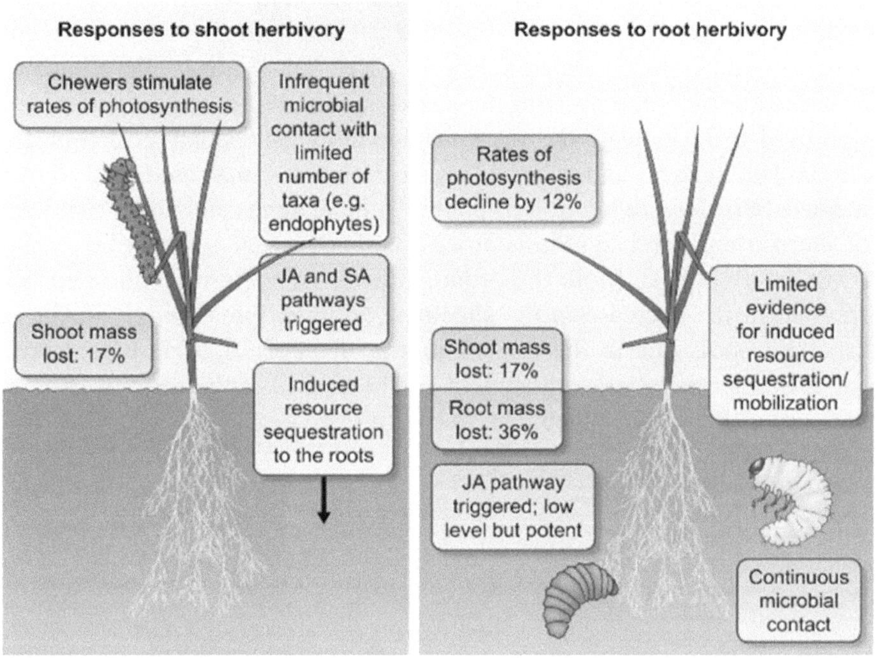

FIGURE 13.5 Relevant and common differences in plant responses to shoot and root herbivory. JA, jasmonic acid; SA, salicylic acid; Global changes in percentage relate to figures testified by Zvereva and Kozlov (2012).
Source: Reprinted with permission from Johnson et al., 2016. © John Wiley.

13.4 ABIOTIC STRESS AND PRIMARY METABOLISM

"A number of conditions, including low and high temperature, inadequate or excess water, high salinity, heavy metals and UV light, are unfriendly to plants. These stressors, called collectively abiotic stresses, present a serious danger to farming and the ecosystem, which leads to significant loss in agricultural yields." (He et al., 2018). The abiotic stresses lead to reactive species damage, protein denaturation, membrane injury, and osmotic stress in the plants. The botanic defenses against these stresses include the production of compatible solutes, reactive species scavengers, and activation of signaling molecules such as H_2O_2, NO, Ca^{2+}, abscisic acid (ABA), polyamines (PAs), and H_2S. Phytochrome B (PHYB) cross talks with the triggered signaling molecules at various stages to initiate a precise directive for the TFs to modify gene expression and activities of protein/enzyme in a definite pattern, thus generating multistress tolerance (He et al., 2018). The compatible solutes are the low molecular weight osmoprotectants including

amino acids, carbohydrates, γ-amino-butyric acid (GABA), amines, and polyols. In response to osmotic stress, carbohydrates such as sucrose, fructose, and trehalose accumulate in cells. Thus, metabolic adjustments are necessary to cope up with the stress conditions. When methyl viologen was used to investigate the relationships amongst oxidative tolerance, the antioxidant system, and cross-tolerance responses in *Arabidopsis thaliana*, it caused the downregulation of photosynthetic genes and allied termination of sucrose and starch synthesis process and activation of catabolic pathways (Dalal et al., 2014). More importantly, the TCA cycle was inhibited which resulted in the decrease in the succinate, malate, fumarate, glutamate, and aspartate pools due to the unavailability of the TCA cycle-linked precursors. Another metabolic adjustment has been seen in the form of re-routing the carbon flux from glycolysis to the oxidative pentose phosphate (OPP) pathway (Fig. 13.6).

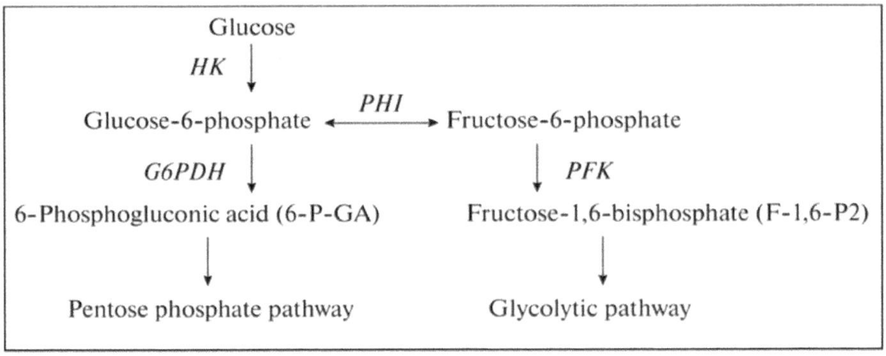

FIGURE 13.6 Glucose-6-phosphate can be irreversibly catalyzed by G6PDH to 6-phosphogluconic acid to enter the pentose phosphate pathway, or can be catalyzed by phosphohexose isomerase to fructose-6-phosphate and then irreversibly catalyzed by PFK to fructose-1,6-bisphosphate to enter the glycolytic pathway.
Source: Reprinted with permission from Yao and Wu (2016). © Springer Nature.

Metabolites of glycolytic pools such as G-6-P and F-6-P decreased and the intermediates of the OPP pathway such as ribulose-5-phosphate and ribose-5-phosphate increased. Any kind of stress factor leading to oxidative stress in plants decreases the reducing power by decreasing the level of reductants. To supply the reducing power, the carbon flux changes to the OPP pathway. This kind of change has been observed in the rice cell cultures that were treated with menadione. The sugar phosphates resulting from the glycolysis process and the TCA cycle organic acids depleted while the OPP intermediates increased (Ishikawa et al., 2010). These variations in the primary metabolism can scavenge the ROS, preserve the osmotic balance, and protect the cellular structure by contributing to the protective functioning of the osmoprotectants. Ultimately, the osmoprotectants produce stress responses to dehydration, salinity, heavy metal stress, drought, and freezing temperatures (Fraire-Velázquez and Balderas-Hernández, 2013). Let us know about the various types of abiotic factors that induce stress and how it affects the primary metabolic (glycolysis and TCA cycle) processes.

13.4.1 RADIATION

Radiations have various effects on the vital physiological processes, including various biosynthetic processes, respiration, photosynthesis, hormonal signaling, and long-distance transport. Ionizing radiations (IR) are capable of inducing reactions, leading to the formation of ROS and hence cause changes in the functionality of plants. As ROS lead to stress conditions in plants, changes in the gene expression levels (transcription and translation) will also be observed to cope up with the stress and bring about stress responses. Both chronic and acute irradiation increase the ROS level and therefore can affect the varied range of physiological processes through genetic and physiological regulation (Caplin and Willey, 2018; Gudkov et al., 2019). Gamma and ion beam irradiation has also been shown to affect the gene expression associated with sugar and starch metabolism in plants (Hwang et al., 2014). Excess exposure to ultraviolet-B radiation (UV-B) leads to the inhibition of photosynthesis as well as the formation of ROS. UV-B induced ROS is scavenged with the assistance of the antioxidant defense systems that include certain enzymes such as ascorbic acid peroxidase, catalase, dehydroascorbate reductase, glutathione reductase, peroxidase, superoxide dismutase; non-enzymatic antioxidant systems include ascorbic acid, α-tocopherol, carotenoids, reduced glutathione, etc. (Kataria et al., 2014). The effect of gamma (γ) radiations on the glucose metabolism and organic acids of the TCA cycle has been studied on the carrot tissue. γ-radiations increased the absorption of glucose and pyruvate; however, the acetate and succinate absorption decreased. The anabolic utilization of all substrates was reduced by the radiation (Bourke and Massey Jr, 1967). Despite the negative effects of radiations, certain studies also proved that radiations of diverse wavelengths can be beneficial for plants by regulating carbohydrate metabolism. Experiments conducted by Liu and coauthors proposed that the soluble sugar present in lettuce could be enhanced by reducing the yellow light fraction and adding an adequate amount of purple light in the plant cultivation light (Liu et al., 2018).

13.4.2 TEMPERATURE

Climate changes leading to variations in temperature can be detrimental for crop plants. High night temperature (HNT) can bring about changes in the pool sizes of intermediates of glycolysis and TCA cycle in crops such as rice. As the intermediate compounds of glycolysis and TCA cycle give rise to different amino acids, stress conditions also lead to the accumulation of those amino acids. This increase in amino acid production can be related to the increase in the glycolysis, giving rise to intermediates that form different amino acids. HNT causes an increase in metabolic flux across the TCA cycle and into biosynthetic pathways branching out from the TCA cycle at oxaloacetic acid and 2-oxoglutaric acid in sensitive rice cultivars (Glaubitz et al., 2015). Succinate dehydrogenase (SDH) is an important enzyme of the Krebs/TCA cycle that is accountable for the conversion of succinate to fumarate. Exposure of Cold Atmospheric Pressure Plasma (CAPP) on germinating seeds of soybean is effective in a positive way effect by increasing the SDH activity and swapping of the seed metabolism from anoxygenic to the oxygenic pathway. Conversely, the longer exposure periods of CAPP inhibited the SDH activity, but stimulated the activities of other enzymes such as lactate and alcohol dehydrogenase activities. The involvement of lactate dehydrogenase suggests anoxygenic metabolism as it converts the end of the product of glycolysis, that is, pyruvate to lactate (Švubová et al., 2021). Latest study on *Arabidopsis* revealed the fluctuations and responses of primary metabolism concerning low night temperature (20°C) in natural light conditions (Annunziata et al., 2018). The researcher worked using three different conditions, that is, controlled light and temperature (LT), naturally variable light and controlled temperature ($L^{VAR}T$), and naturally variable light and variable temperature ($L^{VAR}T^{VAR}$). When the sucrose levels were measured, it was found that sucrose was somewhat higher in $L^{VAR}T^{VAR}$ than in the temperature-controlled treatments, late at night, much higher in the first hours after midnight, and somewhat higher later in the light season, with the variance vanishing at dusk. The phosphorylated intermediates such as F-1,6-P2 and glucose 1,6-bisphosphate, sucrose-6-phosphate, and the TCA cycle intermediates (except isocitrate) increased in $L^{VAR}T^{VAR}$ than in the temperature-controlled systems. Slow accumulation of starch in the light, amid high levels of phosphorylated metabolites and sugars may be an indirect result of the cool night; it is understood that starch accumulation is slower when starch is not completely exhausted in the preceding 24 h period (Mengin et al., 2017). Low temperature, on the other hand, is known to move C allocation away from starch and toward aggregation of sucrose and other cryoprotectants. The influence of temperature change on N metabolism is greater than that of C metabolism. The natural light and variable temperature regimes had higher amino acid levels than the temperature-controlled sinusoidal simulation and, in particular, the temperature-controlled natural light system. It can sometimes be because of the C and N metabolism that are imbalanced due to the harsh effects of warm nights in crop yield as the buffering effect of low night temperature is eliminated. In rice, however, high night temperatures (28°C) are linked with increase in organic acids and amino acids (Glaubitz et al., 2015; Glaubitz et al., 2017). This may be due to variations in habitats or reactions to high and low temperatures. In general conditions (particularly in field), variations in irradiance are usually convoyed by changes in temperature, hence modifying the growth responses. Low temperatures, for instance, decrease maintenance reserve usage and growth during the night (Pilkington et al., 2015).

13.4.3 WATER

Water stress is typically caused by insufficient or excessive water activity in the plant surroundings. The former condition leads to drought or water deficit stress and the latter causes flood stress (Lewitt, 1980). Drought affects plants by reducing the soil water potential, decreases the pressure potential, disturbs membrane integrity, disturbs the physiological process (including photosynthesis, respiration, metabolism, etc.), and ultimately diminishes the plant growth. The major and unavoidable consequences of drought stress are an increase in the production of ROS in mitochondria and chloroplasts. In case of severe water-deficit conditions, the metabolism is disturbed completely leading to plant death (Kaur and Asthir, 2017). The enzymes involved in glycolysis altered differently in cultivars of plants depending on the severity and extent of water-deficit conditions. The specific activities of glycolytic enzymes aldolase, phosphoglycerate kinase (PGK), and NADP linked GADPH decreases with an increase in the severity of drought stress in plants such as chickpea (Khanna et al., 2014), sunflower (Castillejo et al., 2008), and bent grass (Xu and Huang, 2010). In pigeon pea (Singal et al., 1985) and *Sporobolus stapifianus* (Whittaker et al., 2004), a reduction in aldolase activity has been observed during the water stress period. This rule, in contrast, does not apply to plants such as oak (Jorge et al., 2006), poplar (Bogeat-Triboulot et al., 2007), wheat (Ford et al., 2011), and *Sporobolus stapifianus* (Oliver et al., 2011), where the glycolytic enzymes including PGK and aldolase accumulate during drought stress. Enzymes including PGM, glucose-6-phosphate isomerase (GPI), and triosephosphate isomerase (TPI) increase their activity during the stress conditions in chickpea. However, this is not true for the GPI present in mosses where the enzymatic activity of GPI does not get affected during water stress (Stewart and Lee, 1972). The regulatory enzymes of glycolyis, PFK, and hexokinase do not exhibit major changes in chickpea during water-deficit conditions. This proves that during drought, the plants need to maintain the glycolysis process to supply adequate ATP required for the stress responses. In some cases, inhibition of glycolysis can be a mechanism to accumulate sugars that can serve as the energy source for the retrieval and enhanced growth of the plants when the water is made available. From the cited arguments, we can conclude that the enzymes of glycolysis behave differently in diverse plants/crops under drought stress. In essence, the genes associated with glycolysis are upregulated, downregulated, or remain unaffected under drought stress (Khanna et al., 2014).

Flooding or excess water in the surrounding can have negative effects on the plants. Flooding lessens the diffusion of gases and thus reduces the availability of O_2 and CO_2 within plant tissues (Mustroph, 2018). Stress conditions get magnified in plants when the flood water completely submerges the shoots and the roots because it prevents the influx of the carbon dioxide required for photosynthesis. Under these conditions, the plant species begin to synthesize anaerobic proteins that allow O_2-independent energy-producing metabolism to continue where fermentable substrates are accessible. As the oxygen level decreases, the soluble sugars are hastily channeled to fermentative metabolism. The hormonal signaling of the sugars becomes active as the amounts of the soluble sugars are limited. However, few plant species are well adapted to the submergence conditions and they can survive these conditions at growing temperatures (Visser et al., 2003). Let us take an example of a soybean plant and see how the primary metabolism gets affected by flooding (Wang et al., 2018). Glycolysis, fermentation, and TCA cycle are the major pathways of primary metabolism in soybean that gets affected by flood. The soybean plants under flooded conditions show suppressed glycolysis and TCA cycle; however, the fermentation is enhanced. During the initial stress conditions, regulation of glycolytic enzymes such as hexokinase and PFK have a major role in flood tolerance. Later, during the endurance stages, enolase becomes responsible for the balance of glycolysis. As we have already discussed regading the importance of the GAPDH enzyme in stress tolerance, soybean plants also mediate flooding tolerance by regulating the glycolytic pathway. The level of GAPDH usually reduces during abiotic stress; however, in drought-tolerant plants, higher GAPDH activity is required to facilitate the rate of glycolysis. Besides soybean, wheat and mushroom also show increased GAPDH activity to mediate the tolerance mechanism. Due to the accumulation of the pyruvate in the flooded plants of soybean, the fermentation process gets activated due to the unavailability of oxygenic conditions. This happens to compensate for the ATP production by availing NAD^+ which help in the conversion of pyruvate to acetyl-CoA. Here onward, the regulation of the TCA cycle becomes necessary to tolerate the stress conditions. The suppression of the TCA cycle is common in wild-type plants during flooding conditions and mitochondrial enzymes such as aconitase and 2-oxoglutarate dehydrogenase decreases under oxidative stress. These reductions impose flux limitations on the TCA cycle as well as ETC. To avoid metabolite accumulation, enzyme modifications become necessary in flood-tolerant plants. Therefore, aconitase and 2-oxoglutarate dehydrogenase (also known as α-ketoglutarate dehydrogenase) are the two effective enzymes, which, if modified, are responsible for decreasing the metabolite accretion in flooded plants. Aconitase regulates the citrate metabolism and its modification during stress conditions helps in avoiding the accumulation of citrate. 2-oxoglutarate dehydrogenase is an early target during oxidative stress and its inhibition can reduce the respiration rate (Sweetlove et al., 2002). It is an important enzyme concerning the carbon–nitrogen balance ratio (Araújo et al., 2008). Thus, increasing the level of 2-oxoglutarate dehydrogenase will help in directing the flux toward carbohydrate metabolism (Fig. 13.7) (Wang et al., 2018).

In grapevine roots, the waterlogging situations also result in substantial reorganization of primary metabolism through the improvement of the glycolytic and fermentative enzymes and of a subset of enzymes partaking in the TCA cycle (Ruperti et al., 2019). During the hypoxia, G-6-P gets channelized into glycolysis through the upregulation of genes encoding hexokinase and epimerase. On the other hand, downregulation of PGM reduces the flux through gluconeogenesis and the absence of transcripts coding ATP-dependent 6-PFK1 leads to a stable induction of PPi-dependent PFK. The involvement of the PPi-dependent process instead of ATP suggests that an energy-saving metabolic alteration occurs under low oxygen conditions, while safeguarding the recycling of reduced NADH. The rest of the induced genes in the glycolytic process helps in increasing the pyruvate formation which forms a basis for the fermentation process. During waterlogging, the hypoxic conditions redirect sugar metabolism away from the TCA cycle, which is inhibited and reversed, toward an increased glycolytic flux. Studies have reported that reversion of the TCA cycle results in the accumulation of succinate in hypoxic situations and the accumulation continues as the gene encoding the α-subunit of succinyl-CoA synthetase activates. As a remark, we can say that during drought and flood/waterlogging, primary metabolism gets affected and the plants try to cope up with the situation by regulating the genes encoding enzymes involved in glycolysis and TCA cycle and manages to save/produce energy that might be required during the recovery period.

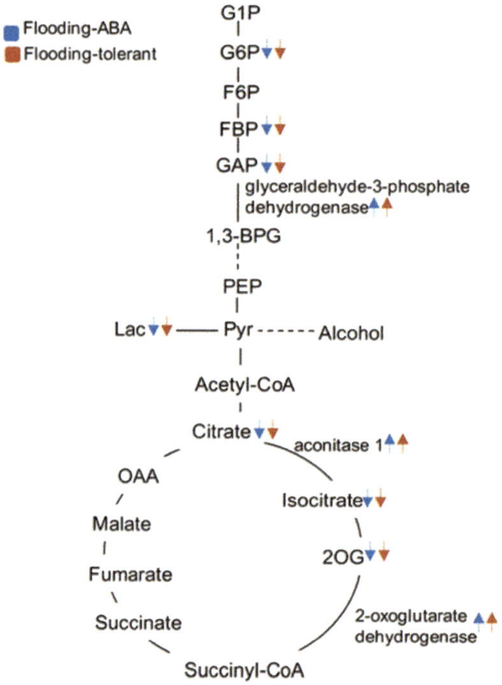

FIGURE 13.7 Schematic summary of essential physiological processes such as glycolysis, fermentation, and tricarboxylic acid cycle that gets affected during flood circumstances in soybean roots. Blue and orange colors present flooded soybean with ABA treatment and flood-tolerant plant, respectively. Protein abundance of aconitase 1, 2-oxoglutarate dehydrogenase, and glyceraldehyde-3-phosphate dehydrogenase, and metabolite accumulation were presented in the ABA-treated soybean and flood-tolerant plant. Upward and downward arrows designate increase and decrease, respectively, in metabolite accretion in ABA-treated soybean and flood-tolerant plant, compared with wide-type soybean under flooding (Wang et al., 2018).

13.4.4 NUTRIENT AND CHEMICAL TOXICITY

Mineral nutrients are an indispensable requirement for the growth and development of plants by playing a vital role at the cellular and molecular levels.

The plants show diverse mineral composition and are influenced by various factors including plant age, genetic makeup, the chemical composition of the soil, and nonetheless climactic conditions. For proper growth of plants, minerals and nutrients are always needed at a critical level. The multifarious phenomenon caused by the low availability of a mineral or by its excess concentrations is called nutrient stress. Excess/deficiency of minerals, heavy metals, or salt concentration disturbs the biochemical homeostasis in plants due to various environmental or anthropogenic activities (Reddy, 2006). The biochemical modifications caused due to salt stress involve changes in the metabolic pathways that play important role in the stress responses. During water stress conditions, we have already seen that it leads to the accumulation of certain metabolites/molecules. Studies on *Arabidopsis thaliana*, maize, and tobacco plants have also shown that exposure to salt stress leads to the accumulation of lactate and sucrose, whereas the levels of glucose decreased. However, different plants adapt to different stress conditions in various ways (Fraire-Velázquez and Balderas-Hernández, 2013). The extensive study carried out by Bandehagh and Taylor (2020) on the impacts of salt stress on carbon metabolism will help us understand the major changes in biochemical pathways important for the plants to cope up with stress environments. The accumulation of salt in soil and water averts plant growth by reducing the water-absorbing capacity. The reduction in plant growth can also be attributed to the Na^+ and Cl^- ion accumulation that primes to toxic effects on the biochemistry of plants. The photosynthetic activity of plants gets restricted due to salinity exposure and henceforth, a change in the respiration can be observed extensively. Both the processes are linked to the carbon balance in plants which gets disturbed during stress conditions. The respiratory metabolism involving processes such as glycolysis and the TCA cycle are required to maintain the NAD(P)H to uphold ATP supply. During long-term exposure to salinity stress, glycolysis and sucrose metabolism have been shown to be co-induced and certain changes in the metabolites of glycolysis have been observed in crops such as wheat, barley, and maize. Saccharides including glucose, sucrose, fructose, and xylose are reduced in the stressed plants because the saccharides serve as the source of glucose required for glycolysis. As the glycolysis proceeds, the abundance of glucose, G-6-P, and F-6-P increases in stressed plants and leads to the accumulation of pyruvate that will be utilized in the TCA cycle. However, in the roots of barley, the metabolites including G-6-P, F-6-P, and 3-phosphoglycerate decrease. Researchers have suggested that the pyruvate utilization/TCA cycle gets reduced under the stress conditions, leading to the accumulation of pyruvate as well as the feedback inhibition of the glycolytic pathway (Che-Othman et al., 2020).

Altogether, the glycolysis process is upregulated to produce more energy that is required to accelerate the stress responses. Apart from glycolysis, the OPP pathway becomes a central pathway during stress conditions in plants such as rice and some halophytic species (Kaur and Asthir, 2017; Lu et al., 2016). OPP acts as an alternative metabolic pathway that catabolizes carbohydrates and produces NADPH that is critical in surviving stress (Huan et al., 2014). Salt stress increases the abundance of OPP pathway metabolites as a response to the stress and suppressess the TCA cycle and amino acid metabolism significantly. This suppression restricts glycolysis and thus diverts the carbon to the OPP pathway. As salt stress increases the energy requirements and ATP production, the collaboration of glycolysis and OPP pathway permits the modification of the abundance of reducing power in the form of NADPH, NADH, and ATP (Fig. 13.8) (Bandehagh and Taylor, 2020).

The glycolysis and OPP pathway are finally linked to the mTCA cycle leading to the conversion of PEP to malate, oxaloacetate, or pyruvate. These organic acids afterward get interconverted in the TCA cycle to generate ATP. Generally, organic acids and the intermediates involved in the TCA cycle are reduced following salt stress. This is attributed to the increase in the biosynthesis of amino acids from the intermediate compounds of the TCA cycle (Fig. 13.9) (Diab and Limami, 2016). Different crop species show differing responses and impacts on the TCA cycle and the changes in the intermediates, for example, the roots of cultivars of rice (salt stress-tolerant) and wheat leaves reduce the abundance of the TCA cycle intermediates but increase the amino acid abundance. On the other hand, intermediates such as aconitate, citrate, isocitrate, malate, α-ketoglutarate, and succinate increase in roots of salt-sensitive cultivar of barley, but there have been no major changes in the salt-tolerant species of Clipper after exposure to salt stress. This increase in the abundance of organic acids is connected with the capacity of the plant to uphold or enhance the yield when stressed. Lower concentration of some organic acids (most remarkably citrate and isocitrate) in salinity-stressed leaves (wheat and barley) and roots (rice) indicates that these metabolites/intermediates which are downstream of pyruvate are depleted probably through the inhibition of pyruvate transport into the mitochondria or its conversion into acetyl-CoA. In wheat plants, 2-oxoglutarate reported to increase during salt stress due to the enhanced activity of isocitrate dehydrogenase or glutamate dehydrogenase, or both (Masclaux-Daubresse et al., 2006). Long-term exposure to salinity stress reduces respiration followed by a reduction in the growth of plants as organic acids decline. Along with this, the decline in the concentration of pyruvate carriers reduces the capability of the TCA cycle to produce NADH (Che-Othman et al., 2020).

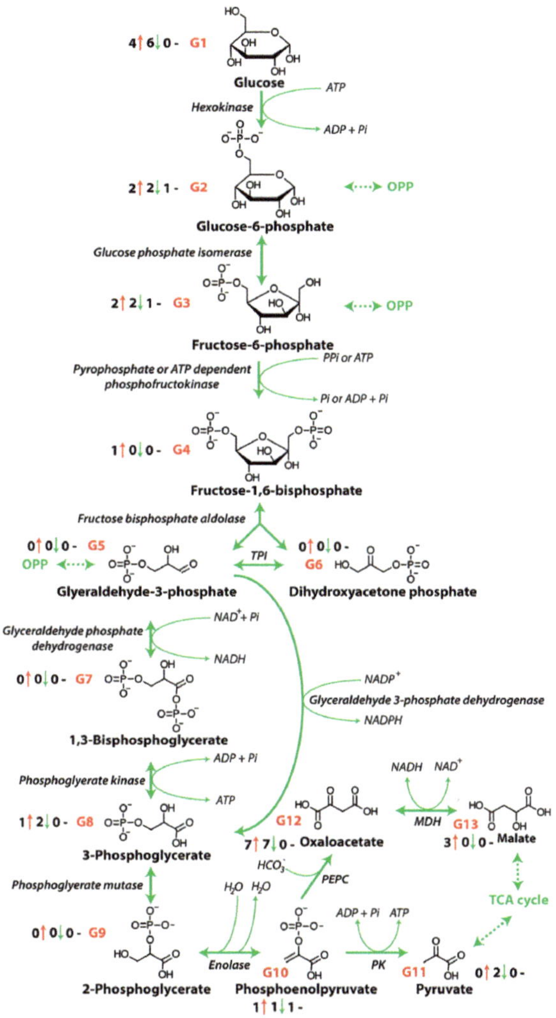

FIGURE 13.8 The figure represents the glycolytic pathway during salt stress conditions where G1-G13 represent each metabolite of glycolysis. The number before the red arrows indicates the number of papers that report an increase in abundance, the number before the green arrows indicates the number of papers that report a decrease in abundance, and the number before the black dashes indicates the number of papers that report no change in abundance.

Abbreviations: ATP, adenosine triphosphate; ADP, adenosine diphosphate; MDH, malate dehydrogenase; NAD^+, nicotinamide adenine dinucleotide (oxidized); NADH, nicotinamide adenine dinucleotide (reduced); $NADP^+$, nicotinamide adenine dinucleotide phosphate (oxidized); NADPH; nicotinamide adenine dinucleotide phosphate (reduced); OPP, Oxidative pentose phosphate pathway; PPi, pyrophosphate, Pi, inorganic phosphate; PEPC, phosphoenolpyruvate carboxylase; PK, pyruvate kinase. TPI, triose phosphate isomerase.

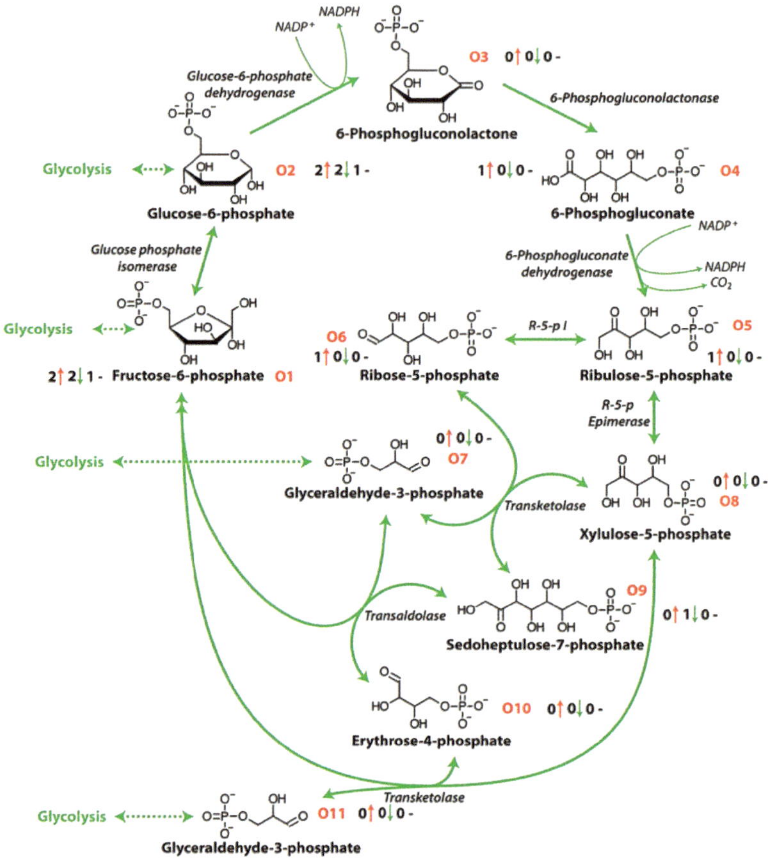

FIGURE 13.9 The figure represents the tricarboxylic acid (TCA) cycle during salt stress where T1-T15 represent each metabolite of the TCA pathway. The number before the red arrows indicates the number of papers that report an increase in abundance, the number before the green arrows indicates the number of papers that report a decrease in abundance, and the number before the black dashes indicates the number of papers that report no change in abundance. Abbreviations: ATP, adenosine triphosphate; ADP, adenosine diphosphate; GDP, guanosine diphosphate; GTP, guanosine triphosphate; NAD^+, nicotinamide adenine dinucleotide (oxidized); NADH, nicotinamide adenine dinucleotide (reduced); Pi, inorganic phosphate; Q, quinone; QH2, dihydroquinone.

Source: Reprinted with permission from Bandehagh and Taylor, 2020.

Putrescine is one of the dominant species of polyamines which is also naturally found in plants. The deficiency of potassium (K^+) ions leads to the accumulation of putrescine in plants, especially in leaves. This helps in monitoring the plant developmental stages during various conditions including abiotic stresses (Cui et al., 2020). The exogenous application of putrescine has positive effects on glycolysis and the TCA cycle when the cucumber leaves are subjected to salt stress. Salinity can inhibit the primary metabolic processes in cucumber leaves. Application of putrescine improves the activities of glycolytic enzymes, that is, PK and PEPC that in turn provide pyruvate and oxaloacetate to be utilized in the TCA cycle. It also helps in improving the enzymes of the TCA cycle that results in the higher production of organic acids. As a result, the TCA cycle can function normally and supply more organic materials for other metabolic pathways. The rise in citric acid content provided more substrate for citric dehydrogenase, alleviated abiotic stress-induced oxidative damage, and controlled mitochondrial oxidative balance. Finally, more carbohydrates join the glycolic pathway and sustain natural metabolism of cucumber seedlings, thus promoting salt tolerance of cucumber seedling (Zhong et al., 2016).

Heavy metals induce oxidative stress in plants and are responsible for the modulation of numerous stress-allied genes in plants. Heavy metal toxicity affects the structural properties of cells, stimulates the formation of ROS and free radicals ($O^{2\bullet-}$, H_2O_2, $OH^\bullet$), suppresses various enzymatic activities, and eventually reduces the rate of transpiration and photosynthesis, for example, the occurrence/presence of Pb (Lead) affects the nutrient (majorly divalent cations: Ca^{2+}, Fe^{2+}, Mg^{2+}, Mn^{2+}, Zn^{2+}) uptake property of plants. Formation of ROS is directly induced by some heavy metals such as copper, lead, chromium, and arsenic (Dubey et al., 2018). The intermediates (organic acids) of primary metabolism such as the TCA cycle are necessary for heavy metal detoxification. During metal stress, organic acids and amino acids can bind to the heavy metals by chelating them and transporting them to the vacuoles. Citrate and histidine chelates Fe and Ni, respectively (Dubey et al., 2018). Let us understand the effects of heavy metal stress on glycolysis by taking the example of cadmium (Cd)-induced stress on pea plants. Cd ions inhibit seedling growth and with an increase in the exogenous Cd ion concentration, the inhibition becomes more distinct (Devi et al., 2007). Cd reduces the activities of sucrose phosphate synthase (SPS) and sucrose synthase in roots and shoots of pea plant due to the inhibition of transport of sucrose from cotyledons to the cells and tissues present under stressed conditions. However, the sucrose synthesizing ability is not affected in cotyledons. SPS is responsible for the conversion of F-6-P to sucrose phosphate and by the action of sucrose phosphate phosphatise, sucrose is formed. Glucose and fructose present in roots/shoots are utilized by glycolysis or OPP pathway or they are converted into sucrose. In cotyledons of pea, Cd ion increases the activity of G-6-P dehydrogenase (G6PDH) and 6-phosphogluconate dehydrogenase (6PGDH), whereas these enzymes have reduced functioning in the stressed roots and shoots of the seedling. This leads to a low rate of NADPH formation in roots and shoots. The process of hexose phosphorylation remains unaffected under cadmium toxicity because the activity of hexokinase increases about 10–15% in the cotyledons, roots, and shoots of stressed pea seedlings. The activity of PGI and fructokinase remains unaffected in cotyledons, which indicates the continued conversion of G-6-P to F-6-P and its final conversion to sucrose-6-phosphate with the help of SPS. The activity of PGI increases at a normal rate during Cd stress indicating that the hexose sugars are generally utilized by glycolysis rather than the OPP pathway. PGM has negligible activity in roots and shoots but maximum in the cotyledons leading to a pathway for the rapid formation of sucrose. Although the reasons are not clear for the changes in enzyme activity, it can be said that the Cd might interact with the sulfhydryl groups of cysteine present in the protein/enzyme (6PGDH) or might be able to disturb the cation equilibrium at the subcellular level (Devi et al., 2007).

Acidic soil limits plant growth in the presence of toxic levels of aluminum (Al^{3+}). The phytotoxicity caused by Al is known to constrain the root growth and cause various changes in the metabolism in leaves of maize as well. Researches have proved that the production of organic acid, as well as carbon metabolism, gets affected during Al phytotoxicity. The key mechanism of Al tolerance is the organic acid exudation. In the leaves of Al-stressed maize grown in acidic soil, enzyme/protein-encoding genes such as citrate synthase and malate dehydrogenase (MDH) are upregulated, whereas citrate transporter is downregulated. This causes an increase in malate and citrate production; however, organic acid transporters are upregulated in hydroponic maize roots. When the gene-encoding MDH overexpresses under a leaf-specific promoter, it leads to the increment in malate content in leaves and stems, as well as the amount of malate exuded by the roots, resulting in greater Al resistance. The organic acid production or the activation of the TCA cycle in Al-stressed leaves increases due to the downregulation of pyruvate dehydrogenase kinase isoform 2 that inhibits pyruvate dehydrogenase. Soil acidity slightly increases the internal carbon concentration. During the stressed conditions, the enzymes that hydrolyze carbohydrates are upregulated due to the reduction in photosynthetic activity and sucrose formation. Genes encoding pyrophosphate F-6-P1-phosphotransferase α-subunit and phosphoglycerate mutase-like protein are seen to be upregulated during soil acidity caused by Al. These changes in the enzyme activities suggest that the plants respond to the stress conditions and try to reimburse for the reduction in the metabolic process and ultimately maintain the carbon skeleton in cells and tissues (Mattiello et al., 2014).

13.5 THE IMPACT OF CLIMATE CHANGES ON PLANT PRIMARY METABOLISM

Climate changes and adverse conditions are the main causes of the abiotic and biotic stresses that have a great impact on crops and plants which ultimately hamper their productivity. The variations in rainfall, heat waves, level of atmospheric CO_2, average temperature, and pests/microbes have negative impacts on worldwide crop production and compromise food security (Raza et al., 2019). An increase in atmospheric CO_2 is likely to be a major factor that can influence disease severity and extent in plants. High levels of CO_2 are likely to increase the fecundity and spread of the fungal pathogens (Luck et al., 2011). Climate change has the potential to impact plant development, thus influencing the quantity and quality of specialized metabolites produced. Flowering time, for example, is highly susceptible to environmental stress, perhaps by a signaling mechanism involving basic sugars including sucrose and glucose (Xu and Weng, 2020). The metabolic adaptations taken up by the plants during the climate changes can be provided by the alternative oxidase (AOX) pathway. Under drought, elevated temperature, and CO_2, the AOX pathway creates plasticity in cellular energy and carbon metabolism, which are the main abiotic pressures threatening current agricultural production as a result of climate change. The ability of AOX to sustain higher respiration in light, which advances chloroplast energy balance and photosynthesis, has been related to the beneficial impact on growth. Nonetheless, the AOX activity in vivo promotes the synthesis of TCA-derived metabolites that play specific roles in defending against salinity stress and high light. AOX overexpression has also been shown to inhibit carbohydrate and energy imbalances in tobacco leaves cultivated in high CO_2 environment (Dahal and Vanlerberghe, 2018; Florez-Sarasa et al., 2020)

13.6 INTERACTION BETWEEN ABIOTIC AND BIOTIC STRESSES

According to recent experiments and studies, plants react to multiple stresses differently than they do to single stresses. Abiotic stress may reduce or increase susceptibility to a biotic pest or pathogen, and vice versa. After plants are exposed to a pest or pathogen, the effects of abiotic stress like water deficiency are often amplified, while long-term abiotic stress can degrade plant defense and increase pathogen susceptibility. Several molecular pathways that function together in a dynamic regulatory network regulate the specificity of numerous stress responses. When biotic and abiotic stresses arise at the same time, the responses are primarily mediated by various hormone signaling pathways that can interfere and inhibit one another, adding to the complication (Atkinson and Urwin, 2012). Interactions of biotic and abiotic stresses such as heat-drought, heat-pathogen infection, drought-fungal infection, heat-salinity, drought-cold, pathogen-water relations, etc. can impart both positive and negative effects on the plant metabolism (Table 13.2).

TABLE 13.2 Effects of Multiple Stress Combinations on Plants.

Multiple stress	
Negative interactive stress	**Effects**
High temperature and drought	Physiological changes
	Affects crop yield and quality
	Replacement of proline by sucrose that acts as osmoprotectants
Drought and salt stress	Imbalance of K^+:Na^+ ratio
Drought and heavy metals	Alteration of xylem structure and hydraulic conductivity
Heavy metal (cadmium) and biotic stress (*Erysiphe cruciferarum*)	Diminishes the plant ability to induce salicylic acid as a defense signal in response to the attack
Heat and water	Decrease in leaf sucrose content
	Decrease photosynthesis and stomatal conductance
High temperature and pathogen infection	Suppression of basal and the resistance-gene-mediated defense responses
Positive interactive stress	**Effects**
Drought and O_3	Lowered stomatal conductance
	Enhances the plant tolerance to O_3 stress
	Decrease in ROS concentrations
	Increase in reduced ascorbate and glutathione
Salt and heat stress	Accumulation of glycine betaine and trehalose
	Maintenance of a lower Na^+: K^+ ratio
	Improved cell water status and photosynthetic performance
High CO_2 and salt/high light	Improves water-use efficiency
	Increase in the biomass of plants
CO_2 and O_3	Decreases stomatal conductance
	Inhibits diffusion of O_3 into leaves
	Enhanced antioxidant capacity

TABLE 13.2 *(Continued)*

Positive interactive stress	Effects
	Decrease in transpiration rate
High temperature and *Puccinia* infection	Enhanced disease resistance in spring wheat
Salinity stress and *Blumeria graminis* (causal agent of powdery mildew)	Increased resistance of barley
Drought stress and plant-parasitic nematode infection	Ameliorated the severity of drought stress in rice plants

Source: The data are collected from Morkunas et al. (2018), Ramegowda and Senthil-Kumar (2015), and Suzuki et al. (2014).

The metabolic responses of maize plants reveal some of the metabolic plasticity when subjected to soil salinity, heat, soil drought, and multiple concurrent stresses. Experiments conducted by Sun and coworkers revealed that maize plants exposed to soil water deficit had a metabolic profile that was more comparable to salt-stressed plants than heat-stressed plants. Compared with those exposed to either stressor separately, drought-stressed maize plants exposed to salt or heat stress had distinct incorporated metabolic profiles. Between maize plants grown in a regulated growth cabinet and those grown naturally, differences in glucose, fructose, malate, alanine, asparagine, aspartate, citrate, proline, and threonine were evident. The TCA cycle and central nitrogen metabolism were linked to these modifications, which may be linked to their numerous roles during plant development. The time-dependent basic metabolic responses to drought, heat, and combined drought and heat stress in maize were shown by the stress-induced trajectory of metabolic alterations (Fig. 13.10). Maize plants are able to acclimate their metabolic ranges in reaction to changing environmental factors because of plasticity at the metabolic stage (Sun et al., 2016). When maize leaves are subjected to soil acidity caused by aluminum (Al), genes encoding pyrophosphate fructose-6-phosphate 1-phosphotransferase α-subunit and phosphoglycerate mutase-like protein are seen to be upregulated. Arabidopsis homolog of this gene is articulated in apical meristems from shoots and roots and is upregulated by parasite invasion, hormones (auxin), nitrate exposure, salt stress, and drought. This advocates that the gene is involved in the reaction to biotic and abiotic stresses, and they upkeep the theory that Al stress alters carbon metabolism pathways (Mattiello et al., 2014).

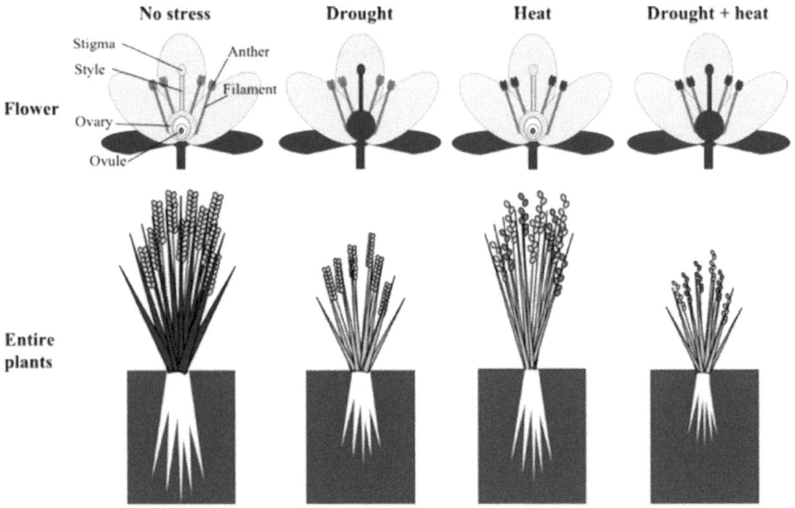

FIGURE 13.10 The figure depicts the effects of drought, heat stress, and their combination on growth and development of maize and wheat. Drought primarily reduces plant height, spike number, and grain weight. Heat stress, on the other hand, produces a rise of aborted spikes and a drop in grain number. Drought has the greatest influence on pistil growth in reproductive tissues, whereas heat stress has the greatest impact on pollen formation. When drought and heat stress are coupled, the separate impacts of both stressors have an influence on plants at the same time. These broad impacts, which have been documented for maize and wheat, may change based on the severity and duration of the stressors, as well as the plant species involved.
Source: Reprinted from Suzuki et al. (2014). © John Wiley and Sons.

Methylglyoxal (MG), $CH_3C(O)CHO$, is a cytotoxic/mutagenic compound and glycolytic by-product formed subsequently via glycolysis in plant cells, but its synthesis increases manifold under stress conditions (Hasanuzzaman et al., 2019). MG is also responsible for impeding root development, seed germination, stomatal opening, and activating a stress response, making it a critical signaling molecule that serves as an essential link between the biotic and abiotic responses (Hoque et al., 2016). Under normal metabolic conditions, plants retain a lower amount of MG, but its content increases in response to abiotic stresses. MG is produced during glycolysis by the fragmentation of GAP and DHAP. The rate-limiting step in glycolysis is PFK, which is allosterically stimulated by F-1,6-P2. The aggregation of MG disrupts cellular functions by damaging DNA, causing point mutations and progressive glycation end products, which eventually lead to death if not detoxified in the early stages. The rate of the glycolysis cycle increases in plants under biotic and abiotic stresses. This abrupt rise in glycolysis rate disrupts the cellular equilibrium, resulting in two- to six-fold overproduction of the MG molecule (Singh et al., 2020).

13.7 ADDITIONAL REGULATING FACTORS OF PRIMARY METABOLISM DURING STRESS CONDITIONS

Primary metabolism in plants amid stress circumstances is sensitive to alterations or inhibitions. However, there are certainly other factors (especially biochemical compounds) that directly or indirectly affect the metabolic pathways and the concentration of the metabolites or the intermediates. The secondary messenger 3',5'-cyclic adenosine monophosphate (cAMP) has diverse roles in plant responses to environmental stimuli or stress. In 2016, Alqurashi and coauthors found certain proteins related to glycolysis and the TCA cycle that were involved in abiotic stress (salinity and cold) as well as biotic stress after *Arabidopsis* was treated with cAMP. cAMP treatment significantly increases the three major glycolytic proteins including mitochondrial MDH, fructose-bisphosphate aldolase 3 (FBA-3), and early response to dehydration 10 (ERD 10), while it decreases glyoxalase I. FBA-3 is known to increase in response to phytohormones (ABA and SA) and also to abiotic stresses, except Cd^{2+} stress where FBA-3 decreases. The increase in FBA-3 after cAMP treatment suggests that cAMP is involved in stress response and energy metabolism. The increase in MDH and ERD10 in response to cAMP suggests that cAMP is involved in the regulation of the TCA cycle and energy metabolism, either directly or indirectly. Therefore, relevant stress responses can be elicited by cAMP alone, which can lead to dynamic changes in cellular energy homeostasis. As a result, cAMP may be a correlation between biotic and abiotic stress responses in stress-related changes in energy metabolism (Alqurashi et al., 2016). During anoxic conditions, accumulation and fermentation of lactate and ethanol is a general phenomenon in plants. Yet, certain studies provided evidence of the accumulation of alanine without injurious consequences under anoxic situations and eventually help in regulating the pH balance inside an anoxic cell (De Sousa and Sodek, 2003). Glycolysis and TCA cycle are connected by alanine aminotransferase, as shown by the importance of Ala accumulation for surviving waterlogging in *L. Japonicus* (Rocha et al., 2010). Alanine and oxoglutarate are formed when pyruvate interacts with glutamine through alanine aminotransferase (AlaAT). This inhibits pyruvate accumulation while concurrently producing oxoglutarate, which can be used by oxoglutarate dehydrogenase and succinate CoA ligase to generate ATP in the mitochondria. The NAD^+ needed for this reaction is produced from NADH by MDH catalyzing the oxaloacetate to malate reaction. Aspartate aminotransferase produces the oxaloacetate that is needed as a substrate for this reaction. Glutamate, a cosubstrate for alanine synthesis, is generated concurrently. The aspartate and glutamine pools deplete to keep the cycle going. Malate either reacts with NAD^+ through ME to produce pyruvate, which can then be used to make alanine, or reacts with fumarate to produce succinate. This mechanism explains why alanine accumulates during hypoxia, as well as why most other TCA cycle-related amino acids drop dramatically. This causes the metabolic flux to be driven from glycolysis to succinate through Ala synthesis and oxoglutarate. The cycle will take place entirely within the mitochondrion and all of the enzymes needed for these reactions are known to have isoforms within the mitochondria. The amount of ATP obtained through this hypoxic pathway is twice that obtained through glycolysis alone (Rocha et al., 2010). The resistance response of plants to abiotic stress is influenced by the glycolytic and OPP pathways. The rate-limiting enzymes of the two pathways, PFK and G6PDH, respectively, regulate carbon flow through the two pathways during drought and bicarbonate stress as predicted in mulberry leaves. Drought and bicarbonate stress have a stimulating influence on G6PDH activity and an inhibitory effect on PFK activity in mulberry leaves at the functional level. Thus, under these two abiotic pressures, carbon flux in the leaves can be redirected from the catabolic glycolytic pathway to the anabolic OPP pathway (Fig. 13.6). The fact that this plant maintains PFK activity under stress means that it prefers to maintain glycolytic metabolism to keep the ATP supply going. When water becomes available, glycolysis inhibition has been proposed as a method for storing sugars as an energy cradle for recovery and rapid development. G6PDH activity rises in response to stress, indicating that the plant needs more biosynthetic precursors and NADPH. Stress responses in plants are complex and varied, with cross-representation among different levels of regulation. Mechanisms of action of different regulators are complex, and it would be useful to see how different types of stresses affect metabolic processes, transcriptional levels, and stress adaptation. Finally, even though mulberry seedlings stop developing at the same time under two different pressures, the association between the PFK activity and the G6PDH activity is not strictly negative, and the variations in mRNA levels of PFK and G6PDH genes, as well as their behaviors, are vastly different. Changes in G6PDH and PFK can vary, meaning that external influences on plant stress responses may also play important roles in glucose and energy metabolism (Yao and Wu, 2016).

13.8 CONCLUSION

In conclusion, abiotic and biotic stresses can influence and hence alter the primary metabolism of the plants. Major metabolic pathways (particularly glycolysis and TCA cycle) and their enzymes and intermediates change frequently in response to stress. Conversely, different plants show different types of regulation and changes during stress conditions. Enzymes of glycolysis such as hexokinase, G6PDH, PFK, PEP, and PK are some of the important enzymes whose activities are affected during stress. Likewise, TCA cycle enzymes including MDH, CS, IDH, KDH, and aconitase are mostly affected during the stress. Intermediates of both glycolysis and TCA cycle increase, decrease, or accumulate in concentration as per the requirement. Accumulation of pyruvate from glycolysis and organic acids from the TCA cycle is a common phenomenon in most stressful conditions. All these changes in the primary metabolism are finally compensated somehow (as in OPP shunt) to keep up the ATP production required to produce responses to various stress factors.

KEYWORDS

- **abiotic stress**
- **biotic stress**
- **glycolysis**
- **glycolytic enzymes**
- **organic acids**
- **primary metabolism**
- **tricarboxylic acid cycle**

REFERENCES

Alqurashi, M.; Gehring, C.; Marondedze, C. Changes in the *Arabidopsis thaliana* Proteome Implicate cAMP in Biotic and Abiotic Stress Responses and Changes in Energy Metabolism. *Int. J. Mol. Sci.* **2016,** *17*, 852.

Amirsadeghi, S.; Robson, C. A.; Vanlerberghe, G. C. The Role of the Mitochondrion in Plant Responses to Biotic Stress. *Physiol. Plant.* **2007,** *129*, 253–266.

Annunziata, M. G.; Apelt, F.; Carillo, P.; Krause, U.; Feil, R.; Koehl, K.; Lunn, J. E.; Stitt, M. Response of Arabidopsis Primary Metabolism and Circadian Clock to Low Night Temperature in a Natural Light Environment. *J. Exp. Bot.* **2018,** *69*, 4881–4895.

Araújo, W. L.; Nunes-Nesi, A.; Trenkamp, S.; Bunik, V. I.; Fernie, A. R. Inhibition of 2-Oxoglutarate Dehydrogenase in Potato Tuber Suggests the Enzyme Is Limiting for Respiration and Confirms Its Importance in Nitrogen Assimilation. *Plant Physiol.* **2008,** *148*, 1782–1796.

Armengaud, P.; Sulpice, R.; Miller, A. J.; Stitt, M.; Amtmann, A.; Gibon, Y. Multilevel Analysis of Primary Metabolism Provides New Insights Into the Role of Potassium Nutrition for Glycolysis and Nitrogen Assimilation in Arabidopsis Roots. *Plant Physiol.* **2009,** *150*, 772–785.

Ashtamker, C.; Kiss, V.; Sagi, M.; Davydov, O.; Fluhr, R. Diverse Subcellular Locations of Cryptogein-Induced Reactive Oxygen Species Production in Tobacco Bright Yellow-2 Cells. *Plant Physiol.* **2007,** *143*, 1817–1826.

Atkinson, N. J.; Urwin, P. E. The Interaction of Plant Biotic and Abiotic Stresses: From Genes to the Field. *J. Exp. Bot.* **2012,** *63*, 3523–3543.

Baalmann, E.; Backhausen, J. E.; Rak, C.; Vetter, S.; Scheibe, R. Reductive Modification and Nonreductive Activation of Purified Spinach Chloroplast NADP-Dependent Glyceraldehyde-3-Phosphate Dehydrogenase. *Arch. Biochem. Biophys.* **1995,** *324*, 201–208.

Babu, M.; Griffiths, J. S.; Huang, T.-S.; Wang, A. Altered Gene Expression Changes in Arabidopsis Leaf Tissues and Protoplasts in Response to Plum Pox Virus Infection. *BMC Genom.* **2008,** *9*, 325.

Bak, A.; Emerson, J. B. Cauliflower Mosaic Virus (CaMV) Biology, Management, and Relevance to GM Plant Detection for Sustainable Organic Agriculture. *Front. Sustain. Food Syst.* **2020,** *4*, 21.

Baldacci-Cresp, F.; Maucourt, M.; Deborde, C.; Pierre, O.; Moing, A.; Brouquisse, R.; Favery, B.; Frendo, P. Maturation of Nematode-Induced Galls in *Medicago truncatula* Is Related to Water Status and Primary Metabolism Modifications. *Plant Sci. Int. J. Exp. Plant Biol.* **2015,** *232*, 77.

Bale, J. S.; Masters, G. J.; Hodkinson, I. D.; Awmack, C.; Bezemer, T. M.; Brown, V. K.; Butterfield, J.; Buse, A.; Coulson, J. C.; Farrar, J. Herbivory in Global Climate Change Research: Direct Effects of Rising Temperature on Insect Herbivores. *Global Change Biol.* **2002,** *8*, 1–16.

Balmer, A.; Pastor, V.; Glauser, G.; Mauch-Mani, B. Tricarboxylates Induce Defense Priming against Bacteria in *Arabidopsis thaliana. Front. Plant Sci.* **2018,** *9*, 1221.

Bandehagh, A.; Taylor, N. L. Can Alternative Metabolic Pathways and Shunts Overcome Salinity Induced Inhibition of Central Carbon Metabolism in Crops? *Front. Plant Sci.* **2020,** *11*.

Baxter, A.; Mittler, R.; Suzuki, N. ROS as key players in plant stress signalling. *J.Exp. Bot.* **2014,** *65*, 1229–1240.

Bilgin, D. D.; Zavala, J. A.; Zhu, J.; Clough, S. J.; Ort, D. R.; Delucia, E. H. Biotic Stress Globally Downregulates Photosynthesis Genes. *Plant Cell Environ.* **2010,** *33*, 1597–1613.

Bogeat-Triboulot, M.-B.; Brosché, M.; Renaut, J.; Jouve, L.; Le Thiec, D.; Fayyaz, P.; Vinocur, B.; Witters, E.; Laukens, K.; Teichmann, T. Gradual Soil Water Depletion Results in Reversible Changes of Gene Expression, Protein Profiles, Ecophysiology, and Growth

Performance in *Populus euphratica*, a Poplar Growing in Arid Regions. *Plant Physiol.* **2007,** *143*, 876–892.
Bolton, M. D.; Kolmer, J. A.; Garvin, D. F. Wheat Leaf Rust Caused by *Puccinia triticina*. *Mol. Plant Pathol.* **2008,** *9*, 563–575.
Bourke, J.; Massey Jr, L. The influence of gamma radiation on carbohydrate and citric acid cycle metabolism in carrot tissue. Radiation Research **1967,** *31*, 783–795.
Caplin, N.; Willey, N. Ionizing radiation, higher plants, and radioprotection: from acute high doses to chronic low doses. *Front. Plant Sci.* **2018,** *9*, 847.
Castillejo, M.; Maldonado, A. M.; Ogueta, S.; Jorrín, J. V. Proteomic Analysis of Responses to Drought Stress in Sunflower (*Helianthus annuus*) Leaves by 2DE Gel Electrophoresis and Mass Spectrometry. *Open Proteom. J.* **2008,** *1*.
Cheng, Y.; Xia, X.; Lin, Y.; Cao, H.; Yao, J.; Li, Z. Ubiquitylome Study Highlights Ubiquitination of Primary Metabolism Related Proteins in Fruit Response to Postharvest Pathogen Infection. *Postharvest Biol. Technol.* **2020,** *163*, 111142.
Che-Othman, M. H.; Jacoby, R. P.; Millar, A. H.; Taylor, N. L. Wheat Mitochondrial Respiration Shifts from the Tricarboxylic Acid Cycle to the GABA Shunt under Salt Stress. *New Phytol.* **2020,** *225*, 1166–1180.
Chojak-Koźniewska, J.; Kuźniak, E.; Linkiewicz, A.; Sowa, S. Primary Carbon Metabolism-Related Changes in Cucumber Exposed to Single and Sequential Treatments with Salt Stress and Bacterial Infection. *Plant Physiol.Plant Physiol. Biochem.* **2018,** *123*, 160–169.
Choudhury, S.; Panda, S. Induction of Oxidative Stress and Ultrastructural Changes in Moss *Taxithelium nepalense* (Schwaegr.) Broth. under Lead and Arsenic Phytotoxicity. *Curr. Sci.*, **2004,** 342–348.
Colell, A.; Ricci, J.-E.; Tait, S.; Milasta, S.; Maurer, U.; Bouchier-Hayes, L.; Fitzgerald, P.; Guio-Carrion, A.; Waterhouse, N. J.; Li, C. W. GAPDH and Autophagy Preserve Survival after Apoptotic Cytochrome C Release in the Absence of Caspase Activation. *Cell* **2007,** *129*, 983–997.
Cui, J.; Pottosin, I.; Lamade, E.; Tcherkez, G. What Is the Role of Putrescine Accumulated under Potassium Deficiency? *Plant Cell Environ.* **2020,** *43*, 1331–1347.
Dahal, K.; Vanlerberghe, G. C. Growth at Elevated CO_2 Requires Acclimation of the Respiratory Chain to Support Photosynthesis. *Plant Physiol.* **2018,** *178*, 82–100.
Dalal, A.; Kumar, A.; Yadav, D.; Gudla, T.; Viehhauser, A.; Dietz, K.-J.; Kirti, P. B. Alleviation of Methyl Viologen-Mediated Oxidative Stress by *Brassica juncea* Annexin-3 in Transgenic Arabidopsis. *Plant Sci.* **2014,** *219*, 9–18.
Danson, J.; Wasano, K.; Nose, A. Infection of Rice Plants with the Sheath Blight Fungus Causes an Activation of Pentose Phosphate and Glycolytic Pathways. *Plant Pathol. Eur.J.Plant Pathol.* **2000,** *106*, 555–561.
De Sousa, C.; Sodek, L. Alanine Metabolism and Alanine Aminotransferase Activity in Soybean (*Glycine max*) during Hypoxia of the Root System and Subsequent Return to Normoxia. *Environ. Exp. Bot.* **2003,** *50*, 1–8.
Devi, R.; Munjral, N.; Gupta, A. K.; Kaur, N. Cadmium Induced Changes in Carbohydrate Status and Enzymes of Carbohydrate Metabolism, Glycolysis and Pentose Phosphate Pathway in Pea. *Environ. Exp. Bot.* **2007,** *61*, 167–174.
Diab, H.; Limami, A. M. Reconfiguration of N Metabolism Upon Hypoxia Stress and Recovery: Roles of Alanine Aminotransferase (AlaAT) and Glutamate Dehydrogenase (GDH). *Plants* **2016,** *5*, 25.

Doehlemann, G.; Wahl, R.; Horst, R. J.; Voll, L. M.; Usadel, B.; Poree, F.; Stitt, M.; Pons-Kühnemann, J.; Sonnewald, U.; Kahmann, R. Reprogramming a Maize Plant: Transcriptional and Metabolic Changes Induced by the Fungal Biotroph *Ustilago maydis*. *Plant J.* **2008,** *56*, 181–195.

Doughari, J. An overview of plant immunity. J. Plant Pathol. Microbiol **2015,** *6*, 10.4172.

Dubey, S.; Shri, M.; Gupta, A.; Rani, V.; Chakrabarty, D. Toxicity and Detoxification of Heavy Metals during Plant Growth and Metabolism. *Environ. Chem. Lett.* **2018,** *16*, 1169–1192.

Dumont, S.; Rivoal, J. Consequences of Oxidative Stress on Plant Glycolytic and Respiratory Metabolism. *Front. Plant Sci.* **2019,** *10*, 166–166.

Eloh, K.; Nicola, S.; Andrea, M.; Pierluigi, C. Untargeted Metabolomics of Tomato Plants after Root-Knot Nematode Infestation. *J. Agric. Food Chem.* **2016.**

Ferrieri, A. P.; Appel, H.; Ferrieri, R. A.; Schultz, J. C. Novel Application of 2-[18F] Fluoro-2-deoxy-D-Glucose to Study Plant Defenses. *Nucl. Med. Biol.* **2012,** *39*, 1152–1160.

Finkemeier, I.; König, A.-C.; Heard, W.; Nunes-Nesi, A.; Pham, P. A.; Leister, D.; Fernie, A. R.; Sweetlove, L. J. Transcriptomic Analysis of the Role of Carboxylic Acids in Metabolite Signaling in Arabidopsis Leaves. *Plant Physiol.* **2013,** *162*, 239–253.

Florez-Sarasa, I.; Fernie, A. R.; Gupta, K. J. Does the Alternative Respiratory Pathway Offer Protection against the Adverse Effects Resulting from Climate Change? *J.Exp. Bot.* 202, *71*, 465–469.

Ford, K. L.; Cassin, A.; Bacic, A. F. Quantitative Proteomic Analysis of Wheat Cultivars with Differing Drought Stress Tolerance. *Front. Plant Sci.* **2011,** *2*, 44.

Forrester, S. J.; Kikuchi, D. S.; Hernandes, M. S.; Xu, Q.; Griendling, K. K. Reactive Oxygen Species in Metabolic and Inflammatory Signaling. *Circ. Res.* **2018,** *122*, 877–902.

Fraire-Velázquez, S.; Balderas-Hernández, V. E. *Abiotic Stress in Plants and Metabolic Responses. Abiotic Stress—Plant Responses and Applications in Agriculture*; InTech: Rijeka, Croatia, 2013; pp 25–48.

Fraire-Velázquez, S.; Rodríguez-Guerra, R.; Sánchez-Calder6n, L. *Abiotic and Biotic Stress Response Crosstalk in Plants. Abiotic Stress Response in Plants—Physiological, Biochemical and Genetic Perspectives*, 2011; pp 3–26.

Gautam, V.; Kaur, R.; Kohli, S. K.; Verma, V.; Kaur, P.; Singh, R.; Saini, P.; Arora, S.; Thukral, A. K.; Karpets, Y. V. *ROS Compartmentalization in Plant Cells under Abiotic Stress Condition, Reactive Oxygen Species and Antioxidant Systems in Plants: Role and Regulation under Abiotic Stress*; Springer, 2017; pp 89–114.

Glaubitz, U.; Erban, A.; Kopka, J.; Hincha, D. K.; Zuther, E. High Night Temperature Strongly Impacts TCA Cycle, Amino Acid and Polyamine Biosynthetic Pathways in Rice in a Sensitivity-Dependent Manner. *J.Exp. Bot.* **2015,** *66*, 6385–6397.

Glaubitz, U.; Li, X.; Schaedel, S.; Erban, A.; Sulpice, R.; Kopka, J.; Hincha, D. K.; Zuther, E. Integrated Analysis of Rice Transcriptomic and Metabolomic Responses to Elevated Night Temperatures Identifies Sensitivity-and Tolerance-Related Profiles. *Plant Cell Environ.* **2017,** *40*, 121–137.

Grau, C. R.; Dorrance, A. E.; Bond, J.; Russin, J. S. Fungal Diseases. *Soybeans: Improvement Production Uses* **2004,** *16*, 679–763–678.

Gudkov, S. V.; Grinberg, M. A.; Sukhov, V.; Vodeneev, V. Effect of Ionizing Radiation on Physiological and Molecular Processes in Plants. *J. Environ. Radioactivity* **2019,** *202*, 8–24.

Hasanuzzaman, M.; Alam, M. M.; Nahar, K.; Mohsin, S. M.; Bhuyan, M. B.; Parvin, K.; Hawrylak-Nowak, B.; Fujita, M. Silicon-Induced Antioxidant Defense and Methylglyoxal

Detoxification Works Coordinately in Alleviating Nickel Toxicity in *Oryza sativa* L. *Ecotoxicology* **2019,** *28*, 261–276.

He, M.; He, C.-Q.; Ding, N.-Z. Abiotic Stresses: General Defenses of Land Plants and Chances for Engineering Multistress Tolerance. *Front. Plant Sci.* **2018,** *9*, 1771.

Henry, E.; Fung, N.; Liu, J.; Drakakaki, G.; Coaker, G. Beyond Glycolysis: GAPDHs Are Multi-Functional Enzymes Involved in Regulation of ROS, Autophagy, and Plant Immune Responses. *PLoS Genet.* **2015,** *11*, e1005199.

Herbers, K.; Takahata, Y.; Melzer, M.; Mock, H. P.; Hajirezaei, M.; Sonnewald, U. Regulation of Carbohydrate Partitioning during the Interaction of Potato Virus Y with Tobacco. *Mol. Plant Pathol.* **2000,** *1*, 51–59.

Hofmann, J.; Abd El Naser El Ashry, S.; Anwar, A. E.; Kopka, J.; Grundler, F. Metabolic Profiling Reveals Local and Systemic Responses of Host Plants to Nematode Parasitism. *Plant J.* **2010,** *62*, 1058–1071.

Hoque, T. S.; Hossain, M. A.; Mostofa, M. G.; Burritt, D. J.; Fujita, M.; Tran, L.-S. P. Methylglyoxal: An Emerging Signaling Molecule in Plant Abiotic Stress Responses and Tolerance. *Front. Plant Sci.* **2016,** *7*, 1341.

Huan, L.; Xie, X.; Zheng, Z.; Sun, F.; Wu, S.; Li, M.; Gao, S.; Gu, W.; Wang, G. Positive Correlation between PSI Response and Oxidative Pentose Phosphate Pathway Activity during Salt Stress in an Intertidal Macroalga. *Plant Cell Physiol.* **2014,** *55*, 1395–1403.

Huang, H.; Ullah, F.; Zhou, D.-X.; Yi, M.; Zhao, Y. Mechanisms of ROS Regulation of Plant Development and Stress Responses. *Front. Plant Sci.* **2019,** *10*.

Hwang, J. E.; Hwang, S. G.; Kim, S. H.; Lee, K. J.; Jang, C. S.; Kim, J. B.; Kim, S. H.; Ha, B. K.; Ahn, J. W.; Kang, S. Y. Transcriptome Profiling in Response to Different Types of Ionizing Radiation and Identification of Multiple Radio Marker Genes in Rice. *Physiol. Plant.* **2014,** *150*, 604–619.

Hýsková, V.; Plisková, V.; Červený, V.; Ryšlavá, H. NADP-Dependent Enzymes Are Involved in Response to Salt and Hypoosmotic Stress in Cucumber Plants. *Gen. Physiol. Biophys.* **2017,** *36*, 247–258.

Ijaz, M.; Ali, Q.; Fahad, S.; Ashraf, S.; Shahid, M.; Ahmad, S.; Hasanuzzaman, M. *Role of Reactive Sulfur Species in the Oxidative Metabolism in Plants. Reactive Oxygen, Nitrogen and Sulfur Species in Plants: Production, Metabolism, Signaling and Defense Mechanisms*, 2019; pp 729–742.

Ishikawa, T.; Takahara, K.; Hirabayashi, T.; Matsumura, H.; Fujisawa, S.; Terauchi, R.; Uchimiya, H.; Kawai-Yamada, M. Metabolome Analysis of Response to Oxidative Stress in Rice Suspension Cells Overexpressing Cell Death Suppressor Bax Inhibitor-1. *Plant Cell Physiol.* **2010,** *51*, 9–20.

Johnson, S. N.; Erb, M.; Hartley, S. E. Roots under Attack: Contrasting Plant Responses to Below-and Aboveground Insect Herbivory. *New Phytol.* **2016,** *2*, 413–418.

Jorge, I.; Navarro, R. M.; Lenz, C.; Ariza, D.; Jorrín, J. Variation in the Holm Oak Leaf Proteome at Different Plant Developmental Stages, between Provenances and in Response to Drought Stress. *Proteomics* **2006,** *6*, S207–S214.

Kataria, S.; Jajoo, A.; Guruprasad, K. N. Impact of Increasing Ultraviolet-B (UV-B) Radiation on Photosynthetic Processes. *J. Photochem. Photobiol. B* 2014; *137*, 55–66.

Kaur, G.; Asthir, B. Molecular Responses to Drought Stress in Plants. *Biol. Plant.* **2017,** *61*, 201–209.

Khanna, S. M.; Taxak, P. C.; Jain, P. K.; Saini, R.; Srinivasan, R. Glycolytic Enzyme Activities and Gene Expression in *Cicer arietinum* Exposed to Water-Deficit Stress. *Appl. Biochem. Biotechnol.* **2014,** *173*, 2241–2253.

Kozlov, M. V. Insect Herbivory on Two Willow Species in Northern Europe Is Independent of Pollution Load. 2015.

Kozlov, M. V.; Zvereva, E. L. Background Insect Herbivory: Impacts, Patterns and Methodology. *Progress Bot.* **2017,** *79*, 313–355.

Krawczyk, C. M.; Holowka, T.; Sun, J.; Blagih, J.; Amiel, E.; DeBerardinis, R. J.; Cross, J. R.; Jung, E.; Thompson, C. B.; Jones, R. G. Toll-Like Receptor–Induced Changes in Glycolytic Metabolism Regulate Dendritic Cell Activation. *Blood* **2010,** *115*, 4742–4749.

Kyndt, T.; Denil, S.; Bauters, L.; Van Criekinge, W.; De Meyer, T. Systemic Suppression of the Shoot Metabolism upon Rice Root Nematode. 2014.

Lata, R.; Chowdhury, S.; Gond, S. K.; White Jr, J. F. Induction of Abiotic Stress Tolerance in Plants by Endophytic Microbes. *Lett. Appl. Microbiol.* **2018,** *66*, 268–276.

Laxalt, A.; Cassia, R.; Sanllorenti, P.; Madrid, E.; Andreu, A.; Daleo, G.; Conde, R.; Lamattina, L. Accumulation of Cytosolic Glyceraldehyde-3-Phosphate Dehydrogenase RNA under Biological Stress Conditions and Elicitor Treatments in Potato. *Plant Mol. Biol.* **1996,** *30*, 961.

Leonard, S.; Hommais, F.; Nasser, W.; Reverchon, S. Plant–Phytopathogen Interactions: Bacterial Responses to Environmental and Plant Stimuli. *Environ. Microbiol.* **2017,** *19*, 1689–1716.

Less, H.; Angelovici, R.; Tzin, V.; Galili, G. Coordinated Gene Networks Regulating Arabidopsis Plant Metabolism in Response to Various Stresses and Nutritional Cues. *Plant Cell* **2011,** *23*, 1264–1271.

Lewitt, J. Responses of Plants to Environmental Stresses, 1980.

Li, K.; Xu, C.; Zhang, J. Proteome Profile of Maize (Zea Mays L.) Leaf Tissue at the Flowering Stage after Long-Term Adjustment to Rice Black-Streaked Dwarf Virus Infection. *Gene* **2011,** *485*, 106–113.

Li, X.; Wei, W.; Li, F.; Zhang, L.; Deng, X.; Liu, Y.; Yang, S. The Plastidial Glyceraldehyde-3-Phosphate Dehydrogenase Is Critical for Abiotic Stress Response in Wheat. *Int. J. Mol. Sci.* **2019,** *20*, 1104.

Liu, H.; Fu, Y.; Hu, D.; Yu, J.; Liu, H. Effect of Green, Yellow and Purple Radiation on Biomass, Photosynthesis, Morphology and Soluble Sugar Content of Leafy Lettuce via Spectral Wavebands "Knock Out". *Sci. Horti.* **2018,** *236*, 10–17.

Llave, C. Dynamic Cross-Talk between Host Primary Metabolism and Viruses during Infections in Plants. *Curr. Opin. Virol.* **2016,** *19*, 50–55.

Lu, X.; Huan, L.; Gao, S.; He, L.; Wang, G. NADPH from the Oxidative Pentose Phosphate Pathway Drives the Operation of Cyclic Electron Flow Around Photosystem I in High-Intertidal Macroalgae under Severe Salt Stress. *Physiol. Plant.* **2016,** *156*, 397–406.

Luck, J.; Spackman, M.; Freeman, A.; Trębicki, P.; Griffiths, W.; Finlay, K.; Chakraborty, S. Climate Change and Diseases of Food Crops. *Plant Pathol.* **2011,** *60*, 113–121.

Ludwig-Müller, J.; Prinsen, E.; Rolfe, S. A.; Scholes, J. D. Metabolism and Plant Hormone Action during Clubroot Disease. *J. Plant Growth Regul.* **2009,** *28*, 229–244.

Machado, R. A.; Ferrieri, A. P.; Robert, C. A.; Glauser, G.; Kallenbach, M.; Baldwin, I. T.; Erb, M. Leaf-Herbivore Attack Reduces Carbon Reserves and Regrowth from the Roots via Jasmonate and Auxin Signaling. *New Phytol.* **2013,** *200*, 1234–1246.

Maheswari, M.; Yadav, S.; Shanker, A. K.; Kumar, M. A.; Venkateswarlu, B. Overview of Plant Stresses: Mechanisms, Adaptations and Research Pursuit, Crop Stress and Its Management: In *Crop Stress and its Management: Perspectives and Strategies*; Springer, 2012; pp 1–18.

Mäkinen, K.; De, S. The Significance of Methionine Cycle Enzymes in Plant Virus Infections. *Curr. Opin. Plant Biol.* **2019,** *50*, 67–75.

Malamy, J.; Klessig, D. F. Salicylic Acid and Plant Disease Resistance. *Plant J.* **1992,** *2*, 643–654.

Masclaux-Daubresse, C.; Reisdorf-Cren, M.; Pageau, K.; Lelandais, M.; Grandjean, O.; Kronenberger, J.; Valadier, M.-H.; Feraud, M.; Jouglet, T.; Suzuki, A. Glutamine Synthetase-Glutamate Synthase Pathway and Glutamate Dehydrogenase Play Distinct Roles in the Sink-Source Nitrogen Cycle in Tobacco. *Plant Physiol.* **2006,** *140*, 444–456.

Mattiello, L.; Begcy, K.; Da Silva, F. R.; Jorge, R. A.; Menossi, M. Transcriptome Analysis Highlights Changes in the Leaves of Maize Plants Cultivated in Acidic Soil Containing Toxic Levels of Al 3+. *Mol. Biol. Rep.* **2014,** *41*, 8107–8116.

Mengin, V.; Pyl, E. T.; Alexandre Moraes, T.; Sulpice, R.; Krohn, N.; Encke, B.; Stitt, M. Photosynthate Partitioning to Starch in *Arabidopsis thaliana* Is Insensitive to Light Intensity But Sensitive to Photoperiod Due to a Restriction on Growth in the Light in Short Photoperiods. *Plant Cell Environ.* **2017,** *40*, 2608–2627.

Mizuno, H.; Kawahigashi, H.; Kawahara, Y.; Kanamori, H.; Ogata, J.; Minami, H.; Itoh, T.; Matsumoto, T. Global Transcriptome Analysis Reveals Distinct Expression among Duplicated Genes during Sorghum-Bipolaris Sorghicolainteraction. *BMC Plant Biol.* **2012,** *12*, 121.

Molassiotis, A.; Fotopoulos, V. Oxidative and Nitrosative Signaling in Plants: Two Branches in the Same Tree? *Plant Signal. Behav.* **2011,** *6*, 210–214.

Morkunas, I.; Woźniak, A.; Mai, V. C.; Rucińska-Sobkowiak, R.; Jeandet, P. The Role of Heavy Metals in Plant Response to Biotic Stress. *Molecules* **2018,** *23*, 2320.

Mustroph, A. Flooding Stress in Plants. 2018.

Mutuku, J. M.; Nose, A. Changes in the Contents of Metabolites and Enzyme Activities in Rice Plants Responding to Rhizoctonia Solani Kuhn Infection: Activation of Glycolysis and Connection to Phenylpropanoid Pathway. *Plant Cell Physiol.* **2012,** *53*, 1017–1032.

Nafees, M.; Fahad, S.; Shah, A. N.; Bukhari, M. A.; Ahmed, I.; Ahmad, S.; Hussain, S. *Reactive Oxygen Species Signaling in Plants, Plant Abiotic Stress Tolerance*; Springer, 2019; pp 259–272.

Nayudu, M. Stress Physiology in Plant Bacterial Diseases. In *Proceedings of the Indian Academy of Sciences-Section B*; Springer, **1969,** pp 146–151.

Nolfi-Donegan, D.; Braganza, A.; Shiva, S. Mitochondrial Electron Transport: Oxidative Phosphorylation, Mitochondrial Oxidant Production, and Methods of Measurement. *Redox Biol.* **2020,** 101674.

Oliver, M. J.; Jain, R.; Balbuena, T. S.; Agrawal, G.; Gasulla, F.; Thelen, J. J. Proteome Analysis of Leaves of the Desiccation-Tolerant Grass, *Sporobolus stapfianus*, in Response to Dehydration. *Phytochemistry* **2011,** *72*, 1273–1284.

Olson, K. R.; Straub, K. D. The Role of Hydrogen Sulfide in Evolution and the Evolution of Hydrogen Sulfide in Metabolism and Signaling. *Physiology* **2015**.

Pál, M.; Janda, T.; Majláth, I.; Szalai, G. Involvement of Salicylic Acid and Other Phenolic Compounds in Light-Dependent Cold Acclimation in Maize. *Int. J. Mol. Sci.* **2020,** *21*, 1942.

Pandey, P.; Sinha, R.; Mysore, K. S.; Senthil-Kumar, M. Impact of Concurrent Drought Stress and Pathogen Infection on Plants. In *Combined Stresses in Plants*; Springer, 2015; pp 203–222.

Pilkington, S. M.; Encke, B.; Krohn, N.; Hoehne, M.; Stitt, M.; PYL, E. T. Relationship between Starch Degradation and Carbon Demand for Maintenance and Growth in *Arabidopsis thaliana* in Different Irradiance and Temperature Regimes. *Plant Cell Environ.* **2015,** *38*, 157–171.

Prasch, C. M.; Sonnewald, U. Simultaneous Application of Heat, Drought, and Virus to Arabidopsis Plants Reveals Significant Shifts in Signaling Networks. 2013.

Rai, M.; Agarkar, G. Plant–Fungal Interactions: What Triggers the Fungi to Switch among Lifestyles? *Crit. Rev. Microbiol.* **2016,** *42*, 428–438.

Ramegowda, V.; Senthil-Kumar, M. The Interactive Effects of Simultaneous Biotic and Abiotic Stresses on Plants: Mechanistic Understanding from Drought and Pathogen Combination. *J. Plant Physiol.* **2015,** *176*, 47–54.

Raza, A.; Razzaq, A.; Mehmood, S. S.; Zou, X.; Zhang, X.; Lv, Y.; Xu, J. Impact of Climate Change on Crops Adaptation and Strategies to Tackle Its Outcome: A Review. *Plants* **2019,** *8*, 34.

Reddy, K. J. *Nutrient Stress, Physiology and Molecular Biology of Stress Tolerance in Plants*; Springer, 2006; pp 187–217.

Robert, C. A.; Ferrieri, R. A.; Schirmer, S.; Babst, B. A.; Schueller, M. J.; Machado, R. A.; Arce, C. C.; Hibbard, B. E.; Gershenzon, J.; Turlings, T. C. Induced Carbon Reallocation and Compensatory Growth as Root Herbivore Tolerance Mechanisms. *Plant Cell Environ.* **2014,** *37*, 2613–2622.

Rocha, M.; Licausi, F.; Araujo, W. L.; Nunes-Nesi, A.; Sodek, L.; Fernie, A. R.; van Dongen, J. T. Glycolysis and the Tricarboxylic Acid Cycle Are Linked by Alanine Aminotransferase during Hypoxia Induced by Waterlogging of *Lotus japonicus*. *Plant Physiol.* **2010,** *152*, 1501–1513.

Rojas, C. M.; Senthil-Kumar, M.; Tzin, V.; Mysore, K. Regulation of Primary Plant Metabolism during Plant-Pathogen Interactions and Its Contribution to Plant Defense. *Front. Plant Sci.* **2014,** *5*, 17.

Roossinck, M. J. Plants, Viruses and the Environment: Ecology and Mutualism. *Virology* **2015,** *479*, 271–277.

Rudrappa, T.; Czymmek, K. J.; Paré, P. W.; Bais, H. P. Root-Secreted Malic Acid Recruits Beneficial Soil Bacteria. *Plant Physiol.* **2008,** *148*, 1547–1556.

Ruperti, B.; Botton, A.; Populin, F.; Eccher, G.; Brilli, M.; Quaggiotti, S.; Trevisan, S.; Cainelli, N.; Guarracino, P.; Schievano, E. Flooding Responses on Grapevine: A Physiological, Transcriptional, and Metabolic Perspective. *Front. Plant Sci.* **2019,** *10*.

Saminathan, T.; Bodunrin, A.; Singh, N. V.; Devarajan, R.; Nimmakayala, P.; Jeff, M.; Aradhya, M.; Reddy, U. K. Genome-Wide Identification of microRNAs in Pomegranate (*Punica granatum* L.) by High-Throughput Sequencing. *BMC Plant Biol.* **2016,** *16*, 122. https://doi.org/10.1186/s12870-016-0807-3.

Savvides, A.; Ali, S.; Tester, M.; Fotopoulos, V. Chemical Priming of Plants against Multiple Abiotic Stresses: Mission Possible? *Trends Plant Sci.* **2016,** *21*, 329–340.

Schneider, M.; Knuesting, J.; Birkholz, O.; Heinisch, J. J.; Scheibe, R. Cytosolic GAPDH as a redox-dependent regulator of energy metabolism. *BMC Plant Biol.* **2018,** *18*, 1–14.

Shao, H.-b.; Chu, L.-y.; Shao, M.-a.; Jaleel, C. A.; Hong-mei, M. Higher Plant Antioxidants and Redox Signaling under Environmental Stresses. *Comptes Rendus Biol.* **2008,** *331*, 433–441.
Singal, H.; Sheoran, I.; Singh, R. Effect of Water Stress on Photosynthesis and in Vitro Activities of the PCR Cycle Enzymes in Pigeonpea (*Cajanus cajan* L.). *Photosynthesis Res.* **1985,** *7*, 69–76.
Singh, U. P.; Chouhan, S. S.; Jain, S.; Jain, S. Multilayer Convolution Neural Network for the Classification of Mango Leaves Infected by Anthracnose Disease. *IEEE Access* **2019,** *7*, 43721–43729.
Singh, Y. J.; Grewal, S. K.; Gill, R. K. Role of Glutathione in Methylglyoxal Detoxification Pathway during Yellow Mosaic Virus (YMV) Infection in Black Gram (*Vigna mungo* (L.) Hepper). *Physiol. Mol. Plant Pathol.* **2020,** *111*, 101513.
Stewart, G.; Lee, J. Desiccation Injury in Mosses. *New Phytol.* **1972,** *71*, 461–466.
Sun, C.; Li, M.; Gao, X.; Liu, L.; Wu, X.; Zhou, J. Metabolic Response of Maize Plants to Multi-Factorial Abiotic Stresses. *Plant Biol.* **2016,** *18*, 120–129.
Suzuki, N.; Rivero, R. M.; Shulaev, V.; Blumwald, E.; Mittler, R. Abiotic and Biotic Stress Combinations. *New Phytol.* **2014,** *203*, 32–43.
Švubová, R.; Slováková, Ľ.; Holubová, Ľ.; Rovňanová, D.; Gálová, E.; Tomeková, J. Evaluation of the Impact of Cold Atmospheric Pressure Plasma on Soybean Seed Germination. *Plants* **2021,** *10*, 177.
Sweetlove, L.; Heazlewood, J.; Herald, V.; Holtzapffel, R.; Day, D.; Leaver, C.; Millar, A. The Impact of Oxidative Stress on *Arabidopsis mitochondria. Plant J.* **2002,** *32*, 891–904.
Tecsi, L. I.; Smith, A. M.; Maule, A. J.; Leegood, R. C. A Spatial Analysis of Physiological Changes Associated with Infection of Cotyledons of Marrow Plants with Cucumber Mosaic Virus. *Plant Physiol.* **1996,** *111*, 975–985.
Terrile, M.; Iglesias, M.; Casalongué, C.; París, R. *Reactive Nitrogen Species Mediated Cross-Stress Tolerance in Plants, Priming-Mediated Stress and Cross-Stress Tolerance in Crop Plants*; Elsevier, 2020; pp 133–148.
Vanlerberghe, G. C. Alternative Oxidase: A Mitochondrial Respiratory Pathway to Maintain Metabolic and Signaling Homeostasis during Abiotic and Biotic Stress in Plants. *Int. J. Mol. Sci.* **2013,** *14*, 6805–6847.
Visser, E.; Voesenek, L.; Vartapetian, B.; Jackson, M. Flooding and Plant Growth. *Ann. Bot.* **2003,** *91*, 107–109.
von Koskull-Döring, P.; Scharf, K.-D.; Nover, L. The Diversity of Plant Heat Stress Transcription Factors. *Trends Plant Sci.* **2007,** *12*, 452–457.
Wang, X.; Sakata, K.; Komatsu, S. An Integrated Approach of Proteomics and Computational Genetic Modification Effectiveness Analysis to Uncover the Mechanisms of Flood Tolerance in Soybeans. *Int. J. Mol. Sci.* **2018,** *19*.
Whittaker, A.; Martinelli, T.; Bochicchio, A.; Vazzana, C.; Farrant, J. Comparison of Sucrose Metabolism during the Rehydration of Desiccation-Tolerant and Desiccation-Sensitve Leaf Material of *Sporobolus stapfianus*. *Physiol. Plant.* **2004,** *122*, 11–20.
Xu, C.; Huang, B. Differential Proteomic Responses to Water Stress Induced by PEG in Two Creeping Bentgrass Cultivars Differing in Stress Tolerance. *J. Plant Physiol.* **2010,** *167*, 1477–1485.
Xu, S. Y.; Weng, J. K. Climate Change Shapes the Future Evolution of Plant Metabolism. *Adv. Genet.* **2020,** *1*, e10022.

Yang, S.; Zhai, Q. Cytosolic GAPDH: A Key Mediator in Redox Signal Transduction in Plants. *Biol. Plant.* **2017,** *61*, 417–426.

Yao, K.; Wu, Y. Phosphofructokinase and Glucose-6-Phosphate Dehydrogenase in Response to Drought and Bicarbonate Stress at Transcriptional and Functional Levels in Mulberry. *Russian J. Plant Physiol.* **2016,** *63*, 235–242.

Yao, N.; Tada, Y.; Sakamoto, M.; Nakayashiki, H.; Park, P.; Tosa, Y.; Mayama, S. Mitochondrial Oxidative Burst Involved in Apoptotic Response in Oats. *Plant J.* **2002,** *30*, 567–579.

Zaffagnini, M.; Fermani, S.; Costa, A.; Lemaire, S. D.; Trost, P. Plant Cytoplasmic GAPDH: Redox Post-Translational Modifications and Moonlighting Properties. *Front. Plant Sci.* **2013,** *4*, 450.

Zeilinger, S.; Gupta, V. K.; Dahms, T. E.; Silva, R. N.; Singh, H. B.; Upadhyay, R. S.; Gomes, E. V.; Tsui, C. K.-M.; Nayak, S. C. Friends or Foes? Emerging Insights from Fungal Interactions with Plants. *FEMS Microbio. Rev.* **2016,** *40*, 182–207.

Zhong, M.; Yuan, Y.; Shu, S.; Sun, J.; Guo, S.; Yuan, R.; Tang, Y. Effects of Exogenous Putrescine on Glycolysis and Krebs Cycle Metabolism in Cucumber Leaves Subjected to Salt Stress. *Plant Growth Regul.* **2016,** *79*, 319–330.

Zhou, S.; Lou, Y.-R.; Tzin, V.; Jander, G. Alteration of Plant Primary Metabolism in Response to Insect Herbivory. *Plant Physiol.* **2015,** *169*, 1488–1498.

Zvereva, E. L.; Kozlov, M. V. Sources of Variation in Plant Responses to Belowground Insect Herbivory: A Meta-Analysis. *Oecologia* **2012,** *169*, 441–452.

CHAPTER 14

Modulation of TCA Cycle and Enzymes During Different Environmental Stresses

SUDIP KUMAR MANDAL[1], DEBAJYOTI MUKHERJEE[2], SINDHUJIT ROY[3], SABYASACHI BANERJEE[4], and SANKHADIP BOSE[5*]

[1]*Dr. B. C. Roy College of Pharmacy and Allied Health Sciences, Bidhannagar, Durgapur 713206, West Bengal, India*

[2]*R&D Head, Herbal Extraction, Keva Fragrances Pvt. Ltd., Mumbai, India*

[3]*Department of Biotechnology, SRM University, Chennai, Tamil Nadu, India*

[4]*Gupta College of Technological Sciences, Ashram More, G.T. Road, Asansol 713101, West Bengal, India*

[5]*Department of Pharmacognosy & Phytochemistry, Bengal School of Technology, A College of Pharmacy, Chuchura, Hooghly 712102, West Bengal, India*

[*]*Corresponding author. E-mail: sankha.bose@gmail.com*

ABSTRACT

Tricarboxylic acid cycle (TCA) is an important metabolic cycle as it facilitates the synthesis of ATP through catabolic pathways. The essential signaling functions of mitochondria include the mitochondrial regulation of autophagy through AMP-activated protein kinase which induces autophagy, thus triggering the release of Ca^{2+} in endoplasmic reticulum which enhances the TCA cycle. The structure and functionality of eight different enzymes

Photosynthesis and Respiratory Cycles during Environmental Stress Response in Plants. Aryadeep Roychoudhury (Ed.)

of TCA cycle inside the mitochondrial matrix, linked with oxidation and reduction products associated with it, facilitate the structures and operations of the TCA cycle. Considering the application of TCA cycle inside the plant body, the carrier for pyruvate into the matrix is unidentified till now. Despite this, this study will broaden the horizons on the interconnectedness of photosynthesis, root development, and various other heterotrophic tissues with TCA cycle. During the study of genomic organization of the enzymes, major genes were found to be subcellular and localized in mitochondria followed by cytosol, peroxisomes, and chloroplasts. Using metabolic control analysis, flux control analysis of respiration was quantified at steady state which associates the quantity of a particular enzyme with respiration rate. Comprehension of complex pathways of metabolism and regulation of TCA activity as well as in vivo regulation requires intensive effort to correlate with photorespiration, nitrate assimilation, and photosynthesis. Environmental effect like high temperature or increased daylight reduces net carbon because it increases the plant respiration, not photosynthesis. Ecological hassle, for example, floods lead to low availability of molecular oxygen, differing levels of enzymes obstruct or stimulate certain processes impacting respiration, or may cause mutation leading to suppressed immunity or cancer, as with the case of HIF-α regulation, and aluminum toxicity which evokes dysfunctionality. These studies show that modulating TCA cycle and its enzymes by environmental and/or intrinsic factors affects various biological processes.

14.1 INTRODUCTION

The most important function of mitochondria in all eukaryote cells is the synthesis of adenosine triphosphate (ATP) along with reduction of metabolites and intermediates for various other biological processes. While mitochondria are the primary site for energy transduction and ATP generation, they also facilitate apoptosis or autophagy, fulfill carbon requirement through anabolic processes, seed germination, and several other vital cellular and developmental processes (Achnine et al., 2004). The mitochondria are said to be important and and are considered to be the essential biosynthetic precursor for growth and maintenance. Mitochondrial metabolism stimulates many light-associated processes as in photorespiration, photosynthesis, and nitrogen metabolism (Bandehagh and Taylor, 2020). The TCA cycle includes consumption and regeneration of citric acid, due to which it is also named as citric acid cycle and is directed by an enzymatic reaction to form water and carbon dioxide. Through the catabolism of fats, sugars, and proteins, the acetyl-CoA that is produced finds entry into the citric acid cycle. The purpose of TCA is to harvest high-energy electrons, but the cycle by itself does not generate a significant quantity of ATP or add oxygen as a reaction. It actually eliminates electrons from acetyl-CoA and utilizes them in the conformation of NADH and $FADH_2$ (Bandehagh and Taylor, 2020). In the method of oxidative phosphorylation, the electrons are released for the re-oxidation of $FADH_2$ and NADH with the help of proton gradients found across membranes in a sequence of membrane proteins primarily referred to as the electron transport chain (Achnine et al., 2004). Then the protons flow through the ATP synthase and produce inorganic phosphate and ATP from adenosine diphosphate (ADP). Majority of the genes responsible for the enzymes were identified, and it was found that more than a single gene contributes to the production of a single enzyme as there is a well-defined distribution of subunits of respective enzymes among the genes (Barone et al., 2008). From the phylogenetic analysis, several proteins were identified as paralogous and the manifestation of these pairs amidst the proteins showed extension of genes after splitting up of these paralogous pairs. Promoters of various genes possess several elements, which are in direct correlation with the environmental temperature and stress (Carrer et al., 2019). These genes directly correlate with root development, flower development, root respiration activity, gametophyte development, etc. Metabolic control analysis (MCA) tool was implemented for processing and evaluating different pathways, it quantizes fluxes and species concentrations. The regulation of TCA cycle is mainly involved with the evaluation of concentrations of both NADH and ATP (Chandel et al., 2000). The significant evaluating markers are mainly the enzymes, α-ketoglutarate dehydrogenase, and isocitrate dehydrogenase. The mitochondrion has evolved in numerous ways to develop strategies to survive with the environment, and associate environmental signals to communicate effectively with the other cells to maintain cellular homeostasis (Carrer et al., 2019). Differences in temperature affect the enzymes of TCA cycle directly, and the pH balance of the environment triggers several other biochemical processes which indirectly affect the TCA cycle (Bandehagh and Taylor, 2020). Inadequate supply of water in certain conditions impact plant growth, mild-to-moderate stress in water balance affects mitochondrial activities and several key enzymes of the TCA. Aluminum affects significantly, as it shows inhibitory behavior toward various TCA enzymatic activities. In case the key enzymes are in a state of disbalance due to environmental factors, they dysregulate the TCA cycle with noticeable changes associated with it (Deng et al., 2009). The succinate is linked with TCA dysfunction as it inhibits the HIF-α-prolyl hydroxylases found in cytosols which stabilizes and activates the HIF-1α. Stem cells are regulated by internal and external mechanisms; interactions with these cellular metabolisms and regulatory mechanisms are identified in multiple systems which are associated with changes in glycolysis or oxidative phosphorylation. In some cancer cells, excessive glycolysis takes place which results in pyruvic acid being converted to lactic acid rather than entering the TCA cycle, which causes the glycolysis products to divert into anabolic pathways instead of going for TCA-mediated total oxidation into substances, such as carbon dioxide and water (Downs et al., 1998).

14.2 ESSENTIAL SIGNALING FUNCTIONS OF THE MITOCHONDRIA

Mitochondria are very important organelles due to their prime role in several biosynthetic reactions. They are actively associated with cell signaling processes, such as initiation of cell signaling, autophagy, innate immunity, cell death signaling, apoptosis signaling, transducers, and effectors (Tan et al., 2007). Various processes, which include mitochondrial dynamics and intermediate metabolites, contribute to the regulation of mitochondrial cell signaling. Failure to control cell signaling and regulation activities may have adverse effects. Apoptosis is crucial for development and homeostasis among tissues, the deregulation of which results in neurodegeneration and cancer. The activation of caspase protease is necessary to rapid apoptosis after activation (Taylor et al, 2008). Mitochondria are responsible for regulating apoptosis and caspase activity through diverse mechanisms (Oberst et al., 2008). Between the intermembrane space, many proteins like cytochrome *c* are responsible for caspase activation after they are released in cytoplasm (Tait and Green, 2010). Mitochondria undergo continuous fission and fusion among themselves (Westermann, 2010). The regulation of apoptosis is conducted through other pathways apart from the mitochondrial outer membrane permeabilization, through caspase-8 regulation. Apart from apoptosis, necrosis-like cell death pathway, death receptor ligation, etc. are also triggered by various mitochondrial processes (Galluzzi et al., 2012). The major cellular source of reactive oxygen species is the mitochondria. Hence, they are vital for commencing necrosis-like cell death pathway (He et al., 2009). The host cells of the immune system detect pathogens and cause damages through pattern recognition receptors (PPRs). There exists a group of PPR such as the retinoic acid-inducible protein-like receptor (RLR) family that reacts to severe viral infection. Mitochondria role for antiviral immunity activated by RLR was confirmed with the help of RLR signal-mediated mitochondrial adaptor protein identification also called the mitochondrial antiviral signaling protein (MAVS) (Xu et al., 2005; Kawai et al., 2005). MAVS in the outer mitochondrial membrane is important for the RLR antiviral signaling from the mitochondria (Seth et al., 2005). The aggregation of mitochondrial antiviral signaling protein on the outer membrane results in the initiation of downstream pathways which are regulated by more mitochondrial proteins (West et al., 2011). There exist positive regulators which are essential for RLR signaling, but are situated in the contact sites between endoplasmic reticulum (ER) and mitochondria (Ishikawa and Barber. 2008). The above findings suggest that the outer mitochondrial membrane is very important as a signaling medium for the antiviral RLR along with mitochondrial dynamics. Damage-associated molecular patterns (DAMPs) are vital for the initiation of signaling cascades in various immune cells. Mitochondria are rich in DAMPs that induce hyperactivation in the innate immunity, resulting in systemic inflammatory response syndrome which is responsible for the shock (Medzhitov, 2007). Mitochondrial reactive oxygen species act as signal molecule for innate immunity which is important for antimicrobial properties during the process of phagocytosis (West et al., 2011). Interaction of certain toll-like receptors results in the signaling adaptors being translocated. Autophagy helps in cellular detoxification and also produces metabolites which are required for energy generation and various anabolic processes. The initiation of autophagy is controlled by mitochondrial signaling processes where specific kinases must be activated for the initiation (Mizushima, 2010). Decrease of cellular ATP results in the activation of adenosine monophosphate (AMP)-activated protein kinase which immediately initiates phosphorylation of the kinase proteins. Mitochondria intakes Ca^{2+} ion released by endoplasmic reticulum during the activation of inositol triphosphate receptors where the intracellular Ca^{2+} is vital for the regulation of autophagy by inhibiting the AMPK activity and efficient oxidative phosphorylation which in turn helps in greater ATP production (Egan et al., 2011). The transformation from pyruvate to lactate or acetyl-CoA throughout TCA cycle activity in the matrix of mitochondria is determined by the amount of oxygen available and the rate of proliferation or cellular growth, which requires ATP and macromolecules to be available. In many cancer cells, this is achieved by the Warburg Effect which is characterized by increased glycolysis under normal oxygen conditions which results in the pyruvate being converted to lactate rather than entering into TCA cycle.

Thus, there occurs diversion of glycolytic products into anabolic biosynthetic pathways rather than opting for complete oxidation into carbon dioxide and water to produce significant amount of ATP. The growth and proliferation can create metabolic changes for the generation of substrates which cause disruption in various other biochemical pathways and mechanisms dependent on mitochondrial TCA cycle (Vander, 2011).

14.3 FUNCTIONAL BEHAVIOR OF THE TCA CYCLE

The citric acid cycle is an important metabolic pathway which links protein, carbohydrate, and fat metabolisms. The cycle reactions are carried out through eight enzymes, which entirely oxidize acetate (a 2-carbon molecule) in the acetyl-CoA form to two molecules of carbon dioxide and water. The two-carbon organic compound acetyl-CoA (a form of acetate) is developed through the catabolism of proteins, sugars, and fats, and joins the citric acid cycle. The cycle also converts three equivalents of nicotinamide adenine dinucleotide (NAD^+) to three equivalents of reduced NAD^+ (NADH), one equivalent of flavin adenine dinucleotide (FAD) to one equivalent of $FADH_2$, and one equivalent each of guanosine diphosphate (GDP) and inorganic phosphate (Pi) to one equivalent of guanosine triphosphate (GTP). The citric acid cycle generates NADH and $FADH_2$, which are then utilized by the oxidative phosphorylation pathway to produce energy-rich ATPs. It acts differently in the case of plants. The synthesis of respiratory substrates such as carbohydrates is carried over to the TCA cycle as the main substrate. Phosphoenolpyruvate (PEP) is the main substrate which due to oxidative pentose phosphate and glycolysis pathway gets metabolized. PEP is further converted to either pyruvate or oxaloacetate (OAA) which further gets converted to malate. The transport of intermediates of metabolic biosynthesis is constricted to citrate with strict interchange with either malate or oxaloacetate (OAA). The TCA cycle is comprised of elementary flux modes for different operation (Schuster, 1999). An elementary flux mode (EFM) is a nominal reaction with consistency to valid steady state. The six EFMs support different operations. EFM A in the TCA cycle undergoes mass–conversion relationship, that is, the total quantity of malate, OAA, citrate, 2-oxoglutarate, and succinate is preserved when pyruvate is transported to the mitochondria and no metabolic intermediates are transported out. Due to this, the cycle does not possess a net synthesis of the aforementioned intermediates due to the inability to catalyze. However, if there is NAD-malic enzyme converting malate to pyruvate, this can lead to instability of the pathway due to depleting metabolic intermediates. For the TCA cycle substrate, EFM B is the malate which is substituted for both OAA and pyruvate. EFM C is the other mode of operation where the total quantity of OAA and citrate is conserved within the mitochondria. This occurs as the TCA cycle provides citrate as a precursor for biosynthesis. EFM D is similar to EFM C with the change being that the malate substitutes for OAA absorption, interchanging with citrate. However, in the existence of NAD-malic enzyme, there will be depletion of intermediates. EFM E and F are very similar to EFM C and EFM D with the change that pyruvate import gets substituted by malate (Fig. 14.1) (Schuster, 2000).

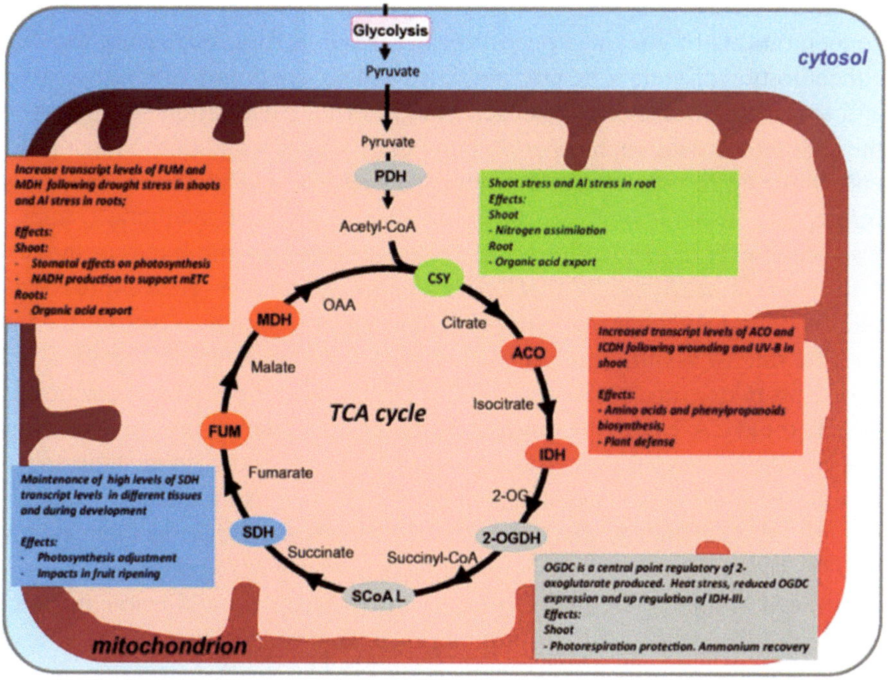

FIGURE 14.1 Enzymes of tricarboxylic acid cycle and their associated processes.

Source: Reprinted with permission from Cavalcanti et al., 2014. © John Wiley and Sons.

14.4 APPLICATION OF TCA CYCLE IN PLANTS

The mitochondria found in plants possess a fully functional tricarboxylic acid cycle that is necessary in producing ATP and supplying carbon for various biosynthetic processes. The acids of organic origin found in plants, specifically the TCA cycle intermediates, are known to help in various other functions beyond cellular metabolism (Fernie and Martinoia, 2009) indicating that these enzymes associated with interconversion are under heavy regulatory controls. A wide range of regulatory mechanisms have been put forward for TCA cycle, such as light-dependent regulation, redox regulation, phosphorylation, phytochrome, and nitric oxide regulation (Nunes-Nesi et al., 2013). Phosphorylation of pyruvate dehydrogenase (PDH) complex is one mechanism through which TCA cycle is downregulated (Isela. et al., 1992). Mitochondrial and cytosolic thioredoxin systems have a very crucial role in regulation (Schmidtmann et al., 2014). The cycle in plants or mammals formulate dynamic enzyme assemblies (Zhang et al., 2020). A few theories are surfaced regarding enzyme assemblies, such as decreasing concentrations of enzymes necessary to sustain a flux, increasing localized concentration of metabolic intermediates and sequestration of highly reactive and toxic intermediates (Jorgensen et al., 2005). Respiration and photosynthesis interact with each other as they share oxygen and carbon as product and substrate or substrate and product, respectively. The TCA cycle is very influential on photosynthetic metabolism and on photosynthetic tissues. It was previously understood that the cycle is completely inhibited for the reaction which is catalyzed by PDH. The plastid and nuclear gene expression of photosynthetic mechanism is upregulated, supported with ascorbate feeding, and the rate of assimilation of carbon is also upregulated (Smirnoff, 2000). These properties are based on the TCA cycle and the circumstances through which the TCA cycle operates. It is being considered that to understand the process of respiration, it is simpler to look at roots rather than leaves. In recent times, it is suggested that the TCA cycle is crucial for producing organic acids which is discharged in the rhizosphere (De la Fuente et al., 1997). To some extent, the mechanisms underlying aluminum tolerance have been studied. The majority of soil in subtropical and tropical regions are acidic because of the accumulation of aluminum cations, which cause nutrient and water disruptions (Nunes-Nesi et al., 2014). Increase in organic acid discharge will enhance tolerance for aluminum. This can be done by increasing the activities of enzymes, viz., pyruvate phosphate dikinase, citrate synthase, malate dehydrogenase, the combinations of citrate synthase and PEP carboxylase, malic enzyme, and oxalate (Wang et al., 2012; Trejo-Tellez et al., 2010; Tesfaye et al., 2001; Han et al., 2009; Ma, 2000). In the process of development for other heterotrophic tissues, it was found that mitochondrial citrate synthase, when downregulated, causes cytoplasmic male sterility. Seed metabolism and development is directly linked to the TCA cycle activity (Heyer et al., 1995; Fait et al., 2006; Zou et al., 1999; Angelovici et al., 2011; Fait et al., 2011).

14.5 GENOMIC ORGANIZATION OF THE ENZYMES OF THE TCA CYCLE

The citric acid cycle primarily consists of eight enzymes which function in metabolic regulation and assist the substrates formed in the cyclic process. Specific genes from the plant genomes are encoded to synthesize the enzymes engaged in the TCA cycle (Liu. et al., 2019). Genes associated with enzyme production are linked to the development of the primary and secondary tissues of plants (Liu. et al., 2019). Enzymes, such as aconitase, citrate synthase, isocitrate dehydrogenase, α-ketoglutarate dehydrogenase complex, succinyl-CoA synthetase, succinate dehydrogenase, malate dehydrogenase, and fumarase are the enzymes present in the TCA cycle (Liu et al., 2019).

In the genome of maize, nearly 91 genes identified are linked to the overall production of the TCA enzyme. Based on gene expression in certain plants, enzyme production is implicated (Liu et al., 2019). Several factors include environmental stress, proper nutrient supplementation. In plants, such as *Solanum lycopersicum* and *Zea mays*, specified genes, such as *ACO* and *IDH* are present in certain regions (roots, shoots, leaves) and express their gene identity to synthesize enzymes depending upon the factors that affect the physiological and metabolic pathway involved (Liu et al., 2019). In *Zea mays*, the genes responsible for enzyme productions are identified through gene mapping corresponding to the protein synthesis. At the initial phase, Zm00001d027558 gene codes for aconitase enzyme, Zm00001d008244 gene encodes the isocitrate dehydrogenases enzyme, Zm00001d025258 gene for alpha-dehydrogenases complexes, and Zm00001d044042 gene for the production of malate dehydrogenase (Liu et al., 2019). Eleven genes are identified for malate dehydrogenase and aconitase production, and 18 genes for alpha-ketoglutarate dehydrogenase enzyme production (Liu et al., 2019). The genes identified are also responsible for several potential developmental processes that include floral development, cotyledon development, and root development (Liu et al., 2019). For *Arabidopsis thaliana*, enzymes, such as citrate synthase, aconitase, isocitrate dehydrogenase are produced by genes present in the mitochondria, peroxisomes, and cytosol, respectively (Wagner, 2012).

14.6 ENZYMES OF THE TCA CYCLE

Citric acid cycle enzymes include isocitrate dehydrogenase, aconitase, citrate synthase, alpha-ketoglutarate dehydrogenase, fumarase, succinyl-CoA synthase, malate dehydrogenase, and succinate dehydrogenase (Martínez-Reyes and Chandel, 2020). The aforementioned enzymes are subjected to numerous mechanisms to produce different substrates in their corresponding phases in the cyclic process. Condensation, isomerization, dehydrogenation, and hydration are the fundamental mechanisms involved in substrate formation (Table 14.1) (Martínez-Reyes and Chandel, 2020).

TABLE 14.1 Enzymes Along with Their Subcellular Localization and Genes Associated with Them.

Enzymes	Localization subcellular	Genes associated
Citrate synthase	Mitochondria and peroxisome	2
Succinate dehydrogenase	Mitochondria	11
Isocitrate dehydrogenase	Cytosol, mitochondria and peroxisome	5
Succinyl-CoA ligase	Cytosol and mitochondria	3
2-oxoglutarate dehydrogenase	Mitochondria	6
Fumarase	Mitochondria and cytosol	2
Aconitase		2
Malate dehydrogenase	Chloroplasts, mitochondria, cytosol, and peroxisomes	4

Acetyl-CoA, formed from the decarboxylation process, is a two-carbon compound, and oxaloacetate undergoes the condensation process to form citrate (Martínez-Reyes and Chandel, 2020). This process performed by the citrate synthase enzyme, acts independently without dependence on the cofactors. Succinyl-CoA inhibits and regulates the function of citrate (Martínez-Reyes and Chandel, 2020). More number of ATP and NADH molecules also inhibit the enzymatic function of citrate synthase. The second most important step in the citric acid cycle is the isomerization of citrate that includes dehydration and hydration process (Martínez-Reyes and Chandel, 2020). Aconitase enzyme conducts the isomerization process. Cis-aconitase is formed as an intermediate and later converted to isocitrate. Isocitrate later is converted to alpha-ketoglutarate by the enzyme isocitrate dehydrogenases. $NADH_2$ is utilized as a cofactor in this hydrogenation process. Oxalosuccinate so formed in the initial stage is a temporary compound that is highly unstable (Tretter and Adam-Vizi, 2005). The oxalosuccinate thus formed undergoes decarboxylation process to form alpha-ketoglutarate (Tretter and Adam-Vizi, 2005). Alpha-ketoglutarate dehydrogenase regulates and converts the alpha-ketoglutarate to the successive product succinyl-CoA. Fe^{2+} acts as a cofactor in this process (Tretter and Adam-Vizi, 2005). The alpha-ketoglutarate dehydrogenase is highly regulated and controls the metabolic fluxes in the Krebs cycle, which can be inhibited by reactive oxygen species (ROS) (Tretter and Adam-Vizi, 2005). CO_2, H^+, and NADH are the products formed.

Substrate-level phosphorylation occurs and GDP is converted to GTP (Jardim-Messeder. et al, 2015). Succinate thiokinase acts on succinyl-CoA to form succinate which is termed as substrate-level phosphorylation (Jardim-Messeder et al, 2015). Thioester bond of succinyl-CoA is cleaved to form succinate in the cyclic process. Succinate dehydrogenase present in the TCA cycle additionally plays a major role in the electron transport chain and is identified as complex II and also regulates the nitrogen assimilation process (Jardim-Messeder et al., 2015). Succinate dehydrogenase converts the succinate to fumarate. Fumarase enzyme present in the Krebs cycle form malate as the product from fumarate. Fumarase carries out hydration process to produce malate catalyzed by the enzyme fumarate hydratase. The fumarate production also takes place in the ornithine cycle (Martínez-Reyes and Chandel, 2020). The final process in the TCA cycle is the formation of oxaloacetate from malate catalyzed by malate dehydrogenase. NADH formation is regulated by malate dehydrogenase (Minárik et al., 2002). Steps, such as reduction of NAD^+ and oxidizing the hydroxyl group are the significant steps involved in the interconversion process, and NAD^+ and $NADP^+$ act as cofactor for the enzymatic activity (Minárik et al., 2002). Finally, the cyclic process continues due to the reaction of oxaloacetate with acetyl-CoA.

14.7 ANALYTICAL APPROACH FOR METABOLIC FLUX CONTROL IN RESPIRATION

Metabolic flux control is a technique, which is mainly used for close examination with respect to the management of the intermediary forming metabolite concentrations and fluxes in a metabolic pathway (Kochian et al., 2007). Metabolic control evaluation is intently associated with a highly specialized engineering area commonly known as sensitivity analysis. Other processes that include the analysis of the peculiar kinetic behavior of a structure, which is made up of multiple enzymes, are flux-orientated principle and biochemical structure theory (Millar et al., 2011). For all and any metabolite, whether primary or secondary, and any flux, one can simply multiply the associated flux control coefficient of the chosen enzyme by implementing its property of elasticity with considerable appreciation to the metabolite in discussion (Park et al., 2016). If someone manages to achieve this for each and every enzyme and provides all the accurate ensuing merchandise, they are supplying a sum which effectively amounts to zero (Patten et al., 2010). This is the overall shape and structure of the connectivity dating. For other substances like metabolites which have the property of elasticity for several other variants of enzymes, the effective sum at the end consists of several phases and is not always particularly beneficial (Ren et al., 2019). Metabolic control evaluation has a peculiar tendency of treating the kinetic components of the enzymes as some black box (Patten et al., 2010). Various studies related to biochemical kinetics have been established for more than a century with the primary focus being molecular mechanisms.

The flux control efficiency and flux control ratio are internal normalizations, describing respiratory flux in a specified state, relative to respiratory flux in a reference state. Whereas flux control ratios express various respiratory states relative to a mutual reference state, they indicate the control of respiration in one step initiated by a particular metabolic control variable. The concept of the flux control efficiency gives a framework for evaluating the impact of an exploratory variable on flux and defines particular expressions, such as the biochemical coupling efficiency. A metabolic control variable is either supplied (stimulation, activation) or withdrawn (reversal of inhibition) to produce a high flux in the reference state, related to flux in the background state. The flux control ratio in the study state associates the fluxes in a diversity of respiratory states, which may be divided by single or multiple variables. Physiological and biochemical aspects of such instances have proved to be the contributor for both the pathway and the enzyme constituents. The reduction (50%) of enzyme activity for citrate synthase promoted in comparative flux within the TCA cycle, and its associated amplified rate of respiration. Various experiments with the wild tomato species *Solanum pennellii* mutant gene like *Aco1*, lacking the manifestation of aconitase, established a 60% reduction with the action of aconitase. The same effect was found to occur with the rate of respiration (Carrari and Fernie, 2006). Greater reductions with the action of succinyl-CoA ligase enzyme have no influence on flux control coefficient outcomes having the values of 0.0008. Decrease of as much as 50% with regard to the action of isocitrate dehydrogenase is dependent on mitochondrial NAD, without any noticeable changes occurring in the respiration rates and produced a flux control coefficient of −0.123. With prominent noticeable change, citrate synthase undoubtedly amplified the respiration rates and also gave a negative value (−0.4) for all flux control coefficient observed. Enzymes, such as succinate dehydrogenase, aconitase, fumarase, 2-oxoglutarate dehydrogenase, and malate dehydrogenase display a prominent effect, with flux control coefficient outcomes of 0.964, 0.786, 0.286, 0.601, and 1.76, respectively. The control coefficient was high, which was an expected result for the branched pathway (Geigenberger, 2011). In defining flux control coefficients from experimental statistics, evidence suggests that the effort is dependent on numerous factors. The data show some significant reduction in actions involving expressions and enzymes, which might be a concern with respect to different distributions in controlling of transgenic plants, as compared with the WT. TCA cycle is associated with substrates and branch points in complex systems (Sweetlove et al., 2010; Rocha et al., 2010). The flux control coefficients could have either optimistic or pessimistic values, making the assessed coefficients less complicated. Finally, it must be considered that the greatest feasible method accessible to us at the moment is the calculative usage of respiration in absence of light as a degree of mitochondrial respiration.

14.8 THE MODULATION OF TCA CYCLE: ENZYMATIC BIOSYNTHETIC APPROACH

Flux control coefficients provide a new approach of quantitative analysis for the control of pathway flux at an enzyme level and do not compulsorily indicate which are physiologically regulated. The reaction undergoing the conversion of sucrose to starch, catalyzed by AGPase, is very crucial for biosynthetic pathways. Plastidial phosphoglucomutase is majorly responsible for starch synthesis in potatoes, and amyloplastidic ATP/ADP transporter occupies a major portion of the regulator (Geigenberger et al., 2011). The advancement for in vivo regulations depends solely upon the evaluation of disequilibrium constants (Geigenberger et al., 2011). The ratio between the concentrations of products and reactants at any given time is called the mass action ratio which is divided by the equilibrium constant of the reaction. This calculation displays the extent of withdrawal of each reaction from equilibrium. Irreversible actions are represented by ratios which have been removed from unity. A prerequisite knowledge of the capacity to decide subcellular metabolite concentrations is necessary to use approach like a pathway which is expanding throughout several components as observed in the TCA cycle. In such unique cases, the most effective component is the trade between the mitochondria and cytosol involving a pool of organic acids with the help of the mitochondrial family proteins. These data indicate the possible equilibrium condition of those metabolites (Palmieri et al., 2011), and thus for a proxy, the data can be used. Mitochondrial stages of ATP, NAD, and malate must take place for the leaf energy balance (Benkeblia et al., 2007; Stitt et al., 1989), whereas the ATP, NAD, malate enzyme, succinate, and citrate enzyme are accessible for the utilization of potato tubers (Farre et al., 2008). As of now, we do not possess enough factual data to perform a scientific analysis for even a single enzyme associated with the TCA cycle. Regardless of the necessity for empirical information which is very much necessary to evaluate the data, it could effectively be presumed that the isocitrate dehydrogenase, citrate synthase, and 2-oxoglutarate dehydrogenase enzyme catalyzed reactions are fundamentally irreversible, taking into account the daily conditions surrounding it. This is not always the trend because it is proved that the isocitrate dehydrogenase response changes unambiguously during the embryonic development of *Brassica napus* (Schwender et al., 2006). The isocitrate dehydrogenase in *Brassica napus* is reversible in vivo. it is presumed that in growing seeds, this may be because of the excessive tissue CO_2 (Goffman and Bergman, 2004), which might change the reaction in the direction of carboxylation, thus disrupting the equilibrium. Although there are insufficient factual data to support the statement that there are regulative enzymes inside the TCA cycle solely dependent on disequilibrium proportions, which is the ratio between mass action ratio and equilibrium constant, the amount of information regarding allosteric and posttranslational regulation is quite large.

Reactions associated with the complex PDH are very similar to gluconeogenesis, glycolysis, and fatty acid production metabolic pathways (Plaxton and Podesta, 2006). As an outcome, a number of allosteric effectors and covalent modification of PDH complex are involved (Plaxton and Podesta, 2006). In vitro, both mitochondrial and TCA cycle dehydrogenases exhibit inhibition of product formation through NADH (McDonald et al., 2006; Tovar-Méndez, 2003; Igamberdiev et al., 2003). NAD^+ is utilized as a co-substrate with the reactions associated with the TCA cycle and the PDH complex, since the ratio of mitochondrial NADH/NAD^+ affects the carbon flux in the TCA cycle. In vivo experiments suggest that NADH/NAD^+ ratio provides stability to the mechanism through the pyruvate oxidation with the help of PDH but at the cost of oxidative phosphorylation (Plaxton and Podesta, 2006).

The enzyme citrate synthase is not found in tissues which transform fatty acids into the sugars where an isoform of glyoxysomal is further operational (Pracharoenwattana et al., 2005). The revocable hydration of cis-aconitate is catalyzed by aconitase to both citrate and isocitrate. In plants, numerous tissues are found to have aconitase and it is proven that a couple of aconitase isoenzymes exist in the intact plant cells (Carrari and Fernie, 2006). The 2-oxoglutarate is degraded in an irreversible manner by 2-oxoglutarate dehydrogenase complex so that the carbon skeletons are available to be integrated with for nitrogen cycle. Metabolic indicators in such cases consist of ATP/ADP, NADH/NAD^+, calcium ions, and thiamine pyrophosphates.

14.9 IN VIVO POTENTIAL OF THE ENZYMES AFFECTING THE TCA CYCLE

The regulations of TCA in living tissues are hard to comprehend as specific intervals of the pathway have one-of-a-kind roles past the renovation of the cycle interest, and evidently, the equilibrium of those functionalities strongly relies upon the physiological context of the cycle. For the special physiological framework of illuminated leaves, this pathway was studied thoroughly, while a few researchers suggested the complete restriction of the TCA cycle activity (Gessler et al., 2009). An inhibitory decarboxylation reaction catalyzed by PDHs in light as opposed to dark was anticipated by identical institutions (Gessler et al., 2009). It was expected that with the high NADPH/NADP ratio of the mitochondria, the isocitrate dehydrogenase would undergo inhibition (Igamberdiev et al., 2003). Therefore, the concerns of respiration during daytime are less as compared with that during dark. It is also true that in the presence of sunlight, the respiratory activities performed play an essential role in metabolizing nitrogen and carbon, and also is able to afford the ATP demands for sucrose production which is an important action in addition to 2-oxoglutaric acid essential for nitrogen assimilation (Raghavendra and Padmasree, 2003; Nunes-Nesi et al., 2014). The generation of the primary amino donors is achieved through nitrogen integration by the glutamate/glutamine synthase pathway. The primary donors, glutamine, and glutamate are essential for the biosynthesis of nitrogenous constituents. The effect of ambient concentrations of O_2 and CO_2 on TCA cycle was studied. Utilizing isotope-tracing strategies, it was proved that TCA cycle and glycolysis activities are inversely proportional to excessive carbon dioxide/oxygen ratio (Gessler et al., 2009). It was found that the respiratory metabolism for a particular duration was upregulated under low carbon dioxide situations, while photorespiration was favored. Reduction of the β-subunit of succinyl-CoA ligase in transgenic tomato (Khan et al., 2021) showed very little alteration in the respiratory control.

14.10 ENVIRONMENTAL IMPACT ON RESPIRATION

Plant respiration is incredibly adaptable due to the abundance of available respiration substrates as well as metabolic pathways, particularly in responses to changing environmental situation. In reality, the light-saturated photosynthesis charge of C3 plants inclusive of rice and wheat is at its peak for temperatures ranging from about 20°C–32°C, whereas the overall respiration suggests a steep nonlinear boom for temperatures between 15°C and 40°C, observed by means of a quick and closely linear decline (Porter and Semenov, 2005). Notwithstanding, the stimulation of C3 photosynthesis at accelerated atmospheric [CO_2] may be forecasted with certainty, the aspect of adjustments in respiration remains uncertain (Leakey et al. 2009b). The main cause for hesitation is that the responses of plant respiration to higher levels of [CO_2] are not completely realized (Leakey et al., 2009; Gifford et al., 2003). In real sense, the results are somewhat contradictory, demonstrating that respiration can increase by $> 37\%$, or indeed cease entirely with increased [CO_2] (Leakey et al., 2009; Gifford et al., 2003). Even if in the leaf, respiration at accelerated [CO_2] might additionally lessen net carbon stability, it is far viable, nonetheless, that such condition may facilitate multiplied yield through supplying more power for the export of photoassimilates from source organ to sink tissue. However, an appropriate function of respiration in enhancing the sink potential remains unclear (Gonzalez-Meler et al., 2004).

Substantial research has also been directed toward the adaptation responses of respiration to lower oxygen concentration. An emergent environment pressure condition that swiftly results in molecular oxygen depletion inside the organs of the plant is water logging or flooding of soil (Bailey-Serres and Voesenek, 2008). The immediate impact of flooding is a decline in the O_2 level and as a result of the reduction in aerobic root breathing, there is a limit in ATP production. Moreover, due to the exceptionally higher resistance to O_2 diffusion through plant tissues, limited availability of O_2 to plant cells could occur even under optimal growth situations (van Dongen et al., 2011). Deep oxygen gradients were discovered in a variety of plant tissues, including stems, roots, tubers, and seeds (Geigenberger et al., 2004; Dongen et al., 2011). Furthermore, limited oxygen concentrations can range, relying on the metabolic condition of the tissues (Geigenberger et al., 2004; Dongen et al., 2011).

It is known that both anatomical and metabolic adjustments are significant ways to permit plant life to cope with temporal and spatial variability of the O_2 availability. The most important structural alternate is an extended development of parenchyma to reduce the resistance to O_2 diffusion in the respirating tissues (Jiang et al., 2010; Colmer, 2003). Instances of certain kind of metabolic diversifications to hypoxia consist of the transfer to sucrose synthase from invertase routes at some stage in sucrose hydrolysis (Huang et al., 2008; Bologa et al., 2006) and inhibiting mitochondrial respiration (Zabalza et al., 2009; Dongen et al., 2011). Such metabolic alterations are vital additives of the surviving technique as they significantly increase the duration of hypoxia that a plant could cope with (Dongen et al., 2011).

Various researchers on the impact of intense drought on respiratory pathways have found out conflicting outcomes, as respiration remains unchanged in soybeans (Ribas-Carbo et al., 2005), enhanced in wheat (Bartoli et al., 2005), and reduced in pepper and bean (Flexas et al., 2005). Nonetheless, the consequences of mild-to-severe water stress have been extraordinarily lower on the mitochondrial TCA cycle key enzymes in two crassulacean acid metabolism (CAM) species (Werner et al., 2000). However, alterations of in vivo actions of the alternative oxidase (AOX) and cytochrome oxidase (COX) pathways, estimated with O_2 isotope fractionation method, have revealed the partitioning of electrons amidst the two primary respirating pathways (Gonzalez-Meler et al., 1999). However, more research is needed to determine whether and to what extent the in vivo respiration pathways, especially the TCA cycle, are impaired under drought stress. Although the responses of respiration are nicely characterized, the unique reaction of the enzymes with intermediates in TCA cycle has been defined in a restricted number of situations. Moreover, the metabolic, molecular, and enzymatic responses have been noted in case of lower nitrogen (Ribas-Carbo et al., 2005), lower carbon (Leakey et al., 2009; Gifford et al., 2003), lower potassium (Ribas-Carbo et al., 2005), minor decrease in temperature (Bartoli et al., 2005) and water scarcity (van Dongen et al., 2011), from which an adaptive mechanism to maintain carbon flux via TCA cycle can be deciphered. Nevertheless, a strong hyperlink within circadian-clock feature and metabolic homeostasis within the TCA cycle is needed (Bartoli et al., 2005). Farther investigations are required to discover the connections of mitochondrial non-phosphorylating pathways with photosynthetic procedures and mobile homeostasis under demanding situations.

14.11 EFFECT OF VARIOUS STRESSES ON TCA CYCLE

14.11.1 METABOLITES AND ENZYMES

14.11.1.1 ACETYL-COA: EFFECT ON CANCER, IMMUNE, AND STEM CELL FUNCTIONS

Increased histone acetylation resulted in excessive acetyl-CoA levels shifting the cells into proliferating conditions, along with enhanced glycolytic enzymes (Sun et al., 2014). As a result, cancerous cells stimulate acetyl-CoA producing enzymes, viz., ACLY to upsurge their proliferative capability, so that inhibiting ACLY subdues tumorigenesis. The signaling pathways that result in an improved histone acetylation in higher acetyl-CoA tiers seemed to be different in distinct contexts. For example, latest discoveries in pancreatic cancer confirmed that higher ranges of acetyl-CoA and their utilization in the mevalonate pathway preferred malignant advancement of KRAS mutant acinar cells via the AKT–ACLY signaling axis (Rui et al, 2011).

14.11.1.2 α-KETOGLUTARATE

α-Ketoglutarate, also referred to as 2-oxoglutarate, is a compulsory co-substrate for 2-oxoglutarate-dependent dioxygenases (2-OGDD), which catalyze hydroxylation reactions of different substrates, inclusive of lipids, nucleic acids, proteins, and metabolic intermediates, generating CO_2 and succinates. 2-OGDD is dependent on the intracellular proportion of α-ketoglutarate to succinates or different inhibitors consisting of 2-hydroxyglutarate (2-HG) or fumarate (Sun et al., 2014). Along with α-ketoglutarate, such hydroxylation reactions necessitate Fe^{2+} as cofactor and O_2 as co-substrate. Ascorbic acid utilizes element in such reactions converting Fe^{3+} to Fe^{2+}. 2-OGDD in humans plays a major role in physiological processes which include response to chromatin and hypoxia adjustments. It is vital to mention that intracellular concentration of substrates consisting of the stages of α-Ketoglutarate surpass the enzyme-binding site affinities. In this way, we advocate the build-up of inhibitors of 2-OGDD, viz., fumarates, succinates or 2-HG above α-ketoglutarate levels as means to moderate 2-OGDD enzymes (Semenza, 2012).

14.11.1.3 PROLINE AND PROLYL HYDROXYLASES

Prolyl hydroxylases (PHD1–3) is a 2-OGDD enzyme essential in regulating transcription of HIF-1, a regulator for oxygen homeostasis. In normoxia, proline accumulated in the O_2-established damaged area of HIF-1α undergoes hydroxylation by means of PHDs. The hydroxylating HIF-1α is the signal for the polyubiquitination of HIF-1α through the VHL protein (von Hippel-Lindau tumor suppressor) that targets the protein for degradations inside the proteasome. Under constrained O_2 circumstances or at decreased ranges of α-ketoglutarate or Fe^{2+}, the action of PHD is restrained, resulting in HIF-1α or -2α accumulations and translocation to nucleus where it promotes alteration in the manifestation of genes associated with erythropoiesis, metabolism, and angiogenesis along with stem and immune cellular functions (Zhou et al., 2010). PHDs are reduced as a result of decreased oxygen levels and mitochondrial ROS production, which causes HIF-1 to increase. Mitochondrial ROS accumulations under normoxia could also prevent PHDs from activating HIFs30. As previously stated, TCA cycle intermediates, such as succinate, fumarate, and L-2-HG could inhibit PHDs under normoxic conditions. Surprisingly, in the reverse manner, the reactivation of PHDs via increased intracellular α-ketoglutarate in hypoxic cancer cells causes excessive metabolic impairment, resulting in cell death (Zatta et al., 1995).

14.11.2 ENVIRONMENTAL STRESS

14.11.2.1 ALUMINUM STRESS

Oxalic acid exhibits a pivotal position regarding the response of the soil microbe *Pseudomonas fluorescence* to aluminum. Its production through oxidizing glyoxylate necessitates a first-rate reconfiguration of the enzymatic reactions concerned within the TCA cycle. The call for glyoxylate seems to increase the interest of isocitrate lyase (Rui et al., 2011). Isocitrate lyase, an enzyme which helps in the breakdown of isocitrate to succinate and glyoxylate incurred a fourfold enhanced activity in aluminum-stressed cells. Whereas the call for oxalate in aluminum-stressed cells influenced negatively the interest of the enzyme α-ketoglutarate dehydrogenase complexes, no discernible trade in the action of malate synthase change was recorded. It emerges that the TCA cycle is able to produce the required precursor for oxalate synthesis attributable to aluminum stress. Aluminum has been diagnosed as a neurotoxic agent in animals for more than a 100 years. It is also recognized to be poisonous to humans and represents an applicable etiological component in pathologies associated with long-term dialysis treatment (Zhou et al., 2010).

14.11.2.2 DROUGHT STRESS

The feedback from the metabolites, which were subjected to drought stress, showed different outcomes between two genotypes of wheat, namely, JD8 and JD17. The activity of TCA cycle was improved in the JD17 variety by succinic acid, fumaric acid, malic acid, and oxalic acid that were amplified, but the same outcome was not found in the JD8 variety. There was a significant change in the responses of metabolites to drought stress between genotypes and shoots. The findings showed that the TCA cycle, sugar, and glycolysis were increased, while the GABA shunt process was hindered in the shoots of drought-tolerant wheat genotypes. TCA cycle, sugar, and glycolysis were interfered in the shoots (Zhou et al., 2010).

14.11.2.3 SALINITY STRESS

As early responses to salt exposures, the relative abundance of several metabolites involved in the production of energy, as well as those engaged in glycolysis, TCA cycle, and OPP pathway have been identified. It has been recorded that salinity stress causes a substantial suppression of the TCA cycle, which limits glycolysis and diverts carbon to other pathways in some cases (Zhou et al., 2010).

14.11.2.4 HEAT STRESS

Plant respiration in general increases with increasing temperature. The regulation of stomata can also increase the rate of transpiration to provide cooling effect to the leaf as a precautionary measure to the increased temperature parameters. In contrast, excessive heat induces the closure of stomata, which results in the reduced activity of photosynthesis, the stomatal size decreases at 45 °C, signifying a possible negative impact on gas exchange process in leaves. In TCA cycle and glycolysis, higher temperature could enhance the pyruvate and sucrose levels, but reduces the sugar phosphates and certain organic acids (Rui et al., 2011).

14.12 CONCLUSION AND FUTURE RESEARCH PROSPECTS

This review broadly covers the various aspects and important notions surrounding the TCA cycle. In terms of respiration, regulation of cellular metabolic pathways, cellular development, seed germination, and other vital process, the TCA cycle affects the outcome directly as well as indirectly. The mitochondria are largely responsible for the signaling and communication functions from various processes to various organelles. One such process is the action of the TCA cycle and its communication with other cells through specific mitochondrial antiviral signaling proteins, retinoic acid-inducible protein-like receptors and other important protein receptors which act as key signaling and communicating molecules. The eight very crucial enzymes which basically govern the whole process of TCA cycle provide us a great insight into the structures and functions, as also the genomic localizations of these aforementioned eight enzymes which are crucial for the understanding of genomics of this biosynthetic process. The use of modern mathematical frameworks to measure the intermediate metabolites is extremely beneficial, one such tool used is the MCA. On further meddling with such mathematical models, one can efficiently figure out most of the data and statistics as opposed to few decades back when such tools were unavailable. One such application of this framework is to measure, control, and analyze the flux generated from the biosynthetic process for more accurate statistics. These tools can be amplified and can be used on a broad variety of aspects.

There still remains plenty of knowledge to be attained with regard to this subject, so that the future research scope is still filled with abundant mystery left to be uncovered and discovered. TCA cycle under different environmental stresses is poorly characterized. Heat stress needs a load of experimentations to support various statements made by specific groups, for salinity and drought stress, there are considerable amounts of data which can be used for guideline purposes but nothing more than that, since the data are not consolidated. The information and data regarding the environmental stress related to cold and heavy metals are also less that it would not even serve as a guide to research groups looking for sufficient information, statistics, and data that can open up a broad field to future researchers to study, experiment and discover new grounds of TCA cycle, not only with respect to stress, but also with respect to the genomics, subcellular localization, and other related aspects.

KEYWORDS

- **apoptosis**
- **hypoxia**
- **normoxia**
- **flux control coefficient**
- **glyoxysomal isoform**
- **plastidial phosphoglucomutase**

REFERENCES

Achnine, L.; Blancaflor, E. B.; Rasmussen, S.; Dixon, R. A. Colocalization of L-Phenylalanine Ammonia-Lyase and Cinnamate 4-Hydroxylase for Metabolic Channeling in Phenylpropanoid Biosynthesis. *Plant Cell.* **2004,** *16* (11), 3098–3109.

Angelovici, R.; Fait, A.; Fernie, A. R.; Galili, G. A Seed High-Lysine Trait Is Negatively Associated with the TCA Cycle and Slows Down Arabidopsis Seed Germination. *New Phytol.* **2011,** *189* (1), 148–519.

Araújo, W. L.; Nunes-Nesi, A.; Nikoloski, Z.; Sweetlove, L. J.; Fernie, A. R. Metabolic Control and Regulation of the Tricarboxylic Acid Cycle in Photosynthetic and Heterotrophic Plant Tissues. *Plant Cell Environ.* **2012,** *35* (1), 1–21.

Bailey-Serres, J.; Voesenek L. A. Flooding Stress: Acclimations and Genetic Diversity. *Annu. Rev. Plant Biol.* **2008,** *59*, 313–339. DOI: 10.1146/annurev.arplant.59.032607.092752

Bandehagh, A.; Taylor, N. L. Can Alternative Metabolic Pathways and Shunts Overcome Salinity Induced Inhibition of Central Carbon Metabolism in Crops? *Front. Plant Sci.* **2020,** *11*, 1072.

Barone, P.; Rosellini, D.; Lafayette, P.; Bouton, J.; Veronesi, F.; Parrott, W. Bacterial Citrate Synthase Expression and Soil Aluminium Tolerance in Transgenic Alfalfa. *Plant Cell Rep.* **2008,** *27* (5), 893–901.

Bartoli, G.; Sarnthein, M.; Weinelt, M.; Erlenkeuser, H.; Garbe-Schönberg, D.; Lea, D. W. Final Closure of Panama and the Onset of Northern Hemisphere Glaciation. *Earth Planet Sci. Lett.* **2005,** *237* (1–2), 33–44.

Benkeblia, N.; Shinano, T.; Osaki, M. Metabolite Profiling and Assessment of Metabolome Compartmentation of Soybean Leaves Using Non-Aqueous Fractionation and GC-MS Analysis. *Metabolomics* **2007,** *3* (3), 297–305. DOI: 10.1007/s11306-007-0078-y

Bologa, C. G.; Revankar, C. M.; Young, S. M.; Edwards, B. S.; Arterburn, J. B.; Kiselyov, A. S.; Parker, M. A.; Tkachenko, S. E.; Savchuck, N. P.; Sklar, L. A.; Oprea, T. I. Virtual and Biomolecular Screening Converge on a Selective Agonist for GPR30. *Nat. Chem. Biol.* **2006,** *2* (4), 207–212. DOI: 10.1038/nchembio775.

Brand, K. Glutamine and Glucose Metabolism during Thymocyte Proliferation. Pathways of Glutamine and Glutamate Metabolism. *Biochem J.* **1985,** *228* (2), 353–361.

Bui, T.; Thompson, C. B. Cancer's Sweet Tooth. *Cancer Cell.* **2006,** *9* (6), 419–420.

Carrari, F.; Fernie, A. R. Metabolic Regulation Underlying Tomato Fruit Development. *J. Exp. Bot.* **2006,** *57* (9), 1883–1897.

Carrer, A.; Trefely, S.; Zhao, S.; Campbell, S. L.; Norgard, R. J.; Schultz, K. C.; Sidoli, S.; Parris, J. L. D.; Affronti, H. C.; Sivanand, S.; Egolf, S.; Sela, Y.; Trizzino, M.; Gardini, A.; Garcia, B. A.; Snyder, N. W.; Stanger, B. Z.; Wellen, K. E. Acetyl-CoA Metabolism Supports Multistep Pancreatic Tumorigenesis. *Cancer Discov.* **2019,** *9* (3), 416–435.

Cavalcanti, J. H. F.; Esteves-Ferreira, A. A.; Quinhones, C. G.; Pereira-Lima, I. A.; Nunes-Nesi, A.; Fernie, A. R.; Araújo, W. L. Evolution and Functional Implications of the Tricarboxylic Acid Cycle as Revealed by Phylogenetic Analysis. *Genome Boil. Evol.* **2014,** *6* (10), 2830–2848.

Chandel, N. S.; McClintock, D. S.; Feliciano, C. E.; Wood, T. M.; Melendez, J. A.; Rodriguez, A. M.; Schumacker, P. T. Reactive Oxygen Species Generated at Mitochondrial Complex III Stabilize Hypoxia-Inducible Factor-1alpha during Hypoxia: A Mechanism of O_2 Sensing. *J. Biol. Chem.* **2000,** *275* (33), 25130–25138.

Ciais, P.; Reichstein, M.; Viovy, N.; Granier, A.; Ogée, J.; Allard, V.; Aubinet, M.; Buchmann, N.; Bernhofer, C.; Carrara, A.; Chevallier, F. Europe-Wide Reduction in Primary Productivity Caused by the Heat and Drought in 2003. *Nature* **2005,** *437* (7058), 529–533. DOI: 10.1038/nature03972

Colmer, T. D. Long-Distance Transport of Gases in Plants: A Perspective on Internal Aeration and Radial Oxygen Loss from Roots. *Plant Cell Environ.* **2003,** *26* (1), 17–36.

de la Fuente, J. M.; Ramírez-Rodríguez, V.; Cabrera-Ponce, J. L.; Herrera-Estrella, L. Aluminium Tolerance in Transgenic Plants by Alteration of Citrate Synthesis. *Science* **1997,** *276* (5318), 1566–1568.

Deng, W.; Luo, K.; Li, Z.; Yang, Y.; Hu, N.; Wu, Y. Overexpression of *Citrus junos* Mitochondrial Citrate Synthase Gene in *Nicotiana benthamiana* Confers Aluminium Tolerance. *Planta* **2009,** *230* (2), 355–365.

Downs, C. A.; Heckathorn, S. A. The Mitochondrial Small Heat-Shock Protein Protects NADH: Ubiquinone Oxidoreductase of the Electron Transport Chain during Heat Stress in Plants. *FEBS Lett.* **1998,** *430.*

Egan, D. F.; Shackelford, D. B.; Mihaylova, M. M.; Gelino, S.; Kohnz, R. A.; Mair, W.; Vasquez, D. S.; Joshi, A.; Gwinn, D. M.; Taylor, R.; Asara, J. M.; Fitzpatrick, J.; Dillin, A.; Viollet, B.; Kundu, M.; Hansen, M.; Shaw, R. J. Phosphorylation of ULK1 (hATG1) by AMP-Activated Protein Kinase Connects Energy Sensing to Mitophagy. *Science* **2011,** *331* (6016), 456–461.

Fait, A.; Angelovici, R.; Less, H.; Ohad, I.; Urbanczyk-Wochniak, E.; Fernie, A. R.; Galili, G. Arabidopsis Seed Development and Germination Is Associated with Temporally Distinct Metabolic Switches. *Plant Physiol.* **2006,** *142* (3), 839–854.

Fait, A.; Nesi, A. N.; Angelovici, R.; Lehmann, M.; Pham, P. A.; Song, L.; Haslam, R. P.; Napier, J. A.; Galili, G.; Fernie, A. R. Targeted Enhancement of Glutamate-to-γ-Aminobutyrate Conversion in Arabidopsis Seeds Affects Carbon-Nitrogen Balance and Storage Reserves in a Development-Dependent Manner. *Plant Physiol.* **2011,** *157* (3), 1026–1042.

Farre, E. M.; Fernie, A. R.; Willmitzer, L. Analysis of Subcellular Metabolite Levels of Potato Tubers (*Solanum tuberosum*) Displaying Alterations in Cellular or Extracellular Sucrose Metabolism. *Metabolomics* **2008,** *4* (2), 161–170. DOI: 10.1007/s11306-008-0107-5

Fernie, A. R.; Martinoia, E. Malate. Jack of All Trades or Master of a Few? *Phytochemistry* **2009,** *70* (7), 828–832.

Flexas, J.; Galmes, J.; Ribas-Carbo, M.; Medrano, H. The Effects of Water Stress on Plant Respiration. In *Plant Respiration*; Springer: Dordrecht, 2005; pp 85–94. DOI: 10.1007/1-4020-3589-6_6

Galluzzi, L.; Vitale, I.; Abrams, J. M.; Alnemri, E. S.; Baehrecke, E. H.; Blagosklonny, M. V.; Dawson, T. M.; Dawson, V. L.; El-Deiry, W. S.; Fulda, S.; Gottlieb, E. Molecular Definitions of Cell Death Subroutines: Recommendations of the Nomenclature Committee on Cell Death. *Cell Death Differ.* **2012,** *19*, 107–120.

Geigenberger, P. Regulation of Starch Biosynthesis in Response to a Fluctuating Environment. *Plant Physiol.* **2011,** *155* (4), 1566–1577. doi:10.1104/pp.110.170399

Geigenberger, P.; Stitt, M.; Fernie, A. R. Metabolic Control Analysis and Regulation of the Conversion of Sucrose to Starch in Growing Potato Tubers. *Plant Cell Environ.* **2004,** *27* (6), 655–673.

Gessler, A.; Tcherkez, G.; Karyanto, O.; Keitel, C.; Ferrio, J. P.; Ghashghaie, J.; Kreuzwieser, J.; Farquhar, G. D. On the Metabolic Origin of the Carbon Isotope Composition of CO_2 Evolved from Darkened Light-Acclimated Leaves in *Ricinus communis*. *New Phytol.* **2009,** *181* (2), 374–386.

Gifford, R. M.; Roderick, M. L. Soil Carbon Stocks and Bulk Density: Spatial or Cumulative Mass Coordinates as a Basis of Expression? *Global Change Biol.* **2003,** *9* (11), 1507–1514.

Goffman, F. D.; Bergman, C. J. Rice Kernel Phenolic Content and Its Relationship with Antiradical Efficiency. *J. Sci. Food Agri.* **2004,** *84* (10), 1235–1240.

Gonzalez-Meler, M. A.; Ribas-Carbo, M.; Giles, L.; Siedow, J. N. The Effect of Growth and Measurement Temperature on the Activity of the Alternative Respiratory Pathway. *Plant Physiol.* **1999,** *120* (3), 765–772. DOI: 10.1104/pp.120.3.765

Gonzalez-Meler, M. A.; Taneva, L. I. N. A.; Trueman, R. J. Plant Respiration and Elevated Atmospheric CO_2 Concentration: Cellular Responses and Global Significance. *Ann. Bot.***2004,** *94* (5), 647–656. DOI: 10.1093/aob/mch189

Han, Y.; Zhang, W.; Zhang, B.; Zhang, S.; Wang, W.; Ming, F. One Novel Mitochondrial Citrate Synthase from *Oryza sativa* L. Can Enhance Aluminum Tolerance in Transgenic Tobacco. *Mol. Biotechnol.* **2009,** *42* (3), 299–305.

Hatzivassiliou, G.; Zhao, F.; Bauer, D. E.; Andreadis, C.; Shaw, A. N.; Dhanak, D.; Hingorani, S. R.; Tuveson, D. A.; Thompson, C. B. ATP Citrate Lyase Inhibition Can Suppress Tumour Cell Growth. *Cancer Cell.* **2005,** *8* (4), 311–321.

He, S.; Wang, L.; Miao, L.; Wang, T.; Du, F.; Zhao, L.; Wang, X. Receptor Interacting Protein Kinase-3 Determines Cellular Necrotic Response to TNF-Alpha. *Cell* **2009,** *137* (6), 1100–1111.

Heidecker, G.; Kölch, W.; Morrison, D. K.; Rapp, U. R. The Role of Raf-1 Phosphorylation in Signal Transduction. *Adv. Cancer Res.***1992,** *58*, 53–73.

Herppich, W. B.; Peckmann, K. Influence of Drought on Mitochondrial Activity, Photosynthesis, Nocturnal Acid Accumulation and Water Relations in the CAM Plants *Prenia sladeniana* (ME-type) and *Crassula lycopodioides* (PEPCK-type). *Ann. Bot.* **2000,** *86* (3), 611–620. DOI: 10.1006/anbo.2000.1229

Heyer, A. G.; Mozley, D.; Landschutze, V.; Thomas, B.; Gatz, C. Function of Phytochrome A in Potato Plants as Revealed through the Study of Transgenic Plants. *Plant Physiol.* **1995,** *109* (1), 53–61.

Huang, S.; Colmer, T. D.; Millar, A. H. Does Anoxia Tolerance Involve Altering the Energy Currency Towards PPi? *Trends Plant Sci.* **2008,** *13* (5), 221–227. DOI: 10.1016/j.tplants.2008.02.007

Igamberdiev, A. U.; Gardeström, P. Regulation of NAD-and NADP-Dependent Isocitrate Dehydrogenases by Reduction Levels of Pyridine Nucleotides in Mitochondria and Cytosol of Pea Leaves. *Biochim. Biophys. Acta Bioenerg. BBA* **2003,** *1606* (1–3), 117–125. DOI: 10.1016/s0005-2728 (03)00106-3

Ishikawa, H.; Barber, G. STING Is an Endoplasmic Reticulum Adaptor That Facilitates Innate Immune Signalling. *Nature* **2008,** *455*, 674–678.

Jardim-Messeder, D.; Caverzan, A.; Rauber, R.; de Souza Ferreira, E.; Margis-Pinheiro, M.; Galina, A. Succinate Dehydrogenase (Mitochondrial Complex II) Is a Source of Reactive Oxygen Species in Plants and Regulates Development and Stress Responses. *New Phytol.* **2015,** *208* (3), 776–789.

Jiang, L.; Pearson, J. C.; Crews, S. T. Diverse Modes of Drosophila Tracheal Fusion Cell Transcriptional Regulation. *Mech. Dev.* **2010,** *127* (5–6), 265–280.

Jørgensen, K.; Rasmussen, A. V.; Morant, M.; Nielsen, A. H.; Bjarnholt, N.; Zagrobelny, M.; Bak, S.; Møller, B. L. Metabolon Formation and Metabolic Channeling in the Biosynthesis of Plant Natural Products. *Curr. Opin. Plant Biol.* **2005,** *8* (3), 280–291.

Kawai, T.; Takahashi, K.; Sato, S.; Coban, C.; Kumar, H.; Kato, H.; Ishii, K. J.; Takeuchi, O.; Akira, S. IPS-1, an Adaptor Triggering RIG-I- and Mda5-Mediated Type I Interferon Induction. *Nat. Immunol.* **2005,** *6* (10), 981–988.

Khan, M. I. R.; Jalil, S. U.; Chopra, P.; Chhillar, H.; Ferrante, A.; Khan, N. A.; Ansari, M. I. Role of GABA in Plant Growth, Development and Senescence. *Plant Gene* **2021,** 100283.

Kochian, L. V.; Hoekenga, O. A.; Pineros, M. A. How Do Crop Plants Tolerate Acid Soils? Mechanisms of Aluminum Tolerance and Phosphorous Efficiency. *Annu. Rev. Plant Biol.* **2004,** *55*, 459–493.

Lambers, H.; Chapin III, F. S.; Pons, T. L. *Plant Physiological Ecology*; Springer Science & Business Media, 2008. DOI: 10.1007/978-0-387-78341-3

Leakey, A. D.; Ainsworth, E. A.; Bernacchi, C. J.; Rogers, A.; Long, S. P.; Ort, D. R. Elevated CO_2 Effects on Plant Carbon, Nitrogen, and Water Relations: Six Important Lessons from FACE. *J. Exp. Bot.* **2009,** *60* (10), 2859–2876. DOI: 10.1093/jxb/erp096

Lee, J. V.; Carrer, A.; Shah, S.; Snyder, N. W.; Wei, S.; Venneti, S.; Worth, A. J.; Yuan, Z. F.; Lim, H. W.; Liu, S.; Jackson, E.; Aiello, N. M.; Haas, N. B.; Rebbeck, T. R.; Judkins, A.; Won, K. J.; Chodosh, L. A.; Garcia, B. A.; Stanger, B. Z.; Feldman, M. D.; Blair, I. A.; Wellen, K. E. Akt-Dependent Metabolic Reprogramming Regulates Tumour Cell Histone Acetylation. *Cell Metab.* **2014,** *20* (2), 306–319.

Liu, Y.; Qu, J.; Zhang, L.; Xu, X.; Wei, G.; Zhao, Z.; Ren, M.; Cao, M. Identification and Characterization of the TCA Cycle Genes in Maize. *BMC Plant Biol.* **2019,** *19*, 592.

Loreto, F.; Centritto, M. Leaf Carbon Assimilation in a Water-Limited World. *Plant Biosyst.* **2008,** *142* (1), 154–161.

Lunt, S. Y.; Vander, H. M. G. Aerobic Glycolysis: Meeting the Metabolic Requirements of Cell Proliferation. *Annu. Rev. Cell Dev. Biol.* **2011,** *27*, 441–464.

Ma, J. F. Role of Organic Acids in Detoxification of Aluminum in Higher Plants. *Plant Cell Physiol.* **2000,** *41* (4), 383–390.

Macfarlane, T. V.; Kenealy, P.; Kingdon, H. A.; Mohlin, B. O.; Pilley, J. R.; Richmond, S.; Shaw, W. C. Twenty-Year Cohort Study of Health Gain from Orthodontic Treatment: Temporomandibular Disorders. *Am. J. Orthod. Dentofac. Orthoped* **2009,** *135* (6), 692–e1. DOI: 10.1016/j.ajodo.2008.10.017

Martínez-Reyes, I.; Chandel, N. S. Mitochondrial TCA Cycle Metabolites Control Physiology and Disease. *Nat. Commun.* **2020,** *11*, 102.

McDonald, A. E.; Vanlerberghe, G. C.; Staples, J. F. Alternative Oxidase in Animals: Unique Characteristics and Taxonomic Distribution. *J. Exp. Biol.* **2009,** *212* (16), 2627–2634.

Medzhitov, R. Recognition of Microorganisms and Activation of the Immune Response. *Nature* **2007,** *449*, 819–826.

Millar, A. H.; Whelan, J.; Soole, K. L.; Day, D. A. Organization and Regulation of Mitochondrial Respiration in Plants. *Annu. Rev. Plant Biol.* **2011,** *62*, 79–104.

Minárik, P.; Tomásková, N.; Kollárová, M.; Antalík, M. Malate Dehydrogenases-Structure and Function. *Gen. Physiol. Biophys.* **2002,** *21* (3), 257–265.

Mizushima, N. The Role of the Atg1/ULK1 Complex in Autophagy Regulation. *Curr. Opin. Cell Biol.* **2010,** *22* (2), 132–139.

Nunes-Nesi, A.; Araújo, W. L.; Obata, T.; Fernie, A. R. Regulation of the Mitochondrial Tricarboxylic Acid Cycle. *Curr. Opin. Plant Biol.* **2013,** *16* (3), 335–343.

Nunes-Nesi, A.; Santos Brito, D.; Inostroza-Blancheteau, C.; Fernie, A. R.; Araújo, W. L. The Complex Role of Mitochondrial Metabolism in Plant Aluminum Resistance. *Trends Plant Sci.* **2014,** *19* (6), 399–407.

Oberst, A.; Bender, C.; Green, D. R. Living with Death: The Evolution of the Mitochondrial Pathway of Apoptosis in Animals. *Cell Death Differ.* **2008,** *15* (7), 1139–1146.

Palmieri, M.; Impey, S.; Kang, H.; di Ronza, A.; Pelz, C.; Sardiello, M.; Ballabio, A. Characterization of the CLEAR Network Reveals an Integrated Control of Cellular Clearance Pathways. *Hum. Mol. Genet.* **2011,** *20* (19), 3852–3866. DOI: 10.1093/hmg/ddr306

Park, J. O.; Rubin, S. A.; Xu, Y. F.; Amador-Noguez, D.; Fan, J.; Shlomi, T.; Rabinowitz, J. D. Metabolite Concentrations, Fluxes and Free Energies Imply Efficient Enzyme Usage. *Nat. Chem. Biol.* **2016,** *12*, 482–489.

Patten, D. A.; Lafleur, V. N.; Robitaille, G. A.; Chan, D. A.; Giaccia, A. J.; Richard, D. E. Hypoxia-Inducible Factor-1 Activation in Nonhypoxic Conditions: The Essential Role of Mitochondrial-Derived Reactive Oxygen Species. *Mol. Biol. Cell.* **2010,** *21* (18), 3247–3257.

Plaxton, W. C.; Podestá, F. E. The Functional Organization and Control of Plant Respiration. *Crit. Rev. Plant Sci.* **2006,** *25* (2), 159–198.

Porter, J. R.; Semenov, M. A. Crop Responses to Climatic Variation. *Phil. Trans. R. Soc. B Biol. Sci.* **2005,** *360* (1463), 2021–2035.

Pracharoenwattana, I.; Cornah, J. E.; Smith, S. M. Arabidopsis Peroxisomal Citrate Synthase Is Required for Fatty Acid Respiration and Seed Germination. *Plant Cell* **2005,** *17* (7), 2037–2048. DOI: 10.1105/tpc.105.031856

Raghavendra, A. S.; Padmasree, K. Beneficial Interactions of Mitochondrial Metabolism with Photosynthetic Carbon Assimilation. *Trends Plant Sci.* **2003,** *8* (11), 546–553. DOI: 10.1016/j.tplants.2003.09.015

Ren, S.; Ma, K. L. Z.; Chen, G.; Cui, J.; Tong, P.; Wang, L.; Teng, N.; Jin, B. Transcriptomic and Metabolomic Analysis of the Heat-Stress Response of *Populus tomentosa* Carr. *Forests* **2019,** *10*, 383.

Ribas-Carbo, M.; Taylor, N. L.; Giles, L.; Busquets, S.; Finnegan, P. M.; Day, D. A.; Lambers, H.; Medrano, H.; Berry, J. A.; Flexas, J. Effects of Water Stress on Respiration in Soybean Leaves. *Plant Physiol.* **2005,** *139* (1), 466–473. DOI: 10.1104/pp.105.065565

Robert, D. H.; Vasu, D. A. Modulation of TCA Cycle Enzymes and Aluminum Stress in *Pseudomonas fluorescens*. *J. Inorg. Biochem.* **2001,** *87* (1–2), 1–8.

Rocha, L. A.; Pinheiro, H. T.; Gasparini, J. L. Description of *Halichoeres rubrovirens*, a New Species of Wrasse (Labridae: Perciformes) from the Trindade and Martin Vaz Island Group, Southeastern Brazil, with a Preliminary mtDNA Molecular Phylogeny of New World Halichoeres. *Zootaxa* **2010,** *2422* (1), 22–30.

Ruderman, N. B.; Saha, A. K.; Vavvas, D.; Witters, L. A. Malonyl-CoA, Fuel Sensing, and Insulin Resistance. *Am. J. Physiol.* **1999,** *276* (1), E1–E18.

Rui, G.; LianXuan, S.; Yang, J.; MingXia, L.; XiuLi, Z.; FengXue, G.; Qi, L.; Xu, X.; HaoRu, L. Metabolic Responses to Drought Stress in the Tissues of Drought-Tolerant and Drought-Sensitive Wheat Genotype Seedlings. *AoB Plants* **2018,** *10* (2).

Schmidtmann, E.; König, A. C.; Orwat, A.; Leister, D.; Hartl, M.; Finkemeier. I. Redox Regulation of Arabidopsis Mitochondrial Citrate Synthase. *Mol. Plant.* **2014,** *7* (1), 156–169.

Schuster, S.; Dandekar, T.; Fell, D. A. Detection of Elementary Flux Modes in Biochemical Networks: A Promising Tool for Pathway Analysis and Metabolic Engineering. *Trends Biotechnol.* **1999,** *17* (2), 53–60.

Schuster, S.; Fell, D. A.; Dandekar, T. A General Definition of Metabolic Pathways Useful for Systematic Organization and Analysis of Complex Metabolic Networks. *Nat. Biotechnol.* **2000,** *18* (3), 326–332.

Schwender, J.; Shachar-Hill, Y.; Ohlrogge, J. B. Mitochondrial Metabolism in Developing Embryos of *Brassica napus*. *J. Biol. Chem.* **2006,** *281* (45), 34040–34047. DOI: 10.1074/jbc.M606266200

Semenza, G. L. Hypoxia-Inducible Factors in Physiology and Medicine. *Cell.* **2012,** *148* (3), 399–408.

Seth, R. B.; Sun, L.; Ea, C. K.; Chen, Z. J. Identification and Characterization of MAVS, a Mitochondrial Antiviral Signaling Protein That Activates NF-kappaB and IRF 3. *Cell* **2005,** *122* (5), 669–682.

Smirnoff, N. Ascorbate Biosynthesis and Function in Photoprotection. *Phil. Trans. R. Soc. Lond. B Biol. Sci.* **2000,** *355* (1402), 1455–1464.

Stitt, M.; Lilley, R. M.; Gerhardt, R.; Heldt, H. W. Metabolite Levels in Specific Cells and Subcellular Compartments of Plant Leaves. *Methods Enzymol.* **1989,** *174*, 518–552.

Sun, L.; Liang, C.; Chen, Z.; Liu, P.; Tian, J.; Liu, G.; Liao, H. Superior Aluminum (Al) Tolerance of Stylosanthes Is Achieved Mainly by Malate Synthesis through an Al-Enhanced Malic Enzyme, SgME1. *New Phytol.* **2014,** *202* (1), 209–219.

Sweetlove, L. J.; Beard, K. F.; Nunes-Nesi, A.; Fernie, A. R.; Ratcliffe, R. G. Not Just a Circle: Flux Modes in the Plant TCA Cycle. *Trends Plant Sci.* **2010,** *15* (8), 462–470. DOI: 10.1016/j.tplants.2010.05.006

Tait, S. W.; Green, D. R. Mitochondria and Cell Death: Outer Membrane Permeabilization and Beyond. *Nat. Rev. Mol. Cell Biol.* **2010,** *11* (9), 621–632.

Tan, F. J.; Fire, A. Z.; Hill, R. B. Regulation of Apoptosis by C. Elegans CED-9 in the Absence of the C-Terminal Transmembrane Domain. *Cell Death Differ.* **2007,** *14* (11), 1925–1935.

Taylor, R. C.; Cullen, S. P.; Martin, S. J. Apoptosis: Controlled Demolition at the Cellular Level. *Nat. Rev. Mol. Cell Biol.* **2008,** *9* (3), 231–241.

Tesfaye, M.; Temple, S. J.; Allan, D. L.; Vance, C. P.; Samac, D. A. Overexpression of Malate Dehydrogenase in Transgenic Alfalfa Enhances Organic Acid Synthesis and Confers Tolerance to Aluminum. *Plant Physiol.* **2001,** *127* (4), 1836–1844.

Tovar-Méndez, A.; Miernyk, J. A.; Randall, D. D. Regulation of Pyruvate Dehydrogenase Complex Activity in Plant Cells. *Eur. J. Biochem.* **2003,** *270* (6), 1043–1049. DOI: 10.1046/j.1432-1033.2003.03469.x

Trejo-Téllez, L. I.; Stenzel, R.; Gómez-Merino, F. C.; Schmitt, J. M. Transgenic Tobacco Plants Overexpressing Pyruvate Phosphate Dikinase Increase Exudation of Organic Acids and Decrease Accumulation of Aluminum in the Roots. *Plant Soil* **2010,** *326*, 187–198.

Tretter, L.; Adam-Vizi, V. Alpha-Ketoglutarate Dehydrogenase: A Target and Generator of Oxidative Stress. *Phil. Trans. R. Soc. Lond. B Biol Sci.***2005,** *360* (1464), 2335–2345.

Van Dongen, H. P.; Belenky, G.; Vila, B. J. The Efficacy of a Restart Break for Recycling with Optimal Performance Depends Critically on Circadian Timing. *Sleep* **2011,** *34* (7), 917–929. doi:10.5665/SLEEP.1128

Wagner, G. P.; Kin, K.; Lynch, V. J. Measurement of mRNA Abundance Using RNA-seq Data: RPKM Measure Is Inconsistent among Samples. *Theor. Biosci.* **2012,** *131* (4), 281–285.

Wang, Q.; Yi, Q.; Hu, Q.; Zhao, Y.; Nian, H.; Li, K.; Yu, Y.; Izui, K.; Chen, L. Simultaneous Overexpression of Citrate Synthase and Phosphoenolpyruvate Carboxylase in Leaves Augments Citrate Exclusion and Al Resistance in Transgenic Tobacco. *Plant Mol. Biol. Rep.* **2012,** *30*, 992–1005.

West, A. P.; Shadel, G. S.; Ghosh, S. Mitochondria in Innate Immune Responses. *Nat. Rev. Immunol.* **2011,** *11* (6), 389–402.

Westermann, B. Mitochondrial Fusion and Fission in Cell Life and Death. *Nat. Rev. Mol. Cell Biol.* **2010,** *11* (12), 872–884.

Xu, L. G.; Wang, Y. Y.; Han, K. J.; Li, L. Y.; Zhai, Z.; Shu, H. B. VISA Is an Adapter Protein Required for Virus-Triggered IFN-Beta Signaling. *Mol. Cell.* **2005,** *19* (6), 727–740.

Zabalza, A.; Van Dongen, J. T.; Froehlich, A.; Oliver, S. N.; Faix, B.; Gupta, K. J.; Schmälzlin, E.; Igal, M.; Orcaray, L.; Royuela, M.; Geigenberger, P. Regulation of Respiration and Fermentation to Control the Plant Internal Oxygen Concentration. *Plant Physiol.* **2009,** *149* (2), 1087–1098. doi:10.1104/pp.108.129288

Zatta, P.; Zambenedetti, P.; Pizziuti, A.; Perazzolo, M. Different Effects of Aluminum Upon Carbonic Anhydrases and Na+/K+-ATPase Activities in Rat. *Neurosci. Lett.***1995,** *197* (1), 65–68.

Zhang, Y.; Fernie, A. R. Metabolons, Enzyme-Enzyme Assemblies That Mediate Substrate Channeling, and Their Roles in Plant Metabolism. *Plant Comm.* **2020,** 100081.

Zhou, H. H.; Chen, Y. N.; Li, W. H.; Chen, Y. P. Photosynthesis of *Populus euphratica* in Relation to Groundwater Depths and High Temperature in Arid Environment, Northwest China. *Photosynthetica* **2010,** *48* (2), 257–68.

Zou, J.; Qi, Q.; Katavic, V.; Marillia, E. F.; Taylor, D. C. Effects of Antisense Repression of an *Arabidopsis thaliana* Pyruvate Dehydrogenase Kinase cDNA on Plant Development. *Plant Mol. Biol.* **1999,** *41* (6), 837–849.

CHAPTER 15

Stress Signaling Dynamics of Mitochondrial Electron Transport Chain and Oxidative Phosphorylation in Plants

ATASI ROUTRAY[1], SHARBANI BAHALI[1], SUCHISMITA PRUSTY[1], SAGAR MAITRA[2], and RANJAN KUMAR SAHOO[1*]

[1]*Department of Biotechnology, Centurion University of Technology and Management, Odisha, India*

[2]*Department of Agronomy and Agroforestry, Centurion University of Technology and Management, Odisha, India*

[*]*Corresponding author. E-mail: ranjan.sahoo@cutm.ac.in*

ABSTRACT

Mitochondria perform a vast range of tasks including physiological as well as metabolic during both biotic and abiotic stresses. They may be directly affected by physicochemical restrictions, and they are frequently implicated in responses to challenges of environment and their essential involvement in cell feeding, respiration, energy balancing, and biosynthesis. Mitochondria from plant cells have a number of biochemical characteristics, such as cyanide-insensitive alternate respiration, and their activity must be coordinated with that of the plastids in addition to integration with the ubiquitous eukaryotic compartments. Furthermore, given sessile lifestyle of plants, relative protection, and challenges due to present climate change, we can assume the plant cell as a suitable model for responses to current environmental stresses. Mitochondria possess a complicated signaling network being involved in

Photosynthesis and Respiratory Cycles during Environmental Stress Response in Plants. Aryadeep Roychoudhury (Ed.)

this response. The mechanism has yet to be studied as very large number of mitochondrial components are involved in stress signaling pathways. This chapter describes the relationship between mitochondrial electron transport chain and the mechanism of oxidative phosphorylation during stress.

15.1 INTRODUCTION

All plant cells have mitochondria, which are responsible for feeding, metabolism, and biosynthesis. One of the biochemical properties of plant mitochondria is the alternative oxidase-mediated cyanide-insensitive respiration (Folgueira et al., 2020). The function of mitochondria is linked to that of the plastids, in light as well as in dark conditions. They derive energy from oxidative phosphorylation system (OXPHOS) and electron transport chain (ETC) which are essential for the survival and development of plants (Wurzinger et al., 2018). Mitochondria are the key organelles for maintaining energy balance when plant responds to abiotic stress (Crawford et al., 2018). Mitochondria function with all of the cell compartments due to a vast area of contact with considerable metabolic interactions with the cytosol. Environmental stress detrimentally impacts on the function of mitochondria, resulting in mitochondrial disruption or stress. On the other hand, disturbances and stresses modify the mitochondrial function with other cell compartments (Van Aken et al., 2016). This complex network of signaling occurring due to stresses initiates a change in the large variety of mitochondrial elements such as metabolites, and regulatory and structural biomolecules are linked to programmed cell death (PCD) signaling pathways (Huang et al., 2019). In this chapter, we have summarized the role of mitochondria during stress and the pathways of OXPHOS and ETC. Different biotic and abiotic stresses can change metabolically the functions which are directly related to OXPHOS and ETC.

15.2 REACTIVE OXYGEN AND NITROGEN SPECIES

Electron transport by means of the ETC can yield reactive oxygen species (ROS), such as superoxide radical and hydrogen peroxide (Murphy, 2009). The ETC blocks like I, II, and III remain the major significant areas for ROS creation (Andreyev et al., 2015). A few metabolic and stress conditions bring about lower ETC state and expanded ROS production (Mller, 2001). At the point when ADP accessibility is diminished because of stress, it can bring about the ETC restraint or easing back. Subsequently, respiratory inhibitors can hinder the ETC, prompting the ETC to overdecrease and cause the overproduction of superoxide (Schwarzländer et al., 2009). Changes in mitochondrial ultrastructure brought about by stress such as those induced by salt stress can influence or disturb the ETC and OXPHOS function (Garcia de la Garma et al., 2015). Environmental stresses are the essential drivers of superoxide overproduction in mitochondria, and superoxide dismutases (SODs) convert superoxide to hydrogen peroxide (Dixit et al., 2002; Turrens, 2003). While superoxide has a short half-life, hydrogen peroxide has extended half-life (Bienert et al., 2007; Bohovych and Khalimonchuk, 2016). ROS flagging components including oxidized intermediates, Ca, and phosphorylation mechanism (Ng et al., 2014) have been connected with ROS-based mitochondrial elements (Noctor and Foyer, 2016), just as transcriptional and posttranscriptional modifications (Ng et al., 2014; Cui et al., 2019). Mitochondrial ROS flagging are connected to auxin-controlled formative cycles (Wang and Auwerx, 2017). In this aspect, it may be stated that *Arabidopsis* FtsH4 freaks are greatly influenced in a mitochondrial protein-handling protease, accumulation of mitochondrial hydrogen peroxide is related to the maintenance of auxin homeostasis, disability of meristem and cell cycle dysregulation (Dolzblasz et al., 2018). Changes in the abscisic acid excessively delicate 8 (ABO8) joining factor, which is fundamental in fitting articulation of the NAD4 segment of complex I, are observed in ROS-expanded mitochondria with decreased auxin collection/flagging with diminished meristem action (Yang et al., 2014). The association among mitochondrial ROS and chemicals indicates that chemical flagging pathways may assume a part in mitochondrial ROS flagging to some extent. During biotic stress, the programmed cell death represses the development and improvement of biotrophic microorganisms (Dangl and Jones, 2001). The hypersensitive response (HR) is associated with an expansion in ROS levels (Mur et al., 2007). Mitochondria are essential cell organelles known for protection against microorganism-mediated oxidative explodes and HR in addition to restricted NADPH oxidases (Suzuki et al., 2011). Programmed cell death might be delivered by expanded ROS creation brought about by ETC complex I flimsiness, as indicated by the *Arabidopsis* mosaic-demise 1 (mod1) freak (Wu et al., 2015). The absence of unsaturated fat blend in the chloroplast causes the ETC complex I destabilization in the mod1 freak, exhibiting that ROS-based mitochondrial flagging could relate interruption in different kinds of subcellular sections to programmed cell death acceptance. Furthermore, ROS-intervened mitochondrial flagging seems to include an assortment of pathways and administrative levels. A change in the mitochondrial porosity could be linked to the early ROS stage (Van Aken and Van Breusegem, 2015; Murik et al., 2014). Incited expansions in ROS creation can intensify mitochondrial breakage, which can buildup ROS levels, finishing in a self-supporting circle of ROS elements in the mitochondria (Amirsadeghi et al., 2007). In the case of biotic stress, plant development controllers like salicylate, pathogenic poisons, for example, harpin (Krause and Durner, 2004) or victorin (Yao et al., 2002) along with other toxic variables could intensify further damages (Amirsadeghi et al., 2007) (Fig. 15.1).

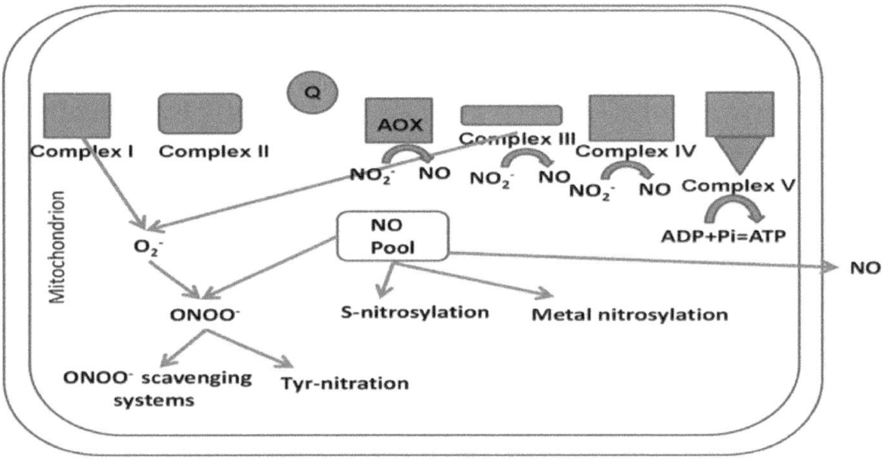

FIGURE 15.1 Mitochondrial dynamics and cellular effects of NO.

15.3 SIGNALING DYNAMICS OF MITOCHONDRIAL ENERGY METABOLISM

There exists a connection in the dynamics of the metabolism of citric acid cycle (CAC), acetyl-coenzyme A, flavin adenine dinucleotide, either as $FADH_2$ (reduced) or as FAD (oxidized), nicotinamide adenine dinucleotide, either as NADH (reduced form) or as NAD^+ (oxidized form), and ADP (Bohovych and Khalimonchuk, 2016). By regulating the ETC along with OXPHOS, the TCA cycle maintains the balance between NAD^+/NADH and FAD/$FADH_2$ (Bohovych and Khalimonchuk, 2016). Mitochondria function as organelles for the detection of stressors by keeping close association with the central metabolites (Crawford et al., 2018). Environmental stress can obstruct the ETC along with OXPHOS function directly or indirectly.; Moreover, the disruption signal is often sent to the nuclear genome, resulting in the synchronization of metabolic as well as stress-related genes (Che-Othman et al., 2017). In situations which produce functional abnormalities or injuries in mitochondria, decreased ATP generation can cause ATP diminution as well as a rise in the level of adenosine mono and diphosphates (Pedrotti et al., 2018). When carbohydrate is limited or depleted, like at night or during prolonged dark period, power constraint develops (Pedrotti et al., 2018). The subsequent decline in the level of ATP may cause mitochondria to send retrograde signals to the genome, prompting alternative respiration pathways, such as aminoalkanoic acid and carboxylic acid catabolism (Hildebrandt et al., 2015). Changes in the ATP levels also can have an indirect effect on other metabolites, which may affect the retrograde signaling. The NADH, Krebs cycle intermediates, as well as acetyl-coenzyme A, have proven to function like retreating signals toward the nucleus through the mitochondria (Wagner et al., 2018). This might cause the proposition that the quantity and diversity of stressed mitochondrial signals have an impact on the effectiveness of stress responses.

15.4 ELECTRON TRANSPORT CHAIN AND EFFECT DURING STRESS

ETC inhibition or slowdown occurs when availability of ADP is diminished due to stress. As a result, respiratory inhibition may block the complex of ETC, leading to the ETC depletion and superoxide overproduction (Schwarzländer et al., 2009). Changes in the mitochondrial ultrastructure generated by stress, such as those that occur during salt stress, can have a direct effect on OXPHOS pathway and the function of ETC, as suggested by Garcia de la Garma et al. (2015). Abiotic stresses are the primary causes of superoxide overproduction in mitochondria (Dixit et al., 2002). The conversion of superoxide to H_2O_2 occurs using superoxide dismutases (SODs) (Turrens, 2003). Superoxide contains a brief half-life compared with H_2O_2, which via aquaporins, may move across the membranes, suggesting that it might act between mitochondria and various cell parts (Bienert et al., 2007; Bohovych and Khalimonchuk, 2016). The reactive oxygen species (ROS) signaling mechanisms involving oxidized intermediates, Ca, and phosphorylation processes (Ng et al., 2014) have been linked to ROS-based mitochondrial dynamics (Noctor and Foyer, 2016) as well as remolding the transcriptional and posttranscriptional function of cell which may cause the regulation of gp130 (Ng et al., 2014). The ROS as well as perturbation of mitochondrial signaling are associated with the auxin-regulated physiological processes (Wang and Auwerx, 2017). The FtsH4 variant of *Arabidopsis* that are compromised in the protein-processing protease of mitochondria indicated the connection of H_2O_2 accumulation with the deterioration of cell cycle, disfigurement of the meristem as well as homeostatic re-regulation of auxin in the cell (Dolzblasz et al., 2018).

15.5 PROLINE REGULATION OF ROS DYNAMICS

Proline serves a variety of functions inside the plant cell, including growth, development, and stress reactions (Szabados and Savouré, 2010). Inside the cytosol, proline is generated from glutamate (P5CR) with the help of enzymes, such as P5CS (pyrroline 5-carboxylate synthetase) and P5CR (pyrroline 5-carboxylate reductase). The oxidative breakdown is mediated through P5CD (pyrroline-5-carboxylate dehydrogenase) and ProDH (proline dehydrogenase), which leads to glutamate production inside mitochondria (P5CDH). In response to stress (high and low temperature, drought, heavy metal, pathogen- mediated infection, deficiency of nutrition, atmospheric pollution, ultraviolet irradiation, anaerobiosis), the upregulation of P5CS (proline synthesis) and downregulation of proline catabolism (ProDH) result in proline accumulation (Verbruggen and Hermans, 2008). Other stressors, on the other hand, have been known to cause a rise in ProDH (proline catabolism) and proline accumulation (Garcia de la Garma et al., 2015). The presence of ProDH in the inner mitochondrial membrane, alongside the ETC, has also been reported by Cabassa-Hourton et al. (2016). ProDH, acting like a flavoenzyme, provides electrons to the ETC system, but when stress-related perturbations cause the ETC to overreduce, reduced $FADH_2$ may instantly diminish the oxygen supply to superoxide along with H_2O_2, adding to ROS production inside mitochondria (Liang et al., 2013). As a result, proline along with ProDH interconnects the stresses condition of ETC with the formation of ROS along with the redox state (Rasmusson and Møller, 2011). Such conversion of proline/P5C to produce ROS via the activity of ProDH is fundamental for the induction of programmed cell death and hypersensitive response under biotic stress (Monteoliva et al., 2014).

As a result, proline mitochondrial metabolism plays a significant role in providing defense against pathogenic oxidative bursts (Verslues and Sharma, 2010). Furthermore, Fabro et al. (2016) suggested that the production of ROS through proline is related to the function of plasma membrane-localized respiratory burst NADPH oxidase homologue D (RBOHD). In response to biotic stress, RBOHD creates apoplastic type ROS, which are included in the oxidative bursts and the immunity mediated by pathogenic attacks (Fabro et al., 2016). In *ProDH* mutants, RBOHD production of ROS was reduced and activation or suppression treatments of ProDH exhibited that ProDH activity is controlled by the activity of RBOHD (Fabro et al., 2016). The association between ProDH, ROS production by mitochondria along with RBOHD suggested the relation between the systemic signaling and mitochondrial stress (Gilroy et al., 2014). Furthermore, the pathway of ProDH/PC5CDH has been linked to glutamate regeneration, suggesting that it may play a role in the dynamic pool of glutamate (Schertl and Braun, 2014). Glutamate has been related to the signal pathways, such as receptor-like channels of glutamate and systemic Ca^{2+} responses to wounds and infections (Toyota et al., 2018). Glutamate could act as a mediator between the systemic signaling and mitochondrial proline dynamics.

15.6 DYNAMICS OF STRESS HORMONE INFORMATION

In the mitochondrial functioning, different growth regulators, such as cytokinins, abscisic acid (ABA), salicylate, ethylene, jasmonate, and other hormones interact directly when plants are grown under biotic or abiotic stresses (Zhu, 2016). Salicylate has a unique function. At lower concentrations, it functions as an uncoupling agent, but it can act on the mitochondrial ETC complexes when its concentration is higher (Czarnocka and Karpiński, 2018). Cytokinins and cytokinin equivalents are able to block the mitochondrial respiration in a number of molecular forms (Alberto et al., 2017). ABA inhibits mitochondrial adenine nucleotide translocators (ANTs) by intervening through ATP/ADP interactions with mitochondria and cytoplasm. Despite the fact that ANTs have a greater affinity for ATP than ABA, increased ABA levels during stress may impede ADP/ATP exchange, resulting in reduced ATP transport to the cytosol and ADP replenishment to the mitochondria (Fig. 15.2). The existence of a direct link between ABA and ATP dynamics was established in vitro (Kharenko et al., 2011). All these interactions to the activation of ROS generation have been related to the assembling of electrons in the ETC induced by a lowering in ADP availability in mitochondria (Turrens, 2003). The enhanced production of ROS in mitochondria is due to retrograde hydrogen peroxide signaling, alteration of membrane permeability, transcriptional and posttranslational regulation, release of cytochrome *c* as well as induction of programmed cell death (Cui et al., 2019). Variations in ATP synthesis or exchange between ATP/ADP may be detected by SnRK1 at the same time. Posttranslational phosphorylation and transcriptional regulation of stress-linked genes lead to the expression of key metabolic enzymes. The hormones, such as salicylate, ABA, and cytokinins have direct effects on inter-organellar stress integration, programmed cell death induction, and mitochondrial ROS signaling. Mitochondria can act as an organelle transducer under the influence on ROS and ATP/ADP proportions, stress hormone signaling as well as organellar integrators for various stress hormone signals. In the feedback regulation of abscisic acid, mitochondria are involved in ROS signaling under salinity stress (Garcia de la Garma et al., 2015). Activation of ROS production in mitochondria is also linked to the salt stress response (Garcia de la Garma et al., 2015). Increase in proline buildup is due to salt stress and it can also cause the stimulation of mitochondrial ProDH and P5CDH activities (Garcia de la Garma et al., 2015). The increased proline level accelerates the electron fluxes through the mitochondrial ETC, resulting in an increased production of mitochondrial ROS (Garcia de la Garma et al., 2015). Phospholipase D (PLD) 1 and P5CS expression in relation to proline biosynthesis appears to be linked to this stress-related regulatory loop through different unknown mechanisms (Garcia de la Garma et al., 2015). On the other hand, this regulation of ROS amplification could cause an increase in programmed cell death. PLD1, on the other hand, has been related to improved salt tolerance. However, its particular role is uncertain or unknown (Garcia de la Garma et al., 2015). In addition, the ABI4 transcription factor is involved in the transmission of ABA signaling pathway and combines mitochondrial and chloroplast retrograde signals (Zhang et al., 2013). The impact of convergence on mitochondrial AOX expression is associated to rheostat the activities of AOX in mitochondrial ROS and RNS dynamics. Several stress hormones signaling systems rely on the ETC-related mitochondrial stress signals including ABA signaling. Such interactions, on the other hand, could be considered as interventions with hormone signaling pathways and mitochondrial hormone signaling regulation mechanisms.

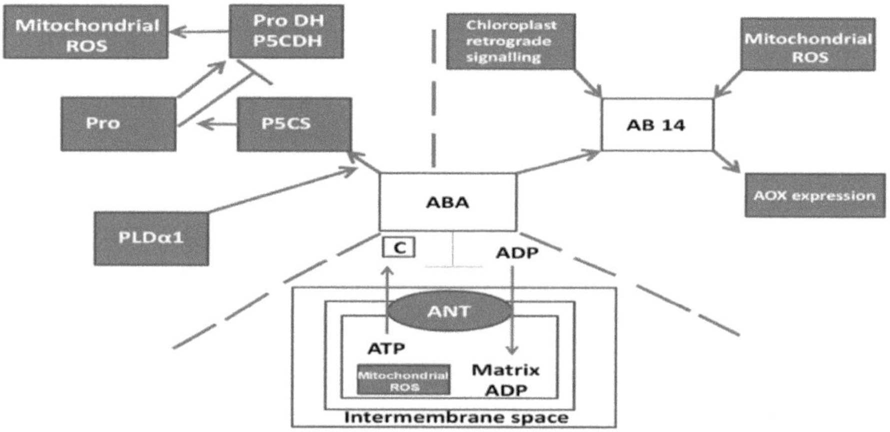

FIGURE 15.2 Signaling interactions between ABA and mitochondrial ROS.

15.7 CONCLUSION

Plant mitochondria are associated with different signaling pathways caused due to different stresses, including abiotic (salinity, heat, cold, prolonged darkness, drought, heavy metals) and biotic stressors (toxins and piercing–sucking insects). Different changes including hormonal, nutritional, and developmental changes cause the activation of different hormones to stabilize the energy level as well as the metabolites. The majority of stresses act as climate change factors influencing plant communities, with varying degrees of frequency and intensity (Komatsu et al., 2019). Mitochondria could potentially act as integrators for different signals derived from climate change and thus associated with energy and redox balance (Munné-Bosch et al., 2013). Synergistic or antagonistic consequences occur during acclimatization against different stresses. OXPHOS and ETC are centrally implicated in this network of mitochondrial signals (Table 15.1). The expression levels of OXPHOS and ETC components regulate the developmental and hormonal environment. However, additional mitochondrial structures and activities that are linked to signaling systems, such as membrane contacts or Ca^{2+} dynamics, should be further examined. Furthermore, mitochondrial signaling hubs may be involved in biogenetic processes, such as protein import and supramolecular assembly, in addition to normal and abnormal functions. As a result, OXPHOS and ETC homeostasis is anticipated to affect not only respiration and energy function, but also ETC and OXPHOS complex assembly.

Finally, more research into the potential connections between mitochondrial signaling hubs, nuclear epigenetic regulation and mitochondrial epigenetics should be conducted (van der Wijst and Rots, 2015). Understanding these connections should lead to new insights into the role of mitochondria in stress memory and priming processes. Regulation of mitochondrial respiration and mitochondria/cell cross talk has long been recognized as key factors in plant vigor and crop productivity. The identification of key genetic and physiological targets will be required to reduce respiratory carbon loss and boost agricultural output (Amthor et al., 2019). The search for such targets, however, should not be limited to metabolic and biosynthetic processes. It should also consider signaling connections that are likely to sustain integrative and adaptive responses, particularly mitochondrial signaling.

TABLE 15.1 Mitochondrial ETC–OXPHOS-Related Signaling in Environmental Stress Responses.

Mitochondrial Signals	Environmental stress	Stress response
ROS	Toxins	PCD
Proline	Flagellin	Immunity
AOX	Sucking insects	Resistance
ATP	Xenobiotics	Alternative respiratory pathways
NO	Hypoxia	Ethylene response
Ascorbate	High light	Heat-shock response
Calcium	Cold	Not characterized

KEYWORDS

- **mitochondria**
- **abiotic and biotic stresses**
- **cell signaling**
- **reactive oxygen species**

REFERENCES

Alberto, D.; Couée, I.; Sulmon, C.; Gouesbet, G. Root-Level Exposure Reveals Multiple Physiological Toxicity of Triazinexenobiotics in *Arabidopsis thaliana*. *J. Plant. Physiol.* **2017,** *212*, 105–114.

Amirsadeghi, S.; Robson, C.; Vanlerberghe, G. The Role of the Mitochondrion in Plant Responses to Biotic Stress. *Physiol. Plant.* **2007,** *129*, 253–266.

Amthor, J. S.; Bar-Even, A.; Hanson, A. D.; Millar, A. H.; Stitt, M.; Sweetlove, L. J.; Tyerman, S. D. Engineering Strategies to Boost Crop Productivity by Cutting Respiratory Carbon Loss. *Plant Cell.* **2019,** *31*, 297–314.

Andreyev, A. Y.; Kushnareva, Y. E.; Murphy, A. N.; Starkov, A. A. Mitochondrial ROS Metabolism: 10 Years Later. *Biochem. (Moscow)* **2015,** *80*, 517–531.

Bienert, G. P.; Møller, A. L.; Kristiansen, K. A.; Schulz, A.; Møller, I. M.; Schjoerring, J. K.; Jahn, T. P. Specific Aquaporins Facilitate the Diffusion of Hydrogen Peroxide across Membranes. *J. Biol. Chem.* **2007,** *282*, 1183–1192.

Bohovych, I.; Khalimonchuk, O. Sending Out an SOS: Mitochondria as a Signaling Hub. *Front. Cell Dev. Biol.* **2016,** *4*, 1–15.

Cabassa-Hourton, C.; Schertl, P.; Bordenave-Jacquemin, M.; Saadallah, K.; Guivarc'h, A.; Lebreton, S.; Planchais, S.; Klodmann, J.; Eubel, H.; Crilat, E.; Lefebvre-De, VD.; Ghelis, T.; Richard, L.; Abdelly, C.; Carol, P.; Braun, H. P.; Savouré, A. Proteomic and Functional Analysis of Proline Dehydrogenase 1 Link Proline Catabolism to Mitochondrial Electron Transport in *Arabidopsis thaliana. Biochem J.* **2016,** *473*, 2623–2634.

Che-Othman, M. H.; Millar, A. H.; Taylor, N. L. Connecting Salt Stress Signalling Pathways with Salinity-Induced Changes in Mitochondrial Metabolic Processes in C3 Plants. *Plant Cell Environ.* **2017,** *40*, 2875–2905.

Crawford, T.; Lehotai, N.; Strand, Å. The Role of Retrograde Signals during Plant Stress Responses. *J. Exp. Bot.* **2018,** *69*, 2783–2795.

Cui, F.; Brosché, M.; Shapiguzov, A.; He, X. Q.; Vainonen, J. P.; Leppälä, J.; Trotta, A.; Kangasjärvi, S.; Salojärvi, J.; Kangasjärvi, J.; Overmyer, K. Interaction of Methyl Viologeninduced Chloroplast and Mitochondrial Signalling in Arabidopsis. *Free Radic. Biol. Med.* **2019,** *134*, 555–566.

Cvetkovska, M.; Alber, N.; Vanlerberghe, G. The Signaling Role of a Mitochondrial Superoxide Burst during Stress. *Plant Signal. Behav.* **2013,** *8*, e22749–166.

Czarnocka, W.; Karpiński, S. Friend or Foe? Reactive Oxygen Species Production, Scavenging and Signaling in Plant Response to Environmental Stresses. *Free Radic. Biol. Med.* **2018,** *122*, 4–20.

Dangl, J.; Jones, J. Plant Pathogens and Integrated Defence Responses to Infection. *Nature* **2001,** *411*, 826–833.

Dixit, V.; Pandey, V.; Shyam, R. Chromium Ions Inactivate Electron Transport and Enhance Superoxide Generation in Vivo in Pea (*Pisumsativum* L. cv. Azad) Root Mitochondria. *Plant Cell Environ.* **2002,** *25*, 687–693.

Dolzblasz, A.; Gola, EM.; Sokolowska, K.; Smakowska-Luzan, E.; Twardawska, A.; Janska, H. Impairment of Meristem Proliferation in Plants Lacking the Mitochondrial Protease AtFTSH4. *Int. J. Mol. Sci.* **2018,** *19*, 1–14.

Fabro, G.; Rizzi, YS.; Alvarez, ME. Arabidopsis proline dehydrogenase contributes to flagellin-mediated PAMP-triggered immunity by affecting RBOHD. MPMI. **2016,** *29*, 620–628.

Folgueira, I., Lamas, J., Sueiro, R. A.; Leiro, J. M. Molecular Characterization and Gene Expression Modulation of the Alternative Oxidase in a Scuticociliate Parasite by Hypoxia and Mitochondrial Respiration Inhibitors. *Sci. Rep.* **2020,** *10*, 1–12.

Garcia de la Garma, J.; Fernandez-Garcia, N.; Bardisi, E.; Pallol, B.; Asensio-Rubio, J. S.; Bru, R.; Olmos, E. New Insights Into Plant Salt Acclimation: The Roles of Vesicle Trafficking and Reactive Oxygen Species Signalling in Mitochondria and the Endomembrane System. *New Phytol.* **2015,** *205*, 216–239.

Gilroy, S.; Suzuki, N.; Miller, G.; Choi, W. G.; Toyota, M.; Devireddy, A. R.; Mittler, R. A Tidal Wave of Signals: Calcium and ROS at the Forefront of Rapid Systemic Signaling. *Trends Plant Sci.* **2014,** *19*, 623–630.

Hildebrandt, T. M.; NunesNesi, A.; Araújo, W. L.; Braun, H. P. Amino Acid Catabolism in Plants. *Mol. Plant* **2015,** *8*, 1563–1579.

Huang, S.; Braun, H. P.; Gawryluk, R. M. R.; Millar, A. H. Mitochondrial Complex II of Plants: Subunit Composition, Assembly, and Function in Respiration and Signaling. *Plant J.* **2019,** *98*, 405–417.

Kharenko, O. A.; Boyd, J.; Nelson, K. M.; Abrams, S. R.; Loewen, M. C. Identification and Characterization of Interactions between Abscisic Acid and Mitochondrial Adenine Nucleotide Translocators. *Biochem. J.* **2011,** *437*, 117–123.

Komatsu, K. J.; Avolio, M. L.; Lemoine, N. P.; Isbell, F.; Grman, E.; Houseman, G. R.; Koerner, S. E.; Johnson, D. S.; Wilcox, K. R.; Alatalo, J. M.; Anderson, J. P.; Aerts, R.; Baer, S. G.; Baldwin, A. H.; Bates, J.; Beierkuhnlein, C.; Belote, R. T.; Blair, J.; Bloor, J. M. G.; Bohlen, P. J.; Bork, E. W.; Boughton, E. H.; Bowman, W. D.; Britton, A. J.; Cahill Jr, J. F.; Chaneton, E.; Chiariello, N. R.; Cheng, J.; Collins, S. L.; Cornelissen J. H. C.; Du, G.; Eskelinen, A.; Firn, J.; Foster, B.; Gough, L.; Gross, K.; Hallett, L. M.; Han, X.; Harmens, H.; Hovenden, M. J.; Jagerbrand, A.; Jentsch, A.; Kern, C.; Klanderud, K.; Knapp, A. K.; Kreyling, J.; Li, W.; Luo, Y.; McCulley, R. L.; McLaren, J. R.; Megonigal, J. P.; Morgan, J. W.; Onipchenko, V.; Pennings, S. C.; Prevéy, J. S.; Price, J. N.; Reich, P. B.; Robinson, C. H.; Russell, F. L.; Sala, O. E.; Seabloom, E. W.; Smith, M. D.; Soudzilovskaia, N. A.; Souza, L.; Suding, K.; Suttle, K. B.; Svejcar, T.; Tilman, D.; Tognetti, P.; Turkington, R.; White, S.; Xu, Z.; Yahdjian, L.; Yu, Q.; Zhang, P.; Zhang, Y. Global Change Effects on Plant Communities Are Magnified by Time and the Number of Global Change Factors Imposed. *PNAS* **2019,** *116*, 17867–17873.

Krause, M.; Durner, J. Harpin Inactivates Mitochondria in Arabidopsis Suspension Cells. *MPMI.* **2004,** *17*, 131–139.

Liang, X.; Zhang, L.; Natarajan, S. K.; Becker, D. F. Proline Mechanisms of Stress Survival. *ARS* **2013,** *19*, 998–1011.

Monteoliva, M. I.; Rizzi, Y. S.; Cecchini, N. M.; Hajirezaei, M-R.; Alvarez, M. E. Context of Action of Proline Dehydrogenase (ProDH) in the Hypersensitive Response of Arabidopsis. *BMC Plant Biol.* **2014,** *14*, 1–21.

Munné-Bosch, S.; Queval, G.; Foyer, C. H. The Impact of Global Change Factors on Redox Signaling Underpinning Stress Tolerance. *Plant Physiol.* **2013,** *161*, 5–19.

Mur, L.; Kenton, P.; Lloyd, A.; Ougham, H.; Prats, E. The Hypersensitive Response; the Centenary Is Upon Us But How Much Do We Know? *J. Exp. Bot.* **2007,** *59*, 501–520.

Murik, O.; Elboher, A.; Kaplan, A. Dehydroascorbate: A Possible Surveillance Molecule of Oxidative Stress and Programmed Cell Death in the Green Alga Chlamydomonas reinhardtii. *New Phytol.* **2014,** *202*, 471–484.

Murphy, M. P. How Mitochondria Produce Reactive Oxygen Species. *Biochem J.* **2009**, *417*, 1–13.

Ng, S.; De Clercq, I.; Van Aken, O.; Law, SR.; Ivanova, A.; Willems, P.; Giraud, E.; Breusegem, F. V.; Whelan, J. Anterograde and Retrograde Regulation of Nuclear Genes Encoding Mitochondrial Proteins during Growth, Development, and Stress. *Mol. Plant.* **2014,** *7*, 1075–1093.

Noctor, G.; Foyer, C. H. Intracellular Redox Compartmentation and ROS Related Communication in Regulation and Signaling. *Plant Physiol.* **2016,** *171*, 1581–1592.

Pedrotti, L.; Weiste, C.; Nägele, T.; Wolf, E.; Lorenzin, F.; Dietrich, K.; Mair, A.; Weckwerth, W.; Teige, M.; Baena-González, E.; Dröge-Laser, W. Snf1-Related Kinase1- Controlled C/S1-bZIP Signaling Activates Alternative Mitochondrial Metabolic Pathways to Ensure Plant Survival in Extended Darkness. *Plant Cell.* **2018,** *30*, 495–509.

Rasmusson, A. G.; Møller, I. M. Mitochondrial Electron Transport and Plant Stress. In *Plant Mitochondria*; Kempken, F., Ed.; Springer: New York, Dordrecht, Heidelberg. London, **2011,** *52*, 357–381.

Schertl, P.; Braun, H. P. Respiratory Electron Transfer Pathways in Plant Mitochondria. *Front. Plant Sci.* **2014,** *5*, 1–12.

Schwarzländer, M.; Fricker, M. D.; Sweetlove, L. J. Monitoring the in Vivo Redox State of Plant Mitochondria: Effect of Respiratory Inhibitors, Abiotic Stress and Assessment of Recovery from Oxidative Challenge. *Biochim Biophys Acta Gen Subj BBA-Gen Subjects* **2009,** *1787*, 468–475.

Suzuki, N.; Miller, G.; Morales, J.; Shulaev, V.; Torres, M.; Mittler, R. Respiratory Burst Oxidases: The Engines of ROS Signaling. *Curr. Opin. Plant Biol.* **2011,** *14*, 691–699.

Szabados, L.; Savouré, A. Proline: A Multifunctional Amino Acid. *Trends Plant Sci.* **2010,** *15*, 89–97.

Toyota, M.; Spencer, D.; Sawai-Toyota, S.; Jiaqi, W.; Zhang, T.; Koo, A. J.; Howe, G. A.; Gilroy, S. Glutamate Triggers Long-Distance, Calcium-Based Plant Defense Signaling. *Science* **2018,** *361*, 1112–1115.

Turrens, J. F. Mitochondrial Formation of Reactive Oxygen Species. *J. Physiol.* **2003,** *552*, 335–344.

Van der Wijst, M. G.; Rots, M. G. Mitochondrial Epigenetics: An Overlooked Layer of Regulation? *Trends Genet.* **2015,** *31*, 353–356.

Van Aken, O.; De Clercq, I.; Ivanova, A.; Law, S. R.; Van Breusegem, F.; Millar, A. H.; Whelan, J. Mitochondrial and chloroplast stress responses are modulated in distinct touch and chemical inhibition phases. Plant Physiol. **2016,** *171*, 2150–2165. .

Van Aken, O.; Van Breusegem, F. Licensed to Kill: Mitochondria, Chloroplasts, and Cell Death. *Trends Plant Sci.* **2015,** *20*, 754–7

Verbruggen, N.; Hermans, C. Proline Accumulation in Plants: A Review. *Amino Acids.* **2008,** *35*, 753–759.

Verslues, P. E.; Sharma, S. Proline Metabolism and Its Implications for Plant-Environment Interaction. *Arabidopsis Book* **2010,** *8*, 1–23.

Wagner, S.; Van Aken, O.; Elsässer, M.; Schwarzländer, M. Mitochondrial Energy Signaling and Its Role in the Low-Oxygen Stress Response of Plants. *Plant Physiol.* **2018,** *176*, 1156–1170.

Wang, X.; Auwerx, J. Systems Phytohormone Responses to Mitochondrial Proteotoxic Stress. *Mol. Cell.* **2017,** *68*, 540–551.

Wu, J.; Sun, Y.; Zhao, Y.; Zhang, J.; Luo, L.; Li, M.; Wang, J.; Yu, H.; Liu, G.; Yang, L.; Xiong, G.; Zhou, J. M.; Zuo, J.; Wang, Y.; Li, J. Deficient Plastidic Fatty Acid Synthesis Triggers Cell Death by Modulating Mitochondrial Reactive Oxygen Species. *Cell Res.* **2015,** *25*, 621–633.

Wurzinger, B.; Nukarinen, E.; Nägele, T.; Weckwerthe, W.; Teige, M. The SnRK1 Kinase as Central Mediator of Energy Signaling between Different Organelles. *Plant Physiol.* **2018,** *176*, 1085–1094.

Yang, L.; Zhang, J.; He, J.; Qin, Y.; Hua, D.; Duan, Y.; Chen, Z.; Gong, Z. ABA-Mediated ROS in Mitochondria Regulate Root Meristem Activity by Controlling Plethora Expression in Arabidopsis. *PLoS Genet.* **2014,** *10*, 1–19.

Yao, N.; Tada, Y.; Sakamoto, M.; Nakayashiki, H.; Park, P.; Tosa, Y.; Mayama, S. et al. Mitochondrial Oxidative Burst Involved in Apoptotic Response in Oats. *Plant J.* **2002,** *30*, 567–579.

Zhu, J. K. Abiotic Stress Signaling and Responses in Plants. *Cell.* **2016,** *167*, 313–324.

CHAPTER 16

Protective Chemicals and Metabolites in Stabilizing Photosynthesis and Respiration Machinery during Abiotic Stresses

ANKUR SINGH and ARYADEEP ROYCHOUDHURY*

Post Graduate Department of Biotechnology, St. Xavier's College (Autonomous), 30, Mother Teresa Sarani, Kolkata 700016, West Bengal, India

**Corresponding author. E-mail: aryadeep.rc@gmail.com*

ABSTRACT

Unfavorable environmental conditions suspend the development and growth of plants and also limit their production that ultimately threatens the food security of the world. Various abiotic factors such as drought, salinity, temperature, heavy metals, and excessive light hamper the photosynthetic and respiratory process in plants. These factors either directly inhibit the activity of enzyme involved in the above-mentioned process or indirectly distort the structure of thylakoids, chloroplast and degrade photosynthetic pigments via the formation of reactive oxygen species that again inhibits the above-mentioned processes and directly lowers the yield of the plants. However, to survive in these hostile environment, plants have developed an extensive network of protective metabolites comprising of osmolytes and secondary metabolites. Additionally, various reports have also demonstrated that exogenous application of certain chemicals such as salicylic acid, nitric

Photosynthesis and Respiratory Cycles during Environmental Stress Response in Plants. Aryadeep Roychoudhury (Ed.)

acid, calcium, melatonin etc. can boost the generation of the metabolites that ultimately lower the formation of cytotoxic by-products in plants. Here, we provide a detailed overview of the research conducted on photosynthetic and respiratory machineries that highlights the negative effects of abiotic stresses on these processes along with the efficacy of protective compounds that restores proper functioning of these pathways. Based on our studies, we tried to present detailed information of the plant responses to unfavorable environmental conditions, and more efficient protective network of metabolites that can improve photosynthetic and respiratory machinery so as to develop stress-resilient crop plants.

16.1 INTRODUCTION

Global climate change has emerged to be a major challenge for human being in the 21st century. According to Singh et al. (2011), the current population of world that is around 7 billion will increase to 8.9 billion by the end of 2050 which will immensely increase the demand of food. However, due to climate change, the soil health and environmental condition have drastically changed which ultimately lowers the yield of plants. Thus, in near future, it will be a huge challenge for our agricultural industry to provide food for such a massive population by overcoming the effects of abiotic stress (drought, salinity, extreme temperature, cold, flood, and metal toxicity) on plants. In recent times, extensive use of chemical fertilizers has been noticed for increasing the yield of plants, but such enormous use of chemical fertilizers further reduces the soil quality and is toxic for human population due to accumulation in edible parts (Banerjee et al., 2020). To overcome such problems, various groups have reported the use of certain beneficial metabolites and chemicals that abrogate the effects of abiotic stresses without jeopardizing the quality and quantity of the yield of plants. These protective metabolites induce the adaptive capability of the plants by scavenging the reactive oxygen species (ROS) generated during abiotic stress, enhancing the formation of other protective metabolites, and maintaining the integrity and homeostatic balance of other cell organelles (Singh et al., 2020). Jakab et al. (2005) showed that the exogenous application of non-proteinaceous amino acid, β-aminobutyric acid (BABA), enhanced the endogenous level of abscisic acid (ABA) in *Arabidopsis thaliana* that resulted in higher survival capability of the plants against drought and salt stress by scavenging the ROS formed and also inducing the formation and activity of protective machineries of plants. Similar work was done by Singh et al. (2021a) where they demonstrated that exogenous application of sodium nitroprusside (SNP), a well-known nitric oxide (NO) donor, abrogated the effects of fluoride-induced damages in *Vigna* sp. by reducing the formation of ROS, protecting the photosynthetic apparatus, and maintaining the integrity of cell membranes. In addition, they also reported that exogenous application of SNP induces the formation of protective metabolites like proline (Pro), glycine-betaine (Gly-bet), and amino acids that in turn maintain the osmotic balance of the cells. Similarly, several other reports have also demonstrated the protective role of various metabolites in ameliorating the toxic effects of abiotic stresses in plants which ultimately maintains the yield of crops without hampering their surrounding environment. Such aspects are discussed in greater detail in this chapter.

Another important process most commonly affected in plants exposed to stressed environment is photosynthesis. All the life form present on earth is directly or indirectly dependent on photosynthesis for food to eat and oxygen (O_2) to breathe. Along with this, fossilized photosynthetic fuels provide ~87% of the global energy (Johnson, 2016). Photosynthesis is a complex photochemical reaction where carbon dioxide (CO_2) and water (H_2O) molecules react in the presence of light to yield complex organic molecules such as carbohydrate and O_2. According to Foyer et al. (2012), over-reduction of electron transport chain during abiotic stress leads to photooxidation. Excess formation of ROS causes damage to the photosynthetic machinery like photosystem II (PSII) complex (Takahashi and Badger, 2011). It was also reported that light-induced damage of oxygen-evolving complex of PSII resulted in higher release of manganese ions. Nishiyama and Murata (2014) stated that the process of photoinhibition enhances exponentially when the amount of light energy absorbed by light-harvesting complex (LHC) II exceeds that of energy consumed. According to Nishiyama and Murata (2014), abiotic stresses do not directly affect the functions of PSII, but they hamper the repair of PSII complex. Different abiotic stresses such as high light, excess salt, and extreme temperature induce the formation of ROS like superoxide radicals and singlet oxygen that can inhibit the translation of *PsbA* mRNA which in turn inhibits the repair of PSII damages (Ohnishi and Murata, 2006; Yang et al., 2007; Allakhverdiev et al., 2002). Thus, maintaining the level of ROS and inhibiting their accumulation in photosynthetic apparatus is highly important for the survival of plant systems. The level of ROS in plant systems is highly dependent on the ROS scavenging system.

Plants have a wide range of protective metabolites that efficiently scavenge these cytotoxic metabolites which induces the adaptive nature of the plants.

Photorespiration is an equally important process which is closely linked with photosynthesis. According to Moroney et al. (2013), photorespiration recycles the phosphoglycolate produced by the oxygenase reaction of RuBisCO, which lowers the carbon loss and also provides some protective regulatory functions for plants. However, Ainsworth and Rogers (2007) reported that at 25°C, around 23–30% of carbon fixed by the plants during photosynthesis is lost due to photorespiration. Thus, photorespiration can regulate the carbon level in plants and is also essential for further strengthening the defense machineries of plants against abiotic stresses (Sunil et al., 2019). In addition to photorespiration, various other protective metabolites and chemicals have been exogenously applied which enhances the survival ability of the plants. Keeping in view these wide ranges of works, this chapter is mainly dedicated to decipher the efficacy of different molecules in abrogating the harmful effects of abiotic stress on plants. This chapter mostly consist of a brief discussion about the importance of photosynthesis and photorespiration in plants, effects of various abiotic stresses in plants, and finally discussing the role of protective metabolites that were reported to abrogate such toxic effects without jeopardizing the quality and yield of plants.

16.2 PROTECTIVE METABOLITES FOR PLANTS

To survive under harsh environmental conditions, plants have specialized mechanisms such as formation of small organic protective metabolites like (proline, glycine betaine, amino acids, etc.), and secondary metabolites (phenolics, ascorbic acid, flavonoids, glutathione, etc.). Osmolytes are small, electrically neutral, and nontoxic compatible osmoprotectants that help to maintain the osmotic balance of the cells (Singh and Roychoudhury, 2021). Secondary metabolites are another major group of protective compounds that scavenge the excessive ROS formed during abiotic stresses. Exogenous application of other protective chemicals such as calcium, SNP, spermine, etc. has been demonstrated to ameliorate the negative effects of abiotic stress in plants. In this section, we will try to shed some light on these diverge groups of protective metabolites formed in plants along with chemicals that are exogenously applied to improve the efficacy of photosynthetic machineries in plants during abiotic stresses.

16.2.1 OSMOLYTES

Osmolytes protect the lipid membrane and proteins present in the cells from denaturation due to harsh effects of abiotic stress. In addition, osmolytes also help in maintaining the osmotic balance of the cells along with the maintenance of turgor pressure by compartmentalizing in the cytoplasm (Rontein et al., 2002). According to Ozturk et al. (2011), osmolytes such as proteins, free amino acids, carbohydrates, proline, and glycine betaine are synthesized by the plants for maintaining the osmotic potential. Of all the above-mentioned osmolytes, proline and glycine betaine are mostly formed in response to abiotic stress (Rontein et al., 2002; Sankar et al., 2007). Several reports have demonstrated the importance of osmolytes in protecting the photosynthetic and respiratory apparatus in plants from negative effects of harsh environmental conditions. Chen and Murata (2011) reported that glycine betaine maintains the assimilation of CO_2 and also protects the proteins and enzymes that optimize the linear electron flow through thylakoid membrane during abiotic stress. Similar work was also reported by Wani et al. (2019) where they showed that exogenous application of proline improved the efficiency of photosystem II (PSII) and other photosynthetic attributes by enhancing the protective machineries of plants to combat the negative effects of salt stress in mustard plants.

16.2.2 SECONDARY METABOLITES

Secondary metabolites formed in plants play a vital role in the plant defence system against environmental stresses and pathogen attacks. Formation and accumulation of secondary metabolites depend on a variety of environmental factors such as soil water and fertility, light, temperature, and salinity (Yang et al., 2018). Secondary metabolites such as terpenoids, steroids, flavonoids, phenolics, and alkaloids are classified according to their chemical structures (Harborne, 1999). Secondary metabolites can be detected in whole plant cells, but they are synthesized in specialized cells and transported via vascular tissues. In addition, they also exhibit specific site of storage. Lipophilic metabolites like terpene-based oils are stored in resin ducts, trichomes, and thylakoid membranes, whereas hydrophilic compounds such as tannins, alkaloids, and glucosinolates are deposited in vacuoles and idioblasts (Acamovic and Brooker, 2005). Along with providing protection to plants, secondary metabolites are also considered as important sources of perfumes and traditional medicines. Several reports have demonstrated that in response to unfavorable environmental conditions, i.e., drought, salinity, extreme temperature, light, irradiation, metal toxicity, and nutritional deficiency, plants stimulate the formation of the above-mentioned metabolites which in turn contributes toward enhancing the survival capability of plants. According to Ahanger et al. (2020), upon exogenous application of kinetin and spermidine, the level of secondary metabolites like phenolics and flavonoids was enhanced which lowered the toxic effects of cadmium stress by lowering the degradation of photosynthetic pigments and also maintaining photosynthetic efficacy of *Vigna angularis*. A similar report was earlier published by Tattini et al. (2015) where they showed that the photosynthetic rate was restored by enhanced formation of isoprenoids and phenylpropanoids in *Platanus* × *acerifolia* plants exposed to drought stress. The role of secondary metabolites in ameliorating the toxic effects of abiotic stress in plants is widely depicted, thus proving their efficacy in enhancing the survival rate of plants by protecting the photosynthetic and respiratory components.

16.2.3 OTHER PROTECTIVE CHEMICALS

In addition to above-mentioned metabolites formed in plants, beneficial role of various chemicals upon exogenous application has been demonstrated by several groups. These chemicals can abrogate the negative effects of abiotic stresses by maintaining the growth and development of plants. They also do not cause harm to the surrounding environment. Ahanger et al. (2020) showed that the exogenous application of salicylic acid and nitric oxide protected the photosynthetic pigments like chlorophyll which in turn maintained the photosynthetic efficiency of *Vigna angularis* exposed to salt stress. Similar work was also reported by Singh et al. (2020) where they showed that priming of rice seeds with calcium compounds such as $Ca(OH)_2$, $Ca(NO_3)_2$, and $CaCl_2$ protected the photosynthetic pigments in seedlings thereby enhancing the survival capacity of plants against fluoride stress. They also showed that treatment of seeds by calcium compounds maintained the efficacy of enzymes of tricarboxylic acid cycle, ultimately inducing the survival capacity of seedlings by providing the amount of energy required to overcome the toxic effects of fluoride stress (Singh et al., 2021b). Efficacy of various other chemicals such as silicon, sodium nitroprusside, spermine, spermidine, melatonin, etc. in ameliorating the negative effects of abiotic stress on photosynthetic and respiratory components has also been reported by various groups that we attempt to explain elaborately in the next section of this chapter.

16.3 PROTECTIVE ROLE OF METABOLITES IN PHOTOSYNTHESIS AND RESPIRATION DURING DROUGHT STRESS

Of all the abiotic stresses, drought stress is one of the major stresses faced by the plants grown in arid or semiarid zones of the earth. Water scarcity occurs in plants when water supply to the roots is limited or if transpiration rate is high in plants. In case both the conditions coincide, survival of plants becomes more difficult. According to Lawlor and Cornic (2002), water scarcity leads to low water potential of plants which reduces the photosynthetic rate. To conserve water, stomatal closure is an effective strategy, but it hampers the photosynthetic reactions by checking the CO_2 uptake and also decreasing the photosynthetic rate in leaf (Reddy et al., 2004). Drought stress also hampers the function of PSII which in turn reduces its quantum yield (Tattini et al., 2015). According to Bertioli et al. (2016), water deficiency results in the degeneration of thylakoid membrane and chlorophyll pigment that lowers the chlorophyll content. Additionally, Jain et al. (2010) and Jaleel et al. (2009) reported higher amount of chlorophyll *b* as compared with that of chlorophyll a in drought affected plants. A similar report was also demonstrated earlier by Ashraf (1994) where they showed that the ratio of chlorophyll *b* to that of chlorophyll α was highly induced in the susceptible cultivars of wheat and *Brassica* as compared with that of tolerant varieties. Teixeira et al. (2016) also demonstrated that drought-mediated lowering of chlorophyll in soybean seeds reduced the nutritional value and oil quantity of seeds. However, in contrast to photosynthesis, drought stress facilitated the process of photorespiration in plants. According to Haupt-Herting and Fock (2002), photosynthetic electron dissipated by CO_2 assimilation decreased in drought-stressed tomato plants, whereas the same was increased due to contribution of photorespiration from 23% to 40%. A similar report was also published by Guan et al. (2004) where they showed that photorespiration increased as a result of moderate water stress in four varieties of grapes that ultimately maintained the efficacy of PSII and helped in rapid induction of photosynthesis on re-watering the plants. Thus, we can conclude that stress imposed due to water deficiency on photosynthetic machineries in plants is slightly reduced due to photorespiration.

To overcome the negative effects of drought stress on photosynthetic machineries, plants regulate the formation of various metabolites. According to Bruggemann and Schnitzler (2002), plants such as *Quercus* sp. induce the formation of isoprene that acts as a short-term thermo-protective agent. Similarly, Beckett et al. (2012) demonstrated that the formation of isoprene unit in *Xerophyta humilis* provides protection to the photosynthetic apparatus from mild water scarcity. In a recent study, Su et al. (2019) demonstrated that exogenous application of melatonin alleviates the symptoms of oxidative damages and protect the chlorophyll content of maize leaves exposed to 20% polyethylene glycol (PEG). Combined treatment of seed and foliar application of methyl jasmonate and salicylic acid induces the formation of osmolytes such as proline that in turn reduces the accumulation of ROS, ultimately reducing the degradation of chlorophyll present in the leaves of drought-stressed maize plants (Tayyab et al., 2020). Similarly, the efficacy of osmolytes and secondary metabolites, i.e., proline, ascorbic acid, and glutathione in reducing the negative effects of moderate and severe drought stress in chickpea plants, upon treating the seeds with these compounds, either singly or in combination, was demonstrated by El-Beltagi et al. (2020). They showed that upon priming the seeds with such compounds for 4 h with above-mentioned compounds, the level of chlorophyll *a* and *b*, along with total chlorophyll, was enhanced which maintained the photosynthetic efficacy of the plants. Thus, we can say that the exogenous application of the above-mentioned metabolites and chemicals strengthened the internal protective network of plants that ultimately maintained the efficiency of both photosynthetic and respiratory machineries which eventually enhanced the yield of plants.

16.4 REGULATION OF PHOTOSYNTHETIC AND RESPIRATORY COMPONENTS IN THE PRESENCE OF SALINITY STRESS

Salt stress negatively affects the physiological and biological processes in plants which lead to stunted growth and poor yield. According to Ahmad et al. (2018), salt stress accounts for around 50% loss in the productivity of crops. Salt stress-mediated osmotic stress reduces the efficiency of photosynthetic machineries via ionic effects on the subcellular organelles (Ahmad et al., 2019). Higher ionic concentration causes significant damages to the thylakoid membrane which in turn causes obstruction to electron transport and irreversible inactivation of photophosphorylation (Omoto et al., 2010; Mittal et al., 2012). Li et al. (2010) demonstrated that higher accumulation of Na^+ ions degrades chlorophyll and also hampers electron transport that ultimately down regulates the photosynthetic activity. In addition to chlorophyll, its precursors such as 5-aminolaevulinic acid and glutamic acid are also remarkably affected in sunflower (Santos, 2004). Along with chlorophyll, salt stress also negatively affects other pigments such as carotenoids as reported by Gomathi and Rakkiyapan (2011) in sugarcane. Agastian et al. (2000) reported that mulberry plants on being exposed to salt stress showed lower assimilation of CO_2, stomatal conductance, and conduction rate, whereas the intercellular level of CO_2 was enhanced. Thus, membrane damage of photosynthetic apparatus, reduction in chlorophyll, and biochemical changes are prime responses in plants exposed to salt stress that ultimately affects the development, growth, and yield of plants. Salt stress also hampers the activity of enzyme involved in tricarboxylic acid cycle. According to the report of Che-Othman et al. (2019), salt stress inhibits the activity of pyruvate dehydrogenase complex and 2-oxoglutarate which impairs tricarboxylic acid cycle (TCA) in wheat plants, reducing the rate of ATP formation. More elaborate work on the enzymes of TCA cycle exposed to salt stress was earlier done by Mittova and Igamberdiev (2000) where they showed that salt stress inhibits the activity of NADP-dependent isocitrate dehydrogenase, malate dehydrogenase, and succinate dehydrogenase in higher plants such as wheat, maize and thallus of small duckweed. Another important process that is induced in plants in response to salt stress is photorespiration. According to Foyer et al. (2009), photorespiration plays a pivotal role in regenerating CO_2 that is utilized during photosynthesis by the plants. In another study, Herschbach et al. (2010) reported that salt stress induces the rate of photorespiration which in turn led to higher accumulation of glycine and serine that are ultimately utilized by the plants during the formation of glutathione, thus demonstrating the protective role played by photorespiration in plants during salt stress. However, Li et al. (2020) reported that salt stress severely repressed the expression of differentially abundant protein related to photorespiration in *Medicago sativa* L. They also reported that metabolites such as glyoxylate, glycolate, glycine, and serine linked to photorespiration were reduced by about 40% on being treated with 200 mM NaCl for 14 days which also checked photosynthesis by around 30%. Thus, based on these earlier reports, prediction of exact effect of salt stress on photorespiration is quite difficult and in the near future, more work needs to be done for accurately determining the role of photorespiration.

Various reports have demonstrated that negative effects of salt stress on photosynthetic and photorespiratory components can be alleviated via exogenous application of metabolites and various chemicals. Elkelisha et al. (2019) demonstrated that exogenous application of 5 μM selenium induces the level of chlorophyll by 13.23% which ultimately results in 22.77% higher photosynthetic rate in wheat seedlings on being exposed to 100 mM NaCl stress. Similarly, Ahanger et al. (2020) reported that on application of 100 mM NaCl, the photochemical efficiency, chlorophyll formation, gas exchange, and rate of photosynthesis were repressed in *Vigna angularis* which were again restored to normal condition after exogenous treatments with nitric acid (100 μM), in combination with salicylic acid (1 mM). In an earlier report, Mimouni et al. (2016) reported that salicylic acid can efficiently induce the formation of osmolytes such as sugar and proline which reduces the damage caused by salt stress and protected the photosynthetic pigments in tomato plants. Protective role of other chemicals in photosynthetic machineries such as acetic acid, melatonin, γ-amino butyric acid, and resveratrol in combination with α-tocopherol has also been demonstrated earlier (Rahman et al., 2019; Shomali et al., 2021; Kostopoulou et al., 2014). An interesting work was demonstrated by Howladar (2014), where they showed that foliar spray of the leaf extract of *Moringa oleifera* can reduce the negative attributes of salt stress on *Phaseolus vulgaris* which in turn maintained the photosynthetic efficiency of the plant and also reduced the degradation of photosynthetic pigment, i.e., chlorophyll. Based on these data, a wide range of work was done to demonstrate the efficacy of these compounds in ameliorating the negative effects of unfavorable environmental conditions.

16.5 NEGATIVE EFFECTS OF LIGHT STRESS ON PHOTOSYNTHETIC AND RESPIRATORY COMPONENTS AND AMELIORATION USING PROTECTIVE CHEMICALS

Light is one of the key components for normal growth and development of plants; however, high light intensity negatively affects plant physiology and growth (Sharma et al., 2020). Low light intensity reduces the efficacy of photosynthesis as well as nitrogen and CO_2 metabolism and stomatal conductance that results in higher formation of ROS, whereas high light intensity causes photoinhibition by disrupting the function of light-harvesting complex (LHC) due to excessive formation of ROS (Wang et al., 2013; Qi-hua et al., 2014; Erickson et al., 2015). A similar report was also demonstrated by Zhang et al. (2016) where they showed that low light reduces the efficacy of stomatal conductance, photosynthetic machineries, quantum efficiency of

PSII, net photosynthetic rate, and transpiration rate. According to Gururani et al. (2015) and Chen et al. (2012), the functional failure of PSII results in photoinhibition that is caused due to poor oxidation rate of D1 protein in the presence of high light intensity. Excess light also reduces photochemical efficiency, mitochondrial activity, and release of excess energy which in turn damages both PSI and PSII (Faseela and Puthur, 2018). High light intensity also induces the activity of chlorophyll *b* reductase enzyme that results in the degradation of chlorophyll *b* (Sato et al., 2015). On the contrary, no sufficient reports are available which can effectively demonstrate the effect of light stress on respiration pathway and further research on this pathway needs to be conducted in the future in the coming time.

As a survival strategy against light stress, plants have developed various mechanisms such as elevated level of secondary metabolites like anthocyanins, carotenoids, and flavonoids. Perez-Lopez et al. (2018) showed that in response to excessive light, red leaf lettuce induces the accumulation of anthocyanins. Similar effects were also demonstrated by Nicole et al. (2016) where they showed higher accumulation of anthocyanins in the tissues of red oak and red salanova lettuce when their photoperiod was increased from 12 h to 18 h a day with a photosynthetic photon flux density 200 $\mu mol/m^2/s$. Upon being exposed to green light, the level of carotenoids was induced in *Lactuca sativa* (Liu et al., 2017). Page et al. (2011) reported that ascorbate and anthocyanins are formed in the tissues of *Arabidopsis* plants indicating a potential relationship between these metabolites. Zhao et al. (2019) demonstrated the role of another protective compound, i.e., fulvic acid in *Haematococcus pluvialis* under high light intensity. They showed that exogenous application of fulvic acid enhances the formation of certain cytoprotective metabolites and signaling molecules such as malic acid, melatonin, D-maltose, succinate, and amino acids which in turn reduces the damaging effects of light intensity on the photosynthetic machineries. Despite such wide negative effects of light intensity on photosynthetic components, very less reports are present demonstrating the role of protective metabolites in abrogating these effects.

16.6 PIVOTAL ROLE PLAYED BY METABOLITES IN LOWERING THE NEGATIVE EFFECTS OF TEMPERATURE STRESS ON PHOTOSYNTHETIC AND RESPIRATORY COMPONENTS

As compared with that of low temperature, high temperature causes more significant damages to the photosynthetic machineries of the plants. According to Hou et al. (2016), low temperature hampers photosynthetic processes such as stomatal conductance, transpiration rate, carbon reduction cycle, and thylakoid electron transport. Rexroth et al. (2011) demonstrated that high temperature reduces the rate of photosynthesis in plants by inhibiting the membrane-associated electron carriers and enzymes, and disrupting the thylakoid membranes. High temperature induces the process of chlorophyll degradation and can also cause irreversible damage to chlorophyll synthesis by inhibiting the activity of 5-aminolevulinate dehydratase, an enzyme actively involved in the pyrrole biosynthetic pathway, or the inhibition of protochlorophyllide biosynthesis (Antoniou et al., 2017; Mathuret al., 2014). Anjana and Allkhverdiev (2017) further demonstrated that heat stress even inhibits the activity of RuBisCO which is the key photosynthetic enzyme. High temperature also affects the structural and functional integrity of PSII and light harvesting complex (Lipova et al., 2010). According to Tikkanen et al. (2012), high-temperature stress is mainly accompanied with light stress in plants. Li et al. (2021) linked photosynthesis with respiration in plants exposed to moderate heat stress and stated that yield loss in plants is primarily linked with respiration. Moderate high temperature causes less damage to photosynthesis, but significantly increases respiration that in turn induces the energy production efficiency, but most of this energy is allocated to the maintenance of respiration which contributes toward yield loss in high-temperature climate.

The pivotal role played by secondary metabolites in ameliorating the toxic effects of temperature stress has been demonstrated by various reports. Rai et al. (2020) reported that exogenous application of nitric acid and salicylic acid in high temperature-stressed *Lablab purpureus* led to enhanced formation of proline, ascorbic acid and glutathione that protected the photosynthetic pigments such as chlorophyll and carotenoids. A similar result was earlier reported by Hu et al. (2016) where they reported that exogenous application of citric acid protected the chlorophyll content and maintained photochemical efficiency in tall fescue plants exposed to heat stress by inducing the activity of enzymatic antioxidants. Hasanuzzaman et al. (2012) also reported that exogenous application of nitric oxide abated the heat-induced damages in wheat cultivars by protecting the chlorophyll content of the seedlings. The pivotal role played by salicylic acid in restoring the photosynthetic efficiency in heat-stressed grapevine leaves was also demonstrated by Wang et al. (2010). Thus, based on such earlier reports, salicylic acid and nitric acid along with other metabolites play an important role in ameliorating the heat-induced damages in photosynthetic components of plants.

16.7 NEGATIVE EFFECTS OF HEAVY METAL STRESS ON PHOTOSYNTHETIC AND RESPIRATORY COMPONENTS AND ALLEVIATING THOSE EFFECTS

Exogenous application of phosphate-based fertilizers, release of industrial wastes, sewage sludges, volcanoes, traffic, and hard water practices lead to the release of heavy metal such as lead (Pb), arsenic (As), copper (Cu), nickel (Ni), cobalt (Co), zinc (Zn), and cadmium (Cd) in the environment (Ghori et al., 2019). When accumulated in plants above the threshold level, heavy metals lead to the production of ROS which in turn hamper the enzymatic activity of photosynthetic and respiratory components, and cause chlorophyll degradation and thylakoid membrane damage. According to Janik et al. (2010), Cd-stressed rye plants showed lower energy dissipation rate, thus indicating the negative effect of Cd toxicity on LHC II that harvests excess light energy. A similar effect was also reported by Ahmed and Tajmir-Riahi (1993) in Pb-stressed plants. Kapoor et al. (2019) demonstrated that Cd stress inhibits the rate of photosynthesis, alters and decreases chloroplast shape and size, degrades chlorophyll, and reduces the size of the thylakoid. Singh et al. (2013) reported that chromium (Cr) reduces the absorption of magnesium and nitrogen which causes reduced formation of chlorophyll. Ni stress affects the photosynthetic apparatus, inhibits the synthesis of pigments, and disturbs the activity of RuBisCO (Shahzad et al., 2018; Kohli et al., 2020). Zn and Cd stress disrupts thylakoid membrane, degrades photosynthetic pigments, reduces stomatal conductance, electron transport, and hampers the activity of enzymes controlling TCA cycle (Vassilev et al., 2011; Paunov et al., 2018). Similar reports demonstrating the negative effects of As stress in rice on the enzymes of TCA cycle, i.e., pyruvate dehydrogenase, citrate synthase, isocitrate dehydrogenase, succinate dehydrogenase, and malate dehydrogenase were presented by Samanta et al. (2020). Thus, heavy metal toxicity severely affects both the photosynthetic and respiratory components of plants.

Extensive work was done in the past, demonstrating the beneficial role played by various metabolites and chemicals on being endogenously generated or when exogenously applied in overcoming the damages caused due to excess metal stress in plants. Jiang et al. (2020) reported that exogenous application of glutamic acid reduces Cd uptake in rice which ultimately restores the chlorophyll level in the seedlings. A similar report was also published by Samanta et al. (2021) where they showed that exogenous application of melatonin restored the chlorophyll level in As-stressed rice seedlings along with the induced expression of RuBisCO. Earlier, they also reported that exogenous application of melatonin in rice could overcome the inhibitory effect of As on the enzymes of TCA cycles which in turn induces energy formation in seedlings, ultimately contributing toward higher tolerance level (Samanta et al., 2020). Soliman et al. (2019) reported that exogenous application of nitric oxide induces the level of osmolytes (proline and glycine betaine), ascorbate, and glutathione which in turn maintained the photosynthetic efficiency and chlorophyll level in Ni-stressed eggplants. In another study, Ulhassan et al. (2019) demonstrated that exogenous application of selenium reduces the uptake of Cr in *Brassica napus* which maintains the photosynthesis-related parameters such as chlorophyll *a*, chlorophyll *b*, carotenoids, net photosynthetic rate, stomatal conductance, and photochemical efficiency of PSII.

16.8 CONCLUSION AND FUTURE PERSPECTIVES

In the current chapter, we tried to summarize the effects of abiotic stress on the photosynthetic and respiratory components of the plants and amelioration based on the earlier reports available. The most common mode of toxicity in plants exposed to unfavorable environmental condition is via the formation of ROS which ultimately causes membrane instability and oxidative damages. Thus, it is necessary for the plants to adapt against those excessive ROS to tolerate the negative impacts. Additionally, surplus ROS also damage photosynthetic and respiratory machineries of plants. Various reports have demonstrated that abiotic stress conditions such as drought, salinity, heat, temperature extremes, and heavy metal toxicity lead to thylakoid membrane damage, distort the structure and shape of chloroplast, and degrade photosynthetic pigments. Evidences also show that the activity of enzymes involved in chlorophyll synthesis and TCA cycle is inhibited due to hostile environment that ultimately leads to reduced energy formation in plants. Despite such wide range of work, there are still some areas that need to be covered to get a clear picture of the damages caused by these hostile conditions on the above-mentioned pathways of plants. The prominent effect on the respiratory system in plants exposed to harsh environment is still somewhat fragmented, and in coming time more work is required to fill this gap.

To survive under these conditions plants have developed various suitable metabolites such as osmolytes and secondary metabolites that help them to scavenge cytotoxic ROS and other metabolites generated due to stressed condition. Along with this, these metabolites also provide protection to the membrane of the subcellular organelles along with maintaining the homeostatic balance of the cells. Additionally, various groups have exogenously applied a wide range of protective chemicals that play a pivotal role in maintaining the growth and development of plants by ameliorating the symptoms caused due to abiotic stress and further strengthening the internal protective machineries of plants. Along with this, more work is required to prove the efficacy of these metabolites and chemicals in real-life conditions, since most of the works are carried out in a controlled manner and challenging plants with only one selected stressed factor. Thus, it is necessary to test the efficacy of the chemicals in field condition where plants will be challenged with multiple stresses at a time.

ACKNOWLEDGMENTS

Financial assistance from Science and Engineering Research Board, Government of India through the grant [EMR/2016/004799] and Department of Higher Education, Science and Technology and Biotechnology, Government of West Bengal, through the grant [264(Sanc.)/ST/P/S&T/1G-80/2017] to Dr. Aryadeep Roychoudhury is gratefully acknowledged.

KEYWORDS

- **abiotic stress**
- **osmolytes**
- **secondary metabolites**
- **photosynthesis**
- **respiratory pathway**
- **photorespiration**

REFERENCES

Acamovic, T.; Brooker J. D. Biochemistry of Plant Secondary Metabolites and their Effects in Animals. *Proc. Nutri. Soc.* **2005,** *64*, 403–412.

Agastian, P.; Kingsley, S. J.; Vivekanandan, M. Effect of Salinity on Photosynthesis and Biochemical Characteristics in Mulberry Genotypes. *Photosynthetica* **2000,** *38*, 287–290.

Ahanger, M. A.; Aziz, U.;Sahli, A. A.; Alyemeni, M. N.; Ahmad, P. Combined Kinetin and Spermidine Treatments Ameliorate Growth and Photosynthetic Inhibition in *Vigna angularis* by Up-regulating Antioxidant and Nitrogen Metabolism under Cadmium Stress. *Biomolecules* **2020,** *10*, 147.

Ahmad, H.; Hayat, S.; Ali, M.; Liu, T.; Cheng, Z. The combination of Arbuscular Mycorrhizal Fungi Inoculation (*Glomus versiforme*) and 28-Homobrassinolide Spraying Intervals Improves Growth by Enhancing Photosynthesis, Nutrient Absorption, and Antioxidant System in Cucumber (*Cucumis sativus* L.) under Salinity. *Ecol. Evol.* **2018,** *8*, 5724–5740.

Ahmad, H.; Hayat, S.; Ali, M.; Liu, H.; Chen, X.; Li, J.; Cheng, Z. The Protective Role of 28-Homobrassinolide and *Glomus versiforme* in Cucumber to Withstand Saline Stress. *Plants* **2019,** *9*, 42.

Ahmed, A.; Tajmir-Riahi, H. A. Interaction of Toxic Metal Ions Cd^{2+}, Hg^{2+}, and Pb^{2+} with Light-Harvesting Proteins of Chloroplast Thylakoid Membranes: An FTIR Spectroscopic Study. *J. Inorg. Biochem.* **1993,** *50*, 235–243.

Ainsworth, E. A.; Rogers, A. The Response of Photosynthesis and Stomatal Conductance to Rising [CO_2]: Mechanisms and Environmental Interactions. *Plant Cell Environ.* **2007,** *30*, 258–270.

Allakhverdiev, S. I.; Nishiyama, Y.; Miyairi, S.; Yamamoto, H.; Inagaki, N.; Kanesaki, Y.; Murata, N. Salt Stress Inhibits the Repair of Photodamaged Photosystem II by Suppressing the Transcription and Translation of *psbA* Genes in *Synechocystis*. *Plant Physiol.* **2002,** *130*, 1443–1453.

Anjana, J.; Allakhverdiev, S. I. High-temperature Stress in Plants: Consequences and Strategies for Protecting Photosynthetic Machinery. In *Plant Stress Physiology*; 2nd ed., Shabala, S., Ed.; CAB International: Oxfordshire, 2017; pp 138–154.

Antoniou, C.; Chatzimichail, G.; Xenofontos, R.; Pavlou, J. J.; Panagiotou, E.; Christou, A.; Fotopoulos, V. Melatonin Systemically Ameliorates Drought Stress-Induced Damage in *Medicago sativa* Plants by Modulating Nitro-oxidative Homeostasis and Proline Metabolism. *J. Pineal. Res.* **2017,** *62*, e12401.

Ashraf, M. Breeding for Salinity Tolerance in Plants. *CRC Crit. Rev. Plant Sci.* **1994,** *13*, 17–42.

Banerjee, A.; Singh, A.; Roychoudhury, A. Fluoride Toxicity Variably Affects Overall Physiology and Grain Development in Three Contrasting Rice Genotypes, Representing a Potential Biohazard. *Environ. Sci. Pollut. Res.* **2020,** *28* (30), 40220–40232. doi:10.1007/s11356-020-10604-7.

Beckett, M.; Loreto, F.; Velikova, V.; Brunetti, C.; Di Ferdinando, M.; Tattini, M.; Calfapietra, C.; Farrant, J. M. Photosynthetic Limitations and Volatile and Non-volatile Isoprenoids in the Poikilochlorophyllous Resurrection Plant *Xerophyta humilis* during Dehydration and Rehydration. *Plant Cell Environ.* **2012,** *35*, 2061–2074.

Bertioli, D. J.; Cannon, S. B.; Froenicke, L.; Huang, G.; Farmer, A. D.; Cannon, E. K. S.; Liu, X.; Gao, D.; Clevenger, J.; Dash, S.; Ren, L.; Moretzsohn, M. C.; Shirasawa, K.; Hunag, W.; Vidigal, B.; Abernathy, B.; Chu, Y.; Niederhuth, C. E.; Umale, P.; Araujo, A. C. G.; Kozik, A.; Do Kim, K.; Burow, M. D.; Varshney, R. K.; Wang, X.; Zhang, X.; Noelle, B.; Guimaraes, P. M.; Isobe, S.; Guo, B.; Liao, B.; Stalker, T. H.; Schmitz, R. J.; Scheffler, B. E.; Leal-Bertioli, S. C. M.; Xun, X.; Jackson, S. A.; Michelmore, R.; Ozias-Akins, P. The

Genome Sequences of *Arachis duranensis* and *Arachis ipaensis*, the Diploid Ancestors of Cultivated Peanut. *Nat. Genet.* **2016,** *48*, 438–446.

Bruggemann, N.; Schnitzler, J. P. Comparison of Isoprene Emission, Intercellular Isoprene Concentration and Photosynthetic Performance in Water-limited Oak (*Quercus pubescens* willd. and *Quercus robur* L.) Saplings. *Plant Biol.* **2002,** *4*, 456–463.

Chen, T. H.; Murata, N. Glycinebetaine Protects Plants Against Abiotic Stress: Mechanisms and Biotechnological Applications. *Plant Cell Environ.* **2011,** *34*, 1–20.

Chen, L.; Jia, H.; Tian, Q.; Du, L.; Gao, Y.; Miao, X.; Liu, Y. Protecting Effect of Phosphorylation on Oxidative Damage of D1 Protein by Down-regulating the Production of Superoxide Anion in Photosystem II Membranes under High Light. *Photosynth. Res.* **2012,** *112*, 141–148.

Che-Othman, M. H.; Jacoby, R. P.; Millar, A. H.; Taylor, N. L. Wheat Mitochondrial Respiration Shifts from the TCA Cycle to the GABA Shunt under Salt Stress. *Phytology* **2019,** *225*, 1166–1180.

El-Beltagi, H. S.; Mohamed, H. I.; Sofy, M. R. Role of Ascorbic Acid, Glutathione and Proline Applied as Singly or in Sequence Combination in Improving Chickpea Plant through Physiological Change and Antioxidant Defense under Different Levels of Irrigation Intervals. *Molecules* **2020,** *25*, 1702.

Elkelisha, A. A.; Solimanb, M. H.; Alhaithlould, H. A.; El-Esawi, M. A. Selenium Protects Wheat Seedlings Against Salt Stress-mediated Oxidative Damage by Up-regulating Antioxidants and Osmolytes Metabolism. *Plant Physiol. Biochem.* **2019,** *137*, 144–153.

Erickson, E.; Wakao, S.; Niyogi, K. K. Light Stress and Photoprotection in *Chlamydomonas reinhardtii*. *Plant J.* **2015,** *82*, 449–465.

Faseela, P.; Puthur, J. T. The Imprints of the High Light and UV-B Stresses in *Oryza sativa* L. 'Kanchana' Seedlings are Differentially Modulated. *J. Photochem. Photobiol. B: Biol.* **2018,** *178*, 551–559.

Foyer, C.; Bloom, A. J.; Queval, G.; Noctor, G. Photorespiratory Metabolism: Genes, Mutants, Energetics, and Redox Signalling. *Annu. Rev. Plant Biol.* **2009,** *60*, 455–484.

Foyer, C. H.; Neukermans, J.; Queval, G.; Noctor, G.; Harbinson, J. Photosynthetic Control of Electron Transport and the Regulation of Gene Expression. *J. Exp. Bot.* **2012,** *63*, 1637–1661.

Ghori, N. H.; Ghori, T.; Hayat, M. Q.; Imadi, S. R.; Gul, A.; Altay, V.; Ozturk, M. Heavy Metal Stress and Responses in Plants. *Int. J. Environ. Sci. Technol.* **2019,** *16*, 1807–1828.

Gomathi, R.; Rakkiyapan, P.; Comparative Lipid Peroxidation, Leaf Membrane Thermostability, and Antioxidant System in four Sugarcane Genotypes Differing in Salt Tolerance. *Int. J. Plant Physiol. Biochem.* **2011,** *3*, 67–74.

Guan, X. Q.; Zhao, S. J.; Li, D. Q.; Shu, H. R. Photoprotective Function of Photorespiration in Several Grapevine Cultivars under Drought Stress. *Photosynthetica* **2004,** *42*, 31–36.

Gururani, M. A.; Venkatesh, J.; Tran, L. S. P. Regulation of Photosynthesis during Abiotic Stress-induced Photoinhibition. *Mol. Plant* **2015,** *8*, 1304–1320.

Harborne, J. B. Classes and Functions of Secondary Products from Plants. In *Chemicals from Plants, Perspectives on Secondary Plant Products*; Walton, N. J., Brown, D. E., Eds.; Imperial College Press: London, UK, 1999; pp 1–25.

Hasanuzzaman, M.; Nahar, K.; Alam, M. M.; Fujita, M. Exogenous Nitric Oxide Alleviates High Temperature Induced Oxidative Stress in Wheat (*Triticum aestivum* L.) Seedlings by Modulating the Antioxidant Defense and Glyoxalase System. *Austr. J. Crop. Sci.* **2012,** *6*, 1314–1323.

Haupt-Herting, S.; Fock, H. P. Oxygen Exchange in Relation to Carbon Assimilation in Water-Stressed Leaves during Photosynthesis. *Ann. Bot.* **2002,** *89*, 851–859.
Herschbach, C.; Teuber, M.; Eiblmeier, M.; Ehlting, B.; Ache, P.; Polle, A.; Schnitzler, J. P.; Rennenberg, H. Changes in Sulphur Metabolism of Grey Poplar (*Populus* × *canescens*) Leaves During Salt Stress: a Metabolic Link to Photorespiration. *Tree Physiol.* **2010,** *30*, 1161–1173.
Hou, W.; Sun, A. H.; Chen, H. L.; Yang, F. S.; Pan, J. L.; Guan, M. Y. Effects of Chilling and High Temperatures on Photosynthesis and Chlorophyll Fluorescence in Leaves of Watermelon Seedlings. *Biol. Plant* **2016,** *60*, 148–154.
Howladar, S. M. A Novel *Moringa oleifera* Leaf Extract can Mitigate the Stress Effects of Salinity and Cadmium in Bean (*Phaseolus vulgaris* L.) Plants. *Ecotoxicol. Environ. Saf.* **2014,** *100*, 69–75.
Hu, L.; Zhang, Z.; Xiang, Z.; Yang, Z. Exogenous Application of Citric Acid Ameliorates the Adverse Effect of Heat Stress in Tall Fescue (*Lolium arundinaceum*). *Front. Plant Sci.* **2016,** *7*, 179.
Jain, M.; Tiwary, S.; Gadre, R. Sorbitol-induced Changes in Various Growth and Biochemical Parameters in Maize. *Plant Soil Environ.* **2010,** *56*, 263–267.
Jakab, G.; Ton, J.; Flors, V.; Zimmerli, L.; Metraux, J. P.; Mauch-Mani, B. Enhancing *Arabidopsis* Salt and Drought Stress Tolerance by Chemical Priming for Its Abscisic Acid Responses. *Plant Physiol.* **2005,** *139*, 267–274.
Jaleel, C. A.; Manivannan, P.; Wahid, A.; Farooq, M.; Al-Juburi, H. J.; Somasundaram, R.; Panneerselvam, R. Drought Stress in Plants: A Review on Morphological Characteristics and Pigments Composition. *Int. J. Agric. Biol.* **2009,** *11*, 100–105.
Janik, E.; Maksymiec, W.; Mazur, R.; Garstka, M.; Gruszecki, W. I. Structural and Functional Modifications of the Major Light-Harvesting Complex II in Cadmium- or Copper-treated *Secale cereale*. *Plant Cell Physiol.* **2010,** *51*, 1330–1340.
Jiang, M.; Jiang, J.; Li, S.; Li, M.; Tan, Y.; Song, S.; Shu, Q.; Huang, J. Glutamate Alleviates Cadmium Toxicity in Rice Via Suppressing Cadmium Uptake and Translocation. *J. Hazard. Mat.* **2019,** *384*, 121319.
Johnson, M. P. Photosynthesis. *Essays Biochem.* **2016,** *60*, 255–273.
Kapoor, D.; Singh, M. P.; Kaur, S.; Bhardwaj, R.; Zheng, B.; Sharma, A. Modulation of the Functional Components of Growth, Photosynthesis, and Antioxidant Stress Markers in Cadmium Exposed *Brassica juncea* L. *Plants* **2019,** *8*, 260.
Kohli, S. K.; Handa, N.; Bali, S.; Khanna, K.; Arora, S.; Sharma, A.; Bhardwaj, R. Current Scenario of Pb Toxicity in Plants: Unraveling Plethora of Physiological Responses. *Rev. Environ. Contamin. Toxicol.* **2020,** *249*, 153–197.
Kostopoulou, Z.; Therios, I.; Molassiotis, A. Resveratrol and Its Combination with α-Tocopherol Mediate Salt Adaptation in Citrus Seedlings. *Plant Physiol. Bochum.* **2014,** *78*, 1–9.
Lawlor, D. W.; Cornic, G. Photosynthetic Carbon Assimilation and Associated Metabolism in Relation to Water Deficits in Higher Plants. *Plant Cell Environ.* **2002,** *25*, 275–294.
Li, T. X. Y. Z.; Zhang, Y.; Liu, H.; Wu, Y. T.; Li, W.; Bin Zhang, H. X. Stable Expression of *Arabidopsis* Vacuolar Na^+/H^+ Antiporter gene *AtNHX1*, and Salt Tolerance in Transgenic Soybean for over Six Generations. *Chi. Sci. Bull.* **2010,** *55*, 1127–1134.
Li, J.; Essemine, J.; Shang, C.; Zhang, H.; Zhu, X.; Yu, J.; Chen, G.; Qu, M.; Sun, D. Combined Proteomics and Metabolism Analysis Unravels Prominent Roles of Antioxidant System in

the Prevention of Alfalfa (*Medicago sativa* L.) against Salt Stress. *Int. J. Mol. Sci.* **2020,** *21*, 909.

Li, G.; Chen, T.; Feng, B.; Peng, S.; Tao, L.; Fu, G. Respiration, Rather than Photosynthesis, Determines Rice Yield Loss under Moderate High-temperature Conditions. *Front Plant Sci.* **2021,** *12*, 678653.

Lipova, L.; Krchnak, P.; Komenda, J.; Ilík, P. Heat-induced Disassembly and Degradation of Chlorophyll-containing Protein Complexes in Vivo. *Biochem. Biophys. Acta Bioenerg.* **2010,** *1797*, 63–70.

Liu, H.; Fu, Y.; Wang, M. Green Light Enhances Growth, Photosynthetic Pigments and CO_2 Assimilation Efficiency of Lettuce as Revealed by 'Knock Out' of the 480–560 nm Spectral Waveband. *Photosynthetica* **2017,** *55*, 144–152.

Mathur, S.; Agrawal, D.; Jajoo, A. Photosynthesis: Response to High Temperature Stress. *J. Photochem. Photobiol. B: Biol.* **2014,** *137*, 116–126.

Mimouni, H.; Wasti, S.; Manaa, A.; Gharbi, E.; Chalh, A.; Vandoorne, B.; Lutts, S.; Ahmed, H. B. Does Salicylic Acid (SA) Improve Tolerance to Salt Stress in Plants? A Study of SA Effects on Tomato Plant Growth, Water Dynamics, Photosynthesis, and Biochemical Parameters. *OMICS: J. Int. Biol.* **2016,** *20*, 180–190.

Mittal, S.; Kumari, N.; Sharma, V.; Differential Response of Salt Stress on *Brassica juncea*: Photosynthetic Performance, Pigment, Proline, D1 and Antioxidant Enzymes. *Plant Physiol. Biochem.* **2012,** *54*, 17–26.

Mittova, V. O.; Igamberdiev, A. U. Effect of Salt Stress on Respiration Metabolism in Higher Plants. *Izv. Akad. Nauk. Ser. Biol.* **2000,** *3*, 322–328.

Moroney, J. V.; Jungnick, N.; Di Mario, R. J.; Longstreth, D. J. Photorespiration and Carbon Concentrating Mechanisms, Two Adaptations to High O_2, Low CO_2 Conditions. *Photosynth. Res.* **2013,** *117*, 121–131.

Nicole, C. C. S.; Charalambous, F.; Martinakos, S.; van de Voort, S.; Li, Z.; Verhoog, M.; Krijn, M. Lettuce Growth and Quality Optimization in a Plant Factory. *Acta Hortic.* **2016,** *1134*, 231–238.

Nishiyama, Y.; Murata, N. Revised Scheme for the Mechanism of Photoinhibition and Its Application to Enhance the Abiotic Stress Tolerance of the Photosynthetic Machinery. *Appl. Microbiol. Biotechnol.* **2014,** *98*, 8777–8796.

Ohnishi, N.; Murata, N. Glycinebetaine Counteracts the Inhibitory Effects of Salt Stress on the Degradation and Synthesis of D1 Protein during Photoinhibition in *Synechococcus* sp. PCC 7942. *Plant Physiol.* **2006,** *141*, 758–765.

Omoto, E.; Taniguchi, M.; Miyake, H. Effects of Salinity Stress on the Structure of Bundle Sheath and Mesophyll Chloroplasts in NAD-Malic Enzyme and PCK Type C4 Plants. *Plant Prod. Sci.* **2010,** *13*, 169–176.

Ozturk, M.; Turkyilmaz, B.; Gucel, S.; Guvensen, A. Proline Accumulation in Some Coastal Zone Plants of the Aegean Region of Turkey. *Eur. J. Plant Sci. Biotechnol.* **2011,** *5*, 54–56.

Page, M.; Sultana, N.; Paszkiewicz, K.; Florance, H.; Smirnoff, N. The Influence of Ascorbate on Anthocyanin Accumulation During High Light Acclimation in *Arabidopsis thaliana*: Further Evidence for Redox Control of Anthocyanin Synthesis. *Plant Cell Environ.* **2011,** *35*, 388–404.

Paunov, M.; Koleva, L.; Vassilev, A.; Vangronsveld, J.; Goltsev, V. Effects of Different Metals on Photosynthesis: Cadmium and Zinc Affect Chlorophyll Fluorescence in Durum Wheat. *Int. J. Mol. Sci.* **2018,** *19*, 787.

Perez-Lopez, U.; Sgherri, C.; Miranda-Apodaca, J.; Micaelli, F.; Lacuesta, M.; Mena-Petite, A.; Quartacci, M. F.; Munoz-Rueda, A. Concentration of Phenolic Compounds Is Increased in Lettuce Grown under High Light Intensity and Elevated CO_2. *Plant Physiol. Biochem.* **2018,** *123*, 233–241.

Qi-hua, L.; Xiu, W.; Bo-Cong, C.; Jia-Qing, M.; Jie, G. Effects of Low Light on Agronomic and Physiological Characteristics of Rice Including Grain Yield and Quality. *Rice Sci.* **2014,** *21*, 243–251.

Rahman, M. M.; Mostofa, M. G.; Rahman, M. A.; Islam, M. R.; Keya, S. S.; Das, A. K.; Miah, M. G.; Kawser, A. Q. M. R.; Ahsan, S. M.; Hashem, A.; Tabassum, B.; Abd Allah, E. F.; Tran, L. P. Acetic Acid: A Cost-effective Agent for Mitigation of Seawater-Induced Salt Toxicity in Mung Bean. *Sci. Rep.* **2019,** *9*, 15186.

Rai, K. K.; Rai, N.; Aamir, M.; Tripathi, D.; Rai, S. P. Interactive Role of Salicylic Acid and Nitric Oxide on Transcriptional Reprogramming for High Temperature Tolerance in *Lablab purpureus* L.: Structural and Functional Insights Using Computational Approaches. *J. Biotechnol.* **2020,** *309*, 113–130.

Reddy, A. R.; Chaitanya, K. V.; Vivekanandan, M. Drought-Induced Responses of Photosynthesis Andantioxidant Metabolism in Higher Plants. *J. Plant Physiol.* **2004,** *161*, 1189–1202.

Rexroth, S.; Mullineaux, C. W.; Ellinger, D.; Sendtko, E.; Rogner, M.; Koenig, F. The Plasma Membrane of the Cyanobacterium *Gloeobacter violaceus* Contains Segregated Bioenergetic Domains. *Plant Cell* **2011,** *23*, 2379–2390.

Rontein, D.; Dieuaide-Noubhani, M.; Dufourc, E. J.; Raymond, P.; Rolin, D. The Metabolic Architecture of Plant Cells Stability of Central Metabolism and Flexibility of Anabolic Pathways during the Growth Cycle of Tomato Cells. *J. Biol. Chem.* **2002,** *277*, 43948–43960.

Samanta, S.; Singh, A.; Banerjee, A.; Roychoudhury, A. Exogenous Supplementation of Melatonin Alters Representative Organic Acids and Enzymes of Respiratory Cycle as Well as Sugar Metabolism During Arsenic Stress in Two Contrasting Indica Rice Cultivars. *J. Biotechnol.* **2020,** *324*, 220–232.

Samanta, S.; Banerjee, A.; Roychoudhury, A. Arsenic Toxicity Is Counteracted by Exogenous Application of Melatonin to Different Extents in Arsenic-Susceptible and Arsenic-Tolerant Rice Cultivars. *J. Plant Growth Regul.* **2021,** doi:10.1007/s00344-021-10432-0.

Sankar, B.; Jaleel, C. A.; Manivannan, P.; Kishorekumar, A.; Somasundaram, R.; Panneerselvam, R. Drought-induced Biochemical Modifications and Proline Metabolism in *Abelmoschus esculentus* (L.). *Moench. Acta Bot. Croat.* **2007,** *66*, 43–56.

Santos, C. V. Regulation of Chlorophyll Biosynthesis and Degradation by Salt Stress in Sunflower Leaves. *Sci. Hortic.* **2004,** *103*, 93–99.

Sato, R.; Ito, H.; Tanaka, A. Chlorophyll *b* Degradation by Chlorophyll *b* Reductase under High-Light Conditions. *Photosynth. Res.* **2015,** *126*, 249–259.

Shahzad, B.; Tanveer, M.; Rehman, A.; Cheema, S. A.; Fahad, S.; Rehman, S.; Sharma, A. Nickel: Whether Toxic or Essential for Plants and Environment—A Review. *Plant Physiol. Biochem.* **2018,** *132*, 641–651.

Sharma, A.; Kumar, V.; Shahzad, B.; Ramakrishnan, M.; Singh Sidhu, G. P.; Bali, A. S.; Handa, N.; Kapoor, D.; Yadav, P.; Khanna, K.; Bakshi, P.; Rehman, A.; Kohli, S. K.; Khan, E. A.; Parihar, R. D.; Yuan, H.; Thukral, A. K.; Bhardwaj, R.; Zheng, B. Photosynthetic Response of Plants under Different Abiotic Stresses: A Review. *J. Plant Growth Regul.* **2020,** *39*, 509–531.

Shomali, A.; Aliniaeifard, S.; Didaran, F.; Lotfi, M.; Mohammadian, M.; Seif, M.; Strobel, W. R.; Sierka, E.; Kalaji, H. M. Synergistic Effects of Melatonin and Gamma-aminobutyric Acid on Protection of Photosynthesis System in Response to Multiple Abiotic Stressors. *Cells* **2021,** *10,* 1631.

Singh, A.; Roychoudhury, A. Salicylic Acid–Mediated Alleviation of Fluoride Toxicity in Rice by Restricting Fluoride Bioaccumulation and Strengthening the Osmolyte, Antioxidant and Glyoxalase Systems. *Environ. Sci. Pollut. Res.* **2021,** doi:10.1007/s11356-021-14624-9.

Singh, J. S.; Pandey, V. C.; Singh, D. P. Efficient Soil Microorganisms: A New Dimension for Sustainable Agriculture and Environmental Development. *Agric. Ecosyst. Environ.* **2011,** *140,* 339–353.

Singh, A.; Banerjee, A.; Roychoudhury, A. Seed Priming with Calcium Compounds Abrogate Fluoride-induced Oxidative Stress by Upregulating Defence Pathways in an Indica Rice Variety. *Protoplasma* **2020,** *257,* 767–782.

Singh, A.; Banerjee, A.; Roychoudhury, A. Differential Responses of *Vigna radiata* and *Vigna mungo* to Fluoride-Induced Oxidative Stress and Amelioration Via Exogenous Application of Sodium Nitroprusside. *J. Plant Growth Regul.* **2021a,** doi:10.1007/s00344-020-10285-z.

Singh, A.; Roychoudhury, A.; Samanta, S.; Banerjee, A. Fluoride Stress-Mediated Regulation of Tricarboxylic Acid Cycle and Sugar Metabolism in Rice Seedlings in Absence and Presence of Exogenous Calcium. *J. Plant Growth Regul.* **2021b,** *40,* 1579–1593.

Singh, H. P.; Mahajan, P.; Kaur, S.; Batish, D. R.; Kohli, R. K. Chromium Toxicity and Tolerance in Plants. *Environ. Chem. Lett.* **2013,** *11,* 229–254.

Soliman, M.; Alhaithloul, H. A.; Hakeem, K. R.; Alharbi, B. M.; El-Esawi, M.; Elkelish, A. Exogenous Nitric Oxide Mitigates Nickel-Induced Oxidative Damage in Eggplant by Upregulating Antioxidants, Osmolyte Metabolism, and Glyoxalase Systems. *Plants* **2019,** *8,* 562.

Su, X.; Fan, X.; Shao, R.; Guo, J.; Wang, Y.; Yang, J.; Yang, Q.; Guo, L. Physiological and iTRAQ-Based Proteomic Analyses Reveal that Melatonin Alleviates Oxidative Damage in Maize Leaves Exposed to Drought Stress. *Plant Physiol. Biochem.* **2019,** *142,* 263–274.

Sunil, B.; Saini, D.; Bapatla, R. B.; Aswani, V.; Raghavendra, A. S. Photorespiration Is Complemented by Cyclic Electron Flow and the Alternative Oxidase Pathway to Optimize Ohotosynthesis and Protect against Abiotic Stress. *Photosynth. Res.* **2019,** *139,* 67–79.

Takahashi, S.; Badger, M. R. Photoprotection in Plants: A New Light on Photosystem II Damage. *Trends Plant Sci.* **2011,** *16,* 53–60.

Tattini, M.; Loreto, F.; Fini, A.; Guidi, L.; Brunetti, C.; Velikova, V.; Gori, A.; Ferrini, F. Isoprenoids and Phenylpropanoids are Part of the Antioxidant Defense Orchestrated Daily by Drought-Stressed *Platanus* × *acerifolia* Plants During Mediterranean Summers. *Phytology* **2015,** *207,* 613–626.

Tayyab, N.; Naz, R.; Yasmin, H.; Nosheen, A.; Keyani, R.; Sajjad, M.; Hassan, M. N.; Roberts, T. H. Combined Seed and Foliar Pre-treatments with Exogenous Methyl Jasmonate and Salicylic Acid Mitigate Drought-Induced Stress in Maize. *PLoS ONE* **2020,** *15,* e0232269.

Teixeira, R. N.; Ligterink, W.; de França-Neto, J. B.; Hilhorst, H. W. M.; da Silva, E. A. A. Gene Expression Profiling of the Green Seed Problem in Soybean. *BMC Plant Biol.* **2016,** *16,* 37.

Tikkanen, M.; Grieco, M.; Nurmi, M.; Rantala, M.; Suorsa, M.; Aro, E. M. Regulation of the Photosynthetic Apparatus under Fluctuating Growth Light. *Philos. Trans. R. Soc. B: Biol. Sci.* **2012,** *367,* 3486–3493.

Ulhassan, Z.; Gill, R. A.; Huang, H.; Ali, S.; Mwamba, T. M.; Ali, B.; Huang, Q.; Hamid, Y.; Khan, A. R.; Wang, J.; Zhou, W. Selenium Mitigates the Chromium Toxicity in *Brassic canapus* L. by Ameliorating Nutrients Uptake, Amino Acids Metabolism and Antioxidant Defense System. *Plant Physiol. Biochem.* **2019,** *145*, 142–152.

Vassilev, A.; Nikolova, A.; Koleva, L.; Lidon, F. Effects of Excess Zn on Growth and Photosynthetic Performance of Young Bean Plants. *J. Phytol.* **2011,** *3*, 58–62.

Wang, L. J.; Fan, L.; Loescher, W.; Duan, W.; Liu, G. J.; Cheng, J. S.; Luo, H. B.; Li, S. H. Salicylic Acid Alleviates Decreases in Photosynthesis Under Heat Stress and Accelerates Recovery in Grapevine Leaves. *BMC Plant Biol.* **2010,** *10*, 34.

Wang, L.; Deng, F.; Ren, W. J.; Yang, W. Y. Effects of Shading on Starch Pasting Characteristics of Indica Hybrid Rice (*Oryza sativa* L.). *PLoS ONE* **2013,** *8*, e68220.

Wani, A. S.; Ahmad, A.; Hayat, S.; Tahir, I. Epibrassinolide and Proline Alleviate the Photosynthetic and Yield Inhibition under Salt Stress by Acting on Antioxidant System in Mustard. *Plant Physiol. Biochem.* **2019,** *135*, 385–394.

Yang, X.; Wen, X.; Gong, H.; Lu, Q.; Yang, Z.; Tang, Y.; Liang, Z.; Lu, C. Genetic Engineering of the Biosynthesis of Glycinebetaine Enhances Thermotolerance of Photosystem II in Tobacco Plants. *Planta* **2007,** *225*, 719–733.

Yang, L.; Wen, K. S.; Ruan, X.; Zhao, Y. X.; Wei, F.; Wang, Q. Response of Plant Secondary Metabolites to Environmental Factors. *Molecules* **2018,** *23*, 762.

Zhang, H.; Zhong, H.; Wang, J. I.; Sui, X.; Xu, N. Adaptive Changes in Chlorophyll Content and Photosynthetic Features to Low Light in *Physocarpus amurensis* Maxim and *Physocarpu sopulifolius* Diabolo. *Peer J.* **2016,** *4*, e2125.

Zhao, Y.; Xing, H.; Li, X.; Geng, S.; Ning, D.; Ma, T.; Yu, X. Physiological and Metabolomics Analyses Reveal the Roles of Fulvic Acid in Enhancing the Production of Astaxanthin and Lipids in *Haematococcus pluvialis* under Abiotic Stress Conditions. *J. Agric. Food Chem.* **2019,** *67*, 12599–12609.

CHAPTER 17

Carbon Concentration Mechanism during Environmental Stress in Plants

LOPAMUDRA NAYAK[1], SARANGADHAR NAYAK[1], AWADHESH KUMAR[1], and MILAN KUMAR LAL[2*]

[1]*ICAR-National Rice Research Institute, Cuttack, Odisha, India*

[2]*ICAR-Central Potato Research Institute, Shimla, Himachal Pradesh, India*

[*]*Corresponding author. E-mail: milan.lal@icar.gov.in*

ABSTRACT

The present scenario of the enhanced human population has led to an increase in the demand for grain production to feed billions. On the other hand, climate change due to anthropogenic activities also significantly affected the production and productivity especially of cereals and tubers crops. Photosynthesis is an integral metabolism of the plant which fixes CO_2 by the process of carboxylation of ribulose 1,5-bisphosphate (RuBP), catalyzed by ribulose 1,5-bisphosphate carboxylase–oxygenase (Rubisco), to yield two molecules of 3-phosphoglycerate (3PGA). Photosynthesis is significantly affected due to change in climate that might include heat, drought, and other kinds of abiotic stresses. Plant has developed different strategies of C_3 and C_4 mechanisms to enhance carbon concentration mechanism in the vicinity of RuBiCo enzyme to have photosynthetically efficiency. Photorespiration is another mechanism of recycling of CO_2 which is a necessary evil mainly in C_3 plants. In this chapter, we have tried to provide insight into the carbon concentration mechanism under different abiotic stresses in plants.

Photosynthesis and Respiratory Cycles during Environmental Stress Response in Plants. Aryadeep Roychoudhury (Ed.)

17.1 INTRODUCTION

The anthropogenic activities have significantly affected the global climatic scenario that has led to climate change and reduced global agricultural productivity (Ortiz-Bobea et al., 2021). It was reported that reduced global agricultural productivity of approximately 21% since 1961 was due to anthropogenic climate change (ACC) and has a greater impact for warm region Africa (−34%) than for cooler regions Europe and Central Asia (−7.1%). In the next few decades, climate change is expected to affect more the world's supply of cereal grains, impacting their quantity and quality due to the complex effects of elevated atmospheric (CO_2) and changing temperature on crops (Macabuhay et al., 2018). Approximately 0.85°C temperature increased in air during the last two decades (Pachauri et al., 2014). The most probable outcome of climate resembles model projections foresees increases of 1.8 to 4.0°C by the end of the 21st century (2090–2099) compared to the period 1980–1999.

The common form of photosynthesis is C_3 as CO_2 gets fixed using the enzyme ribulose-1,5-bisphosphate carboxylase/oxygenase (RuBisCO) into a three-carbon compound known as 3-phosphoglycerate (Bassham et al., 1954). As RuBisCO is unable to distinguish between CO_2 and O_2, it leads to competing between the carboxylation and oxygenation reactions in the leaf thereby inhibiting the process of photosynthesis in C_3 plants. Oxygenation is the process of utilization of oxygen as a substrate by the RuBisCo enzyme. The process of oxygenation is elevated in hot, dry environments and yields phosphoglycolate, which has to be recycled in an energy-consuming process termed photorespiration (Long et al., 2015). Both temperature and drought stress were reported to increase photorespiration and decrease photosynthetic efficiency in C_3 plants. A variety of strategies to concentrate CO_2 around RuBisCO have been deciphered in the recent past to reduce photorespiration that can be found in nature. These include the carbon-concentrating mechanisms of many cyanobacteria and green algae, and both crassulacean acid metabolism and C_4 photosynthesis in flowering plants (Raven et al., 2008). As compared to the C_3 plant, the photosynthetic efficiency is higher in the C_4 plant which is agronomically important in terms of productivity which include crops such as maize, sorghum, and sugarcane are C_4, as are many weeds (Sales et al., 2021). The most common form of the C_4 pathway involves the spatial separation of metabolic reactions between two cell types organized within characteristic leaf anatomy known as Kranz Anatomy (Häusler et al., 2002; Slewinski et al., 2012).

CO_2 is initially fixed by an oxygen-insensitive carboxylase into a four-carbon compound (hence the C_4 pathway) in the outer mesophyll cells of the leaf, and then the oxaloacetate is formed after carboxylation followed by malate formation is transferred to inner vascular sheath cells. It is then decarboxylated by NADP malic enzyme to release CO_2 for refixation in bundle sheath cells (Maeda and Fernie, 2021). This intercellular C_4 shuttle concentrates CO_2 at the site of RuBisCO, leading to low levels of photorespiration and enhanced photosynthetic efficiency relative to C_3 plants, particularly in hot and dry environments. The CO_2 concentration surrounding RubisCO is crucial for proper carbon fixation to sustain plant growth and ultimately all life on earth (Sage et al., 2010). During evolution, especially during periods of low CO_2 in the atmosphere, different biochemical and even anatomical strategies have independently emerged that overcome limiting carbon fixation that is designated as inorganic "carbon-concentrating mechanisms" (CCMs). An alternative strategy to installing a C_4-like biochemical pump into C_3 plants to increase photosynthetic efficiency is to mimic the inorganic CO_2-concentrating mechanism (CCM) present in cyanobacteria and algae. This biophysical mechanism has the advantage that only a small subset of genes needs to be transferred to C_3 crop species and specialized anatomy and morphology may not be required. In addition, the energy costs of this mechanism may be inherently lower than those of the C_4 pathway (Nowicka et al., 2018). A plasma membrane or chloroplast envelope localized CCM has never been observed in a terrestrial plant. However, our knowledge of CCMs in cyanobacteria and micro-algae indicates that there are clear opportunities to extend this approach based on CCMs that feature the active accumulation of CO_2 (Taniguchi et al., 1991; Zhu et al., 2010; Schuler et al., 2016; Nowicka et al., 2018).

17.2 IMPACT OF ENVIROMENTAL STRESS ON PRODUCTION AND PRODUCTIVITY OF C_3 AND C_4 CROPS

Numerous effects of elevated atmospheric (CO_2) on plants have been documented through a photosynthesis-mediated CO_2 fertilization effect, including increased carbon (C) assimilation, growth, yield, and C content (Ainsworth and Ort, 2010; Ainsworth and Bush, 2011). Thus, elevated (CO_2) could enhance the concentration of photosynthesis-derived carbohydrates in grains, starch being the major component (Cure et al., 1988; Ferrise et al., 2011; Pandey et al., 2018; Lal et al., 2019). Since grains are predominantly composed of carbohydrates (mostly in the form of starch), it has been suggested that increases in starch concentrations can cause a dilution effect on other nutrients, including proteins, lipids, vitamins, and minerals (Kumar et al., 2020c; Lal et al., 2020; Tiwari et al., 2020a, 2021). In addition, adjustments in the photosynthetic apparatus and later on the redistribution from senescing leaves to grains must be considered the key mechanisms (Kumar et al., 2021b; Lal et al., 2021). Due to the different biochemistry of C_3 and C_4 photosynthesis, the positive effect of elevated (CO_2) on photosynthesis is more pronounced in C_3 crops such as wheat and rice but less notable in C_4 crops such as maize (Crafts-Brandner and Salvucci, 2002). Elevated (CO_2) has been reported to decrease mineral concentrations in barley grains (−6.9%), rice grains (−7.2%), and wheat grains (−7.6%), increasing the ratio of nonstructural carbohydrates (TNC) to protein by 6–47% in grains and tubers (Loladze, 2014).

Elevated (CO_2) affects nutrients beyond protein and minerals that are essential to humans, such as vitamins and carotenoids (Dorais et al., 2008). The decrease in mineral concentrations is notable in C_3 plants but less so in C_4 plants (Loladze, 2014) and is consistent with differences in physiology; the simulation of carbohydrate production by elevated (CO_2) is stronger in C_3 plants, whereas reduced transpiration is present in both C_3 and C_4 plants. Heat stress is a major constraint to sustainable cereal production, with reductions in grain yield being associated with high temperatures during the reproductive or grain-filling stages in wheat (Li et al., 2013), rice (Bahuguna et al., 2017; Kumar et al., 2020a; Singh et al., 2020), and other starch crops (Kumar et al., 2020b; Thakur et al., 2020; Chourasia et al., 2021; Raigond et al., 2021). High-temperature impacts on grain filling can vary enormously, depending on timing (days after anthesis) and duration (Nahar et al., 2010). Both chronic moderately high temperatures (25–35°C) and heat shocks (>35°C) during the grain-filling phase are frequently associated with an increase in grain protein concentration in wheat (Guzmán et al., 2016) and rice (Kumar et al., 2021c). Indeed, high temperature primarily impacts the accumulation of starch in wheat grain, with accumulation beginning earlier than under cooler temperatures, the duration of its accumulation also being reduced, and the result is a greater concentration of protein in the grain (Jain et al., 2014; Tiwari et al., 2020b; Kumar et al., 2021a).

Grain protein and mineral composition are quality characteristics that can change due to high temperature, and they respond to changes in enzymes involved in starch and protein synthesis. The activity of glutamate synthase was enhanced by heat stress, whereas sucrose phosphate synthase, sucrose synthase, and soluble starch synthase were significantly decreased during grain filling (Murakawa et al., 2014). However, Monjardino et al. (2005) found that protein concentration was negatively affected by heat stress during the early stage of endosperm development. They found that among the protein fractions, zeins are the most affected by heat stress. In fact, zein accumulation was repressed under high temperatures rather than being degraded in the early developmental stages. In rice, elevated temperature also alters grain protein and mineral nutrient composition (Ihsan et al., 2019; Chaturvedi et al., 2021).

Terrestrial higher plants use carbon as the substrate in the photosynthesis. About 800–1000 cm^3 CO_2/m^3 is needed for photosynthesis in C_3 plants getting adequate light. In current atmospheric CO_2 concentrations, C_4 plants growing inadequate light is almost saturated (Cundiff and Mankin, 2013; Ngoma, 2019). Therefore, increased CO_2 will increase the rate of photosynthesis and dry matter production of C_3 plants substantially but affect C_4 plants little. A wide range of conditions show that doubling atmospheric CO_2 concentrations from 330 to 650 cm^3 CO_2/m^3 increases the productivity of a large number of C_3 crop plants on average by 33%. Thus, dry matter production increases by approximately 0.1% per cm^3 in CO_2 concentration. However, dry matter production of C_4 plants increases by only 10% with a doubling of CO_2 (Berry and Roderick, 2002).

17.3 IMPACT OF ENVIROMENTAL STRESS ON PHOTOSYNTHESIS, RESPIRATION, AND TRANSPIRATION OF C_3 AND C_4 PLANTS

For the growth and development of crop (Ferrise et al., 2011), such as respiration (Atkin and Tjoelker, 2003), photosynthesis (Sage and Zhu, 2011); Sage et al., 2010), transpiration (Crawford et al., 2012), the temperature plays an important role. There are several studies on the influence of high-temperature stress on the biochemical, physiological, morphological, and molecular aspects of plants (O'Sullivan et al., 2017; Szymańska et al., 2017). The ambient temperature is the vital conditions for development of plants (Criddle et al., 1997); growth and development rates will decrease with higher or lower temperatures (Sánchez et al., 2014). The photosynthesis will increase in the range of 10–30°C and will decrease in increasing temperature. Respiration generally increases at 30–35°C and slows with an increasing temperature. In rice plants, photosynthesis would be inhibited by a temperature of approximately 45°C (Crafts-Brandner and Salvucci, 2002). Drought and heat stress significantly decrease stomatal conductance, photosynthetic efficiency, water-use efficiency, and leaf area of maize and wheat (*Zea mays* L.) (Zandalinas et al., 2018; Farooq et al., 2019). The global temperature increase of every 1°C affects the yield of major crops such as *Zea mays*, *Triticum aestivum*, *Oryza sativa*, *Glycine max* which is declined by 7.4%, 6.4%, 3.2%, and 3.1%, respectively (Zhao et al., 2017).

The plants are extremely sensitive to heat stress, which inhibits the process of photosynthesis by disturbing the photosynthetic pigments (Camejo et al., 2006), reducing activity of photosystem II (Camejo et al., 2005) and damaging the regeneration capacity of RuBP (Wise et al., 2004), reduction in chlorophyll biosynthesis (Dutta et al., 2009). The high-temperature stress damages the thylakoid, which is the main site for light-dependent chemical reaction and C metabolism. The photosystem II (PSII) is very much reactive to temperature and its activity is partially inhibited under heat stress (Camejo et al., 2005). Various mechanisms of PSII were damaged under high-temperature stress in barley and wheat (Tóth et al., 2005). The regeneration capacity of RuBP and electron transport chain was reduced, which was limited by the photosynthesis in the cotton (Wise et al., 2004). Net photosynthesis rate is limited due to low affinity of Rubisco toward binding with CO_2 (Morales et al., 2003). Net photosynthesis (Pn) and Stomatal conductance (gs) are inhibited by high-temperature stress in many crops due to reductions in the activation state of rubisco (Wahid et al., 2007).

Temperature is one of the most important environmental parameters influencing mitochondrial respiration. The balance between photosynthetic gains and respiration losses is determined as the net assimilation of crop (Amthor, 2000). High-temperature effects on respiration are driven by changes in enzyme activity (Brooks and Farquhar, 1985). Mostly, the respiration affected by global warming rather than photosynthesis is the reason for yield loss on maize. Some of respiration protein complexes, such as NADH dehydrogenase (Complex I), succinate dehydrogenase (Complex II), cytochrome bc1 (Complex III), cytochrome *c* oxidase (Complex IV), ATPase (Complex V), and alternative oxidase (AOX), which also help to drive the ATP production are generally affected by high temperature (Rashid et al., 2020). The Q10 temperature coefficient of respiration declines linearly from 3 to 1, which is linked with reduced light harvesting in photosystem II (PSII) that results from cyclic electron flow, thylakoid membrane instability (Crafts-Brandner and Salvucci, 2002). Mostly respiration affects on high temperature due to respiration enzymes' Q10 temperature coefficient (Ryan, 1991). It has been stated that the rice and wheat are affected by high night temperature which is mainly due to higher dark respiration, and this enhances the consumption of photoassimilates (Impa et al., 2021; Xu et al., 2021).

Transpiration is the primary mediator of energy and also is a mechanism of heat that can be avoided (Zhao et al., 2013). The rate of transpiration depends on the vapor pressure deficit (VPD) around leaves and the surrounding air (Seversike et al., 2013). Plant highly requires transpiration mechanism in response to cooling of leave tissues with increasing temperature (Seginer, 1994). Transpiration efficiency is an important component of WUE. Higher wind speed, as well as moderate watered conditions (Machado and Paulsen, 2001), is positively correlated with transpiration increases with increase in temperature.

17.4 CCM AS AFFECTED BY ENVIRONMENTAL STRESS CONDITION IN PLANTS

Global environmental change like high temperature and dry condition is having significant effects on the CO_2-concentrating mechanisms (CCM) also known as inorganic CCM (Reinfelder, 2011). The acute peripheral concentration of inorganic carbon differs among organisms with the kinetic properties of Rubisco of an organism. However, CCM is influenced by PAR (photosynthetically active radiation, 400–700 nm), UVA (ultraviolet A, 320–400 nm), UVB (ultraviolet B, 280–320 nm), and the availability of nitrogen, phosphorus, and iron (Raven, 2013). Inorganic CCM facilitate the elevation of CO_2 near the active sites of RuBisCO. RuBisCO activity is the most necessary photosynthetic enzyme and is reduced by heat stress (Crafts-Brandner and Law, 2000). The derived to gradually decrease in light-saturated CO_2-exchange rate (CER) from 33 to 45°C decrease in response to decline in Rubisco activity (Law and Crafts-Brandner, 1999). High temperature decreases the activation of RuBisCO by inhibiting the Rubisco activase (Crafts-Brandner and Salvucci, 2002). The presence of a CCM in C_4 plants has facilitated more water and nitrogen use efficiency than the C_3 plants (Singh et al., 2014; Schuler et al., 2016). The carbon fixation process is severely inhibited in C_3 plants as compared with C_4 plants in drought conditions (Schuler et al., 2016). Implementation of C_4-like CCM in C_3 crops may be one of the great solutions to increase photosynthetic capacity under stressful climatic conditions. Photorespiration becomes more particular under heat and drought conditions, which favor oxygenation reaction relative to CO_2 fixation (Levey et al., 2019). In C_4 photosynthesis, the CCM serve to enable the uptake of CO_2 at low concentrations which furthermore decreases the oxygenation reaction of Rubisco. Implementation of C_4-like CCM in C_3 crops may be one of the great solutions to increase photosynthetic capacity under stressful climatic conditions. Photorespiration becomes more particular under heat and drought conditions, which favor oxygenation reaction relative to CO_2 fixation (Levey et al., 2019). In C_4 photosynthesis, the CCM serve to enable the uptake of CO_2 at low concentrations which furthermore decreases the oxygenation reaction of RuBisCO.

17.5 OTHER PHYSIOLOGICAL AND BIOCHEMICAL PARAMETERS AFFECTING CCM UNDER ENVIRONMENTAL STRESS CONDITION

RuBisCO stimulates the oxygenase reaction at higher temperature, which induces the photorespiratory pathway and generates H_2O_2 as a by-product of the pathway (Sharkey, 2005). All these factors expressively decrease the photosynthetic efficiency under moderate high-temperature stress. Reactive oxygen species (ROS) are generally produced during photosynthesis in chloroplast (Foyer and Shigeoka, 2011). Some of ROS act as signaling molecules in plant growth and development under heat and drought stresses (Tripathy and Oelmüller, 2012; Baxter et al., 2014). Drought and heat stresses increase the production of these ROS (Singh et al., 2014). In two ways, the ROS generation affects plant metabolism either through the extremal ROS accumulation level or through redox signaling pathways (Foyer and Noctor, 2015). The excess amounts of ROS are scavenged by enzymatic components such as glutathione reductase, catalase, peroxidase, SOD, and nonenzymatic antioxidant components such as ascorbic acid, tocopherol, carotenoids, phenolics, glutathione, flavonoids (Kasote et al., 2015). High temperature will shorten the crop cycles, reduce spikelet fertility, decrease grain yield, and induce decreased crop production as well as effects on meiosis and flowering stages (Hatfield et al., 2011). Exposure of plants to abiotic stresses primarily causes oxidative damage by the synthesis of ROS and these ROS result in damaging lipids and proteins (Fahad et al., 2017). Heat stress under 33°C caused oxidative stress which damage protein degradation, membrane properties, enzyme deactivation in wheat. High temperature also highly enhanced the membrane peroxidation and decreased the membrane thermostability in wheat (Savicka et al., 2010). ROSs are highly damaging to photosynthetic membranes and proteins, including the oxygen-evolving complex polypeptides and the photosystem II (PSII) reaction center D1 protein.

17.6 MOLECULAR RESPONSE OF CCM UNDER ENVIRONMENTAL STRESS CONDITION

Under high temperature the modification of physiological and biochemical processes by gene expression changes gradually leads to acclimation or adaptation on stress condition (Moreno and Orellana, 2011). The heat stress-responsive genes and proteins are divided into two groups such as protein kinases, transcription factors and the other group like heat shock proteins (HSPs) and catalase (CAT). Heat shock proteins are generally expressed and protecting cells against abiotic stress conditions (Barua et al., 2003). HSP turns as a molecular chaperonin and helps to prevent incorrect folds of protein as well as to achieve the correct folds of collected protein. This mechanism facilitates proper cell activity at heat stress. Chaperons function primarily only in case of misfolding in protein and do not function in protein that has normal biological functions (Nakamoto et al., 1999). Different plants make heat shock proteins of various sizes according to rising temperatures. There are five different structures of heat shock proteins such as HSP 100, HSP 90, HSP 70, HSP 60, Small HSP (SmHSP) that are found in the chloroplast, mitochondria, endoplasmic reticulum, nucleus, and cytosol (Vierling, 1991). HSP 100 generally acts to prevent protein denaturation and protein formation (Wang et al., 2004). HSP 104 has an important role in Arabidopsis and yeast to protect against heat stress (Maestri et al., 2002). HSP 101 induces growth throughout the recovery period and is also essential for thermotolerance by genetic analysis (Vinocur and Altman, 2005; Myouga et al., 2006). The formation of smHSP depends on the duration of the temperature stress period (Sun et al., 2006). Small HSPs could help to protect oxygen evolution and oxygen-evolving complex proteins of PSII against heat stress.

17.7 CONCLUSION

In conclusion, there are many approaches (conventional breeding/modern transgenic approaches) that have shown that the CCM might be increased which leads to enhanced production and productivity. Moreover, increased CO_2 assimilation in different studies has revealed improved growth in terms of both vegetative and reproductive stages. Researchers are involved in the reduction of the photorespiration rate in different plant species which has direct correlation with agricultural productivity. The future studies should be focused on both cereal and tuber crops to enhance the CCM that will provide not only food security but also nutritional security.

KEYWORDS

- **photosynthesis**
- **Kranz Anatomy**
- **C_3 plants**
- **C_4 plants**

REFERENCES

Ainsworth, E. A.; Ort, D. R. How Do We Improve Crop Production in a Warming World? *Plant Physiol.* **2010,** *154*, 526–530.

Ainsworth, E. A.; Bush, D. R. Carbohydrate Export from the Leaf: A Highly Regulated Process and Target to Enhance Photosynthesis and Productivity. *Plant Physiol.* **2011,** *155*, 64–69. doi:10.1104/pp.110.167684.

Amthor, J. S. The McCree–de Wit–Penning de Vries–Thornley Respiration Paradigms: 30 Years Later. *Ann. Bot.* **2000,** *86*, 1–20. doi:10.1006/ANBO.2000.1175.

Atkin, O. K.; Tjoelker, M. G. Thermal Acclimation and the Dynamic Response of Plant Respiration to Temperature. *Trends Plant Sci.* **2003,** *8*, 343–351. doi:10.1016/S1360-1385(03)00136-5.

Bahuguna, R. N.; Solis, C. A.; Shi, W.; Jagadish, K. S. V. Post-flowering Night Respiration and Altered Sink Activity Account for High Night Temperature-Induced Grain Yield and Quality Loss in Rice (*Oryza sativa* L.). *Physiol. Plant.* **2017,** *159*, 59–73. doi:10.1111/ppl.12485.

Barua, D.; Downs, C. A.; Heckathorn, S. A. Variation in Chloroplast Small Heat-shock Protein Function is a Major Determinant of Variation in Thermotolerance of Photosynthetic Electron Transport among Ecotypes of Chenopodium Album. *Funct. Plant Biol.* **2003,** *30*, 1071–1079. doi:10.1071/FP03106.

Bassham, J. A.; Benson, A. A.; Kay, L. D.; Harris, A. Z.; Wilson, A. T.; Calvin, M. The Path of Carbon in Photosynthesis. XXI. The Cyclic Regeneration of Carbon Dioxide Acceptor1. *J. Am. Chem. Soc.* **1954,** *76*, 1760–1770.

Baxter, A.; Mittler, R.; Suzuki, N. ROS as Key Players in Plant Stress Signalling. *J. Exp. Bot.* **2014,** *65*, 1229–1240. doi:10.1093/jxb/ert375.

Berry, S. L.; Roderick, M. L. CO_2 and Land-Use Effects on Australian Vegetation over the Last Two Centuries. *Aust. J. Bot.* **2002,** *50*, 511–531.

Brooks, A.; Farquhar, G. D. Effect of Temperature on the CO_2/O_2 Specificity of Ribulose-1,5-bisphosphate Carboxylase/Oxygenase and the Rate of Respiration in the Light—Estimates from Gas-exchange Measurements on Spinach. *Planta* **1985,** *165*, 397–406. doi:10.1007/BF00392238.

Camejo, D.; Rodríguez, P.; Morales, M. A.; Dell'Amico, J. M.; Torrecillas, A.; Alarcón, J. J. High Temperature Effects on Photosynthetic Activity of Two Tomato Cultivars with Different Heat Susceptibility. *J. Plant Physiol.* **2005,** *162*, 281–289. doi:10.1016/J.JPLPH.2004.07.014.

Camejo, D.; Jiménez, A.; Alarcón, J. J.; Torres, W.; Gómez, J. M.; Sevilla, F.; et al. Changes in Photosynthetic Parameters and Antioxidant Activities Following Heat-Shock Treatment in Tomato Plants. *Funct. Plant Biol.* **2006,** *33*, 177–187. doi:10.1071/FP05067.

Chaturvedi, P.; Wiese, A. J.; Ghatak, A.; Záveská Drábková, L.; Weckwerth, W.; Honys, D. Heat Stress Response Mechanisms in Pollen Development. *Phytol* **2021,** *231* (2), 571–585.

Chourasia, K. N.; Lal, M. K.; Tiwari, R. K.; Dev, D.; Kardile, H. B.; Patil, V. U.; et al. Salinity Stress in Potato: Understanding Physiological, Biochemical and Molecular Responses. *Life* **2021,** *11*, 545. doi:10.3390/life11060545.

Crafts-Brandner, S. J.; Law, R. D. Effect of Heat Stress on the Inhibition and Recovery of the Ribulose-1,5-bisphosphate Carboxylase/Oxygenase Activation State. *Planta* **2000,** *2121* (212), 67–74. doi:10.1007/S004250000364.

Crafts-Brandner, S. J.; Salvucci, M. E. Sensitivity of Photosynthesis in a C_4 Plant, Maize, to Heat Stress. *Plant Physiol.* **2002,** *129*, 1773–1780. doi:10.1104/pp.002170.

Crawford, A. J.; McLachlan, D. H.; Hetherington, A. M.; Franklin, K. A. High Temperature Exposure Increases Plant Cooling Capacity. *Curr. Biol.* **2012,** *22*, R396–R397. doi:10.1016/J.CUB.2012.03.044.

Criddle, R. S.; Smith, B. N.; Hansen, L. D. A Respiration Based Description of Plant Growth Rate Responses to Temperature. *Planta* **1997,** *2014* (201), 441–445. doi:10.1007/S004250050087.

Cundiff, J. S.; Mankin, K. R. Basics of Plant Growth. *Dynamics of Biological Systems*; American Society of Agricultural and Biological Engineers: St. Joseph, MI, 2013, Chapter 5; pp 5.1–5.43. doi:10.13031/2013.34920.

Cure, J. D.; Rufty, T. W.; Israel, D. W. Phosphorus Stress Effects on Growth and Seed Yield Responses of Nonnodulated Soybean to Elevated Carbon Dioxide. *Agron. J.* **1988,** *80*, 897–902. doi:10.2134/agronj1988.00021962008000060012x.

Dorais, M.; Ehret, D. L.; Papadopoulos, A. P. Tomato (*Solanum lycopersicum*) Health Components: From the Seed to the Consumer. *Phytochem. Rev.* **2008,** *7*, 231.

Dutta, S.; Mohanty, S.; Tripathy, B. C. Role of Temperature Stress on Chloroplast Biogenesis and Protein Import in Pea. *Plant Physiol.* **2009,** *150*, 1050–1061. doi:10.1104/PP.109.137265.

Fahad, S.; Bajwa, A. A.; Nazir, U.; Anjum, S. A.; Farooq, A.; Zohaib, A.; et al. Crop Production under Drought and Heat Stress: Plant Responses and Management Options. *Front. Plant Sci.* **2017,** *8*, 1147. doi:10.3389/fpls.2017.01147.

Farooq, M.; Hussain, M.; Ul-Allah, S.; Siddique, K. H. M. Physiological and Agronomic Approaches for Improving Water-Use Efficiency in Crop Plants. *Agric. Water Manage.* **2019,** *219*, 95–108. doi:10.1016/J.AGWAT.2019.04.010.

Ferrise, R.; Moriondo, M.; Bindi, M. Probabilistic Assessments of Climate Change Impacts on Durum Wheat in the Mediterranean Region. *Nat. Hazards Earth Syst. Sci.* **2011,** *11*, 1293–1302. doi:10.5194/NHESS-11-1293-2011.

Foyer, C.; Noctor, G. Stress-Triggered Redox Signalling: What's in Prospect? *Plant, Cell, Environ.* **2015,** *39*, 951–964. doi:10.1111/pce.12621.

Guzmán, C.; Autrique, J. E.; Mondal, S.; Singh, R. P.; Govindan, V.; Morales-Dorantes, A.; et al. Response to Drought and Heat Stress on Wheat Quality, with Special Emphasis on Bread-making Quality, in Durum Wheat. *Food Crop. Res.* **2016,** *186*, 157–165.

Hatfield, J. L.; Boote, K. J.; Kimball, B. A.; Ziska, L. H.; Izaurralde, R. C.; Ort, D.; et al. Climate Impacts on Agriculture: Implications for Crop Production. *Agron. J.* **2011,** *103*, 351–370. doi:10.2134/agronj2010.0303.

R

S

Häusler, R. E.; Hirsch, H. J.; Kreuzaler, F.; Peterhänsel, C. Overexpression of C_4-Cycle Enzymes in Transgenic C_3 Plants: A Biotechnological Approach to Improve C_3-Photosynthesis. *J. Exp. Bot.* **2002,** *53*, 591–607. doi:10.1093/jexbot/53.369.591.

Ihsan, M. Z.; Daur, I.; Alghabari, F.; Alzamanan, S.; Rizwan, S.; Ahmad, M.; et al. Heat Stress and Plant Development: Role of Sulphur Metabolites and Management Strategies. *Acta Agric. Scand. Sect. B: Soil Plant Sci.* **2019,** *69*, 332–342. doi:10.1080/09064710.2019.1569715.

Impa, S. M.; Raju, B.; Hein, N. T.; Sandhu, J.; Prasad, P. V. V.; Walia, H.; et al. High Night Temperature Effects on Wheat and Rice: Current Status and Way Forward. *Plant Cell Environ.* **2021,** *44*, 2049–2065. doi:10.1111/pce.14028.

Jain, N.; Singh, G. P.; Singh, P. K.; Ramya, P.; Krishna, H.; Ramya, K. T.; et al. Molecular Approaches for Wheat Improvement under Drought and Heat Stress. *Indian J. Genet. Plant Breed.* **2014,** *74*, 578–583. doi:10.5958/0975-6906.2014.00893.1.

Kasote, D. M.; Katyare, S. S.; Hegde, M. V.; Bae, H. Undefined Significance of Antioxidant Potential of Plants and its Relevance to Therapeutic Applications. *Int. J. Biol. Sci.* **2015,** *11* (8), 982–991.

Kumar, A.; Dash, G. K.; Barik, M.; Panda, P. A.; Lal, M. K.; Baig, M. J.; et al. Effect of Drought Stress on Resistant Starch Content and Glycemic Index of Rice (*Oryza sativa* L.). *Starch/Staerke* **2020a,** *72*, 1900229. doi:10.1002/star.201900229.

Kumar, A.; Panda, P. A.; Lal, M. K.; Ngangkham, U.; Sahu, C.; Soren, K. R.; et al. Addition of Pulses, Cooking Oils, and Vegetables Enhances Resistant Starch and Lowers the Glycemic Index of Rice (*Oryza sativa* L.). *Starch/Staerke* **2020b,** 1900081. doi:10.1002/star.201900081.

Kumar, A.; Sahu, C.; Panda, P. A.; Biswal, M.; Sah, R. P.; Lal, M. K.; et al. Phytic Acid Content May Affect Starch Digestibility and Glycemic Index Value of Rice (*Oryza sativa* L.). *J. Sci. Food Agric.* **2020c,** *100*, 1598–1607. doi:10.1002/jsfa.10168.

Kumar, A.; Nayak, S.; Ngangkham, U.; Sah, R. P.; Lal, M. K.; Azharudheen, T. P.; et al. A Single Nucleotide Substitution in the SPDT Transporter Gene Reduced Phytic Acid and Increased Mineral Bioavailability from Rice Grain (*Oryza sativa* L.). *J. Food Biochem.* **2021a,** *45*, e13822. doi:10.1111/jfbc.13822.

Kumar, A.; Singh, B.; Raigond, P.; Sahu, C.; Nandan Mishra, U.; Sharma, S.; et al. Phytic Acid: Blessing in Disguise, a Prime Compound Required for Both Plant and Human Nutrition. *Food Res. Int.* **2021b,** *142*, 110193. doi:10.1016/j.foodres.2021.110193.

Kumar, R. R.; Dubey, K.; Arora, K.; Dalal, M.; Rai, G. K.; Mishra, D.; et al. Characterizing the Putative Mitogen-Activated Protein Kinase (MAPK) and Their Protective Role in Oxidative Stress Tolerance and Carbon Assimilation in Wheat under Terminal Heat Stress. *Biotechnol. Rep.* **2021c,** *29*, e00597. doi:10.1016/j.btre.2021.e00597.

Lal, M. K.; Vengavasi, K.; Pandey, R. Interactive Effects of Low Phosphorus and Elevated CO_2 on Root Exudation and Nutrient Uptake in Wheat Is Modified under Sulphur Nutrition. *Indian J. Plant Physiol.* **2019,** *24*, 63–73. doi:10.1007/s40502-019-0433-9.

Lal, M. K.; Kumar, A.; Raigond, P.; Dutt, S.; Changan, S. S.; Chourasia, K. N.; et al. Impact of Starch Storage Condition on Glycemic Index and Resistant Starch of Cooked Potato (*Solanum tuberosum*) Tubers. *Starch/Stärke* **2020,** 1900281. doi:10.1002/star.201900281.

Lal, M. K.; Tiwari, R. K.; Kumar, R.; Naga, K. C.; Kumar, A.; Singh, B.; et al. Effect of Potato Apical Leaf Curl Disease on Glycemic Index and Resistant Starch of Potato (*Solanum tuberosum* L.) Tubers. *Food Chem.* **2021,** *359*, 129939. doi:10.1016/j.foodchem.2021.129939.

Law, R. D.; Crafts-Brandner, S. J. Inhibition and Acclimation of Photosynthesis to Heat Stress Is Closely Correlated with Activation of Ribulose-1,5-bisphosphate Carboxylase/ Oxygenase. *Plant Physiol.* **1999,** *120*, 173–181. doi:10.1104/pp.120.1.173.

Levey, M.; Timm, S.; Mettler-Altmann, T.; Borghi, G. L.; Koczor, M.; Arrivault, S.; et al. Efficient 2-Phosphoglycolate Degradation Is Required to Maintain Carbon Assimilation and Allocation in the C_4 Plant Flaveria Bidentis. *J. Exp. Bot.* **2019,** *70*, 575–587. doi:10.1093/ jxb/ery370.

Li, Y. F.; Wu, Y.; Hernandez-Espinosa, N.; Peña, R. J. Heat and Drought Stress on Durum Wheat: Responses of Genotypes, Yield, and Quality Parameters. *J. Cereal Sci.* **2013,** *57*, 398–404. doi:10.1016/j.jcs.2013.01.005.

Loladze, I. Hidden Shift of the Ionome of Plants Exposed to Elevated CO_2 Depletes Minerals at the Base of Human Nutrition. *Elife* **2014,** *3*, e02245.

Long, S. P.; Marshall-Colon, A.; Zhu, X. G. Meeting the Global Food Demand of the Future by Engineering Crop Photosynthesis and Yield Potential. *Cell* **2015,** *161*, 56–66. doi:10.1016/j.cell.2015.03.019.

Macabuhay, A.; Houshmandfar, A.; Nuttall, J.; Fitzgerald, G. J.; Tausz, M.; Tausz-Posch, S. Can Elevated CO_2 Buffer the Effects of Heat Waves on Wheat in a Dryland Cropping System? *Environ. Exp. Bot.* **2018,** *155*, 578–588. doi:10.1016/j.envexpbot.2018.07.029.

Maeda, H. A.; Fernie, A. R. Evolutionary History of Plant Metabolism. *Annu. Rev. Plant Biol.* **2021,** *72*, 185–216.

Maestri, E.; Klueva, N.; Perrotta, C.; Gulli, M.; Nguyen, H. T.; Marmiroli, N. Molecular Genetics of Heat Tolerance and Heat Shock Proteins in Cereals. *Plant Mol Biol.* **2002,** *48*, 667–681. doi:10.1023/A:1014826730024.

Monjardino, P.; Smith, A. G.; Jones, R. J. Heat Stress Effects on Protein Accumulation of Maize Endosperm. *Crop Sci.* **2005,** *45*, 1203–1210.

Morales, D.; Rodríguez, P.; Dell'Amico, J.; Nicolás, E.; Torrecillas, A.; Sánchez-Blanco, M. J. High-Temperature Preconditioning and Thermal Shock Imposition Affects Water Relations, Gas Exchange and Root Hydraulic Conductivity in Tomato. *Biol. Plant.* **2003,** *47*, 203–208. doi:10.1023/B:BIOP.0000022252.70836.fc.

Moreno, A. A.; Orellana, A. The Physiological Role of the Unfolded Protein Response in Plants. *Biol. Res.* **2011,** *44*, 75–80. doi:10.4067/S0716-97602011000100010.

Murakawa, M.; Shimojima, M.; Shimomura, Y.; Kobayashi, K.; Awai, K.; Ohta, H. Monogalactosyldiacyl Glycerol Synthesis in the Outer Envelope membrane of Chloroplasts Is Required for Enhanced Growth under Sucrose Supplementation. *Front. Plant Sci.* **2014,** *5*, 1–13. doi:10.3389/fpls.2014.00280.

Myouga, F.; Motohashi, R.; Kuromori, T.; Nagata, N.; Shinozaki, K. An Arabidopsis Chloroplast-Targeted HSP 101 Homologue, APG6, Has an Essential Role in Chloroplast Development as Well as Heat-Stress Response. *Plant J.* **2006,** *48*, 249–260. doi:10.1111/j.1365-313X.2006.02873.x.

Nahar, K.; Ahamed, K. U.; Fujita, M. Phenological Variation and Its Relation with Yield in Several Wheat (*Triticum aestivum* L.) Cultivars under Normal and Late Sowing Mediated Heat Stress Condition. *Not. Sci. Biol.* **2010,** *2*, 51–56.

Nakamoto, H.; Marcel, T. H. H. Heat Shock Protein and thermal stress in chicken; 1999.

Ngoma, J. Climate-change Effects on the Zambezi Teak Forests Productivity in Zambia. 2019.

Nowicka, B.; Ciura, J.; Szymańska, R.; Kruk, J. Improving Photosynthesis, Plant Productivity and Abiotic Stress Tolerance—Current Trends and Future Perspectives. *J. Plant Physiol.* **2018,** *231*, 415–433. doi:10.1016/j.jplph.2018.10.022.

Wang, W.; Vinocur, B.; Shoseyov, O.;Altman, A. Role of Plant Heat-Shock Proteins and Molecular Chaperones in the Abiotic Stress Response. *Trends Plant Sci.* **2004,** *9*, 244–252. doi:10.1016/j.tplants.2004.03.006.

Wise, R. R.; Olson, A. J.; Schrader, S. M.; Sharkey, T. D. Electron Transport Is the Functional Limitation of Photosynthesis in Field-Grown Pima Cotton Plants at High Temperature. *Plant. Cell Environ.* **2004,** *27*, 717–724. doi:10.1111/J.1365-3040.2004.01171.X.

Xu, J.; Misra, G.; Sreenivasulu, N.; Henry, A. What Happens at Night? Physiological Mechanisms Related to Maintaining Grain Yield under High Night Temperature in Rice. *Plant Cell Environ.* **2021,** *44*, 2245–2261. doi:10.1111/PCE.14046.

Zandalinas, S. I.; Mittler, R.; Balfagón, D.; Arbona, V.; Gómez-Cadenas, A. Plant Adaptations to the Combination of Drought and High Temperatures. *Physiol. Plant.* **2018,** *162*, 2–12. doi:10.1111/ppl.12540.

Zhao, J.; Hartmann, H.; Trumbore, S.; Ziegler, W.; Zhang, Y. High Temperature Causes Negative Whole-Plant Carbon Balance under Mild Drought. *Phytol* **2013,** *200*, 330–339. doi:10.1111/NPH.12400.

Zhao, C.; Liu, B.; Piao, S.; Wang, X.; Lobell, D. B.; Huang, Y.; et al. Temperature Increase Reduces Global Yields of Major Crops in Four Independent Estimates. *Proc. Natl. Acad. Sci. U. S. A.* **2017,** *114*, 9326–9331. doi:10.1073/pnas.1701762114.

Zhu, X. G.; Shan, L.; Wang, Y.; Quick, W. P. C_4 Rice—An Ideal Arena for Systems Biology Research. *J. Integr. Plant Biol.* **2010,** *52*, 762–770. doi:10.1111/j.1744-7909.2010.00983.x.

Sharkey, T. D. Effects of Moderate Heat Stress on Photosynthesis: Importance of Thylakoid Reactions, Rubisco Deactivation, Reactive Oxygen Species, and Thermotolerance Provided by Isoprene. *Plant Cell Environ.* **2005,** *28,* 269–277. doi:10.1111/j.1365-3040.2005.01324.x.

Singh, J.; Pandey, P.; James, D.; Chandrasekhar, K.; Achary, V. M. M.; Kaul, T.; et al. Enhancing C_3 Photosynthesis: An Outlook on Feasible Interventions for Crop Improvement. *Plant Biotechnol. J.* **2014,** *12,* 1217–1230. doi:10.1111/pbi.12246.

Singh, A.; Raigond, P.; Lal, M. K.; Singh, B.; Thakur, N.; Changan, S. S.; et al. Effect of Cooking Methods on Glycemic Index and in Vitro Bioaccessibility of Potato (*Solanum tuberosum* L.) Carbohydrates. *LWT—Food Sci. Technol.* **2020,** *127,* 109363. doi:10.1016/j.lwt.2020.109363.

Slewinski, T. L.; Anderson, A. A.; Zhang, C.; Turgeon, R. Scarecrow Plays a Role in Establishing Kranz Anatomy in Maize Leaves. *Plant Cell Physiol.* **2012,** *53,* 2030–2037. doi:10.1093/pcp/pcs147.

Sun, A. Q.; Yi, S. Y.; Yang, J. Y.; Zhao, C. M.; Liu, J. Identification and Characterization of a Heat-inducible ftsH Gene from Tomato (*Lycopersicon esculentum* Mill.). *Plant Sci.* **2006,** *170,* 551–562. doi:10.1016/J.PLANTSCI.2005.10.010.

Szymańska, R.; Ślesak, I.; Orzechowska, A.; Kruk, J. Physiological and Biochemical Responses to High Light and Temperature Stress in Plants. *Environ. Exp. Bot.* **2017,** *139,* 165–177. doi:10.1016/J.ENVEXPBOT.2017.05.002.

Taniguchi, Y.; Ohkawa, H.; Masumoto, C.; Fukuda, T.; Tamai, T. Er Det Etisk Forsvarligt? *Sygeplejersken* **1991,** *91,* 3. doi:10.1093/jxb/ern016.

Thakur, N.; Raigond, P.; Singh, Y.; Mishra, T.; Singh, B.; Lal, M. K.; et al. Recent Updates on Bioaccessibility of Phytonutrients. *Trends Food Sci. Technol.* **2020,** *97,* 366–380. doi:10.1016/j.tifs.2020.01.019.

Tiwari, R. K.; Kumar, R.; Sharma, S.; Sagar, V.; Aggarwal, R.; Naga, K. C.; et al. Potato Dry Rot Disease: Current Status, Pathogenomics and Management. *3 Biotech* **2020a,** *10,* 503. doi:10.1007/s13205-020-02496-8.

Tiwari, R. K.; Lal, M. K.; Naga, K. C.; Kumar, R.; Chourasia, K. N.; et al. Emerging Roles of Melatonin in Mitigating abiotic and Biotic Stresses of Horticultural Crops. *Sci. Hortic. (Amsterdam).* **2020b,** *272,* 109592. doi:10.1016/j.scienta.2020.109592.

Tiwari, R. K.; Lal, M. K.; Kumar, R.; Chourasia, K. N.; Naga, K. C.; Kumar, D.; et al. Mechanistic Insights on Melatonin-Mediated Drought Stress Mitigation in Plants. *Physiol. Plant.* **2021,** *172,* 1212–1226. doi:10.1111/ppl.13307.

Tóth, S. Z.; Schansker, G.; Kissimon, J.; Kovács, L.; Garab, G.; Strasser, R. J. Biophysical Studies of Photosystem II-Related Recovery Processes after a Heat Pulse in Barley Seedlings (*Hordeum vulgare* L.). *J. Plant Physiol.* **2005,** *162,* 181–194. doi:10.1016/J.JPLPH.2004.06.010.

Tripathy, B. C.; Oelmüller, R. Reactive Oxygen Species Generation and Signaling in Plants. *Plant Signal. Behav.* **2012,** *7.* doi:10.4161/PSB.22455.

Vierling, E. The Roles of Heat Shock Proteins in Plants. *Annu. Rev. Plant Physiol. Plant Mol. Biol.* **1991,** *42,* 579–620. doi:10.1146/ANNUREV.PP.42.060191.003051.

Vinocur, B.; Altman, A. Recent Advances in Engineering Plant Tolerance to Abiotic Stress: Achievements and Limitations. *Curr. Opin. Biotechnol.* **2005,** *16,* 123–132. doi:10.1016/J.COPBIO.2005.02.001.

Wahid, A.; Gelani, S.; Ashraf, M.; Foolad, M. R. Heat Tolerance in Plants: An Overview. *Environ. Exp. Bot.* **2007,** *61,* 199–223. doi:10.1016/j.envexpbot.2007.05.011.

O'Sullivan, O.; Heskel, M. A. Thermal Limits of Leaf Metabolism across Biomes. *Wiley Online Libr.* **2017,** *23*, 209–223. doi:10.1111/gcb.13477.

Ortiz-Bobea, A.; Ault, T. R.; Carrillo, C. M.; Chambers, R. G.; Lobell, D. B. Anthropogenic Climate Change Has Slowed Global Agricultural Productivity Growth. *Glob. Change Biol.* **2021,** *11*, 306–312.

Pachauri, R. K.; Allen, M. R.; Barros, V. R.; Broome, J.; Cramer, W.; Christ, R.; et al. *Climate Change 2014: Synthesis Report. Contribution of Working Groups I, II and III to the Fifth Assessment Report of the Intergovernmental Panel on Climate Change*; IPCC, 2014.

Pandey, R.; Lal, M. K.; Vengavasi, K. Differential Response of Hexaploid and Tetraploid Wheat to Interactive Effects of Elevated [CO_2] and Low Phosphorus. *Plant Cell Rep.* **2018,** *37*, 1231–1244. doi:10.1007/s00299-018-2307-4.

Raigond, P.; Parmar, V.; Thakur, A.; Lal, M. K.; Changan, S. S.; Kumar, D.; et al. Composition of Different Carbohydrate Fractions in Potatoes: Effect of Cooking and Cooling. *Starch/Stärke* **2021,** *2100015*, 2100015. doi:10.1002/star.202100015.

Rashid, F. A. A.; Crisp, P. A.; Zhang, Y.; Berkowitz, O.; Pogson, B. J.; Day, D. A.; et al. Molecular and Physiological Responses During Thermal Acclimation of Leaf Photosynthesis and Respiration in Rice. *Plant. Cell Environ.* **2020,** *43*, 594–610. doi:10.1111/PCE.13706.

Raven, J. A. Rubisco: Still the Most Abundant Protein of Earth? *New Phytol.* **2013,** *198*, 1–3. doi:10.1111/nph.12197.

Raven, J. A.; Cockell, C. S.; De La Rocha, C. L. The Evolution of Inorganic Carbon Concentrating Mechanisms in Photosynthesis. *Philos. Trans. R. Soc. B Biol. Sci.* **2008,** *363*, 2641–2650.

Reinfelder, J. R. Carbon Concentrating Mechanisms in Eukaryotic Marine Phytoplankton. *Ann. Rev. Mar. Sci.* **2011,** *3*, 291–315. doi:10.1146/ANNUREV-MARINE-120709-142720.

Ryan, M. G. Effects of Climate Change on Plant Respiration. *Ecol. Appl.* **1991,** *1*, 157–167. doi:10.2307/1941808.

Sage, R. F.; Zhu, X. G. Exploiting the Engine of C 4 Photosynthesis. *J. Exp. Bot.* **2011,** *62*, 2989–3000. doi:10.1093/jxb/err179.

Sage, R. F.; Kocacinar, F.; Kubien, D. S. C_4 Photosynthesis and Temperature. *Advances in Photosynthesis and Respiration*; 2010, Chapter 10; pp 161–195. doi:10.1007/978-90-481-9407-0_10.

Sales, C. R. G.; Wang, Y.; Evers, J. B.; Kromdijk, J. Improving C_4 Photosynthesis to Increase Productivity under Optimal and Sub-optimal Conditions. *J. Exp. Bot.* **2021,** *72* (17), 5942–5960.

Sánchez, B.; Rasmussen, A.; Porter, J. R. Temperatures and the Growth and Development of Maize and Rice: A Review. *Glob. Chang. Biol.* **2014,** *20*, 408–417. doi:10.1111/GCB.12389.

Savicka, M.; Ekologija, N. Š. Effects of High Temperature on Malondialdehyde Content, Superoxide Production and Growth Changes in Wheat Seedlings (*Triticum aestivum* L.). *Ekologija* **2010,** *56* (1). doi:10.2478/v10055-010-0004-x.

Schuler, M. L.; Mantegazza, O.; Weber, A. P. M. Engineering C_4 Photosynthesis into C_3 Chassis in the Synthetic Biology Age. *Plant J.* **2016,** *87*, 51–65. doi:10.1111/tpj.13155.

Seginer, I. Transpirational Cooling of a Greenhouse Crop with Partial Ground Cover. *Agric. For. Meteorol.* **1994,** *71*, 265–281. doi:10.1016/0168-1923(94)90015-9.

Seversike, T. M.; Sermons, S. M.; Sinclair, T. R.; Carter, T. E.; Rufty, T. W. Temperature Interactions with Transpiration Response to Vapor Pressure Deficit Among Cultivated and Wild Soybean Genotypes. *Physiol. Plant.* **2013,** *148*, 62–73. doi:10.1111/J.1399-3054.2012.01693.X.

Index

U

V

W

X